Clean Energy Production Technologies

Series Editors

Neha Srivastava, Department of Chemical Engineering and Technology
IIT (BHU) Varanasi
Varanasi, India

P. K. Mishra, Department of Chemical Engineering and Technology
IIT (BHU) Varanasi
Varanasi, India

The consumption of fossil fuels has been continuously increasing around the globe and simultaneously becoming the primary cause of global warming as well as environmental pollution. Due to limited life span of fossil fuels and limited alternate energy options, energy crises is important concern faced by the world. Amidst these complex environmental and economic scenarios, renewable energy alternates such as biodiesel, hydrogen, wind, solar and bioenergy sources, which can produce energy with zero carbon residue are emerging as excellent clean energy source. For maximizing the efficiency and productivity of clean fuels via green & renewable methods, it's crucial to understand the configuration, sustainability and techno-economic feasibility of these promising energy alternates. The book series presents a comprehensive coverage combining the domains of exploring clean sources of energy and ensuring its production in an economical as well as ecologically feasible fashion. Series involves renowned experts and academicians as volume-editors and authors, from all the regions of the world. Series brings forth latest research, approaches and perspectives on clean energy production from both developed and developing parts of world under one umbrella. It is curated and developed by authoritative institutions and experts to serves global readership on this theme.

Mudasir A. Dar • Hossain M. Zabed
Mohd Shahnawaz
Editors

Recent Trends in Lignocellulosic Biofuels and Bioenergy

Advancements and Sustainability Assessment

Springer

Editors
Mudasir A. Dar
Biofuels Institute
School of Environment and Safety Engineering
Jiangsu University
Zhenjiang, Jiangsu, China

Hossain M. Zabed
School of Life Sciences
Guangzhou University
Guangzhou, Guangdong Province, China

Mohd Shahnawaz
Department of Botany
Govt. Degree College Drass A Constituent College of University of Ladakh
Ladakh, India

ISSN 2662-6861 ISSN 2662-687X (electronic)
Clean Energy Production Technologies
ISBN 978-981-96-4635-7 ISBN 978-981-96-4636-4 (eBook)
https://doi.org/10.1007/978-981-96-4636-4

This Springer imprint is published by the registered company Springer Nature Singapore Pte Ltd.
The registered company address is: 152 Beach Road, #21-01/04 Gateway East, Singapore 189721, Singapore

All editors dedicate this book to their respective parents, spouses, kids and family members for their unconditional love and support.

Preface

To find solutions to the challenges of climate change and the search for energy to replace fossil fuels, the global energy is currently transitioning. From the array options on the table, lignocellulosic biofuels and bio-energy are found to be one of the potential strategies toward a sustainable future free from environmental impacts including degradation and scarcity of resources. Thus, the book *Recent Trends in Lignocellulosic Biofuels and Bioenergy: Advancements and Sustainability Assessment* is an important attempt to represent the newest approaches in the field, the latest innovations and technologies, policies, and sustainability management that shape the current and future development of this quickly evolving industry.

Chapters from different researchers around the globe were invited and only 15 best chapters were included in the current volume and are aimed at discussing the different aspects of lignocellulosic biofuels and bioenergy.

The opening chapter, "Biofuels as Emerging and Sustainable Alternatives to Fossil Fuels: Environmental, Energetic, and Economic Perspectives," begins by explaining how biofuels can be used to support the development of energy for any country. The authors present the environmental, economic, and energetic aspects, pointing out resource and energy efficiency as important factors for satisfying the energy demands of an increasingly industrialized society. The second chapter "Life Cycle Assessment (LCA) of Biofuel Production from Lignocellulosic Biomass" details the life cycle assessment of biofuel production. In conducting the LCA methodologies, the authors assess the sustainability of the biofuel supply chains, and also they proffer ways in which those carbon footprints can be reduced and the effectiveness can be improved. Chapter 3 is all about the analyses of the environmental impact of biofuels. It is a critical look at the effects of biofuels on the environment, biodiversity, agriculture, and water resources. It not only outlines certain risks affecting the environment but also suggests the next steps to control these problems' impacts and sustain biofuel implementation. Chapters 4–8 are dedicated to the developments in the sphere of technologies and bioengineering. This is by employing extremophilic microorganisms, engineering the rumen microbial ecosystem, and adapting biological systems from insects and termites for efficient biorefineries. Altogether, the individual chapters of this book highlight biomimicry and genetic adjustments in improving the utilization efficiency and sustainability of lignocellulosic biofuels. More details on the technological aspect are presented in Chaps. 9 and 10 exploring the contribution of computational methodologies, machine learning, and artificial intelligence in enhancing bioenergy systems. These state-of-the-art, advanced techniques allow for accurate modeling, process improvement, and decision-making that pave the way to a future of successful bioenergy production. The policies and economic aspects of bioenergy are explored in Chaps. 11 and 12 of the book. These chapters focus on the global and regional policy for the promotion of biofuels and use the role played by the developed countries as well as Asian countries in the determination of biofuels markets and policies. The observations made in this contribution are essential in explaining the socioeconomic drivers of bioenergy consumption. The roles, abilities, and use of machine learning, artificial intelligence, and other computational approaches to biofuels are described in Chaps. 13 and 14. These tools develop not only the design and operating parameters of the biofuel systems but also give a vision of how data will play an important role in energy sustainability in the future. Finally, Chap. 15, "Fourth-Generation Biofuels: Genetic Engineering for Sustainable Energy," defines the next chapter of biofuel development. It concentrates on economically exploiting the existing potential in genetic engineering to break the existing barriers and pave the way for newer forms of efficient environmentally sustainable biofuels.

Zhenjiang, Jiangsu, China — Mudasir A. Dar
Guangzhou, Guangzhou, China — Hossain M. Zabed
Ladakh, India — Mohd Shahnawaz

Acknowledgments

First, we want to thank the Supreme Power Almighty, whose blessings are beyond measure. We sincerely thank all the authors for their valuable contributions to this endeavor. We are highly thankful to the Book Series Editors, Neha Srivastava and P.K. Mishra, Springer production team, particularly Nupur Verma (Publishing Editor) and Aishwarya Thyagarajan (Production Editor), for their support and guidance in making this project successful. We hope readers will like this book and become aware of the advances in biofuel production from waste lignocellulosic biomass. Last but not least, the editors are thankful to their respective families for their kind support, trust, time, and blessing to complete this project successfully.

Contents

Editors and Contributors

About the Editors

Mudasir A. Dar is an Associate Professor at Jiangsu University, China. Dr. Dar has conducted extensive research at the Biofuels Institute (China), the National Center for Microbial Resource, and the University of Pune (India). He has more than 15 years of research experience during which he harnessed high-throughput sequencing technologies to better understand the bio-conversion of lignocellulose in animal gut-systems. He has received several grants including Foreign Outstanding Young Scientist (NFSC) grant and China Postdoctoral Fund and Foreign Expert fund from the Ministry of Science and Technology (MoST) of PR China. He has published over 90 publications, including 69+ Scopus-indexed papers and book chapters with over 1100 citations. He is serving as an active editorial board member of several peer-reviewed international journals. He is also an active regular reviewer of over 70 peer-reviewed high quality international journals. He is also a Visiting Scientist at the National Research and Innovation Agency (BRIN), in Indonesia. Dr. Dar has edited/authored several books of international repute. He has also served as an international examiner for PhD theses from Beni Suef University-Egypt, Durban University-South Africa, Lovely Professional University, and Annamalai University-India, to name a few. He has several national and international awards to his credit such as Maulana Azad National Senior Research Fellow, DST PURSE fellow, UGC-UPE II fellow, and CSIR-UGC-NET (All India Rank 41).

Hossain M. Zabed is currently working as a Professor at the School of Life Sciences, Guangzhou University, China. He also previously worked at Jiangsu University, China (as a Professor), Primeasia University, Bangladesh, and Incepta Pharmaceuticals Ltd., Bangladesh. He completed his B.Sc. (Honours) and M.S. in Microbiology from the University of Chittagong, Bangladesh, and his Ph.D. in Biotechnology from the University of Malaya, Malaysia. He was selected as an "Outstanding Hundred Talent" of Guangzhou University (2023), a "Doctor of Innovation and Entrepreneurship" of the Jiangsu province of China (2021), and a "Foreign Young Talent" of the Ministry of Science and Technology of China (2020 and 2022). His main research area is synthetic biology-mediated conversion of waste and biomass into biofuels and high-value chemicals. In particular, he works on metabolic engineering, protein engineering, bioprocess engineering, fermentation technology, and circular bioeconomy. He received research grants from the National Natural Science Foundation of China (NSFC), the Ministry of Science and Technology of China, the China Postdoctoral Science Foundation, the Jiangsu Provincial Science & Technology Department, and the Guangzhou Municipal Government. He has published over 110 publications, including 90+ SCI/ISI-cited papers and 10+ book chapters. His publications received >5200 citations with an h-index of 31. He has been recognized as one of the world's top 2% highly-cited scientists consecutively for the past 4 years (2021–2023). He presented his research outcomes at a good number of international conferences as a plenary/keynote/invited speaker. Prof. Zabed has served as a guest editor/editorial board member of several international journals and as a reviewer for over 50 journals. He serves/served as a member of several international societies and organizations.

Mohd Shahnawaz is currently working as an Assistant Professor (contractual) at the Department of Botany, Govt. Degree College Drass, A Constituent College of University of Ladakh, Dras, Ladakh, India. In 2022–23 session, he worked as an Assistant Professor (contractual) at the Department of Botany, Kargil Campus, University of Ladakh, Khumbathang, Ladakh, India. In 2021, he was selected as an Associate Professor under invited talent program at Biofuel Institute, School of the Environment and Safety Engineering, Jiangsu University, Zhenjiang, China. In the year, he was selected as a postdoc fellow at Yeungnam University, South Korea. However, due to the COVID-19 scenario, he was not able to join. Dr. Shahnawaz got selected for DST-SERB National Postdoctoral Fellowship and worked at Plant Biotechnology Division, CSIR-Indian Institute of Integrative Medicine, Canal Road, Jammu, Jammu and Kashmir, India. He has several years of teaching and postdoctoral research experience. Dr. Shahnawaz worked on bioremediation of polythene at the Botany Department, SP Pune University, during his doctoral research with Prof. A. B. Ade. He is recipient of various fellowships awarded by the SPPU, UGC-BSR, UGC-MANF, and DST-SERB. His research interest is in ecology, biofuels, microbiology, bioremediation, plant biotechnology, and integrative medicine. Currently, he is focused on the mitigation of microplastic using bioremediation technology. He has served as a reviewer and editor for several international journals and had received the Young Scientist Award (2019) from VDGood Professional Association at Vishakhapatnam, India. He has published more than 25 research articles in peer-reviewed journals, authored 24 book chapters and authored/edited 16 books of international repute with a total of 1360 citations.

Contributors

Waha Ismail Yahia Abdelmula International Joint Laboratory on Synthetic Biology and Biomass Biorefinery, School of Emergency Management, School of the Environment and Safety Engineering, Biofuels Institute, Jiangsu University, Zhenjiang, People's Republic of China

Jiangsu Collaborative Innovation Center of Technology and Material of Water Treatment, Suzhou University of Science and Technology, Suzhou, People's Republic of China

Kenneth Kennedy Adama Department of Chemical Engineering, Faculty of Engineering, Edo State University, Uzairue, Edo State, Nigeria

Farhan Ahmad Biofuels Institute, School of Environment and Safety Engineering, Jiangsu University, Zhenjiang, China

International Joint Laboratory on Synthetic Biology and Biomass Biorefinery, School of Emergency Management, School of the Environment and Safety Engineering, Biofuels Institute, Jiangsu University, Zhenjiang, People's Republic of China

Jiangsu Collaborative Innovation Center of Technology and Material of Water Treatment, Suzhou University of Science and Technology, Suzhou, People's Republic of China

Sameh S. Ali Biofuels Institute, School of Environment and Safety Engineering, Jiangsu University, Zhenjiang, China

Shehbaz Ali Biofuels Institute, School of Environment and Safety Engineering, Jiangsu University, Zhenjiang, China

Wajeeha Ali Department of Biotechnology, University of Sargodha, Sargodha, Pakistan

Wajid Ali Department of Biotechnology, University of Sargodha, Sargodha, Pakistan

Osikemekha Anthony Anani Laboratory for Hydrobiology and Ecotoxicology, Department of Animal and Environmental Biology, Delta State University, Abraka, Delta State, Nigeria

Laboratory for Ecotoxicology and Hydrobiology, Department of Animal and Environmental Biology, Delta State University Abraka, Abraka, Nigeria

Trisanti Anindyawati Research Center for Applied Microbiology, National Research and Innovation Agency (BRIN), Cibinong Bogor, Indonesia

Fatima Atif Department of Biotechnology, University of Sargodha, Sargodha, Pakistan

Anjuman Ayub School of Bioengineering and Biosciences, Lovely Professional University, Jalandhar, Punjab, India

Swastika Banerjee Department of Science, Alliance University, Bangalore, Karnataka, India

Mukul Machhindra Barwant Department of Botany, Sanjivani Arts Commerce and Science College, Kopargaon, India

Showkeen Muzamil Bashir Biochemistry and Molecular Biology Lab, Division of Veterinary Biochemistry, Faculty of Veterinary Sciences and Animal Husbandry, Sher-e-Kashmir University of Agricultural Sciences and Technology, Srinagar, Jammu and Kashmir, India

Sunil P. Bhagat Department of Botany, ADTs Sharadabai Pawar Mahila Arts, Commerce and Science College, Shardanagar Baramati, Maharashtra, India

Mudasir A. Dar Biofuels Institute, School of Environment and Safety Engineering, Jiangsu University, Zhenjiang, Jiangsu, China

International Joint Laboratory on Synthetic Biology and Biomass Biorefinery, School of Emergency Management, School of the Environment and Safety Engineering, Biofuels Institute, Jiangsu University, Zhenjiang, People's Republic of China

Jiangsu Collaborative Innovation Center of Technology and Material of Water Treatment, Suzhou University of Science and Technology, Suzhou, People's Republic of China

School of Environment and Safety Engineering, Biofuels Institute of Jiangsu University, Zhenjiang, Jiangsu, China

Apridah Cameliawati Djohan Research Center for Applied Microbiology, National Research and Innovation Agency (BRIN), Cibinong Bogor, Indonesia

Kossivi Fabrice Dossa Faculty of Forestry, Geography and Geomatics, Laval University, Quebec, QC, Canada

Action-Research for Sustainable Development NGO, Department of Research Project, Cotonou, Benin

Department of Agricultural Economics, Faculty of Agriculture, University of Nigeria, Nsukka, Nigeria

Liqin Du Guangxi Microorganism and Enzyme Research Center of Engineering Technology, College of Life Science and Technology, Guangxi University, Nanning, Guangxi, China

Zahra Eslamifar Department of Laboratory Sciences, School of Allied Medical Sciences, Dezful University of Medical Sciences, Dezful, Iran

Arthur Efeoghene Essaghah Department of Urban and Regional Planning, Delta State University of Science and Technology (DSUST), Ozoro, Delta State, Nigeria

Peter Mudiaga Etaware Department of Plant Science and Biotechnology, Faculty of Science, Delta State University of Science and Technology, Ozoro, Delta State, Nigeria

Uduenevwo Francis Evuen Department of Biochemistry, Faculty of Science, Delta State University of Science and Technology, Ozoro, Delta State, Nigeria

Marcelo Franco Department of Exact Science, State University of Santa Cruz, Ilheus, Brazil

Pratiksha Fulari Department of Botany, Sanjivani Arts Commerce and Science College, Kopargaon, India

Arezoo Gheibipour Department of Physics, Faculty of Sciences, Urmia University, Urmia, Iran

Fateme Gheibipour Department of Humanities, Payam Noor Andimeshk University, Andimeshk, Iran

Maryam Gheibipour Environmental Technologies Research Center, Ahvaz Jundishapur University of Medical Sciences, Ahvaz, Iran

Department of Animal Science, Faculty of Agriculture, University of Birjand, Birjand, Iran

Center for Climate Change and Health Research (CCCHR), Dezful University of Medical Sciences, Dezful, Iran

Babbiker Mohammed Taher Gorish International Joint Laboratory on Synthetic Biology and Biomass Biorefinery, School of Emergency Management, School of the Environment and Safety Engineering, Biofuels Institute, Jiangsu University, Zhenjiang, People's Republic of China

Jiangsu Collaborative Innovation Center of Technology and Material of Water Treatment, Suzhou University of Science and Technology, Suzhou, People's Republic of China

Mahesh B. Gunjal Department of Botany, Sahakar Maharshi Bhausaheb Santuji Thorat Arts, Science and Commerce College, Sangamner, Ahilyanagar, Maharashtra, India

Md Muzammel Hossain Biofuels Institute, School of Environment and Safety Engineering, Jiangsu University, Zhenjiang, China

Prosper Monday Ohien Ikhazuangbe Department of Chemical Engineering, Faculty of Engineering, Edo State University Uzairue, Uzairue, Edo State, Nigeria

Muhammad Irfan Department of Biotechnology, University of Sargodha, Sargodha, Pakistan

Research Center for Applied Microbiology, National Research and Innovation Agency (BRIN), Cibinong Bogor, Indonesia

Festus Oamen Isaac Department of Mechanical Engineering, Faculty of Engineering, Edo State University, Uzairue, Edo State, Nigeria

Sahand Jorfi Environmental Technologies Research Center, Ahvaz Jundishapur University of Medical Sciences, Ahvaz, Iran

Department of Environmental Health Engineering, School of Health, Ahvaz Jundishapur University of Medical Sciences, Ahvaz, Iran

Shreya K. Kasar Department of Botany, S.N. Arts, D.J.M. Commerce and B.N.S. Science College (Autonomous), Sangamner, Ahilyanagar, Maharashtra, India

Ika Rahmatul Layly Research Center for Applied Microbiology, National Research and Innovation Agency (BRIN), Cibinong Bogor, Indonesia

Zhenchong Li Guilin University of Technology, Guilin, Guangxi, China

Jiankai Lin Guangxi Microorganism and Enzyme Research Center of Engineering Technology, College of Life Science and Technology, Guangxi University, Nanning, Guangxi, China

Debasree Lodh Department of Botany, Holy Cross College, Agartala, Tripura, India

S. M. Hasan Mahmud Department of Computer Science, American International University-Bangladesh (AIUB), Dhaka, Bangladesh

Saira Mansab Department of Environmental Sciences, The Women University Multan, Multan, Pakistan

Nida Maqsood Department of Biotechnology, University of Sargodha, Sargodha, Pakistan

Yann Emmanuel Miassi Faculty of Forestry, Geography and Geomatics, Laval University, Quebec, QC, Canada

Action-Research for Sustainable Development NGO, Department of Research Project, Cotonou, Benin

Department of Agricultural Economics, Faculty of Agriculture, Çukurova University, Adana, Turkey

Ali Abdulmawjood Mohammed College of Science, Mustansiriyah University, Baghdad, Iraq

Naiem Harun Nadaf Department of Microbiology, Shivaji University, Kolhapur, Maharashtra, India

Salama Harun Nadaf Department of Biotechnology, Vivekananda College, Kolhapur, Maharashtra, India

Sahar Nasreen Department of Biotechnology, Fatima Jinnah Women University, Rawalpindi, Pakistan

Saima Nasreen Department of Environmental Sciences, The Women University Multan, Multan, Pakistan

Chukwudi Ogwu Department of Environmental Management and Toxicology, Delta State University of Science and Technology, Ozoro, Delta State, Nigeria

Shiyou Pan Guangxi Microorganism and Enzyme Research Center of Engineering Technology, College of Life Science and Technology, Guangxi University, Nanning, Guangxi, China

Maryam Panahizade Department of Animal Science, Faculty of Animal Science and Food Industry, Ramin Khuzestan University of Agriculture and Natural Resources, Molasani, Iran

Kousar Parveen Department of Environmental Sciences, The Women University Multan, Multan, Pakistan

Juliana Heloisa Pinê Américo Pinheiro Department of Forest Science, Soils and Environment, School of Agronomic Sciences, São Paulo State University (UNESP), Botucatu, São Paulo, Brazil

Brazil University, São Paulo, São Paulo, Brazil

Xianghui Qi School of Life Sciences, Guangzhou University, Guangzhou, Guangdong Province, China

Farida Rahayu Research Center for Genetic Engineering, National Research and Innovation Agency, Bogor, Indonesia

Niloofar Rahimi Babadi Department of Midwifery, Faculty of Medical Sciences, Karaj Branch, Islamic Azad University, Karaj, Iran

Praveen B. Roylawar Department of Botany, S.N. Arts, D.J.M. Commerce and B.N.S. Science College (Autonomous), Sangamner, Ahilyanagar, Maharashtra, India

Savita M. Satpute Department of Botany, Shankarrao Mohite Mahavidyalaya, Akluj, Solapur, Maharashtra, India

Mohd Shahnawaz Department of Botany, Govt. Degree College Drass, A Constituent College of University of Ladakh, Ladakh, India

Shalagha A. B. Sharma School of Biological Engineering, Shobhit Deemed to be University, Meerut, Uttar Pradesh, India

Anjuvan Singh School of Bioengineering and Biosciences, Lovely Professional University, Jalandhar, Punjab, India

Reena Singh School of Bioengineering and Biosciences, Lovely Professional University, Jalandhar, Punjab, India

Jianzhong Sun Biofuels Institute, School of Environment and Safety Engineering, Jiangsu University, Zhenjiang, China

International Joint Laboratory on Synthetic Biology and Biomass Biorefinery, School of Emergency Management, School of the Environment and Safety Engineering, Biofuels Institute, Jiangsu University, Zhenjiang, People's Republic of China

Jiangsu Collaborative Innovation Center of Technology and Material of Water Treatment, Suzhou University of Science and Technology, Suzhou, People's Republic of China

Jamila A. Tuly School of Life Sciences, Guangzhou University, Guangzhou, Guangdong Province, China

Balasaheb D. Wagh Department of Botany, Sahakar Maharshi Bhausaheb Santuji Thorat Arts, Science and Commerce College, Sangamner, Ahilyanagar, Maharashtra, India

Atif Khurshid Wani School of Bioengineering and Biosciences, Lovely Professional University, Jalandhar, Punjab, India

Yutuo Wei Guangxi Microorganism and Enzyme Research Center of Engineering Technology, College of Life Science and Technology, Guangxi University, Nanning, Guangxi, China

Rongrong Xie Biofuels Institute, School of Environment and Safety Engineering, Jiangsu University, Zhenjiang, China

International Joint Laboratory on Synthetic Biology and Biomass Biorefinery, School of Emergency Management, School of the Environment and Safety Engineering, Biofuels Institute, Jiangsu University, Zhenjiang, People's Republic of China

Jiangsu Collaborative Innovation Center of Technology and Material of Water Treatment, Suzhou University of Science and Technology, Suzhou, People's Republic of China

Mahta Yousefipour Center for Climate Change and Health Research (CCCHR), Dezful University of Medical Sciences, Dezful, Iran

Department of Educational Sciences, Farhangian University, Tehran, Iran

Junhua Yun School of Life Sciences, Guangzhou University, Guangzhou, Guangdong Province, China

Hossain M. Zabed School of Life Sciences, Guangzhou University, Guangzhou, Guangdong Province, China

Guoyang Zhang School of Life Sciences, Guangzhou University, Guangzhou, Guangdong Province, China

Daochen Zhu Biofuels Institute, School of Environment and Safety Engineering, Jiangsu University, Zhenjiang, China

International Joint Laboratory on Synthetic Biology and Biomass Biorefinery, School of Emergency Management, School of the Environment and Safety Engineering, Biofuels Institute, Jiangsu University, Zhenjiang, People's Republic of China

Jiangsu Collaborative Innovation Center of Technology and Material of Water Treatment, Suzhou University of Science and Technology, Suzhou, People's Republic of China

Chapter 1
Biofuels as Emerging and Sustainable Alternatives to Fossil Fuels: Environmental, Energetic, and Economic Perspectives

Saima Nasreen, Sahar Nasreen, Kousar Parveen, and Saira Mansab

Abstract This chapter highlights the importance of biofuels in multiple perspectives to explore them as a best suitable eco-friendly and sustainable alternative to fossil fuels. Rapid industrialization leads to the overconsumption of fossil fuels in order to fulfill the need for energy. However, due to the adverse environmental impacts of fossil fuels, a replacement energy source is required, one that is eco-friendly, sustainable, and economical, as well as capable of fulfilling the energy demands of industrial sectors. Therefore, researchers are working on the development of alternative renewable sources of energy that fulfill the criteria of an eco-friendly, sustainable, and economical energy source. This chapter begins by exploring the environmental and health hazards inflicted by fossil fuels and how immensely the planet needs a sustainable alternative to rectify issues such as fuel inefficiency, fluctuation in oil prices, and future energy demand. The chapter further investigates the production process, types, and specifications of biofuels. Hydrolysis, saccharification, transesterification, and fermentation are the main production processes of biofuels. Biofuels are classified into four types based on the natural source of feedstock: first-, second-, third-, and fourth-generation biofuels. They have versatile applications in different fields, such as transportation, power generation, industrial processes, etc., along with potential drawbacks and future prospects. The energy balance of biofuels emphasizes the importance of a positive energy input-to-output ratio. The important role of biofuels in the circular economy is highlighted by exploring their potential to create jobs and promote rural development. The

S. Nasreen (✉) · K. Parveen · S. Mansab
Department of Environmental Sciences, Faculty of Life Sciences, The Women University Multan, Multan, Pakistan
e-mail: saima.6333@wum.edu.pk

S. Nasreen
Department of Biotechnology, Faculty of Science and Technology, Fatima Jinnah Women University, Rawalpindi, Pakistan

M. A. Dar et al. (eds.), *Recent Trends in Lignocellulosic Biofuels and Bioenergy*, Clean Energy Production Technologies,
https://doi.org/10.1007/978-981-96-4636-4_1

chapter concludes by addressing the limitations and discussing the future prospects of adopting biofuels as a potential green energy source for mankind.

Keywords Fossil fuels · Biofuels · Sustainable alternative · Eco-friendly · Economically feasible

1.1 Introduction

Energy resources are the most crucial factor for the long-term sustainability of mankind. It is predicted that energy consumption between 2005 and 2030 will rise 2.6 times (Sun 2024), leading to the drastic depletion of fossil fuels and environmental degradation. Therefore, the fulfillment of energy demand and environmental protection are the two supreme concerns of industrialists and environmentalists (Ambaye et al. 2021; Malik et al. 2024). Any substance that can interact with other substances and produce heat energy is termed fuel. Fuel has three physical states, i.e., solid (wood, charcoal, coal, etc.), liquid (diesel, gasoline, ethanol, LPG, etc.), or gas (coal gas, propane, methane, etc.). Fossil fuels are hydrocarbons formed through fossilization, where plant and animal biomass is compressed over long periods under elevated pressure and temperature. The extracted derivatives of fossil fuels include propane, petrol, and kerosene. Humans have been utilizing fossil fuels for cooking, heating, electricity generation, construction, transportation, and production of industrially important chemicals (Muhammad et al. 2018). It has not only powered modern economies but also brought up significant environmental and health concerns. To overcome this situation, an alternative approach is adopted through the development of biofuels as a sustainable energy source. It is reported that 80% of the world's energy demands are met using fossil fuels (Khan et al. 2022). The United Nations Environment Emission, 2015, reported the release of almost 54 gigatons of CO_2 equivalent by the road transport sector only, and this is expected to elevate to about 87 gigatons by 2050 (Ambaye et al. 2021; Khan et al. 2022). The burning of fossil fuels leads to the release of toxic greenhouse gases like carbon dioxide, SOx, NOx, carbon monoxide, unburned hydrocarbons, particulate matter, etc. The impacts of these toxic gases and particulate matter result in the entrapment of greenhouse gases (GHGs) in Earth's atmosphere, causing global climate change, acid rain formation, ozone depletion, heavy metal contamination, etc.

1.2 Alternative Energy Source: Biofuels

The uncontrolled inflation of oil prices and the negative environmental impacts of fossil fuels have diverted the research scope toward biofuels. Biofuels are produced from biological substrates (Ambaye et al. 2021). They possess distinctive characteristics, such as being biodegradable, nontoxic, and economically feasible. As a renewable energy source, biofuels minimize the release of air pollutants such as GHGs, NOx, SOx, and unburnt hydrocarbons into the atmosphere. Biofuel synthesis and utilization are considered carbon-neutral processes (Ambaye et al. 2021; Malla et al. 2023). Biofuels can also be synthesized locally, consequently reducing the dependence on imported oil (Bhattarai et al. 2011). The biofuel industry can create job opportunities in diverse fields such as agriculture, engineering, and transportation. The bioenergy sector has systemized around 2.8 million jobs in 2016 (Ambaye et al. 2021). However, certain challenges emerge when switching to biofuels, such as infrastructure development, political risks, and high production costs, which require reliable solutions that ensure their sustainability and environmental friendliness (Malla et al. 2023). Despite various constraints, the biofuel industry has been flourishing well, with a significantly increasing share in global energy consumption.

This chapter aims to highlight the significance of biofuels as a sustainable alternative to biofuels. It focuses on achieving the following objectives:

- Describe the characteristics of biofuels.
- Specification and importance of biofuels.
- Types and classification of biofuels.
- Applications of biofuels.
- Assess environmental impact.
- Explore economic perspective.
- Analyze energy perspective.

By addressing these areas, the chapter aims to present biofuels as a viable solution to the environmental and economic challenges posed by fossil fuel consumption.

1.3 Biofuel Characteristics

Biofuels offer prospective environmental benefits and energy security. The physicochemical characteristics of biofuels play a prominent part in evaluating their appropriateness as an alternative to fossil fuels. Physicochemical qualities such as viscosity, calorific value, density, flash point, acid number, cloud point, cetane number, sulfur content, and oxygen content determine the performance, combustion characteristics, and environmental effects of biofuels.

1.3.1 Physical Characteristics

(a) *Density*: Density affects the energy content and combustion properties of biofuels. Biodiesel generally has a density of roughly 0.88 g/cm^3, slightly more than that of petroleum diesel, which is around 0.85 g/cm^3 (Bajpai and Tyagi 2006). Meanwhile, bioethanol has a density ranging from 0.80 to 0.82 kg/L at 15 °C, which is greater than gasoline's density of 0.75 kg/L (Yasin et al., 2019.

(b) *Viscosity*: Viscosity is critical for fuel injection and atomization in engines. Biodiesel has a kinematic viscosity at 40 °C ranging from 1.9 to 6.0 mm^2/s, which is within the permissible range for diesel engines (McCormick and Moriarty, 2023). Bioethanol has flow properties equivalent to gasoline (Yasin et al., 2019).

(c) *Calorific value*: The calorific value of biodiesel is around 37 MJ/kg, which is lower than the calorific value of petroleum diesel (41.3 MJ/kg), and the calorific value of bioethanol is lower than that of biodiesel, i.e., 26.4 MJ/kg (Yasin et al., 2019).

(d) *Flash point*: The flash point of biodiesel is high, often between 100 °C and 170 °C, making it safer for handling and storage compared to petroleum diesel, which has a flash point of 52 °C. Bioethanol has a flash point of 15 °C, showing greater volatility compared to biodiesel (Yasin et al., 2019).

(e) *Cloud point and pour point*: The cloud point of biodiesel spans from −3 to 15 °C, whereas it has a pour point ranging from −5 to 10 °C. The cloud and pour points fluctuate within the range depending on the kind of feedstock utilized as a source. These qualities are examined for cold weather performances and flow difficulties of the fuel.

1.3.2 Chemical Characteristics

(a) *Acid number*: The acid number assesses the free fatty acid level in biodiesel, which may affect engine corrosion and fuel stability. A typical acid number for biodiesel is about 0.497 mg KOH/g, which is somewhat higher than that of petroleum diesel (0.35 mg KOH/g) but still within safe limits for engine operation (Karmakar et al. 2018).

(b) *Cetane number*: The cetane number is the quality of combustion the diesel fuel possesses throughout the process of compression/ignition. The cetane number of biodiesel is normally between 47 and 65, making it greater than that of petroleum diesel (45–50). This high cetane number of biodiesels gives them with superior ignition qualities and smoother engine performance (McCormick and Moriarty, 2023).

(c) *Sulfur content*: Less content of sulfur (0.0–0.0015%) in biofuel is regarded as a sustainable, environment-friendly, and affordable option to petroleum diesel, which has substantial sulfur content. Low sulfur concentration in biodiesel minimizes sulfur dioxide emissions, leading to reduced air pollution.

(d) *Oxygen content*: Biofuels like bioethanol and biodiesel have oxygen content, which boosts combustion efficiency and minimizes the emission of unburned hydrocarbons. Bioethanol has an oxygen concentration of 34.8% by weight (Yasin et al., 2019).

(e) *Ash content and carbon residue*: The ash content and carbon residue of biofuels are indications of the existence of noncombustible elements that deposit in engines. Biodiesel usually contains low ash content and carbon residue.

1.4 Biodiesel Specifications

There are two standards designed by the American Society for Testing and Materials (ASTM) and the European Committee for Standardization (CEN) for the quality evaluation of biodiesel.

1.4.1 ASTM D6751 Standard

ASTM D6751 standard defines pure biodiesel (B100) used as a blend with conventional diesel fuel (ASTM D6751-20a 2020). Key characteristics include:

- Ester content: minimum 96.5% (m/m).
- Density: 860–900 kg/m^3 at 15 °C.
- Viscosity: 1.9–6.0 mm^2/s at 40 °C.
- Flash point: minimum 93 °C (closed cup).
- Sulfur content: maximum 0.0015% (m/m) or 0.05% (m/m) for various grades.
- Cetane number: minimum 47.

1.4.2 EN 14214 Standard

The European standard EN 14214 also defines biodiesel quality, with certain limitations (EN 14214:2012+A2:2019 2019). Key characteristics include:

- Ester content: minimum 96.5% (m/m).
- Density: 860–900 kg/m^3 at 15 °C.
- Kinematic viscosity: 3.5–5.0 mm^2/s at 40 °C.
- Flashpoint: minimum 101 °C.
- Sulfur content: maximum 10 mg/kg.
- Cetane number: minimum 51.

1.5 Bioethanol Specifications

There are two standards designed by the American Society for Testing and Materials (ASTM) and the European Committee for Standardization (CEN) for the quality evaluation of bioethanol.

1.5.1 ASTM D4806 Standard

ASTM D4806 standard defines the denaturation of bioethanol (ASTM D4806-22 2022). Key characteristics include:

- Ethanol content: minimum 92.1% (v/v).
- Density: 0.7877–0.8033 g/mL at 20 °C.
- Acidity: maximum 0.007% (m/m) as acetic acid.
- Water content: maximum 1.0% (v/v).
- Sulfur content: maximum 30 mg/kg.

1.5.2 EN 15376 Standard

The European standard EN 15376 also defines the sets of identical standards for bioethanol content and blend with gasoline (EN 15376:2014+A1:2021 2021). Key characteristics include:

- Ethanol content: minimum 98.7% (m/m).
- Density: 0.7938–0.8053 g/mL at 15 °C.
- Acidity: maximum 0.0025% (m/m) as acetic acid.
- Water content: maximum 0.3% (m/m).
- Sulfur content: maximum 10 mg/kg.

1.5.3 Other Standards

In addition to biodiesel and bioethanol, other alternative biofuels have been studied, which contain distinct qualities and standards.

- *Renewable diesel*: It is derived from biomass using hydro-treating or other procedures and fulfills the ASTM D975 or EN 590 criteria for conventional diesel fuel (Pearlson 2011).
- *Biobutanol*: It covers qualities such as vapor pressure, distillation range, and corrosion resistance. It fulfills the standards of ASTM D7862 (Dernotte et al. 2010).

- *Pyrolysis bio-oils*: They are derived from the thermochemical conversion of biomass and have unique characteristics including water content, viscosity, and heating value (Oasmaa and Meier 2005; Lehto et al. 2014).

By complying with these criteria, biofuel producers and consumers can ensure the dependable and sustainable use of these renewable energy sources. The basic properties of biofuels mentioned earlier, i.e., environment-friendliness (low carbon monoxide and particulate matter emissions in the environment, no smoke or sulfur production), renewability, high degradability, sustainability, zero toxicity, non-flammability, portability, and no release of hydrocarbons during combustion, are already promising enough to replace synthetic diesel as an energy source in heavily populated cities and industrialized areas. Using biofuels as an energy source leads to a 78% less production of CO_2 over a life cycle, compared to conventional fossil fuels. Additionally, biofuels contribute to high oxygen production (10–11%), which ultimately leads to complete combustion and reduced greenhouse gas emissions. The lubricating properties and cetane ratings of biofuels also make them suitable alternatives for low-sulfur diesel fuels, as they increase lubrication in engines, enhancing their efficiency and decreasing engine wear. The calorific value of biodiesels is around 37.27 MJ/kg, which is 9% less than that of regular petroleum diesels. The higher flash point of biodiesels (100–170 °C) makes them convenient for transportation, utilization, and storage, compared to petroleum diesels, which have a flash point of 60–80 °C (Mishra and Goswami 2018).

1.6 Biomass as a Biofuel Feedstock

Biomass as biofuel feedstock has been investigated and split into four types:

(a) Crop-based biomass (herbaceous, woody, industrial, agricultural, and aquatic crops).
(b) Agricultural residue and waste (rice, wheat straw, etc.)
(c) Forestry waste and residue (mill wood waste, logging residues, trees, shrubs, etc.)
(d) Industrial and municipal waste (municipal solid waste, sewage sludge, industrial waste).

Direct burning of biomass has been a matter of conflict for environmentalists, as it has contributed to GHG emissions and climate change at a significant level. Therefore, the utilization of biomass as biofuel has great potential for global energy security. In total, the global installed renewable energy capacity reached some 3.9 terawatts in 2023, up by almost 14 percent from the previous year (Fernández 2024). The renewable energy market has experienced an explosion over recent decades due to decreasing prices in renewable technology, as well as concerns about the environmental impact of more traditional sources (Fig. 1.1).

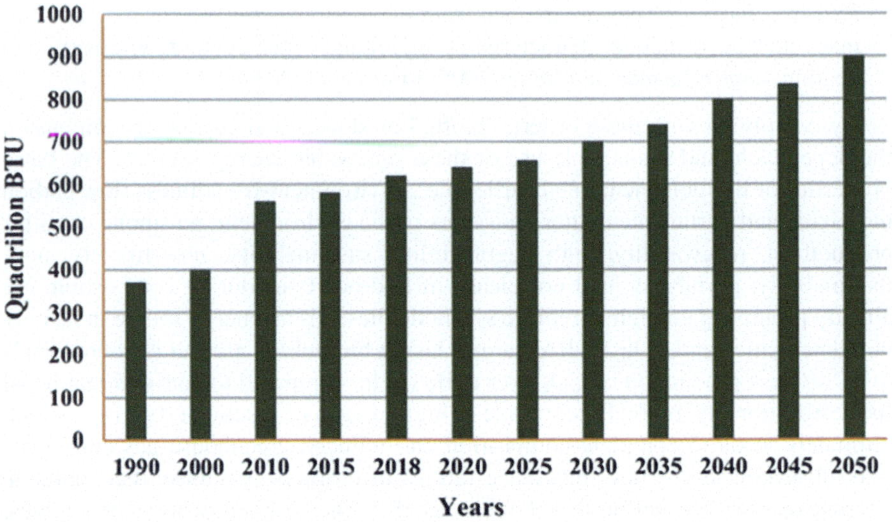

Fig. 1.1 Worldwide Energy consumption from 1990 to 2050 (El-Araby, 2024)

1.7 Types of Biofuels

Biofuels have two types, i.e., conventional and advanced biofuels (Fatma et al. 2018; Rai et al. 2022).

1.7.1 Conventional Biofuels

Conventional biofuels are produced from planting crops (sugar, starch, grain, maize, and vegetable oil), which contribute only 4% of total transport fuel consumption due to limited feedstock availability, low CO_2 mitigation, and low cost. They are carbon-neutral biofuels with low GHG emissions such as bioethanol, biodiesel, and biogas (Muhammad et al. 2018; Malik et al. 2024).

1.7.2 Advanced Biofuels

Advanced biofuels utilize waste or nonfood crops (woody biomass, grasses, agricultural by-products, seaweeds, algae) as feedstock sources. These biofuels are more sustainable and have low GHG emissions. These fuels are still in the research and development or pilot phase and have yet to be commercialized (Fig. 1.2).

A comparative analysis is given below:

	Advance Biofuel			Conventional Biofuel	
	Basic & Applied Research	Demonstration	Early Commercial Production	Commercial	
Bioethanol		Cellulosic ethanol		Sugar and Starch crop-based ethanol	
Biodiesel type Biofuel	Biodiesel (microalgae) Hydrocarbon (sugar-based)	Biomass to liquid-diesel (gasification+Fisher+tropsch)	Hydro treated Vegetable oil	Biodiesel by transesterification	Liquid Biofuel
Others fuels and additives	Novel fuels (Furanics)	Dimethyl ether, Biobutanol, Pyrolysis based fuels	Methanol		
Biomethane		Biosynthetic Gas		Biogas (anaerobic digestion)	
Biohydrogen	Other Biogas Novel routes	Gasification			Gaseous Biofuel

Fig. 1.2 Overview of conventional and advanced biofuels. Source: Malik et al. (2024)

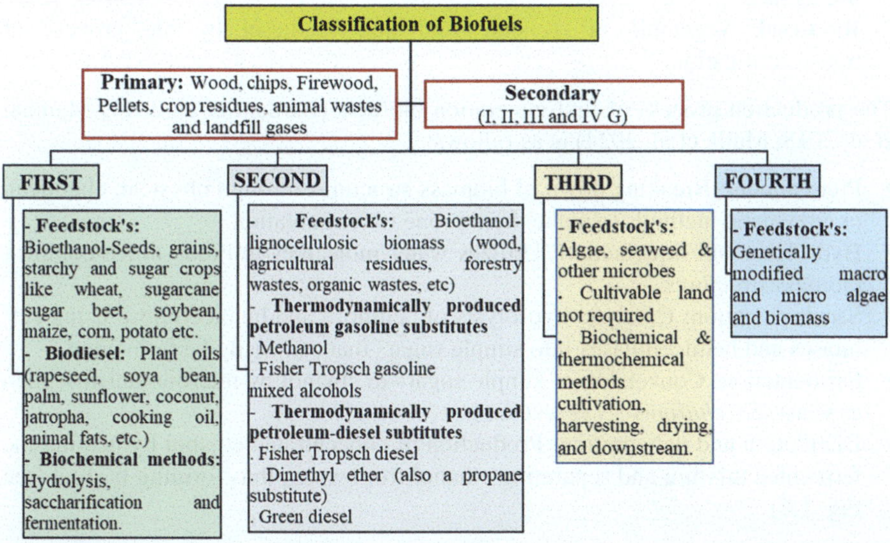

Fig. 1.3 A generalized classification of biofuels and their production sources. Source: Malik et al. (2024)

1.8 Classification of Biofuels

Biofuels are distinguished into different generations based on their class, nature, type of feedstock, production processes, and type of catalysts as shown in Fig. 1.3.

1.8.1 First-Generation Biofuels

When biofuels are synthesized using organic sources such as food crops rich in sugar, starch, and vegetable oils, e.g., soybean, rapeseed, or palm oil, they are termed first-generation biofuels (Demirbas 2009a, b).

1. **Bioethanol**: Carbohydrates are converted into simple sugars, which are then fermented by microorganisms, typically yeast, known as bioethanol (Balat and Balat 2009). Bioethanol is an alcohol fuel made from biological matter, such as crops or waste products, through a fermentation process. It is produced by fermenting sugars and starches found in biomass, such as corn and sugarcane, using microorganisms like yeast. The fermentation process breaks down the sugars in the biomass into alcohol (ethanol), which can then be distilled and purified for use as fuel.
2. **Biodiesel**: Vegetable oil is converted into biodiesel by the process of transesterification.

The production process of first-generation biofuels (Bardant et al. 2017; Manmai et al. 2018; Malik et al. 2024) is as follows:

- Pretreatment: Breaking down of biomass structures through physical, chemical, or enzymatic methods to make them prone to fermentation.
- Hydrolysis and liquefaction: Utilizes water molecules to break down complex biomass structures.
- Saccharification: Catalytic hydrolysis of cellulose and hemicellulose using cellulases and hemicellulases into simple sugars that can easily be fermented.
- Fermentation: Conversion of simple sugars to ethanol by microorganisms, such as yeast (*Saccharomyces cerevisiae*).
- Distillation and dehydration: Production of concentrated ethanol by heating the fermented mixture and separating ethanol from water, thus forming biofuel (see Fig. 1.4).

While first-generation biofuels have contributed to the development of the biofuel industry, they face criticism due to their potential adverse impacts on food security and land-use changes (Naik et al. 2010).

1.8.2 Second-Generation Biofuels

Biofuels synthesized using nonfood biomass sources, such as agricultural residues, forestry waste, and energy crops like switchgrass or miscanthus, are termed second-generation biofuels (Sims et al. 2010).

- **Cellulosic ethanol**: The process requires pretreatment to fractionate the complex biomass structure, followed by enzymatic hydrolysis, which involves the

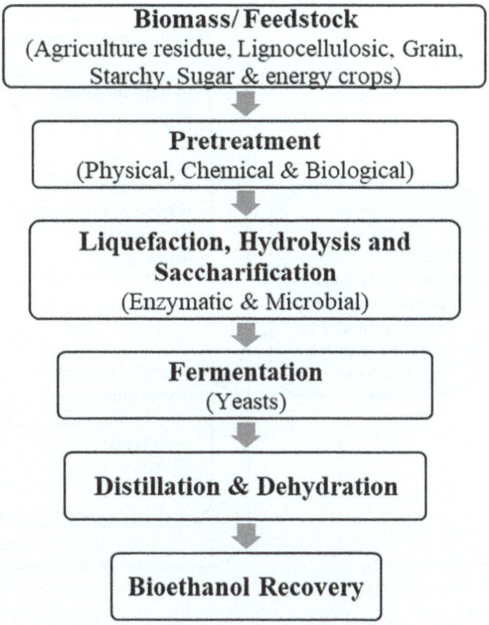

Fig. 1.4 The major stages involved in the production of second-generation/ lignocellulosic biofuels. Source: Malik et al. (2024)

conversion of cellulose and hemicellulose into sugars that may be fermented to make ethanol (Balat 2011).

- **Biomass-to-liquid fuels**: The process of thermochemical transformation of lignocellulosic biomass into biomass-to-liquid (BTL) fuels begins with gasification of biomass, for synthesis gas (syngas) production, which is then followed by the catalytic conversion of syngas into liquid hydrocarbons (Bridgwater 2012). BTL fuels include renewable diesel and gasoline.

The two basic approaches used to produce second-generation biofuels are as follows:

1. **Thermochemical processes**: Thermochemical processes such as combustion, gasification, liquefaction, hydrogenation, and pyrolysis are used for biofuel production. Pyrolysis is the most beneficial and commonly opted method because it converts biomass into any physical state (solid, liquid, or gas) through the thermal decomposition of biomass in anaerobic conditions (Fig. 1.5). This process may or may not require a catalyst. Important products of pyrolysis include pyro gas, pyrolytic oil, and char.
2. **Biochemical processes**: Biochemical processes such as anaerobic digestion and alcoholic fermentation are used for biofuel production through biological actions, converting semisolid or liquid-state biomass sources into biogas or liquid biofuels (Muhammad et al. 2018).

Biological digestion or microbial fermentation products are obtained in the presence of biological catalysts.

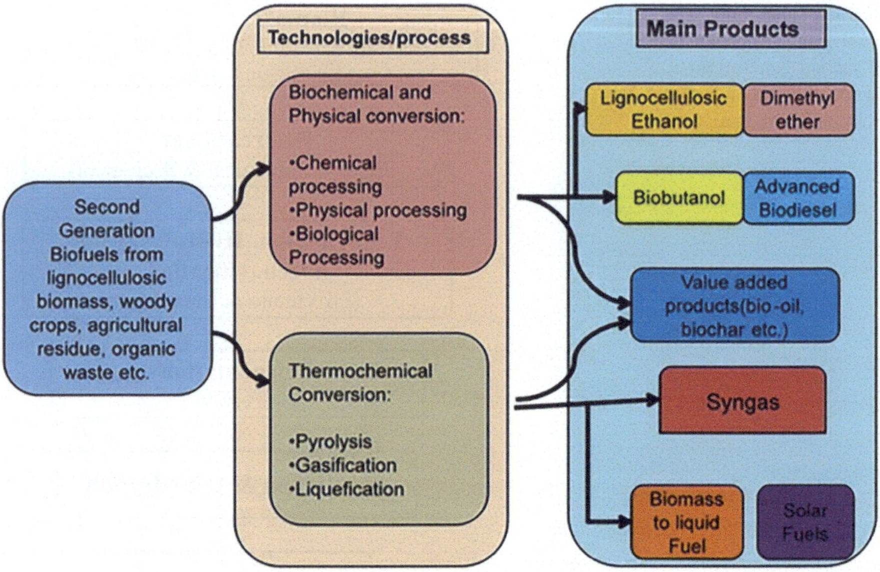

Fig. 1.5 First-generation biofuel production (Malik et al. 2024)

Second-generation biofuels offer improved sustainability and reduced competition with food production.

1.8.3 Third-Generation Biofuels

Biofuels synthesized using microalgae are termed third-generation biofuels (Brennan and Owende 2010).

- **Algal biodiesel**: The process involves cultivating and harvesting the algae to extract lipids, which are then converted into biodiesel through transesterification (Chisti 2007).
- **Algal bioethanol**: The process involves the fermentation of sugars through carbohydrate hydrolysis (Harun et al. 2010).

Third-generation biofuel production involves cultivation, cell separation, transesterification, and biodiesel harvesting, as shown in Fig. 1.6:

Third-generation biofuels offer improved sustainability and reduced competition with food production.

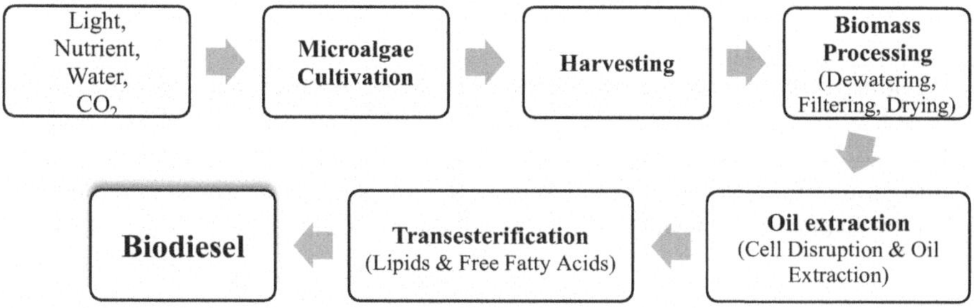

Fig. 1.6 Third-generation biofuel production. Source: Malik et al. (2024)

1.8.4 Fourth-Generation Biofuels

Fourth-generation biofuels involve the use of genetically modified organisms (GMOs) and synthetic biology approaches to produce advanced biofuels (Liao et al. 2016).

- **Metabolically engineered microorganisms**: Metabolically engineered microorganisms, such as bacteria, yeast, or algae, can be designed to produce specific biofuel molecules directly from biomass (Liao et al. 2016). This approach aims to improve the efficiency and sustainability of biofuel production by optimizing metabolic pathways and reducing the number of processing steps.
- **Synthetic biology**: Synthetic biology involves the design and construction of artificial biological systems or pathways to produce biofuels (Liao et al. 2016).

Fourth-generation biofuels are still in the early stages of research and development, but they hold the potential to revolutionize biofuel production by addressing some of the limitations of earlier generations (Liao et al. 2016). These biofuels leverage advances in genetic engineering and synthetic biology to create novel biofuel production pathways.

1.9 Applications of Biofuel

The versatile applications of biofuels in different sectors, such as transportation, power generation, heating, industrial processes, etc. are listed below:

Transportation: This sector consumes the largest percentage of biofuel produced globally, both bioethanol and biodiesel, in different vehicles, and presents a cleaner alternative to gasoline and LPG.

- Blending bioethanol with gasoline produces E10 and E85, with different ratios, and these combinations can be used in conventional engines that operate on the principle of combustion (Renewable Fuels Association 2021). Flex-fuel vehicles (FFVs) are particularly programmed to operate on higher ethanol blends, for instance, E85, which offers greater flexibility and reduces dependence on fossil fuels (U.S. Department of Energy 2021).

- Blending biofuel with petroleum diesel in different ratios, such as B5 (5% biodiesel, 95% diesel) and B20 (20% biodiesel, 80% diesel), allows these combinations to be utilized in engines operating on diesel with minimal alterations, mitigating the release of particulate matter, CO, and hydrocarbons into the atmosphere (National Renewable Energy Laboratory 2021).

1.9.1 Power Generation

The power generation sector predominantly utilizes biofuels for energy production, which is a reliable and sustainable energy source for lighting houses and running mills. The combined heat and power systems utilize biofuels for the production of both heat and electricity, which ultimately leads to reduced GHG emissions into the atmosphere.

- **Biomass power plants**: Solid biofuels, i.e., wood chips, energy crops, and agricultural wastes, are used by biomass power plants to produce electricity. These plants can serve as independent facilities or be co-fired with coal in preexisting power plants, lowering the carbon footprint of electricity production (International Renewable Energy Agency 2020).
- **Combined heat and power (CHP) systems**: CHP systems generate electricity and usable heat from biofuels. These systems are very efficient, trapping and using the heat that would otherwise be squandered in traditional power production. CHP systems may employ different biofuels, including biogas, biodiesel, and bioethanol, to supply energy for industrial operations, district heating, and residential uses (U.S. Environmental Protection Agency 2021).

1.9.2 Heating

The heating units in residential and commercial areas can utilize biofuels for heating applications. Biofuels are comparatively cleaner, more reliable, and more sustainable alternatives to conventional fuels that have been used for heating for decades, e.g., propane, natural gas, and heating oil.

- **Residential heating**: Biofuels such as wood pellets, biodiesel, and biogas are used for space heating and water heating. Pellet stoves and boilers are popular options for using wood pellets, providing efficient and low-emission heating solutions (Biomass Energy Centre 2021). Biodiesel can be used in oil-fired furnaces and boilers, offering a cleaner burning alternative to traditional heating oil (U.S. Department of Energy 2021).

- **Industrial Heating**: Biofuels are used for process heating, providing energy for manufacturing, drying, and other industrial processes. Biomass boilers and furnaces can use various biofuels, including wood chips, agricultural residues, and biogas, to generate heat for industrial applications (International Renewable Energy Agency 2020).

1.9.3 Industrial Processes

Biofuels in industrial processes play a role in the development of a bio-based economy, minimizing the dependence on fossil resources and lowering the environmental impact of industrial activities.

- **Bio-based chemicals and materials**: Biofuels serve as feedstock for the synthesis of bio-based chemicals and materials, such as bioplastics, bio-lubricants, and bio-solvents. These products offer renewable alternatives to conventional petrochemical-based products, contributing to sustainability and reducing the carbon footprint of industrial processes (European Commission 2021).
- **Biorefineries**: Biorefineries integrate the production of biofuels with the synthesis of bio-based chemicals, materials, and other products. This integrated approach maximizes the value of biomass feedstock, improving sustainability and economic feasibility for biofuel production.

1.10 Biofuel Environmental Consideration

The ecological footprint of biofuels encompasses various environmental impacts, including land use, water use, biodiversity, soil health, and pollution. These impacts vary, depending on the types of biofuel, feedstocks, and production methods used.

(a) **Land use**: The cultivation of biofuel feedstocks requires significant land resources, which can lead to land-use changes and associated environmental impacts. Land-use change has contributed to the addition of around 600 ± 290 Gt of carbon dioxide into the atmosphere from 1750 to 2011. The conversion of natural ecosystems, such as forests and grasslands, into agricultural land for biofuel production can result in habitat loss, reduced biodiversity, and increased carbon emissions due to the release of stored carbon in vegetation and soil (Stoeglehner and Narodoslawsky 2009; Sahin 2011; Jeswani et al. 2020).

(b) **Water use**: Biofuel production, particularly first-generation biofuel production, is water-intensive, requiring substantial amounts of water for crop irrigation, processing, and refining. The water footprint of biofuels varies depending on the feedstocks and production methods. For example, the water footprint of corn ethanol is significantly higher than that of sugarcane ethanol due to the irrigation requirements of corn (Gheewala 2023). Excessive water use for biofuel production can lead to water scarcity and water pollution, affecting aquatic ecosystems and human health (Sahin 2011). With the increase in biofuel synthesis from agricultural biomass, global water consumption by agricultural biomass might increase significantly by 2050, adding more stress to agricultural areas with decreased water availability. As for the water footprints of biofuels, they are 55–246 times higher than that of fossil fuels (Jeswani et al. 2020).

(c) **Biodiversity**: The replacement of diverse natural ecosystems with monoculture biofuel crops can lead to a decline in biodiversity. Monocultures are more susceptible to pests and diseases, requiring increased use of chemical inputs, which can further harm biodiversity (Demirbas 2009a, b). However, the cultivation of energy crops on marginal lands can still pose risks to local flora and fauna if not managed sustainably (Sahin 2011). As per studies, second-generation biofuels damage biodiversity to a reduced extent than first-generation biofuels. Similarly, using agricultural and forest residues as biomass for biofuel production has fewer negative impacts on biodiversity (Jeswani et al. 2020).

(d) **Soil health**: The cultivation of biofuel feedstocks can affect soil health through practices such as intensive tillage, mono-cropping, and the use of chemical fertilizers and pesticides. These practices can lead to soil erosion, nutrient depletion, and reduced soil organic matter, affecting soil fertility and productivity (Demirbas 2009a, b).

(e) **Pollution**: The production and use of biofuels can result in various forms of pollution, including air, water, and soil pollution. The use of fertilizers and pesticides in biofuel crop cultivation can lead to nutrient runoff and water pollution, causing eutrophication and harming aquatic ecosystems (Sahin 2011).

1.11 Life Cycle Assessment of Biofuel Production

Life cycle assessment (LCA) is a tool widely used to assess environmental impacts. LCA considers all phases of biofuel production, including feedstock cultivation, processing, transportation, and combustion, providing a comprehensive assessment of their ecological footprint (Reijnders 2021; Gheewala 2023).

- **Methodological challenges**: LCA studies of biofuels face several methodological challenges, including defining system boundaries, selecting functional units, and accounting for land-use changes and co-products. These challenges can lead to variations in LCA outcomes and uncertainties in the assessment of biofuel impacts (Gheewala 2023; Reijnders 2021).
- **Key findings**: LCA studies have shown that GHG emissions can be effectively reduced by adopting first-generation biofuels, such as corn ethanol and soybean biodiesel, as energy sources, compared to fossil fuels.

1.12 Energy Considerations of Biofuels

The use of energy in the life cycles of biofuel production has been quantified using different indicators, such as fossil energy consumption, net energy ratio, and primary, secondary, or cumulative energy demand. Fossil energy consumption has been the major indicator, since the key objectives of biofuel production are energy security, reducing dependence on fossil fuels, and mitigating rapidly changing climate conditions. The amount of energy produced from biomass must be greater than the amount of energy used in the production of biomass and the production of energy from it (Wasiak and Orynycz 2017). For the first- and second-generation biofuel production, the fossil energy consumption is less than 0.5 MJ MJ^{-1}. There is also wide variation in fossil energy consumption across different types of biofuels for both first and second generation based on the energy credited to co-products, such as electricity and heat. The variations occur due to certain factors such as fluctuation in feedstock productivity, agricultural practices, conservation technologies, and methods of allocation. Many research studies have reported that there is no energy viability in algal biofuel production because of high energy expenditure in processes such as pumping, dewatering, lipid extraction, and thermal drying (Jeswani et al. 2020).

1.13 Economic Considerations of Biofuels

Biofuels, derived from renewable biological resources, are a valuable alternative to fossil fuels in the global energy landscape. They offer potential benefits in terms of energy security, environmental sustainability, and economic development. Multiple sustainability and economic constraints exist in the production and adaptation of biofuels as an energy source. This section explores the economic consideration and the role of biofuels in the circular economy.

- **Economic benefits**: Biofuel production offers several economic benefits. It plays a role in energy security by reducing dependence on fossil fuels, enhances national security, and promotes stability in the energy market (Sivarathnakumar et al. 2021). The biofuel industry generates versatile job opportunities, fostering economic growth and prosperity. The production, processing, and distribution of biofuels create jobs in rural areas, contributing to rural development and income generation (Sivarathnakumar et al. 2021). It can drive technological innovation and investment in renewable energy infrastructure, further boosting economic growth (Moschini et al. 2012; Lundgren et al. 2008).
- **Economic challenges**: Along with the versatile benefits that biofuels offer, there also exist certain challenges. One major concern is the competition for land resources, which can lead to increased food prices and affect agricultural productivity. The "food versus fuel" debate highlights the potential conflict between using crops for biofuel production and for food supply (Hassan and Kalam 2013;

Petrou and Pappis 2009). Government policies also play an important role in determining the economic viability of biofuel production (Sivarathnakumar et al. 2021).

- **Market dynamics**: The biofuel market is affected by various factors, such as feedstock prices, energy prices, government policies, and technological advancements. Fluctuations in these factors can impact the competitiveness of biofuels relative to fossil fuels and affect investment decisions in the biofuel industry (Moschini et al. 2012).

1.14 Challenges and Future Prospects in Biofuel Production

The gradual decline in energy availability due to the increasing oil prices and energy demands, continuous depletion of natural resources, and environmental destruction is eventually jeopardizing human existence. The idea of switching to biofuels as an energy source is fascinating enough in terms of sustainability and economic feasibility. Upscaling biofuel production requires integrated research and experimentation in the domains of biology and genetic engineering. Even though the feedstock source of biofuels is readily and abundantly available in the environment, manipulating these organic substrates requires a great deal of effort and alteration until they become suitable for commercialization. The major constraints in the process of product commercialization are the high energy requirements for downstream processes like pretreatment and purification. There is a dire need for future research on innovative technologies that can synthesize sustainable and value-added chemicals, e.g., biofuels, optimize cultural conditions for maximum yield, develop novel microbial strains, and also decipher the genetic engineering world for enhanced and commercialized production of biofuels, which can meet a maximum of the world's energy demand.

1.15 Conclusions

Biofuels provide a promising and sustainable alternative to fossil fuels, addressing the critical challenges of energy security and environmental degradation. Produced from biological sources, they are renewable and biodegradable and significantly reduce greenhouse gas emissions compared to conventional fossil fuels. Biofuels also contribute to economic development by creating jobs in agriculture, energy, and related industries while reducing dependence on imported oil. Despite these benefits, challenges such as high production costs, competition for land and water, and the need for advanced infrastructure limit their widespread use. The food versus fuel debate and the environmental impacts of large-scale biofuel production require careful management and sustainable practices. Advanced research in biofuel technologies including genetic engineering and process optimization is essential to enhance efficiency, lower cost, and minimize environmental impacts.

Acknowledgments I would like to acknowledge all the authors whose data is cited in this book chapter. I also acknowledge the support of my colleagues and sister for their encouragement and constructive feedback. I am thankful to the Department of Environmental Sciences, Women University Multan, for their affiliation.

Declarations The authors declare that they have no conflict of interest.

References

Ambaye TG, Vaccari M, Bonilla-Petriciolet A, Prasad S, van Hullebusch ED, Rtimi S (2021) Emerging technologies for biofuel production: a critical review on recent progress, challenges, and perspectives. J Environ Manag 290:112627

ASTM D4806-22 (2022) Standard specification for denatured fuel ethanol for blending with gasolines for use as automotive spark-ignition engine fuel. ASTM International, West Conshohocken, PA

ASTM D6751-20a (2020) Standard specification for biodiesel fuel blend stock (B100) for middle distillate fuels. ASTM International, West Conshohocken, PA

Bajpai D, Tyagi VK (2006) Biodiesel: source, production, composition, properties, and its benefits. J Oleo Sci 55(10):487–502

Balat M (2011) Production of bioethanol from lignocellulosic materials via the biochemical pathway: a review. Energy Convers Manag 52(2):858–875

Balat M, Balat H (2009) Recent trends in global production and utilization of bio-ethanol fuel. Appl Energy 86(11):2273–2282

Bardant TB, Dahnum D, Amaliyah N (2017) Simultaneous or separated; comparison approach for saccharification and fermentation process in producing bio-ethanol from EFB. In: AIP conference proceedings, vol 1904, 1. AIP Publishing

Bhattarai K, Stalick WM, McKay S, Geme G, Bhattarai N (2011) Biofuel: an alternative to fossil fuel for alleviating world energy and economic crises. J Environ Sci Health Part A 46(12):1424–1442

Biomass Energy Centre (2021) Residential heating with wood pellets. https://www.biomassenergycentre.org.uk

Brennan L, Owende P (2010) Biofuels from microalgae—a review of technologies for production, processing, and extractions of biofuels and co-products. Renew Sust Energ Rev 14(2):557–577

Bridgwater AV (2012) Review of fast pyrolysis of biomass and product upgrading. Biomass Bioenergy 38:68–94

Chisti Y (2007) Biodiesel from microalgae. Biotechnol Adv 25(3):294–306

Demirbas A (2009a) Biofuels securing the planet's future energy needs. Energy Convers Manag 50(9):2239–2249

Demirbas A (2009b) Political, economic and environmental impacts of biofuels: a review. Appl Energy 86:S108–S117

Dernotte J, Mounaïm-Rousselle C, Halter F, Seers P (2010) Evaluation of butanol–gasoline blends in a port fuel-injection, spark-ignition engine. Oil Gas Sci Technol 65(2):345–351

Olivier JGJ, Bouwman AF, Van der Maas CWM, Berdowski JJM (1994). Emission database for global atmospheric research (EDGAR). Environ Mon Asses 31:93–106.

EN 14214:2012+A2:2019 (2019) Liquid petroleum products—fatty acid methyl esters (FAME) for use in diesel engines and heating applications—requirements and test methods. European Committee for Standardization, Brussels, Belgium

EN 15376:2014+A1:2021 (2021) Automotive fuels—ethanol as a blending component for petrol—requirements and test methods. European Committee for Standardization, Brussels, Belgium

Yasin MHM, Ali MA, Mamat R, Yusop AF, Ali MH (2019) Physical properties and chemical composition of biofuels. In Second and Third Generation of Feedstocks (pp 291-320). Elsevier

European Commission (2021) Bio-based chemicals and materials. https://ec.europa.eu/growth/sectors/biotechnology/bio-based-products_en

Fatma S, Hameed A, Noman M, Ahmed T, Shahid M, Tariq M, Sohail I, Tabassum R (2018) Lignocellulosic biomass: a sustainable bioenergy source for the future. Protein Pept Lett 25(2):148–163

Fernández L (2024) Global cumulative renewable capacity installed 2010–2023. Retrieved from https://www.statista.com/statistics/1094331/global-renewable-capacity-cumulative/

Gheewala SH (2023) Life cycle assessment for sustainability assessment of biofuels and bioproducts. Biofuel Res J 37:1810–1815

Harun R, Danquah MK, Forde GM (2010) Microalgal biomass as a fermentation feedstock for bioethanol production. J Chem Technol Biotechnol 85(2):199–203

Hassan MH, Kalam MA (2013) An overview of biofuel as a renewable energy source: development and challenges. Procedia Eng 56:39–53

International Renewable Energy Agency (2020) Renewable power generation costs in 2020. Retrieved from https://www.irena.org/publications/2021/Jun/Renewable-Power-Costs-in-2020

Jeswani HK, Chilvers A, Azapagic A (2020) Environmental sustainability of biofuels: a review. Proc R Soc A 476(2243):20200351

Karmakar R, Kundu K, Rajor A (2018) Fuel properties and emission characteristics of biodiesel produced from unused algae grown in India. Petrol Sci 15:385–395

Khan S, Naushad M, Iqbal J, Bathula C, Ala'a H (2022) Challenges and perspectives on innovative technologies for biofuel production and sustainable environmental management. Fuel 325:124845

Lehto J, Oasmaa A, Solantausta Y, Kytö M, Chiaramonti D (2014) Review of fuel oil quality and combustion of fast pyrolysis bio-oils from lignocellulosic biomass. Appl Energy 116:178–190

Liao JC, Mi L, Pontrelli S, Luo S (2016) Fueling the future: microbial engineering for the production of sustainable biofuels. Nat Rev Microbiol 14(5):288–304

Lundgren T, Marklund PO, Brännlund R, Kriström B (2008) The economics of biofuels. Int Rev Environ Resour Econ 2(3):237–280

Malik K, Capareda SC, Kamboj BR, Malik S, Singh K, Arya S, Bishnoi DK (2024) Biofuels production: a review on sustainable alternatives to traditional fuels and energy sources. Fuels 5(2):157–175

Malla FA, Bandh SA, Wani SA, Hoang AT, Sofi NA (2023) Biofuels: potential alternatives to fossil fuels. In: Biofuels in circular economy. Springer Nature Singapore, pp 1–15

Manmai N, Bautista K, Unpaprom Y, Ramaraj R, Jayabalan R (2018) Bioethanol production from pretreated-corn stalk waste by separate hydrolysis and fermentation process. In: Proceeding of the 2nd GCIC, 46th national and 9th international graduate research conference, Chiang Mai, Thailand, pp 383–391

McCormick R, Moriarty K (2023) Biodiesel Handling and Use Guide (No. NREL/TP-4A00-86939; CRD-15-00593). National Renewable Energy Laboratory (NREL), Golden, CO (United States)

Mishra VK, Goswami R (2018) A review of production, properties and advantages of biodiesel. Biofuels 9(2):273–289

Moschini G, Cui J, Lapan H (2012) Economics of biofuels: an overview of policies, impacts and prospects. Bio-based Appl Econ 1(3):269–296

Muhammad UL, Shamsuddin IM, Danjuma A, Musawa RS, Dembo UH (2018) Biofuels as the starring substitute to fossil fuels. Petrol Sci Eng 2:44–49

Naik SN, Goud VV, Rout PK, Dalai AK (2010) Production of first and second generation biofuels: a comprehensive review. Renew Sust Energ Rev 14(2):578–597
National Renewable Energy Laboratory (2021) Biodiesel basics. Retrieved from https://www.nrel.gov/docs/fy21osti/78972.pdf
Oasmaa A, Meier D (2005) Norms and standards for fast pyrolysis liquids: 1. Round robin test. J Anal Appl Pyrolysis 73(2):323–334
Pearlson MN (2011) A techno-economic and environmental assessment of hydroprocessed renewable distillate fuels. Doctoral dissertation. Massachusetts Institute of Technology
Petrou EC, Pappis CP (2009) Biofuels: a survey on pros and cons. Energy Fuel 23(2):1055–1066
Rai AK, Al Makishah NH, Wen Z, Gupta G, Pandit S, Prasad R (2022) Recent developments in lignocellulosic biofuels, a renewable source of bioenergy. Fermentation 8(4):161
Reijnders L (2021) Life cycle assessment of biofuels. In: Biofuels and Biodiesel. Springer, US, pp 53–67
Renewable Fuels Association (2021) Ethanol facts: energy security. Retrieved from https://ethanolrfa.org/ethanol-facts/energy-security
Sahin Y (2011) Environmental impacts of biofuels. Energy Educ Sci Technol Part A-Energy Sci Res 26:2
Sims RE, Mabee W, Saddler JN, Taylor M (2010) An overview of second-generation biofuel technologies. Bioresour Technol 101(6):1570–1580
Sivarathnakumar S, Adhitiyan T, Gubendhiran S, Praveenkumar R (2021) Ecological and economic impacts on biofuel production. In: Int Conf Adv Innov Recycling Eng. Springer Nature Singapore, pp 211–231
Stoeglehner G, Narodoslawsky M (2009) How sustainable are biofuels? Answers and further questions arising from an ecological footprint perspective. Bioresour Technol 100(16):3825–3830
Sun Y (2024) Technology research and development prospects of biofuels. Technol 7(1):2024
U.S. Department of Energy (2021) Biofuels for transportation. Retrieved from https://www.energy.gov/eere/bioenergy/biofuels
U.S. Environmental Protection Agency (2021) Combined heat and power (CHP) partnership. Retrieved from https://www.epa.gov/chp
United Nations Environment Programme (UNEP) (2015).The Emission Gap Report 2015-Executive Summary. Available online at: http://uneplive.unep.org/theme/index/13#indcs
Wasiak A, Orynycz O (2017) Energy efficiency of a biofuel production system. Manag Prod Eng Rev 8(1):60–68

Chapter 2
Life Cycle Assessment (LCA) of Biofuel Production from Lignocellulosic Biomass

Shiyou Pan, Jiankai Lin, Zhenchong Li, Liqin Du, and Yutuo Wei

Abstract Nowadays, global nations express concerns regarding fossil fuel supplies, environmental pollution, and greenhouse gas emissions, increasing the demand for a shift towards renewable energy. The debate between food and fuel has prompted research into second-generation (2G) biofuels that rely on nonfood biomass. Lignocellulosic biomass (LCB) is an abundant resource, and it is a highly attractive renewable source of biofuels. They help reduce competition for land by utilizing nonfood materials, thus preventing biomass from ending up in landfills. The generation of biofuels from LCB involves a series of steps, encompassing the feedstock collection, pretreatment, conversion processes, and the separation and purification of targeted products. Evaluating the environmental implications of 2G biofuel production is of great significance. Factors such as feedstock choices, product generation, and economic viability require consideration. The sustainability of 2G biofuel production is evaluated by employing the life cycle assessment (LCA). As per ISO 14040/14044 standards, LCA consists of four key components: goal and scope, inventory analysis, impact assessment, and interpretation. The assessment of lignocellulosic biofuel production through LCA is significantly impacted by methodological factors, such as functional unit, system boundaries, allocation methods, and more. This chapter aims to synthesize and analyze the existing and latest knowledge regarding the LCA of lignocellulosic biofuel production. Additionally, it uses bioethanol production as an example to evaluate its environmental impacts in comparison with conventional sugar milling byproduct disposal methods, such as landfill and combustion. It is intended to identify possible hotspots and suggest improvement options to enhance the effectiveness of the biofuel production process. Future research directions could focus on developing standardized data collection

S. Pan · J. Lin · L. Du · Y. Wei (✉)
Guangxi Microorganism and Enzyme Research Center of Engineering Technology, College of Life Science and Technology, Guangxi University, Nanning, Guangxi, China
e-mail: weiyutuo@gxu.edu.cn

Z. Li
Guilin University of Technology, Guilin, Guangxi, China

M. A. Dar et al. (eds.), *Recent Trends in Lignocellulosic Biofuels and Bioenergy*, Clean Energy Production Technologies,
https://doi.org/10.1007/978-981-96-4636-4_2

methods, improving modeling techniques, and promoting the application of sustainable practices across the biofuel production chain based on LCB.

Keywords Bioethanol · LCA · Lignocellulose · Second-generation biofuel · Sugarcane bagasse · Renewable energy

2.1 Introduction

In the past few decades, global research undertakings remain constantly dedicated to seeking alternative energy sources. The main justifications are as follows: (a) the augmentation of carbon emissions; (b) the exhaustion of nonrenewable energy resource; and (c) the intensification of environmental pollutions. In 2019, the world's energy consumption attained 418 EJ. It is forecast that this will further increase by 23%, reaching 516 EJ by 2040 [1]. Currently, over 80% of energy comes from fossil fuels. Among them, coal, oil, and natural gas account for 26.8%, 30.9%, and 23.2%, respectively (IEA 2024a). This heavy reliance on fossil fuels has significant implications for our environment and future energy security. Relying on traditional energy sources is concerning due to limited reserves and the link between fossil fuel utilization and CO_2 emissions. Annually, 32 Gt of CO_2 equivalent greenhouse gases are released into the atmosphere due to the combustion of coal, oil, and natural gas (Siegelman et al. 2019). Even though the amount of CO_2 on Earth dropped by 5.8% since World War II as a result of the COVID-19 pandemic, this amount rebounded by 4.8% in 2021 with the revival of global economic activities and transportation services (IEA 2024b). To adhere to the goal of net zero CO_2 emissions by 2050 proposed in the Paris Agreement and Sustainable Development Goal, it is of urgent necessity to conduct research and development on sustainable energy sources. The International Energy Agency (IEA) predicts that in the next five years (2025), global renewable energy capacity will nearly double, surpassing coal to become the largest source of power generation (IEA 2024b). This shift towards clean, affordable, and sustainable fuel is not only vitally important for addressing climate change but also brings numerous economic and environmental advantages. For instance, it might bring about new employment opportunities in the clean energy field and decrease reliance on limited and polluting fossil fuel resources.

Biomass, which refers to organic substances originating from plants and animals, presently makes up 9.4% of the global energy provision and 10% of chemical raw materials (IEA 2024a). It is regarded as the earliest fuel in human history and remained the primary energy source until the mid-nineteenth century. Recently, biomass resources have drawn renewed interest as a significant source of organic carbon for the production of an array of fuels and a diverse range of chemicals. Lignocellulose biomass (LCB) stands out as the most promising biomass feedstock due to its nonedible character, abundance, low cost, and accessibility. LCB has been identified as the most abundant biomass on Earth, with an annual production of ~180 billion tons. It possesses the potential to function as a substitute for petroleum in the manufacture of materials, chemicals, and fuels. It is estimated that LCB can generate ~6 MJ of energy or ~ 0.8 kg of chemicals per kg of raw material (Siegelman et al. 2019). Although the utilization of LCB still encompasses a carbon cycle, which leads to CO_2 emissions during the process, the carbon footprint is counterbalanced by the carbon absorbed during plant growth through photosynthesis. Given that human activities persistently rely on the carbon cycle, LCB offers a vital alternative for future energy solutions. Furthermore, ongoing research and technological advancements are expected to enhance the LCB-based biofuel production and the scalability of its applications, rendering it an even more attractive option in the future. For instance, innovative conversion techniques and catalyst developments could potentially increase the yield and quality of the derived materials, chemicals, and fuels from LCB, thereby accelerating its widespread adoption and contribution to global energy transitions (Cai et al. 2021; Gao et al. 2023).

Estimating environmental footprints from cradle to grave is crucial for formulating the most optimal biofuel strategies. In this context, life cycle assessment (LCA) functions as a comprehensive methodology that integrates the economic and environmental effects across the entire production chain and has been extensively employed in diverse renewable energy systems (Osman et al. 2021; Chouinard-Dussault et al. 2011). It offers valuable insights into the interactions between environmental factors and human activities, facilitating successful strategic planning. LCA encompasses inputs and related emissions during the bioproduction process. Meanwhile, it considers the ultimate destination of the product to identify potential development directions in the manufacturing process. Typically, the production of LCB-based biofuels includes the entire life cycle stages, from raw material collection, pretreatment, and conversion processes to the final usage of products (Fig. 2.1). It analyzes the environmental effects of different technical routes and processes, including aspects like greenhouse gas emissions, energy consumption, and water resource utilization. Additionally, it discusses how to enhance the sustainability of biofuel production through optimizing processes and technologies. It is of great significance for understanding the role of LCB in the bioproduction process and conducting sustainability assessments (Prasad et al. 2016). This chapter briefly describes the importance of LCB as a raw material for producing biofuels. Special attention has been directed towards investigating the entire life cycle of the bioproducts (Bilal and Iqbal 2020). An example of LCA application is comparing the environmental impacts of three methods for disposing of sugar production byproducts: landfill, electricity generation by combustion, and ethanol production.

2.2 Source and Gathering of Lignocellulosic Biomasses

The sources of LCB are varied and include a broad range of agricultural residues, forestry byproducts, dedicated energy crops, and industrial wastes (Fig. 2.2).

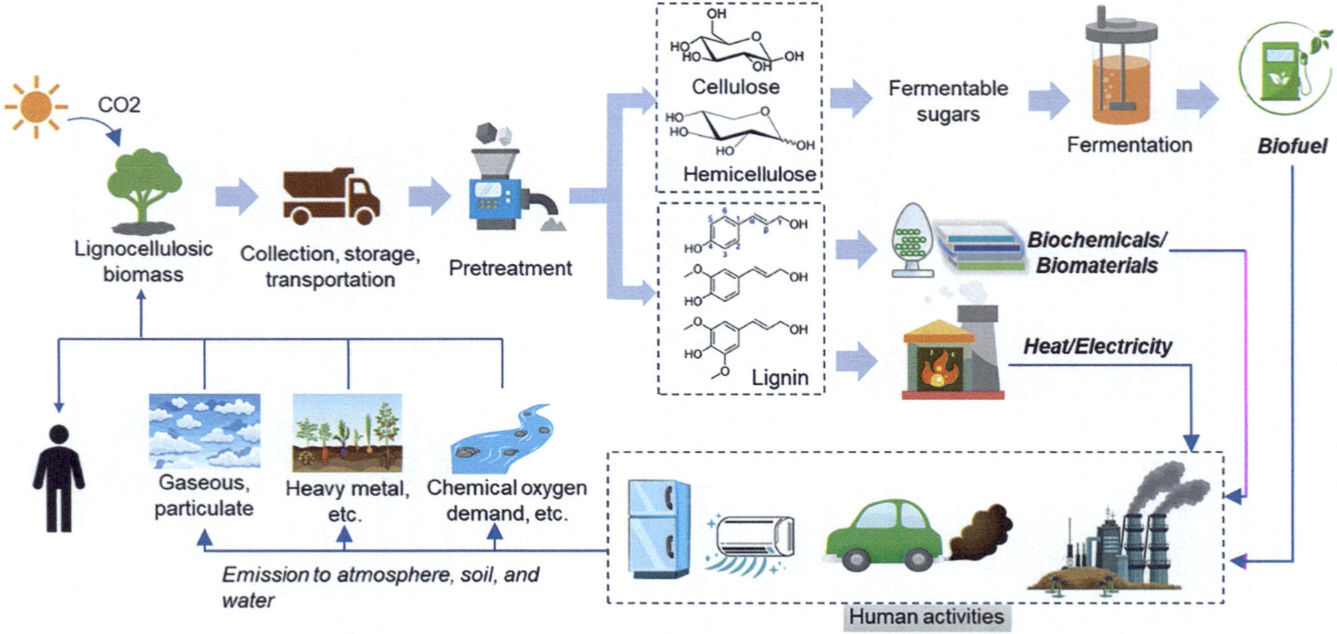

Fig. 2.1 Life cycle of producing high-value products using lignocellulosic biomass

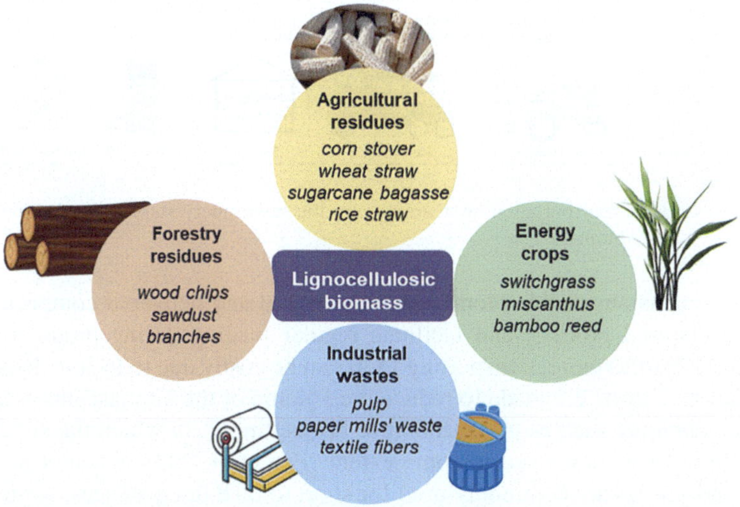

Fig. 2.2 Source of lignocellulosic biomasses

1. Agricultural residues: These include leftover materials from crop production, such as wheat straw, rice straw, corn stover, and sugarcane bagasse. Generally, agricultural residues have relatively higher contents of cellulose or hemicellulose.
2. Forestry residues: These are the byproducts of forestry operations like wood chips, sawdust, and branches. In a managed forest, when trees are harvested for timber, the leftover branches and smaller pieces can be collected and used as LCB. Compared to some agricultural residues, forestry residues might have higher lignin content.
3. Energy crops: These are crops specifically grown for their biomass production, such as switchgrass and miscanthus. Switchgrass, for instance, is a perennial grass that is often cultivated for its high biomass yield and suitability as a source of lignocellulosic material. These crops are designed to thrive on marginal lands, promoting environmental benefits such as soil conservation and biodiversity. Their fast growth rates and high adaptability make them an attractive source of LCB for the bioenergy industry.
4. Industrial wastes: These are materials from industrial operations, such as waste from pulp and paper mills and textile fibers. Some pulp and paper mills yield large quantities of lignin-rich residues that can be considered LCB.

Prior to biorefinery processing, LCB needs to be efficiently gathered, transported, and stored (Fig. 2.3). LCB sources are often dispersed and spread over large areas, necessitating efficient and organized collection methods. For agricultural residues like corn stover or rice straw, collection usually takes place after the main crop harvest. Specialized equipment like balers or forage harvesters is employed to gather and compress the biomass into manageable forms. The regular geometric shape and

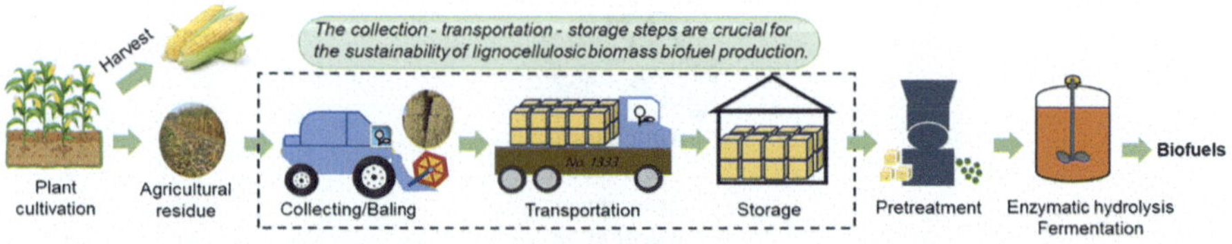

Fig. 2.3 Collection, transportation, and storage are the preliminary steps in biofuel production from lignocellulosic biomass

standard dimensions enable biomasses to be stored and delivered compactly, simplify the disposal process, and facilitate regular feeding during production (Cui et al. 2021). Furthermore, transporting LCB can be costly due to its low density and bulky nature. Efforts are made to reduce the volume of the biomass through densification technique, such as pelletization or briquetting [10], in which the structure of lignocellulose was destroyed to some extent (Pan et al. 2024; Chen et al. 2021). Trucks and trailers are commonly used for short to medium distances, while rail or shipping might be employed for longer hauls. Logistical planning is crucial to ensure smooth transportation, considering factors such as road conditions, distance, and availability of transportation infrastructure (Sun et al. 2017). LCB needs to be stored properly to prevent spoilage, moisture absorption, and degradation. Open-air storage is common, but it can expose the biomass to weather conditions, increasing the risk of mold growth and quality deterioration. Covered storage facilities or silos can provide better protection. Moisture control is essential, as excessive moisture can lead to microbial activity and loss of energy content. Storage for longer periods may require additional measures like chemical treatment or controlled ventilation to maintain the quality of the biomass.

In reality, not all biomass residues are suitable for energy utilization. In some instances, their extensive dispersion or low bulk densities render collection, transportation, and storage extremely costly. In a straw biogas facility, the expenses associated with straw gathering, transporting, and storing constitute 30–50% of the total energy usage [11], which limited the sustainability of biofuel production from LCB. Material characteristics, including moisture content, particle dimensions, and bulk density, exert a substantial influence on the modification and design of storage bins, hoppers, feeders, and pretreatment machines. Overall, optimizing the collection, transportation, and storage of LCB is crucial to ensure its efficient utilization and economic viability in various applications, such as bioenergy production and biorefining.

2.3 Constituent Structure of the Lignocellulosic Biomass

The LCB is predominantly composed of cellulose (35–50%), hemicellulose (15–25%), and lignin (10–15%) (Zheng et al. 2014) by dry weight calculation and also contains a small amount of soluble sugars, proteins, and ashes (Dar et al. 2021). Nevertheless, the percentage of these three major components can change considerably according to the species, types, and sources of LCB. Table 2.1 shows the lignocellulose contents in common LCB.

Cellulose and hemicellulose are undoubtedly the crucial compounds in LCB. These two substances are of great significance because they possess the potential to be converted into biofuels through the action of various microbes. Cellulose is a macromolecular polysaccharide composed of numerous glucose molecules linked by β-1,4-glycosidic bonds. Its long-chain structure is interwoven, possessing high mechanical strength and stability, and it plays a crucial role in the formation of plant cell walls. Hemicellulose is a heteropolysaccharide mainly composed of pentose units such as xylose, mannose, galactose, and arabinose. These sugar units are connected by different glycosidic bonds to form a shorter branched-chain structure and coexist with cellulose in plant cell walls. The structure of hemicellulose is relatively complex and has a certain degree of diversity, unlike cellulose, which has a highly regular long-chain structure. Lignin is a complex polymer with aromatic characteristics. It has a highly complex three-dimensional network structure and exerts a vital function in strengthening plant cell walls. Lignin is formed by the three distinct phenylpropane monomers, namely, p-coumaryl alcohol, coniferyl alcohol, and sinapyl alcohol (dos Santos et al. 2019). These three monomer units are combined in an irregular way, and their ratio in biomass varies according to the species and sources of LCB.

Table 2.1 Major constituents of lignocellulose biomasses

Lignocellulosic biomass	Cellulose (TS%) [a]	Hemicellulose (TS%)	Lignin (TS%)	Ash (TS%)	References
Banana pseudo-stem	26.2	16.7	5.0	17.0	Pan et al. (2020)
Chestnut shell	26.8	24.5	36.8	2.5	Bianco et al. (2021)
Corn stover	37–42	20–28	18–22	– [b]	Kumar et al. (2009); Biswas et al. (2015)
Corn cob	30–40	30–35	10–20	–	Liang et al. (2023)
Grape stalk	17.1–24.1	4.3–6.4	19.6–26.3	6.5–9.0	Spigno et al. (2013)
Rice straw	26.7	20.6	8.8	12.0	Pan et al. (2024)
Sugarcane bagasse	42.6	29.5	18.9	1.3	Pan et al. (2023)
Switch grass	33.5	32.2	3.5	–	Zhao et al. (2017)
Wheat straw	35.1	25.6	7.5	11.1	Chandra et al. (2012)

[a]Dry weight basis
[b]Not determined

Lignin's recalcitrant aromatic-phenolic structure and antimicrobial properties often render it an undesirable component in bioprocessing. Hence, the delignification procedure is often proposed as a way to produce biofuels from natural LCB (Radhika et al. 2022). However, with advancing research and the expansion of bio-economic strategies, lignin has emerged as a strategic feedstock in lignocellulosic biorefineries (Agarwal et al. 2018). Lignin is isolated from LCB and used to produce high-value biopolymers and nutraceuticals/drugs (Zhou et al. 2022). It shows great prospect to serve as a sustainable alternative to traditional petroleum-based materials. Overall, the three polymers in lignocellulose—cellulose, hemicellulose, and lignin—are intricately and mutually cross-linked. This creates a highly complex and extremely stable structure. Lignocellulose degradation mechanisms are truly pivotal for the existence of many microorganisms. This is because these microorganisms must have the ability to degrade plant polymers in order to obtain essential nutrients. As a result, pretreatment is generally an essential step for the delignification and release of sugars.

2.4 The Importance of Pretreatment Technologies

Pretreatment is aimed at enhancing the hydrolysis of polysaccharides, namely, cellulose and hemicellulose, and transforming them into saccharides that are readily utilized by microorganisms. The pretreatment procedure of LCB is the technical difficulty for its conversion into biofuels and is crucial for the economic benefits of the production process. Biomass pretreatment processes typically constitute ≥20% of total biofuel production costs, representing the costliest operational phase in lignocellulosic biorefinery systems (Amin et al. 2017). Pretreatment causes the disruption of the recalcitrant structure of lignocellulose, and carbohydrates are exposed in a form that is readily accessible to microorganisms or enzymes. In recent decades, various pretreatment methods are devised specifically to tackle the recalcitrance of LCB (Zheng et al. 2014). These techniques work to release sugars that are trapped within the complex structure, degrade lignin, and dissolve it to make the polysaccharides more accessible to enzymatic breakdown. Diverse pretreatment approaches may involve physical, chemical, or biological processes (Table 2.2). Each technique has its respective merits and drawbacks. The selection of pretreatment method is contingent upon factors, including the type of biomass, the target product, and economic considerations.

To achieve commercial success in biofuel production, one needs to have a basic understanding of the commercial feasibility of the pretreatment step. The efficacy of pretreatment methods can be evaluated based on specific precautions (Zheng et al. 2014): preventing the loss of reducing sugars, avoiding harmful byproducts, e.g., furfural and 5-hydroxymethylfurfural (HMF); minimizing solid waste while ensuring full material utilization; and maintaining economic feasibility. Previous investigations have reported diverse pretreatment approaches like acid treatment, alkali treatment, organic solvent treatment, and steam explosion and their combinations in order to expedite substrate hydrolysis. They have demonstrated efficacy in

Table 2.2 Pretreatment techniques employed for lignocellulosic biomass, with their merits and demerits. Adapted from Pan et al. (2022a, b)

Pretreatment methods		Merits	Demerits
Physical	Milling	• Reducing particle size, increasing surface area • Reduction of cellulose crystallinity • Ease of substrate management	• High electricity demand
	Irradiation	• Reduction of cellulose crystallinity • Increase surface area • Fast heat transfer, short reaction time	• High electricity demand • Scalability issues
	Extrusion	• Destruction of cellulose structure • Increase surface area	• High energy consumption
Chemical	Acid	• Hydrolysis of hemicellulose • Alteration of cellulose structure	• High cost of acids • Formation of inhibitors
	Alkali	• Dissolving of lignin • Alteration of cellulose structure	• Cost of alkali, formation of inhibitors
	Oxidative	• Removal of hemicellulose and lignin	• Chemical cost • Formation of inhibitors
	Organo-solvent	• Dissolving cellulose/hemicellulose/lignin	• Solvent cost • Solvent removal steps
Physicochemical	Steam explosion	• Hydrolysis of hemicellulose and lignin • Destruction of cellulose structure	• High energy demand • Recalcitrant compound formation
	Hydrothermal	• Hydrolysis of hemicellulose and lignin • Destruction of cellulose structure	• High water consumption • High energy input
Biological	Microbial	• Hydrolysis of cellulose, hemicellulose, lignin • No any inhibitor formation • Low energy consumption	• Slow process • Carbon losses • Necessity of a large sterile area
	Enzyme	• Hydrolysis of cellulose, hemicellulose, lignin • Fast process • Low energy demand	• High cost of enzymes • Continuous addition may require

enhancing enzymatic hydrolysis efficiency. However, these methods require demanding infrastructure, incur substantial chemical/energy inputs, and generate detrimental environmental impacts, significantly constraining their industrial scalability. In contrast, biological methods using microorganisms consume less energy and are nonpolluting; however, they tend to have long action cycles and low enzymatic activity of lignin-degrading enzymes (Sharma et al. 2017). Therefore, much research is needed to develop economical and environmentally friendly methods that facilitate complete delignification and 100% biomass conversion into high-value products. Achieving the concept of sustainable biorefineries is contingent upon the creation of efficient and eco-friendly biomass pretreatment techniques.

2.5 Bioconversion Process of Lignocellulosic Biomass

After pretreatment, LCB will undergo biotransformation, including fermentation or enzymatic hydrolysis. The ultimate goal is still to effectively convert cellulose/hemicellulose into high-value products. Specialized enzymes such as cellulases and hemicellulases are of crucial importance in the decomposition process. Through their enzymatic activity, cellulose and hemicellulose are converted into simple sugars like glucose and xylose, respectively. The efficacy of this enzymatic process is influenced by aspects such as enzyme type and amount, reaction circumstances, and the properties of pretreated biomass (Zhang et al. 2021). The resulting sugars are then fermented by microorganisms. Ethanol-producing microorganisms like *Saccharomyces cerevisiae* (Stanley et al. 2010) or bacteria such as *Zymomonas mobilis* (Xia et al. 2019) convert the sugars into ethanol through fermentation. It is worth noting that the enzymatic hydrolysis step can sometimes be omitted during ethanol production, which mainly depends on the nature of the pretreated hydrolysate. For example, dilute sulfuric acid hydrolysate does not require enzymatic hydrolysis because it is rich in monosaccharides, which can be directly used by microorganisms (Jung et al. 2013).

The bioavailability of LCB is boosted by pretreatment. However, the present production of biofuels is far from satisfactory. Take the conversion of sugarcane bagasse (SCB) as an illustration. In actuality, the output of biomethane is typically less than 150 L/kg dry SCB, and the yield of biomethane falls within the range of 200–300 L/kg dry SCB (Zhang et al. 2021). On the contrary, in theory, the production yields of biohydrogen and biomethane from SCB per kilogram of dry weight are 336 L and 440 L, respectively. In reality, with the exception of bioethanol, the output of biofuels produced from SCB is less than half of its theoretical value, which implies that there is still tremendous undeveloped potential for SCB-based biofuels (Zhang et al. 2021).

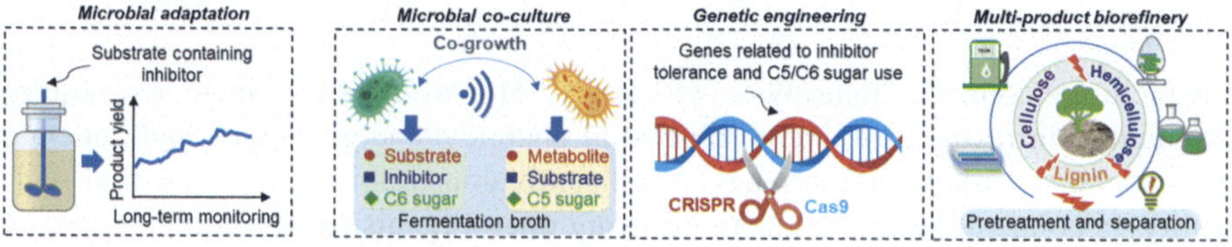

Fig. 2.4 Strategies to enhance biofuel production from lignocellulosic biomass. Adapted from our previous publication (Pan et al. 2022b).

To enhance the biochemical conversion efficiency, various strategies are being explored, as shown in Fig. 2.4. Microbial adaptation makes it possible to obtain higher levels of biofuel production from lignocellulosic hydrolysate containing inhibitors. In this way, the complex detoxification process can be effectively avoided, thereby reducing production costs (Chandel et al. 2012). Nevertheless, this approach is laborious as well as inefficient. Co-culture engineering offers an extremely promising way forward. The reason lies in its outstanding capacity to integrate the procedures of detoxification and microbial fermentation. Moreover, it is capable of conducting the simultaneous fermentation of hexose and pentose. This characteristic significantly expands the range of available feedstocks, providing more options for the production of biofuels. The compatibility of microorganisms is of utmost importance, as it determines whether they can coexist under the same fermentation conditions. Factors such as pH, temperature, and aeration need to be carefully considered to ensure a harmonious coexistence (Oliva-Rodriguez et al. 2019; Tondro et al. 2020). Only when these conditions are met can a successful co-culture be created, leading to more efficient biofuel production processes. Genetic engineering, a powerful technique, is commonly used in microbiology to modify targeted microorganisms. The aim is to enhance their toxicity resistance and improve their assimilation capabilities, enabling them to better withstand harmful substances and environmental stressors. However, recombinant strains often encounter limitations in industrial applications due to being unstable and prone to various stress factors such as pH and temperature. Moreover, most recombinant strains are still sensitive to inhibitors. In the past decade, substantial endeavors have been exerted to mitigate this sensitivity and enhance the robustness of microorganisms. Some emerging genome editing tools, like CRISPR-Cas9 (Javed et al. 2019), have gained attention for their precision and ease of use. It offers an efficient way to introduce genetic changes in non-model microorganisms that are hard to manipulate with traditional methods. This strategy holds great potential, as it will offer more appealing host platforms for achieving increased biofuel yields. In the future, genetic engineering holds great promise and is highly anticipated. Multiproduct biorefinery holds promise for achieving "zero waste" and making biofuels cost-competitive. It combines production of biofuels and other products, such as co-production of biofuel and bioelectricity (Furlan et al. 2013). However, it is still in the early stages and uninformative for SCB. Barriers include finding a feasible fractionation method and ensuring proximity of production units. Therefore, further studies on the separation of LCB components and plant design are required prior to large-scale implementation.

2.6 Biofuel Product Separation and Purification

The separation of valuable compounds from other elements is an extremely crucial step in the biofuel production process. For instance, during biomass bioconversion, extracting fuel molecules from other components like water and nutrients is vital, as

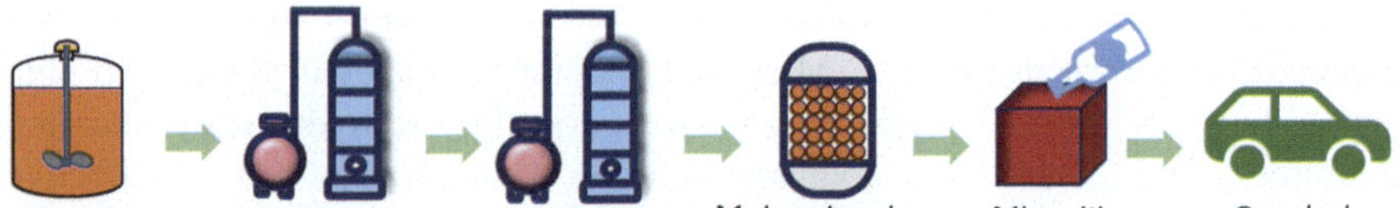

Fig. 2.5 Bioethanol production and purification

it purifies the fuel and ensures its quality and efficiency (Huang and Ramaswamy 2013). Product purity holds immense significance for the economic viability of biofuel production. As the demand for biofuels increases, so does the need for large-scale production technologies. The ultimate aim is to introduce biofuels into the market more rapidly and efficiently. However, currently, biofuels often come with a higher price tag compared to fossil fuels. The US Department of Energy's Bioenergy Technologies Office (BETO) states that enhancing separation technology can potentially lead to a significant reduction of at least 50% or more in the minimum price of biofuels. There are various biofuel separation technologies available, including adsorption, extraction, membrane tech, pertraction, gas stripping, and some innovative ones (Li et al. 2019). The development of these technologies provides strong support for improving the production efficiency of biofuels, reducing costs, and reducing environmental pollution.

However, the separation and purification of biofuel products is a complex and challenging task. One major challenge resides in the diversity and complexity of the LCB. Biomass, such as agricultural straw, wood chips, and miscanthus, contains various substances (e.g., lipids, carbohydrates, proteins, and minerals.), with vastly different properties. This heterogeneity makes it challenging to selectively extract and purify the desired biofuel components (Kiss et al. 2016). Another significant challenge is the low concentration of the target biofuel in the initial mixture. In the case of bioethanol production (Fig. 2.5), the biofuel in fermentation broth is present in relatively small amounts, and separating it from the bulk of the raw material requires extensive processing and large-scale equipment. This not only increases the operational costs but also demands efficient separation methods to achieve economically viable production. Currently, the separation procedure consumes a large amount of energy and is also carbon-intensive and expensive, with separation accounting for 70% of the cost of biomass processing (Burmahl 2023). Furthermore, the impurities present in biofuel mixtures can have adverse effects on the fuel properties and engine performance. Trace metals, sulfur compounds, and other contaminants need to be removed to meet the criterion for use in transportation and other applications. To overcome these challenges, a combination of different separation and purification techniques is often necessary. Process optimization, the development of novel materials and membranes, and the integration of multiple unit operations can lead to more efficient and cost-effective biofuel production.

2.7 Life Cycle Assessment (LCA)

Concerns regarding environmental pollution and the scarcity of energy have unequivocally spurred the development of environment-centered life cycle product approaches. In the case of biofuels originating from LCB, their environmental sustainability is highly significant. This must be evaluated from various aspects including the source of feedstocks, the procedure of product generation, and the economic returns. Life cycle assessment (LCA), as a comprehensive tool, makes every effort to assess the environmental impact of products throughout their entire life cycle (Prasad et al. 2016). The LCA of biofuel production from LCB is gaining prominence due to several factors. Firstly, technological advancement is crucial for enhancing efficiency and sustainability. Secondly, bioconversion of diverse feedstocks like agricultural residues and energy crops into value-added products reduces fossil fuel reliance and promotes renewable resource use. Thirdly, evaluating biorefinery environmental performance in terms of emissions, water use, and land use is essential for sustainability. Finally, validating sustainable conversion processes is crucial for biofuel's long-term viability, ensuring economic feasibility, environmental friendliness, and social acceptability. In this context, the purpose of LCA is to determine the environmental impacts associated with the production process and assist in minimizing negative consequences. Specifically, the LCA study provides a reference in identifying "process hotspots" in the process value chain.

Attributional (ALCA) and consequential (CLCA) are two significant types of LCA (Bamber et al. 2020). ALCA is employed to assess the environmental influences that are related to a product, process, or service throughout its entire life cycle. Its purpose is to attribute the environmental burdens and benefits to different stages and components of the system being analyzed. For example, in the production of a car, ALCA would focus on raw material acquisition, component manufacturing, assembly, use phase, and end-of-life disposal and attribute the environmental impacts to each of these steps. CLCA is an approach within the field of LCA that focuses on evaluating the potential consequences of decisions or changes in a system. Rather than simply accounting for the existing flows and impacts as in ALCA, CLCA considers the dynamic and systemic effects that might occur due to a particular decision or action. It is intended to comprehensively understand and precisely predict how the market, along with the more extensive economic and environmental systems, reacts to diverse changes. For instance, if a company decides to use a new raw material for its product, a CLCA would evaluate not only the direct environmental impacts of using that material but also the potential indirect effects. These include changes in the supply and demand for that material, resulting price fluctuations, and subsequent shifts in production patterns in related industries. CLCA often incorporates economic models and market analyses to anticipate these cascading effects. This makes it more suitable for informing strategic decisions that have the potential to significantly influence market dynamics and resource allocation. For instance, if a large manufacturer decides to increase the production of a certain product, a CLCA would look at how this increase might lead to expanded

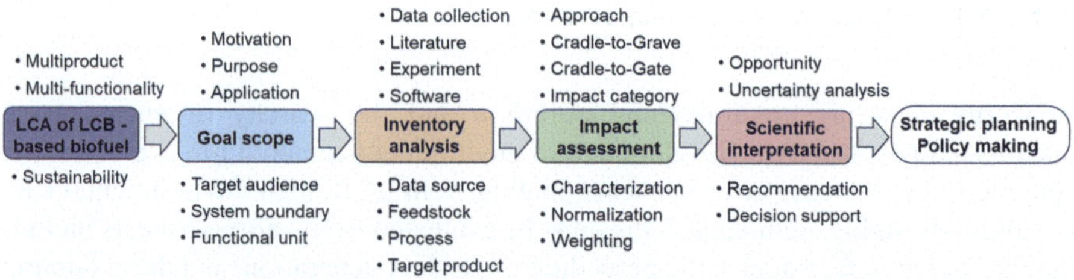

Fig. 2.6 Life cycle assessment (LCA) methodology and framework of biofuel production from lignocellulosic biomass (LCB)

production capacity, changes in the availability and cost of raw materials, and potential impacts on competing products and industries.

Since the first LCA model was proposed by The Coca-Cola Company in the 1960s, notable progress has been made in both its methodology and applications (Klein et al. 2015). The tool is extremely diverse and is reliant on the choices made during each phase of the assessment process. As per ISO 14040/14044, LCA consists of four key stages: goal and scope definition, inventory analysis, impact assessment, and interpretation (Koffler 2012). The primary steps and key elements of the LCA for biofuel production from LCB are presented in Fig. 2.6.

2.7.1 *Goal and Scope Definition*

Goal setting is the first and most important stage of LCA, as it lays the foundation for the entire assessment. The goal is linked to the motivation of the LCA study, and it reveals the target audience and the potential application. However, the importance of this initial stage is often overlooked. The reason is that people are more inclined to pay attention to the stages of data collection and analysis. The determination of goals and scopes not only introduces the LCA process but also determines the specific methods to be followed later. Notably, as new data and information emerge, goals and scopes can be adjusted during the work process. For instance, it might be found that the byproduct allocation principle put forward is not feasible, or there might be a lack of data to assemble a full life cycle inventory. Such changes need to be explicitly presented in the LCA result report. The ISO 14040/14044 standards feature deliberately lenient regulations to enable diverse options when defining the goal and scope. In this stage, the system boundary, functional unit, and LCA approach must be clearly defined.

2.7.1.1 System Boundary

The system boundary of an LCA states the assessment scope by determining which processes and activities are included or excluded. Performing a system boundary analysis is essential for an accurate LCA. Here are the key steps:

1. Clarify the research purpose and problem, for example, determining that the purpose of LCA is to be evaluated as either assessing the environmental impact of a product for design improvement or comparing the sustainability of different products. Different research purposes result in diverse choices of boundary ranges.
2. Consider all life cycle stages. Energy, materials, and emissions involved in all stages, i.e., raw material collection and transportation, biomass pretreatment, product production and use, and waste treatment and reuse, should be considered.
3. Conduct a sensitivity analysis to identify significant unit processes and iteratively adjust the system boundary. Sensitivity analysis involves observing changes in the overall environmental impact by modifying the inclusion or exclusion of specific production processes. Processes with significant impacts must be incorporated into the system boundaries. For those with less impact, the decision on whether to include them can be made according to the actual situation.
4. Consider data availability. When determining system boundaries, take into account the availability of required data. If data for some process is difficult to collect or is inaccurate, its inclusion in the system boundaries may need to be reconsidered. In cases of missing data, reasonable assumptions or approximations can be made, but the uncertainty of the data should be clearly stated in the evaluation report.

By comprehensively considering these factors, the accuracy and reliability of LCA can be improved. However, even for simple products, setting the system boundary is rather complex. This is because the life cycle process flow may be unbounded or unclear, resulting in numerous related processes and making data collection difficult (Li et al. 2014). A balance must be struck between comprehensiveness and manageability. Current ISO 14040/14044-compliant frameworks encounter operational constraints due to their stringent documentation requirements and exhaustive system boundary specifications. According to ISO 14040 and 14,044 (Souza et al. 2021), the system boundary is determined through an iterative process. However, decisions regarding inclusion or exclusion are often not scientific. Determining excluded processes is difficult, as their negligibility cannot be guaranteed. Because of personal experiences or being confined by individual cases and temporal or geographical limitations, different people may choose different boundaries. This may cause important processes to be overlooked. Selecting boundaries presents a challenge for public comparative claims. Consequently, in current LCA research, the issue of boundary selection remains a significant weakness and lacks scientific methods (Li et al. 2014).

2.7.1.2 Functional Unit

The functional unit serves as a reference for collecting and calculating input values during the second stage (e.g., inventory analysis) of the LCA. It defines the performance provided by the LCA framework and serves as a benchmark for comparing different alternatives. It provides a common reference point for all the elements being evaluated in the LCA (Souza et al. 2021). For instance, when comparing the environmental impacts of two kinds of light bulbs, the functional unit might be set as "offering 1,000 hours of illumination with a specific light intensity." This makes sure that the comparison is fair and founded on the same level of functionality. By clearly defining the functional unit, LCA studies can accurately quantify and compare the environmental impacts related to achieving a specific function. It is beneficial as it helps avoid comparing things that are completely different (e.g., apples to oranges), thereby enabling the generation of meaningful and comparable results (Turconi et al. 2013).

The functional unit is essential in LCA calculations, as it is defined in light of the target product and the intended objective (Cossutta et al. 2017). This fundamental element can take on various manifestations such as product mass, which could be measured in kilograms or tons. Another form might be calorific value, expressed in joules or kilojoules. Surface area is also a possible expression for the functional unit, perhaps in square meters or square centimeters. For the evaluation of different pathways to produce bioethanol, a megajoule is usually considered as the functional unit. The reason for this is that it offers a standardized measurement to compare the energy output of different production methods. In reality, when it comes to defining the functional unit for integrated biorefineries, there are two main approaches. One is to consider biomass input, which can be measured in tons or kilograms of raw materials. The other is to focus on the targeted product, which may be in liters or kilograms based on product attribute. A balanced consideration of biomass input and the targeted product is essential. This is because such consideration can result in a more thorough perception of the overall performance of the biorefinery.

2.7.1.3 LCA Approach

LCA typically has several common assessment scopes, each with its own characteristics and applications:

1. Cradle-to-grave: This is the most comprehensive scope. It encompasses the entire life cycle of a product or process, starting from the extraction of raw materials (the "cradle") and ending with the final disposal, recycling, or reuse (the "grave"). It includes all stages such as production, distribution, use, and waste management. For instance, when assessing a smartphone, it would cover the mining of minerals for components, manufacturing, shipping to stores, usage by the consumer, and eventual disposal or recycling.

2. Cradle-to-gate: This scope stops at the factory gate. It considers the environmental impacts from raw material extraction up to the point when the product departs from the manufacturing facility. It excludes the distribution, use, and end-of-life stages. For instance, in the case of a car tire, it covers the production of rubber and the making of the tire within the factory. But it excludes its transportation to dealerships and its performance while in use on a vehicle.
3. Gate-to-gate: This focuses only on a specific part of the whole process, typically from one gate of a facility to another. For example, it might assess the environmental impacts of converting raw materials into intermediate products within a single factory.
4. Cradle-to-cradle: This approach emphasizes the possibility of continuously recycling or reusing materials in a closed-loop system. Doing so minimizes waste and reduces the demand for virgin resources.

The assessment scope is selected according to the specific goals and context of the LCA study. A narrower scope like cradle-to-gate might be sufficient for some initial analyses or comparisons within a specific industry sector. However, for a more holistic understanding of the environmental implications, a cradle-to-grave scope is often preferred (Ibbotson and Kara 2012). Nevertheless, due to the lack of available data, a cradle-to-gate approach is also employed.

2.7.2 *Multifunctionality Issues*

Handling the issue of multifunctionality is a key point in the LCA of LCB biofuels (Gnansounou et al. 2009). Multifunctionality refers to the situation where a product or process serves more than one function or provides multiple outputs. For example, a biorefinery might produce not only bioethanol but also other co-products such as biogas and animal feed. In this case, it is essential to allocate environmental effects to co-products. Generally, the integrated biorefineries have multifunctionality, and this requires that the environmental impacts be distributed between the different outputs (Ijassi et al. 2021). However, this is complicated and challenging in the LCA.

Allocation is the process of distributing the environmental burdens and benefits among the different functions or outputs when a system is multifunctional (Obydenkova et al. 2021). There are several methods for allocation, each with its own pros and cons. One common allocation method is based on physical properties like mass or surface area. For instance, if a process produces two products in specific ratios by mass, the environmental impacts could be allocated based on those mass ratios. Another approach is the economic allocation, in which the impacts are distributed according to the economic worth of the different outputs.

However, allocation can introduce uncertainties and potential biases in the LCA results. So, ISO 14044 does not recommend allocation (Moretti et al. 2020). For a more accurate handling of multifunctionality, some prefer using system expansion or substitution methods. The system expansion indicates that the system boundary is preferably reconstructed in a way that it can effectively incorporate the diverse functions of all the byproducts. Instead of allocating impacts among multiple outputs of a system, it considers the avoided or additional impacts elsewhere in the system (Heijungs et al. 2021). For example, if a biorefinery produces both biofuel and a valuable co-product, the system expansion would take into account the avoided production of the equivalent fossil-based fuel. It would also take into account the impacts of any additional processes required to handle and utilize the co-product. This helps to account for the full environmental implications of the multifunctional system. The substitution method assesses the environmental impact by considering what would happen if the product or process under study was replaced by an alternative (Heijungs et al. 2021). For example, when a company is considering switching to a more energy-efficient production process, the substitution method would compare the impacts of the new process to the impacts of the existing process that would be displaced. Both system expansion and substitution methods are intended to assess the environmental impacts of multifunctional processes in LCA in a comprehensive and accurate manner, thereby reducing the uncertainties and limitations associated with traditional allocation methods.

2.7.3 Inventory Analysis

Based on the goals of LCA research, the inventory analysis aims to collect and quantify input and output data within the system boundary. This involves detailed data collection and assessment of various stages, for example:

1. Raw materials acquisition: Including the energy and resources needed to mine ores or harvest agricultural products.
2. Manufacturing and production: Including the consumption of energy and materials (e.g., chemicals) and the emissions produced during these processes.
3. Distribution and transportation: Accounting for the fuel consumption and emissions from moving the product from the factory to the market.
4. Use and consumption phase: Considering factors like energy consumption during the product's usage.
5. End-of-life: Including disposal (like landfill, incineration), recycling, and reuse or potential environmental impacts of the product after its useful life.

Accurate and comprehensive inventory analysis is essential for a reliable LCA. Errors or omissions in this stage can lead to incorrect conclusions and potentially misguided decisions based on the LCA results (Chen and Chen 2024). For example, in the event that the inventory of a particular pollutant is underestimated during the production process, it is highly probable that one would erroneously assume that the environmental impact of the product is less than the actual situation. In an effective inventory analysis, data sources include collection directly from the studied process or system as well as from existing databases and literature.

Sophisticated software tools and models are also employed to handle and analyze the large amounts of data involved.

2.7.4 Impact Assessment

The output of the inventory analysis acts as an input to the life cycle impact assessment (LCIA). The purpose of the LCIA is to evaluate the environmental impacts associated with the product, process, or service being analyzed (Hauschild et al. 2013). It quantifies and characterizes the emissions, resource consumption, and other potential effects involved in the studied system. This assessment helps decision-makers understand the environmental hotspots and trade-offs, enabling them to make more informed and sustainable choices. For example, in the LCA of a plastic bottle, the LCIA would consider the environmental impacts of obtaining the petroleum needed for making plastic, the energy used in the manufacturing process, the transportation emissions during distribution, the potential pollution during its use, and the challenges in disposing of or recycling it when it reaches the end of its life. The typical LCIA involves several steps:

1. Characterization: First, inventory data is translated into the impact categories, such as global warming potential, human toxicity, and resource depletion. Here, the classified inventory data is quantified in terms of its potential contribution to each impact category. For instance, CO_2 emissions are converted into their equivalent contribution to global warming potential.
2. Normalization: Then, these impacts are quantified and normalized to make comparisons possible. This helps in understanding the significance of the impacts in a broader context.
3. Weighting: Finally, weighting factors may be applied to combine the various impact categories into a single score or indicator, but this is a controversial step as the assignment of weights is subjective.

For example, in the LCIA of a mobile phone, the classification step might group the emissions from manufacturing into categories like greenhouse gases and heavy metals. The characterization would calculate the equivalent warming impact of the greenhouse gases and the potential hazard from the heavy metals. Normalization could show how the impact of this phone compares to the average impact of similar products.

2.7.5 Interpretation

Interpretation is a critical and final step that entails analyzing the LCA results derived from the preceding steps. It aims to draw meaningful conclusions, identify significant issues and hotspots, evaluate the reliability and completeness of the study, and make recommendations for improvement or decision-making (Laurent et al. 2020). In the LCA of LCB-based biofuels, interpretation could include:

1. Comparative analysis: Making a comparison between the environmental consequences of LCB-based biofuel and the conventional utilization of LCB. Additionally, it would be advantageous to assess the sustainability of the transition from fossil fuel-based systems to LCB-based biofuel systems, such as examining the carbon emission, energy consumption, and land use requirements of both and determining if the biofuel offers a significant advantage.
2. Hotspot identification: Identifying stages or processes within the biofuel production life cycle that have the greatest contribution to the overall environmental impacts. This could be the plant cultivation, bioconversion process, or the transportation of materials.
3. Sensitivity analysis: Evaluating the impact of changes in key assumptions or parameters on the LCA results, for instance, varying the yield of the feedstock or the efficiency of the conversion technology, to understand their influence on the environmental profile.
4. Uncertainty analysis: Considering the uncertainties related to the employed data and models. This helps in understanding the reliability of the conclusions and provides a margin of error.
5. Decision context: Linking the LCA results to the broader decision-making context. For example, if the goal is to reduce carbon emissions, interpreting whether the biofuel production meets the targeted reduction and what additional measures might be needed.
6. Recommendations: Based on the findings, suggesting potential improvements in the biofuel production process, such as optimizing feedstock cultivation methods, improving conversion efficiency, or implementing better waste management strategies.

Scientific interpretation helps identify opportunities and offers room for improvement in LCB biorefineries. For instance, the LCA reveals that the transportation of feedstock over long distances is a notable cause of the total environmental consequences. The interpretation could recommend sourcing feedstock from closer locations or exploring more efficient transportation methods. Or if the uncertainty analysis shows that the data on the energy consumption during a particular stage is highly variable, it might suggest conducting more detailed measurements to improve the accuracy of the assessment.

2.7.6 Uncertainty Analysis

The LCA of biofuel production from LCB is notably impacted by methodological characteristics like the functional unit, system boundaries, allocation methods, and LCA approach. Several factors can affect the accuracy and dependability of LCA studies, among which are inadequate data, some assumptions, and geographical disparities. Uncertainty analysis is a vital aspect that helps to understand the dependability and robustness of the LCA outcomes (Deng et al. 2011). It involves assessing and quantifying the possible fluctuations and errors in the LCA data, models, and key assumptions. Uncertainties can arise from multiple sources throughout the LCA process.

1. One major source of uncertainty is data quality and availability. For example, data on emissions from a particular industrial process might be based on limited measurements or estimates, leading to potential inaccuracies. The precision and representativeness of the data collected for different stages, such as raw material acquisition, manufacturing, and disposal, can all contribute to uncertainty.
2. Model uncertainties also play a role. The models used to translate inventory data into impact indicators may have inherent limitations and simplifications. Different models might give slightly different results for the same input data, adding to the uncertainty.
3. Assumptions made during the LCA, such as the lifetime of a product or the allocation methods for shared processes, can introduce uncertainty.

To conduct an uncertainty analysis, various techniques can be employed. Monte Carlo simulation is commonly used, where random values within the defined uncertainty ranges of input parameters are sampled repeatedly to generate a distribution of possible LCA results (Igos et al. 2019). The findings of the uncertainty analysis are presented in the form of confidence intervals or probability distributions. This helps decision-makers understand the scope of potential results and the degree of trust they can have in the main LCA findings. For instance, LCA of photovoltaic module manufacturing frequently reveal significant variability in energy return on investment metrics, attributable to parameter uncertainties in device conversion efficiencies and process energy intensities. This indicates that more precise data collection and model refinement might be necessary for more reliable conclusions.

2.8 LCA of Biofuel Production from Sugar Industry Byproducts

2.8.1 Background

Since the 1980s, China's sugar industry has witnessed rapid development. Currently, it produces around 13 million tons of sugar annually (Li and Yang 2015). Over 90% of the sugar output comes from sugarcane (Wang et al. 2022). Sugarcane is mostly cultivated in Guangxi, Yunnan, and Guangdong. Among them, Guangxi is the largest sugar-producing region in China, contributing over 60% (Li et al. 2016). Generally, crushing one ton of sugarcane yields about 0.125 tons of sugar. As for byproducts, bagasse, molasses, and filter mud are present in quantities of 0.3 tons, 0.041 tons, and 0.03 tons, respectively (Huang et al. 2020). Bagasse is straw yellow in color at about 50% humidity (Wang et al. 2022), which comprises 35–50% cellulose, 20–27% hemicellulose, 15–25% lignin, and 1–6% ash on the basis of dry weight (Pan et al. 2022a, b). The amount of its constituent elements differs depending on the types of sugarcane. Moreover, it is also influenced by the age of the sugarcane, the harvesting method, and the manufacturing technique of sugar mills (Dibazar et al. 2023). In China, around 80% of bagasse is burned in boilers to generate electricity for sugar mills. Molasses is composed of 40–60% carbohydrates, amino acids, vitamins, and inorganic salt, which can be utilized in alcohol and feed industry (Wang et al. 2022). Filter mud is usually burned along with bagasse. Otherwise, it is employed in fertilizer manufacturing because it is rich in organic materials and inorganic salts.

As mentioned earlier, byproducts of the sugar industry, particularly bagasse, contain substantial quantities of cellulose and hemicellulose. These substances possess the potential to produce biofuels like ethanol, butanol, and methane, etc. Therefore, bagasse can be utilized to provide combined heat and/or electricity and biofuel within sugar refineries. Bioethanol is an attractive alternative fuel that is receiving great attention worldwide. According to the Renewable 2019 published report, in 2018, the global biofuel output rose by 10% and reached 154 billion liters; it is believed that it will expand by 25% by 2024. Biofuels like bioethanol make up 65% of the total biofuels and hold a crucial position in ensuring global energy and economic security (Raj et al. 2022). In China, the annual bioethanol output grew from 5 billion liters in 2005 to 10 billion liters in 2018, ranking it third after the United States and Brazil (Wang et al. 2023). The superiority of 2G biofuels over 1G ones can be affirmed by considering the issue of producing bioethanol from bagasse. This is mainly because biofuels do not have the controversy of "competing with people for food." Moreover, the production of bioethanol from bagasse holds great potential for further expansion. As research and technology progress, it can lead to more efficient conversion methods. This action will not only boost energy security but also contribute to a more environmentally friendly world by reducing reliance on fossil fuels.

2.8.2 *Goal and Scope*

The LCA framework was constructed following ISO 14040/14044 principles. This study is conducted based on the conditions in the Guangxi Zhuang Autonomous Region (GZAR) of China. The framework includes three integrated parts: sugarcane farming, sugar milling, and bagasse utilization. In accordance with the cradle-to-gate assessment, the system boundaries of three scenarios were defined, as depicted in Fig. 2.7. The three scenarios correspond to the three processing methods of bagasse: sanitary landfill, co-generation, and combined co-generation and bioethanol production. The goal of this LCA was to examine the environmental consequences of the three ways for treating sugar industry byproducts, with the intention of identifying chances for environmental enhancements. This LCA set its functional

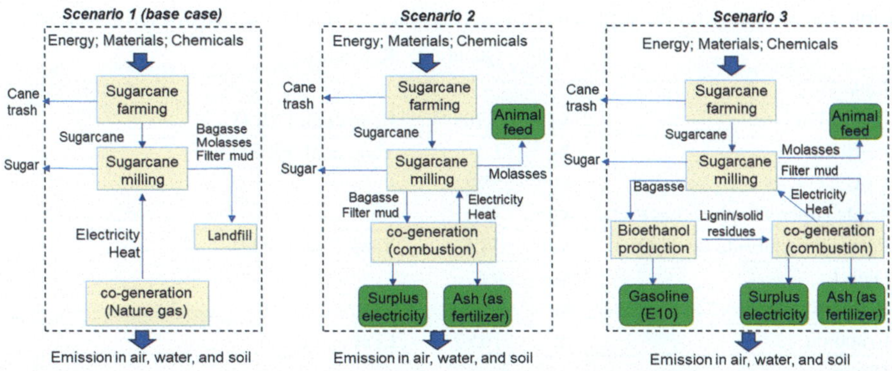

Fig. 2.7 System boundaries of the base case (landfill) and the two examined alternative cases (combustion and bioethanol production)

unit as producing 1 ton of cane sugar. All input (raw materials, chemical reagents, and energy) and output data (air emissions and wastewater) for the system are calculated. The impacts related to capital equipment and infrastructure are excluded, since they only provided minor impacts. Cane trash is left on the fields to decompose and outside the system boundaries.

2.8.3 Data Source

The primary data was acquired from the ecoinvent database 3.0. Data not available from this database was collected from literature or based on actual estimations. The details of the technical description and inventory analysis are as follows.

2.8.4 Technical Description and Inventory Analysis

2.8.4.1 Sugarcane Farming

In Guangxi, sugarcane fields produce around 80 tons of sugarcane per hectare. After harvest, there are around 26.7 tons of cane trash. After being harvested, the sugarcane is transported to the sugar refinery by trucks over a distance of 50 km. The transport truck—"Transport, freight, lorry, 16–32, metric ton"—was selected from ecoinvent v3 database of the LCA software. The truck employed complies with Euro 3 emission standards. Table 2.3 shows the inventory related to sugarcane cultivation.

Table 2.3 Inventory for the sugarcane farming stage based on the production of 1 ton of cane sugar

	Unit	Quantity	References
Input			
Use of land	ha	0.107	Peng et al. (2014)
Pesticide	kg	1.1	Yang et al. (2022)
Urea	kg	58.8	
Triple superphosphate (as P_2O_5)	kg	45.1	
Potassium chloride (as K_2O)	kg	58.6	Yang et al. (2022)
Diesel	kg	9.6	
Plastic film	kg	0.3	
Irrigation	m^3	105.7	Li et al. (2022)
Output			
Cane	kg	8000	The National Bureau of Standards (2023)
Cane trash	kg	2666.7	Li et al. (2022)
Emissions			
N_2O	kg	0.0198	Contreras et al. (2009)
N_{total} to water	kg	0.0014	
Pesticides to water	kg	0.00001	
Pesticides to soil	kg	0.0001	

2.8.4.2 Sugarcane Milling

Typically, in Chinese sugarcane mills, 8 kg of sugarcane are required to produce 1 kg of sugar. The byproducts generated from sugarcane processing include bagasse, molasses, and filter mud, which account for 30%, 4.1%, and 3.0% of the total mass of fresh sugarcane, respectively. In Scenario 1, sugar industry byproducts are sent to landfills for disposal. In alternative scenarios, they are used to make bioproducts or generate bioelectricity through a combined heat and power generation (CHP) unit. The electricity needed in Chinese sugar mills is 0.34 kWh/kg of sugar, and the heat required is 3.13 kWh/kg of sugar (Mohammadi et al. 2020), respectively. This demand is preferentially met by the CHP unit, and the remaining is supplied from the national grid. Table 2.4 lists the inventory data related to sugar milling.

2.8.4.3 Scenario 1: Landfill (Base Case)

In the landfill scenario, the requirements for electricity and heat are primarily met via a CHP unit fueled by natural gas. By referring to the ecoinvent database and literature data, it is considered that the heat efficiency and power efficiency of the CHP unit stand at 55% and 30%, respectively. Additionally, for 1 kWh of electricity produced, 0.03 m^3 of natural gas is consumed. Based on the ecoinvent database, the emissions from this procedure are determined and then adjusted to conform to the presumed natural gas lower heating value (LHV) of 36 MJ/m^3. Meanwhile, it is supposed that the usage of lubricating oil is 0.03 g/MJ of natural gas. Here, the

Table 2.4 Inventory for the sugarcane processing stage based on the production of 1 ton of cane sugar

	Unit	Quantity	References
Input			
Cane	kg	8000.0	Peng et al. (2014)
Cane transport	tkm	50	Estimated value
Lime	kg	3.0	Mohammadi et al. (2020)
Sulfur dioxide, liquid	kg	2.7	
Sulfuric acid	kg	3.9	
Chlorine, gaseous	kg	0.004	
Sodium chloride, powder	kg	0.051	
Chemical, inorganic	kg	0.740	
Chemical, organic	kg	0.580	
Electricity	kWh	340	
Heat	kWh	3130	
Output			
Sugar	kg	1000	Functional unit
Bagasse	kg	2400	Wang et al. (2022)
Molasses	kg	328	
Filter mud	kg	240	
Emissions			
Wastewater	m^3	10.0	Mohammadi et al. (2020)
COD	kg	19.1	Chauhan et al. (2011)
Total suspended particles	kg	1.7	
CO_2	kg	160.0	
CH_4	kg	0.0	
NO_x	kg	1.3	
SO_2	kg	1.2	
CO	kg	1.26	
Total soluble salt	kg	13.1	

byproducts of the sugar industry are discharged as waste and must be dealt with by disposal in landfills. The emission data for landfills is derived from the research of Zhou et al. (2018). Table 2.5 lists the inventory for the landfill of the byproduct of sugar milling.

2.8.4.4 Scenario 2: Combustion

The mixture of bagasse and filtered mud is delivered to the CHP unit. It is estimated that the combustor efficiency covers 30% for electricity generation and 55% for heat production. It is assumed that sugarcane bagasse (SCB) consists of 18.9% lignin, 42.6% cellulose, and 29.5% hemicelluloses; the ash content totals up to 1.3% (Pan et al. 2022a, b). The LHV of bagasse and filter mud on a wet basis are 10.0 MJ/kg and 8.8 MJ/kg, respectively (Mohammadi et al. 2020). The generated electricity is

Table 2.5 Inventory of the sanitary landfill of the byproduct of 1 ton of sugar production

	Unit	Quantity	References
Input			
SCB	kg	2400	–
Molasses	kg	328	–
Filter mud	kg	240	–
Natural gas	m^3	113.33	Mohammadi et al. (2020)
Lubricating oil	kg	0.12	
Heat	kWh	2506.7	Suppl. heat energy
Products			
Electricity	kWh	340.0	Surplus electricity
Emissions			
CO_2	kg	353.6	Mohammadi et al. (2020)
CO	kg	0.0140	
CH_4	kg	0.0068	
N_2O	kg	0.0068	
NO_x	kg	0.1700	
SO_2	kg	0.0034	

initially utilized to fulfill its internal heat and electricity requirements. In the event of excess electricity, it is provided to the public grid. It is assumed that the available waste heat is released into the air. The combustion emission data are calculated according to the ecoinvent database and adjusted to fit the characterization of Chin's sugar milling byproducts. The ashes resulting from combustion are utilized to substitute synthetic fertilizers, for example, triple superphosphate and potassium chloride. Table 2.6 lists the inventory for the combustion of the sugar milling byproduct.

2.8.4.5 Scenario 3: Bioethanol Production

In Scenario 3, bagasse is used to produce ethanol, while filter mud is burned in a CHP unit. According to reference (Peng et al. 2014), in China, it is estimated that producing 1 kg of ethanol requires 14 kg of bagasse (50% moisture). The simultaneous saccharification and fermentation approach is chosen for ethanol fermentation, as it is time-saving and efficient (Unrean and Ketsub 2018). The reaction occurs at 35 °C for 48 hours. A detailed description of this scenario can be found in the literature (Pan et al. 2023). It is estimated that based on dry solid loading, 0.02% cellulase, 0.02% hydrolase, and 0.5% yeast are used, and 0.1 kilogram of ammonium per kilogram of yeast is added as a nutrient. Due to a lack of data, the ecoinvent dataset "enzymes" is adjusted for Chinese conditions as referred to in literature (Pan et al. 2023), and the "Cellulase" and "Hydrolase" were replaced. Ethanol with a purity of 99.6% is regarded as an avoided product because it can replace 10% of gasoline

Table 2.6 Inventory of the combustion of the byproduct of 1 ton of sugar production

	Unit	Quantity	Ref.
Input			
Bagasse	kg	2400	–
Molasses	kg	328	–
Filter mud	kg	240	–
Diesel	kg	6.36	Estimated value
Products			
Electricity	kWh	1674	Surplus electricity
Ash (triple superphosphate, as P_2O_5)	kg	2.2	–
Ash (potassium chloride, as K_2O)	kg	4.6	–
Emissions			
CO_2	kg	797.54	Peng et al. (2023)
CO	kg	0.6525	
CH_4	kg	0.0225	
NO_x	kg	0.5873	
N_2O	kg	0.0058	
SO_2	kg	0.0138	
NMVOC	kg	0.0370	
Particulates	kg	0.4423	

consumption. The solid residue from bioethanol production, mainly consisting of lignin and vinasse, which is about 30% of the mass of the initial raw materials, is also burned with filter mud. As reported in the literature, the LHV of the solid residue is supposed to be 19 MJ/kg (Peng et al. 2023). In this scenario, ashes are considered for use as fertilizer, similar to Scenario 2. Table 2.7 lists the inventory for the bioethanol produced from the byproduct of sugar milling.

2.8.5 *Multifunctionality and Allocation Method*

In this research, a broad set of allocation methods has been applied to track the relevant environmental impacts. Three multifunctional situations are identified at different process stages, and their solutions are as follows.

The first situation occurs during the sugarcane harvest stage and involves sugarcane and sugarcane trash. In this context, both economic benefit and energy value can serve as allocation factors. However, sugarcane is utilized to manufacture sugar for sale. Cane trash is considered biowaste and remains in the field where it decays. The resulting greenhouse gases are offset by photosynthesis. Therefore, economic allocation is applied in this situation, and sugarcane is burdened with all environmental impacts.

Table 2.7 Inventory associated with bioethanol production from the byproduct, based on 1 t of sugar production

	Unit	Quantity	References
Input			
Bagasse	kg	2400	–
Molasses	kg	328	–
Filter mud	kg	240	–
NaOH	kg	96	Li et al. (2022)
Yeast	kg	0.9	Pan et al. (2023)
Cellulase	kg	0.017	
Hydrolase	kg	0.4	
Ammonium	kg	0.1	
Electricity	kWh	59.5	Li et al. (2022)
Water	kg	4718.4	Kadam (2002)
Heat	kWh	1759.9	Suppl. Thermal energy
Diesel	kg	2.4	Estimated value
Products			
Ethanol	kg	342.9	–
Electricity	kWh	407.3	Surplus electricity
Ash (triple superphosphate, as P_2O_5)	kg	0.87	–
Ash (potassium chloride, as K_2O)	kg	1.80	–
Emissions from bioethanol generation			
COD	kg	0.51	Li et al. (2022)
BOD_5	kg	0.10	
Suspended solids	kg	0.15	
Ammonia	kg	0.02	
NO_x	kg	0.63	
SO_x	kg	1.36	
Particulates	kg	0.98	
Emissions from combustion			
CO_2	kg	295.94	Peng et al. (2023)
CO	kg	0.2421	
CH_4	kg	0.0083	
NO_x	kg	0.2179	
N_2O	kg	0.0022	
SO_2	kg	0.0051	
NMVOC	kg	0.0137	
Particulates	kg	0.1641	

The second occurs during the operation of the sugar mill when it separates the streams of sugar, bagasse, molasses, and filtered mud. In previous studies, economic allocation, system expansion, or a combination of them has been a common method (Mohammadi et al. 2020). In Scenario 1, the economic allocation factors are assumed to be 91.4% for sugar, 3.3% for bagasse, 5.3% for molasses, and 0% for filtered mud due to their current treatment in China. In Scenario 2 and Scenario 3, molasses can compensate for animal feed elsewhere based on system expansion. It is assumed that molasses can replace maize grain as animal feed based on the protein content of each product, with a substitution rate of 0.66 (Yang et al. 2022).

The third occurs in the processes of converting byproducts into bioenergy (bioethanol, electricity, and biofertilizer). Here, a system expansion approach is used, assuming byproducts replace similar market products. Specifically, combustion residues serve as synthetic fertilizer substitutes contingent upon their plant-available phosphorus (P), potassium (K), and nitrogen (N) contents, following characterization protocols outlined in the manual of industry sugar derivates (Contreras et al. 2009). Surplus electricity is exported to the national grid, reducing coal power output. In China, E10 (containing 10% ethanol and 90% gasoline by volume) is the only vehicle ethanol fuel that has received approval. Ethanol has a lower heating value but higher combustion efficiency. It should be noted that the calorific value of ethanol is 26.7 MJ/kg, while that of gasoline is 44.3 MJ/kg. After estimation, 1 kg of ethanol is equivalent to 0.603 kg of gasoline (Khatiwada et al. 2016). Additionally, cane sugar is the main product and has an impact on the total environmental consequences. Therefore, the system boundaries are expanded to incorporate the avoided products, and their environmental load is deducted from the total.

Table 2.8 The categories of impact assessment

Number	Impact category	Unit
1	Global warming	kg CO_2 eq
2	Stratospheric ozone depletion	kg CFC_{11} eq
3	Ionizing radiation	kBq Co-60 eq
4	Ozone formation, human health	kg NO_x eq
5	Fine particulate matter formation	kg $PM_{2.5}$ eq
6	Ozone formation, terrestrial ecosystems	kg NO_x eq
7	Terrestrial acidification	kg SO_2 eq
8	Freshwater eutrophication	kg P eq
9	Marine eutrophication	kg N eq
10	Terrestrial ecotoxicity	kg 1,4-DCB
11	Freshwater ecotoxicity	kg 1,4-DCB
12	Marine ecotoxicity	kg 1,4-DCB
13	Human carcinogenic toxicity	kg 1,4-DCB
14	Human noncarcinogenic toxicity	kg 1,4-DCB
15	Land use	m^2a crop eq
16	Mineral resource scarcity	kg Cu eq
17	Fossil resource scarcity	kg oil eq
18	Water consumption	m^3

2.8.6 Impact Assessment

The life cycle impact assessment (LCIA) was performed in SimaPro 9.0 (PR'e Sustainability, Netherlands) with the use of the ReCiPe method (ReCiPe 2016 Midpoint (H) V1.08/World (2010) H). The default 18 categories are presented in Table 2.8.

2.8.7 Results and Interpretation

2.8.7.1 Characterization

The LCIA methods are developed through in-depth modeling based on natural sciences to establish a connection between inventory results and environmental issues. They provide distinct characterization factors. These factors show the influence that a certain amount of pollutant has on specific impact categories such as global warming, acidification, and human toxicity (Levasseur et al. 2010). The function of the characterization indicator is to transform life cycle inventory results into common impact categories, thereby quantifying the LCA results. Figure 2.8 shows the contribution of each scenario to 18 environmental impact categories. A positive value means an impact contribution; a negative value indicates "avoided impact" contribution. For instance, it is shown that Scenario 2 and Scenario 3 are shown to have lower environmental burdens in terms of mineral resource scarcity compared to the base case, since they avoid coal and fuel oil-based electricity production.

As shown in Fig. 2.8, Scenario 2 has positive impacts on nearly all categories, while Scenario 1 is the opposite. In Scenario 1, the landfill of sugar milling

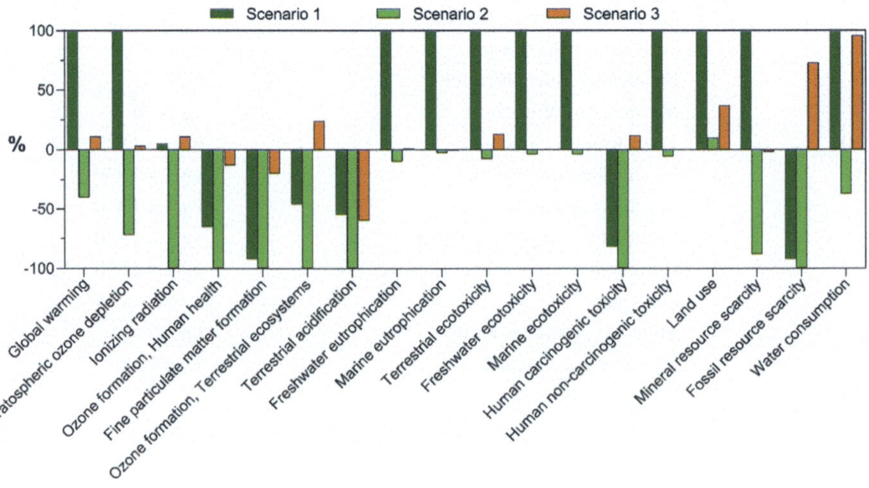

Fig. 2.8 Life cycle impact assessment in characterization indicator

byproducts could result in the release of leachate and particulate matter from manufacturing facilities. This would contribute to air pollution and potentially have negative impacts on human respiratory health and the aquatic environment (Postacchini et al. 2018). Bioelectricity from the CHP unit in Scenario 2 conserves significant amounts of fossil fuels. In reality, with regard to greenhouse gas emissions, many LCA studies conclude that combustion is more favorable than landfilling because of the significant energy recovery offsets (Anshassi et al. 2022). Moreover, the CHP plant occupies a considerably smaller space when compared with landfill sites. This makes site selection much easier and more flexible (Nabavi-Pelesaraei et al. 2017). In contrast, Scenario 3 shows neutral environmental benefits. Although it produces bioethanol as an alternative to gasoline, the ethanol production process requires more chemicals and enzymes. Nevertheless, in Scenario 3, the substitution of gasoline by alcohol brings about a positive reduction in the impact on the category of mineral resource scarcity. In summary, Scenario 2 may be a better choice for processing sugar milling byproducts.

2.8.7.2 Normalization

The normalization aims to help understand the magnitude of environmental consequences by comparing them with a reference value. Normalized values can assist in identifying which impact categories are relatively more significant for the targeted product or process. When the normalized value for a product's contribution to acidification is considerably higher than its normalized value for ozone depletion, it implies that acidification may be a more crucial environmental concern for that product. Figure 2.9 presents the results of applying normalization factors for the three scenarios. The impacts on freshwater eutrophication, freshwater ecotoxicity,

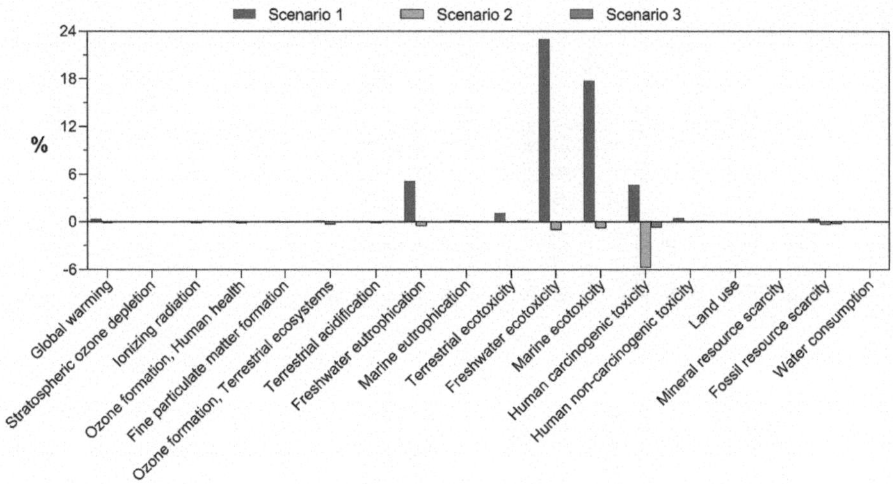

Fig. 2.9 Life cycle impact assessment in normalization indicator

Table 2.9 Analysis of the largest contributor to the total impact

Categories	Process	Unit	Contribution					
			Scenario 1		Scenario 2		Scenario 3	
			Value	Percentage	Value	Percentage	Value	Percentage
Freshwater eutrophication	Municipal solid waste {GLO}I treatment of municipal solid waste, open burning I Cutoff, U	kg P eq	0.019	23.2%	−2.36E-06	0.01%	−9.12E-07	−0.01%
Freshwater ecotoxicity	Municipal solid waste {GLO}I treatment of municipal solid waste, open burning I cutoff, U	kg 1,4-DCB	2.48	51.96%	−3.01E-04	0.04%	−1.02E-04	0.02%
Marine ecotoxicity	Municipal solid waste {GLO}I treatment of municipal solid waste, open burning I Cutoff, U	kg 1,4-DCB	2.97	40.11%	−3.06E-04	0.03%	−1.40E-04	0.03%
Human carcinogenic toxicity	Hard coal ash {RoW}I treatment of hard coal ash, residual material landfill I Cutoff, U	kg 1,4-DCB	−0.36	−1.92%	−6.28	56.82%	−1.46	73.27%

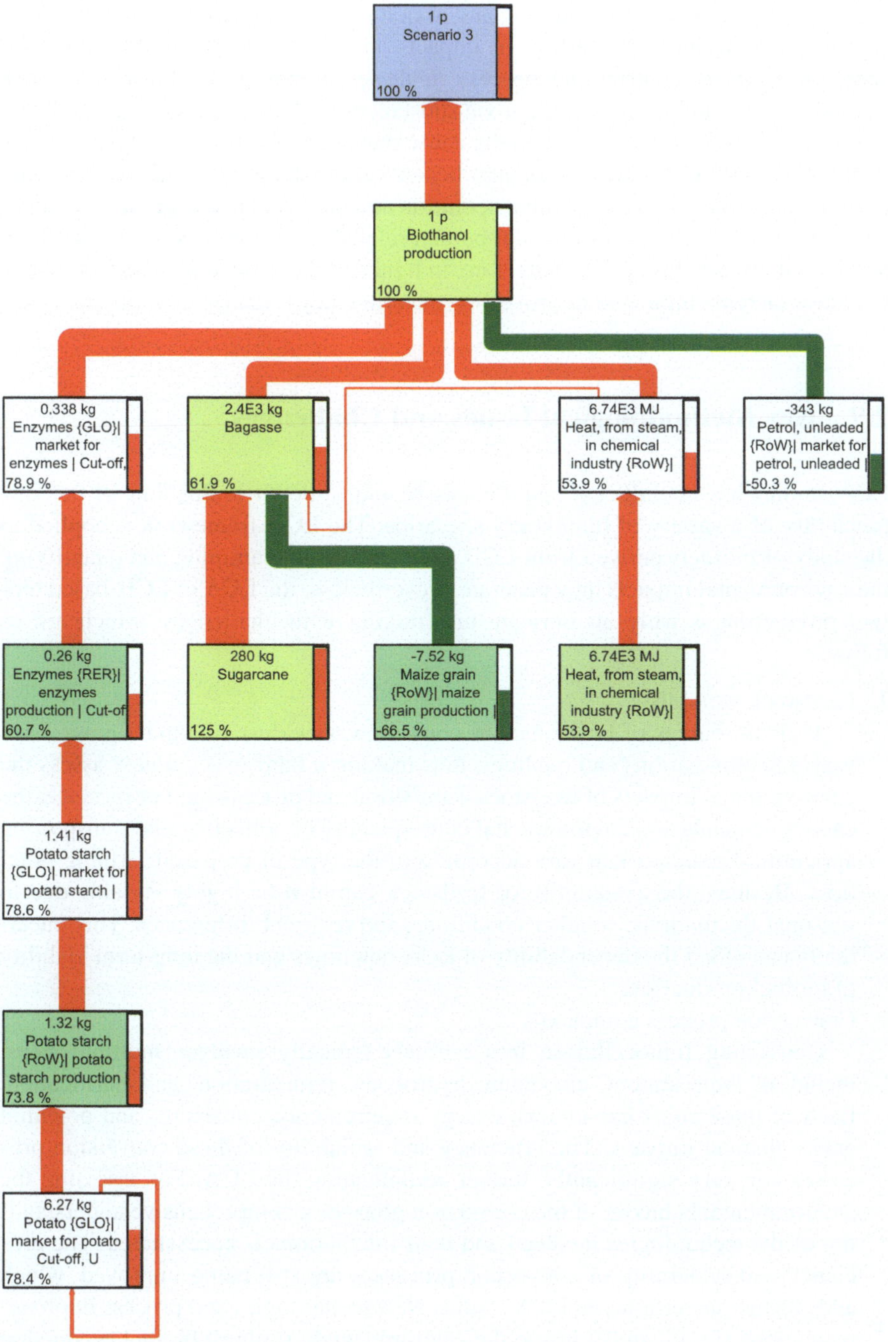

Fig. 2.10 Tree diagrams of LCA for Scenario 3. Node cutoff in 30%

marine ecotoxicity, human carcinogenic toxicity, and fossil resource scarcity have significant contributions, whereas the impacts on the right are negligible. Table 2.9 shows the largest contributing process in these categories. It shows that when byproducts are landfilled, specific toxic substances such as mercury, lead, and cadmium are released. This can potentially cause damage to freshwater ecosystems. By contrast, in Scenario 2, environmental benefits (human carcinogenic toxicity) are attributed to reduced hard coal mine operation and hard coal preparation. In particular, in Scenario 3, the environmental burden related to enzymes contributes 78.5% to the total impact (Fig. 2.10). Therefore, in lignocellulosic ethanol production, how to reduce enzyme input and improve enzyme efficiency remains an important issue.

2.9 Key Methodological Issues and Challenges

LCA studies have conducted comprehensive and in-depth evaluations of the sustainability of a variety of biorefinery scenarios. The LCA framework is applied to the study of biofuels derived from LCB and is aimed at examining and quantifying the environmental impacts they generate. Nevertheless, the LCA of LCB-based biofuel production is truly an intricate undertaking with challenges, which are as follows:

1. Feedstock variability

   Various sources of LCB, such as corn straw, saw dust, and giant grass, have varying compositions and qualities, thus making it hard to accurately assess the environmental impacts of feedstock acquisition and processing. For instance, the energy demands and environmental consequences of collecting and transporting agricultural residues can vary according to the type of crop and farming practices. Besides, the availability of feedstock can also be highly variable due to seasonal fluctuations, weather conditions, and regional differences. This uncertainty can affect the dependability of LCA outcomes and the long-term viability of biofuel production.
2. Conversion process complexity

   Converting lignocellulose into biofuels typically involves multiple steps, including pretreatment, enzymatic hydrolysis, fermentation, and purification. Each of these steps has its own energy requirements, emissions, and potential environmental impacts. The efficiency and scalability of these conversion processes can vary significantly, further complicating the LCA. For assessing the environmental behavior of the conversion process, a comprehensive understanding of the technologies involved and their interactions is necessitated. The efficiency and scalability of conversion processes are still being improved, which adds further uncertainty to LCA results. New technologies and process improvements may significantly change the environmental profile of biofuel production over time.
3. Co-product allocation

Biofuel production from lignocellulose often generates co-products such as electricity, heat, and valuable chemicals. Allocating the environmental burdens and benefits between the biofuel and co-products is a complex methodological issue. Different allocation methods can lead to very different LCA results, making it difficult to compare and evaluate different biofuel production pathways. Defining the system boundaries for co-product allocation can also be challenging. For example, should the environmental impacts of producing the enzymes used in hydrolysis be allocated to the biofuel or considered a separate process? Therefore, the wide variability in system boundaries requires more in-depth studies.

4. Indirect land use change (ILUC)

   The potential for indirect land use change (ILUC) adds another layer of complexity (Schmidt et al. 2015). When substantial amounts of land are employed for cultivating lignocellulosic feedstocks, there is a risk of displacing other land uses and potentially causing indirect environmental impacts, such as the conversion of natural habitats or displacement of food production. Accounting for ILUC is difficult as it involves complex economic and environmental interactions and is often subject to high uncertainty. However, ILUC can bring about notable environmental consequences, including loss of biodiversity, carbon emissions, and changes in water and soil quality. Incorporating ILUC into LCA is essential for enhancing the reliability of the sustainability assessment of lignocellulosic biofuels.

5. Data uncertainty and availability

   Conducting a reliable LCA requires accurate and detailed data on all aspects of biofuel production, from feedstock acquisition to end use. However, data on many aspects of lignocellulosic biofuel production are limited or uncertain, particularly for emerging technologies and processes. Gathering data on environmental impacts such as global warming, resource depletion, and land use change can be challenging and time-consuming. Additionally, different sources of data may vary in quality and reliability, further complicating the LCA process. This makes it essential for researchers and practitioners to carefully evaluate and validate data sources. Rigorous data verification procedures can help mitigate uncertainties and enhance the credibility of LCA studies.

2.10 Conclusion

Lignocellulosic biomass (LCB) is an attractive feedstock, as it is readily available, abundant, and cheaper and can potentially replace petro-fuels. Recently, LCB-derived biofuels are supported globally for multiple goals like mitigating climate change, ensuring energy security, and promoting rural economic development. They can address first-generation biofuel hurdles and are an attractive raw material for various bioproducts. LCA is crucial for recognizing risks and opportunities in biorefining, helping policymakers consider biofuel production technology and logistics while accounting for environmental impacts. It empowers policymakers to make decisions that promote sustainable development and a shift to a more eco-friendly energy future. Nevertheless, the LCA of biofuel production from LCB is truly a complex process fraught with challenges. Despite these challenges, efforts must be made to improve the LCA of biofuel production from LCB. This calls for collaboration among researchers, policymakers, and industry stakeholders to develop standardized data collection methods, improve modeling techniques, and promote the implementation of sustainable practices across the biofuel chain. Only through such collective action can we hope to fully understand and manage the environmental impacts of biofuel production from LCB and head towards a more sustainable energy future.

Acknowledgments The authors acknowledge the financial support from the National Natural Science Foundation of China (Grant No. 52270121) and the Guangxi Natural Science Foundation of China (Grant No. 2023JJD160010). They also thank Prof. Zabed for providing language assistance.

Declaration of Competing Interest The authors declare no competing interest.

References

Agarwal A, Rana M, Park JH (2018) Advancement in technologies for the depolymerization of lignin. Fuel Process Technol 181:115–132. https://doi.org/10.1016/j.fuproc.2018.09.017

Amin FR, Khalid H, Zhang H, Rahman SU, Zhang R, Liu G, Chen C (2017) Pretreatment methods of lignocellulosic biomass for anaerobic digestion. AMB Express 7. https://doi.org/10.1186/s13568-017-0375-4

Anshassi M, Smallwood T, Townsend TG (2022) Life cycle GHG emissions of MSW landfilling versus incineration: expected outcomes based on US landfill gas collection regulations. Waste Manag 142:44–54. https://doi.org/10.1016/j.wasman.2022.01.040

Bamber N, Turner I, Arulnathan V, Li Y, Zargar Ershadi S, Smart A, Pelletier N (2020) Comparing sources and analysis of uncertainty in consequential and attributional life cycle assessment: review of current practice and recommendations. Int J Life Cycle Assess 25:168–180. https://doi.org/10.1007/s11367-019-01663-1

Bianco F, Senol H, Papirio S (2021) Enhanced lignocellulosic component removal and biomethane potential from chestnut shell by a combined hydrothermal-alkaline pretreatment. Sci Total Environ 762:144178. https://doi.org/10.1016/j.scitotenv.2020.144178

Bilal M, Iqbal HMN (2020) Recent advancements in the life cycle analysis of lignocellulosic biomass. Curr Sustain Renew Energy Rep 7:100–107. https://doi.org/10.1007/s40518-020-00153-5

Biswas R, Uellendahl H, Ahring BK (2015) Wet explosion: a universal and efficient pretreatment process for lignocellulosic biorefineries. Bioenergy Res 8:1101–1116. https://doi.org/10.1007/s12155-015-9590-5

Burmahl B (2023) Separations technology critical to converting biomass to low-carbon biofuel. Argonne National Laboratory. Available at: https://www.anl.gov/article/separations-technology-critical-to-converting-biomass-to-lowcarbon-biofuel

Cai CG, Xu ZX, Zhou HR, Chen ST, Jin MJ (2021) Valorization of lignin components into gallate by integrated biological hydroxylation, O-demethylation, and aryl side-chain oxidation. Sci Adv 7:eabg4585. https://doi.org/10.1126/sciadv.abg4585

Chandel AK, da Silva SS, Singh OV (2012) Detoxification of lignocellulose hydrolysates: biochemical and metabolic engineering toward white biotechnology. Bioenergy Res 6:388–401. https://doi.org/10.1007/s12155-012-9241-z

Chandra R, Takeuchi H, Hasegawa T, Kumar R (2012) Improving biodegradability and biogas production of wheat straw substrates using sodium hydroxide and hydrothermal pretreatments. Energy 43:273–282. https://doi.org/10.1016/j.energy.2012.04.029

Chauhan MK, Varun, Chaudhary S, Kumar S, Samar (2011) Life cycle assessment of sugar industry: a review. Renew Sust Energ Rev 15:3445–3453. https://doi.org/10.1016/j.rser.2011.04.033

Chen CF, Chen SK (2024) Carbon footprint inventory using life cycle energy analysis. Mrs Energy Sustain 11:107–122. https://doi.org/10.1557/s43581-023-00074-y

Chen X, Yuan X, Chen S, Yu J, Zhai R, Xu Z, Jin M (2021) Densifying lignocellulosic biomass with alkaline chemicals (DLC) pretreatment unlocks highly fermentable sugars for bioethanol production from corn Stover. Green Chem 23:4828–4839. https://doi.org/10.1039/d1gc01362a

Chouinard-Dussault P, Bradt L, Ponce-Ortega JM, El-Halwagi MM (2011) Incorporation of process integration into life cycle analysis for the production of biofuels. Clean Technol Environ 13:673–685. https://doi.org/10.1007/s10098-010-0339-8

Contreras AM, Rosa E, Pérez M, Van Langenhove H, Dewulf J (2009) Comparative life cycle assessment of four alternatives for using byproducts of cane sugar production. J Clean Prod 17:772–779. https://doi.org/10.1016/j.jclepro.2008.12.001

Cossutta M, McKechnie J, Pickering SJ (2017) A comparative LCA of different graphene production routes. Green Chem 19:5874–5884. https://doi.org/10.1039/c7gc02444d

Cui XY, Yang JH, Wang ZY, Shi XY (2021) Better use of bioenergy: a critical review of co-pelletizing for biofuel manufacturing. Carbon Capture Sci Technol 1:100005. https://doi.org/10.1016/j.ccst.2021.100005

Dar RA, Parmar M, Dar EA, Sani RK, Phutela UG (2021) Biomethanation of agricultural residues: potential, limitations and possible solutions. Renew Sustain Energy Rev 135:110217. https://doi.org/10.1016/j.rser.2020.110217

Deng L, Babbitt CW, Williams ED (2011) Economic-balance hybrid LCA extended with uncertainty analysis: case study of a laptop computer. J Clean Prod 19:1198–1206. https://doi.org/10.1016/j.jclepro.2011.03.004

Dibazar AS, Aliasghar A, Behzadnezhad A, Shakiba A, Pazoki M (2023) Energy cycle assessment of bioethanol production from sugarcane bagasse by life cycle approach using the fermentation conversion process. Biomass Convers Bioref. https://doi.org/10.1007/s13399-023-04288-5

dos Santos AC, Ximenes E, Kim Y, Ladisch MR (2019) Lignin–enzyme interactions in the hydrolysis of lignocellulosic biomass. Trends Biotechnol 37:518–531. https://doi.org/10.1016/j.tibtech.2018.10.010

Furlan FF, Filho RT, Pinto FH, Costa CB, Cruz AJ, Giordano RL, Giordano RC (2013) Bioelectricity versus bioethanol from sugarcane bagasse: is it worth being flexible? Biotechnol Biofuels 6:12. https://doi.org/10.1186/1754-6834-6-142

Gao JQ et al (2023) Engineering co-utilization of glucose and xylose for chemical overproduction from lignocellulose. Nat Chem Biol 19:1524–1531. https://doi.org/10.1038/s41589-023-01402-6

Gnansounou E, Dauriat A, Villegas J, Panichelli L (2009) Life cycle assessment of biofuels: energy and greenhouse gas balances. Bioresour Technol 100:4919–4930. https://doi.org/10.1016/j.biortech.2009.05.067

Hauschild MZ, Goedkoop M, Guinée J, Heijungs R, Huijbregts M, Jolliet O, Pant R (2013) Identifying best existing practice for characterization modeling in life cycle impact assessment. Int J Life Cycle Assess 18:683–697. https://doi.org/10.1007/s11367-012-0489-5

Heijungs R, Allacker K, Benetto E, Brandão M, Guinée J, Schaubroeck S, Zamagni A (2021) System expansion and substitution in LCA: a lost opportunity of ISO 14044 Amendment 2. Front Sustain 2. https://doi.org/10.3389/frsus.2021.692055

Huang HJ, Ramaswamy S (2013) Overview of biomass conversion processes and separation and purification technologies in biorefineries. Separation And Purification Technologies in Biorefineries, 1–36. https://doi.org/10.1002/9781118493441.ch1

Huang JF, Khan MT, Perecin D, Coelho ST, Zhang MQ (2020) Sugarcane for bioethanol production: potential of bagasse in Chinese perspective. Renew Sust Energ Rev 133:110296. https://doi.org/10.1016/j.rser.2020.110296

Ibbotson S, Kara S (2012) LCA case study. Part 1: cradle-to-grave environmental footprint analysis of composites and stainless steel I-beams. Int J Life Cycle Assess 18:208–217. https://doi.org/10.1007/s11367-012-0452-5

International Energy Agency (IEA) (2024a) Key world energy statistics. https://www.iea.org/reports/key-world-energy-statistics-2021. Accessed 16 July 2024

IEA (2024b) IEA's electricity mid-year update. https://www.iea.org/. Accessed 16 July 2024

Igos E, Benetto E, Meyer R, Baustert P, Othoniel B (2019) How to treat uncertainties in life cycle assessment studies? Int J Life Cycle Assess 24:794–807. https://doi.org/10.1007/s11367-018-1477-1

Ijassi W, Ben Rejeb H, Zwolinski P (2021) Environmental impact evaluation of co-products: decision-aid tool for allocation in LCA. Int J Life Cycle Assess 26:2199–2214. https://doi.org/10.1007/s11367-021-01984-0

Javed MR, Noman M, Shahid M, Ahmed T, Khurshid M, Rashid MH, Khan F (2019) Current situation of biofuel production and its enhancement by CRISPR/Cas9-mediated genome engineering of microbial cells. Microbiol Res 219:1–11. https://doi.org/10.1016/j.micres.2018.10.010

Jung YH, Kim IJ, Kim HK, Kim KH (2013) Dilute acid pretreatment of lignocellulose for whole slurry ethanol fermentation. Bioresour Technol 132:109–114. https://doi.org/10.1016/j.biortech.2012.12.151

Kadam KL (2002) Environmental benefits on a life cycle basis of using bagasse-derived ethanol as a gasoline oxygenate in India. Energ Policy 30:371–384. https://doi.org/10.1016/S0301-4215(01)00104-5

Khatiwada D, Venkata BK, Silveira S, Johnson FX (2016) Energy and GHG balances of ethanol production from cane molasses in Indonesia. Appl Energy 164:756–768. https://doi.org/10.1016/j.apenergy.2015.11.032

Kiss AA, Lange JP, Schuur B, Brilman DWF, van der Ham AG, Kersten SR (2016) Separation technology-making a difference in biorefineries. Biomass Bioenergy 95:296–309. https://doi.org/10.1016/j.biombioe.2016.05.021

Klein D, Wolf C, Schulz C, Weber-Blaschke G (2015) 20 years of life cycle assessment (LCA) in the forestry sector: state of the art and a methodical proposal for the LCA of forest production. Int J Life Cycle Assess 20:556–575. https://doi.org/10.1007/s11367-015-0847-1

Koffler C (2012) The critical review of life cycle assessment studies according to ISO 14040 and 14044-origin, purpose and practical performance. Int J Life Cycle Assess 17:1087–1093. https://doi.org/10.1007/s11367-012-0426-7

Kumar R, Mago G, Balan V, Wyman CE (2009) Physical and chemical characterizations of corn stover and poplar solids resulting from leading pretreatment technologies. Bioresour Technol 100:3948–3962. https://doi.org/10.1016/j.biortech.2009.01.075

Laurent A, Weidema BP, Bare J, Liao X, Maia de Souza D, Pizzol M et al (2020) Methodological review and detailed guidance for the life cycle interpretation phase. J Ind Ecol 24:986–1003. https://doi.org/10.1111/jiec.13012

Levasseur A, Lesage P, Margni M, Deschenes L, Samson R (2010) Considering time in LCA: dynamic LCA and its application to global warming impact assessments. Environ Sci Technol 44:3169–3174. https://doi.org/10.1021/es9030003

Li YR, Yang LT (2015) Sugarcane agriculture and sugar industry in China. Sugar Tech 17:1–8. https://doi.org/10.1007/s12355-014-0342-1

Li T, Zhang HC, Liu ZC, Ke QD, Alting L (2014) A system boundary identification method for life cycle assessment. Int J Life Cycle Assess 19:646–660. https://doi.org/10.1007/s11367-013-0654-5

Li YR, Song XP, Wu JM, Li CN, Liang Q, Liu XH et al (2016) Sugar industry and improved sugarcane farming technologies in China. Sugar Tech 18:603–611. https://doi.org/10.1007/s12355-016-0480-8

Li P, Sakuragi K, Makino H (2019) Extraction techniques in sustainable biofuel production: A concise review. Fuel Process Technol 193:295–303. https://doi.org/10.1016/j.fuproc.2019.05.009

Li J, Xiong F, Fan M, Chen Z (2022) The role of nonfood bioethanol production in neutralizing China's transport carbon emissions: an integrated life cycle environmental-economic assessment. Energy Sustain Dev 70:68–77. https://doi.org/10.1016/j.esd.2022.06.002

Liang M, Lin Y, Sun L, Pang H, Wei H, Huang R et al (2023) Discovery of a novel β-xylosidase with xylanase activity and its application in the production of xylitol from corncob xylan. Fermentation-Basel 9:606. https://doi.org/10.3390/fermentation9070606

Mohammadi F, Roedl A, Abdoli MA, Amidpour M, Vahidi H (2020) Life cycle assessment (LCA) of the energetic use of bagasse in Iranian sugar industry. Rene Energ 145:1870–1882. https://doi.org/10.1016/j.renene.2019.06.023

Moretti C, Corona B, Edwards R, Junginger M, Moro A, Rocco M, Shen L (2020) Reviewing ISO compliant multifunctionality practices in environmental life cycle modeling. Energies 13:3579. https://doi.org/10.3390/en13143579

Nabavi-Pelesaraei A, Bayat R, Hosseinzadeh-Bandbafha H, Afrasyabi H, Chau KW (2017) Modeling of energy consumption and environmental life cycle assessment for incineration and landfill systems of municipal solid waste management—a case study in Tehran Metropolis of Iran. J Clean Prod 148:427–440. https://doi.org/10.1016/j.jclepro.2017.01.172

Obydenkova SV, Kouris PD, Smeulders DMJ, Boot MD, van der Meer Y (2021) Modeling life-cycle inventory for multi-product biorefinery: tracking environmental burdens and evaluation of uncertainty caused by allocation procedure. Biofuels Bioprod Biorefin 15:1281–1300. https://doi.org/10.1002/bbb.2214

Oliva-Rodriguez AG et al (2019) Clostridium strain selection for co-culture with Bacillus subtilis for butanol production from agave hydrolysates. Bioresour Technol 275:410–415. https://doi.org/10.1016/j.biortech.2018.12.085

Osman AI et al (2021) Conversion of biomass to biofuels and life cycle assessment: a review. Environ Chem Lett 19:4075–4118. https://doi.org/10.1007/s10311-021-01273-0

Pan S, Chi Y, Zhou L, Li Z, Du L, Wei Y (2020) Evaluation of squeezing pretreatment for improving methane production from fresh banana pseudo-stems. Waste Manag 102:900–908. https://doi.org/10.1016/j.wasman.2019.12.011

Pan SY, Zabed HM, Li ZC, Qi XH, Wei YT (2022a) Enrichment and balancing of nutrients for improved methane production using three compositionally different agro-livestock wastes: process performance and microbial community analysis. Bioresour Technol 357:127360. https://doi.org/10.1016/j.biortech.2022.127360

Pan S, Zabed HM, Wei Y, Qi X (2022b) Technoeconomic and environmental perspectives of biofuel production from sugarcane bagasse: current status, challenges, and future outlook. Ind Crop Prod 188:115684. https://doi.org/10.1016/j.indcrop.2022.115684

Pan SY, Zabed HM, Zhao M, Qi XH, Wei YT (2023) Techno-economic and life cycle assessments for bioenergy recovery from acid-hydrolyzed residues of sugarcane bagasse in the biobased xylose production platform. J Clean Prod 400:136718. https://doi.org/10.1016/j.jclepro.2023.136718

Pan S, Guo F, Li X, Feng L, Li Z, Du L, Wei Y (2024) Briquetting as a source pretreatment strategy to improve energy recovery from the anaerobic digestion of agricultural straw: experimental and economic evaluation. Energy 294:130918. https://doi.org/10.1016/j.energy.2024.130918

Peng L, Jackson PA, Li QW, Deng HH (2014) Potential for bioenergy production from sugarcane in China. Bioenergy Res 7:1045–1059. https://doi.org/10.1007/s12155-013-9403-7

Peng J, Xu Y, Zhu B, Yu H, Li B, Xu H (2023) Life cycle assessment of bioethanol production by two methods of pretreatment of rice straw based on process simulation. Ind Crop Prod 191:115810. https://doi.org/10.1016/j.indcrop.2022.115810

Postacchini L, Ciarapica FE, Bevilacqua M (2018) Environmental assessment of a landfill leachate treatment plant: impacts and research for more sustainable chemical alternatives. J Clean Prod 183:1021–1033. https://doi.org/10.1016/j.jclepro.2018.02.219

Prasad A, Sotenko M, Blenkinsopp T, Coles SR (2016) Life cycle assessment of lignocellulosic biomass pretreatment methods in biofuel production. Int J Life Cycle Assess 21:44–50. https://doi.org/10.1007/s11367-015-0985-5

Radhika NL, Sachdeva S, Kumar M (2022) Lignin depolymerization and biotransformation to industrially important chemicals/biofuels. Fuel 312:122935. https://doi.org/10.1016/j.fuel.2021.122935

Raj T, Chandrasekhar K, Kumar AN, Banu JR, Yoon JJ, Bhatia SK et al (2022) Recent advances in commercial biorefineries for lignocellulosic ethanol production: current status, challenges and future perspectives. Bioresour Technol 344:126292. https://doi.org/10.1016/j.biortech.2021.126292

Schmidt JH, Weidema BP, Brandao MA (2015) A framework for modelling indirect land use changes in life cycle assessment. J Clean Prod 99:230–238. https://doi.org/10.1016/j.jclepro.2015.03.013

Sharma HK, Xu C, Qin W (2017) Biological pretreatment of lignocellulosic biomass for biofuels and bioproducts: an overview. Waste Biomass Val 10:235–251. https://doi.org/10.1007/s12649-017-0059-y

Siegelman RL, Milner PJ, Kim EJ, Weston SC, Long JR (2019) Challenges and opportunities for adsorption-based CO2 capture from natural gas combined cycle emissions. Energy Environ Sci 12:2161–2173. https://doi.org/10.1039/c9ee00505f

Souza HHD et al (2021) Functional unit influence on building life cycle assessment. Int J Life Cycle Assess 26:435–454. https://doi.org/10.1007/s11367-020-01854-1

Spigno G, Maggi L, Amendola D, Dragoni M, De Faveri DM (2013) Influence of cultivar on the lignocellulosic fractionation of grape stalks. Ind Crop Prod 46:283–289. https://doi.org/10.1016/j.indcrop.2013.01.034

Stanley D, Bandara A, Fraser S, Chambers PJ, Stanley GA (2010) The ethanol stress response and ethanol tolerance of. J Appl Microbiol 109:13–24. https://doi.org/10.1111/j.1365-2672.2009.04657.x

Sun Y, Cai W, Chen B, Guo X, Hu J, Jiao Y (2017) Economic analysis of fuel collection, storage, and transportation in straw power generation in China. Energy 132:194–203. https://doi.org/10.1016/j.energy.2017.05.077

The National Bureau of Standards (2023) China Statistical Yearbook 2023 (China statistics press). https://www.stats.gov.cn/sj/ndsj/2023/indexch.htm. Accessed 25 August 2024

Tondro H, Musivand S, Zilouei H, Bazarganipour M, Zargoosh K (2020) Biological production of hydrogen and acetone-butanol-ethanol from sugarcane bagasse and rice straw using co-culture of Enterobacter aerogenes and clostridium acetobutylicum. Biomass Bioenergy 142:105818. https://doi.org/10.1016/j.biombioe.2020.105818

Turconi R, Boldrin A, Astrup T (2013) Life cycle assessment (LCA) of electricity generation technologies: overview, comparability and limitations. Renew Sust Energ Rev 28:555–565. https://doi.org/10.1016/j.rser.2013.08.013

Unrean P, Ketsub N (2018) Integrated lignocellulosic bioprocess for co-production of ethanol and xylitol from sugarcane bagasse. Ind Crop Prod 123:238–246. https://doi.org/10.1016/j.indcrop.2018.06.071

Wang Q, Xie CF, Shi CR, Li Y, Shang YH, Hou CX et al (2022) Sugar product diversification and its opportunities in China. Sugar Tech 24:1093–1106. https://doi.org/10.1007/s12355-022-01134-2

Wang X et al (2023) Process design, modeling and life cycle analysis of energy consumption and GHG emission for jet fuel production from bioethanol in China. J Clean Prod 389:136027. https://doi.org/10.1016/j.jclepro.2023.136027

Xia J, Yang Y, Liu CG, Yang S, Bai FW (2019) Engineering Zymomonas mobilis for robust cellulosic ethanol production. Trends Biotechnol 37:960–972. https://doi.org/10.1016/j.tibtech.2019.02.002

Yang L, Zhou Y, Meng B, Li H, Zhan J, Xiong H et al (2022) Reconciling productivity, profitability and sustainability of small-holder sugarcane farms: a combined life cycle and data envelopment analysis. Agric Syst 199:103392. https://doi.org/10.1016/j.agsy.2022.103392

Zhang H, Han L, Dong H (2021) An insight to pretreatment, enzyme adsorption and enzymatic hydrolysis of lignocellulosic biomass: experimental and modeling studies. Renew Sust Energ Rev 140:110758. https://doi.org/10.1016/j.rser.2021.110758

Zhao X, Liu J, Liu J, Yang F, Zhu W, Yuan X et al (2017) Effect of ensiling and silage additives on biogas production and microbial community dynamics during anaerobic digestion of switchgrass. Bioresour Technol 241:349–359. https://doi.org/10.1016/j.biortech.2017.03.183

Zheng Y, Zhao J, Xu F, Li Y (2014) Pretreatment of lignocellulosic biomass for enhanced biogas production. Prog Energ Combust Sci 42:35–53. https://doi.org/10.1016/j.pecs.2014.01.001

Zhou Z, Tang Y, Dong J, Chi Y, Ni M, Li N, Zhang Y (2018) Environmental performance evolution of municipal solid waste management by life cycle assessment in Hangzhou, China. J Environ Manag 227:23–33. https://doi.org/10.1016/j.jenvman.2018.08.083

Zhou N, Thilakarathna WPDW, He QS, Rupasinghe HPV (2022) A review: Depolymerization of lignin to generate high-value bio-products: opportunities, challenges, and prospects. Front Energy Res 9:758744. https://doi.org/10.3389/fenrg.2021.758744

Chapter 3
Impact of Biofuels on the Environment, Biodiversity, Agriculture, and Water Resources: Potential Solutions

Mukul Machhindra Barwant, Shalagha A. B. Sharma, Swastika Banerjee, Debasree Lodh, and Pratiksha Fulari

Abstract Biofuels are promoted as a low-carbon substitute for fossil fuels due to their potential benefits in lowering greenhouse gas (GHG) emissions and the consequent climate change impact of the automotive industry. However, with their expanding use comes the possibility of unintended environmental impacts. Biofuels' impacts on climate change and other environmental issues have been the subject of numerous life cycle assessment (LCA) studies. Nevertheless, the estimations vary widely, and their outcomes are frequently incoherent. The biofuels impact biodiversity will depend on the type of biofuel crop and the prior land use. If suitable plants are cultivated in appropriate settings, biofuels may promote biodiversity. Additionally, they may indirectly maintain biodiversity worldwide if they contribute to reducing the effects of climate change. The biodiversity of that region is already known to be at risk, regardless of whether natural ecosystems are abruptly changed to biofuels or non-degraded land is subtly converted to biofuels. When biofuel production spread into the tropics, it destroyed wetlands and tropical forests and pushed into set-aside lands in temperate zones. Based on current data, first-generation biofuels can have lower greenhouse gas emissions than fossil fuels in the absence of land-use change (LUC); however, for most feedstocks, the reductions are not high enough to achieve the GHG savings required by the European Renewable Energy Directive (RED). Second-generation biofuels have a greater chance of lowering emissions if there is no LUC. Since third-generation biofuels release more green-

M. M. Barwant (✉) · P. Fulari
Department of Botany, Sanjivani Arts Commerce and Science College, Kopargaon, India

S. A. B. Sharma
School of Biological Engineering, Shobhit Deemed to be University, Meerut, Uttar Pradesh, India

S. Banerjee
Department of Science, Alliance University, Bangalore, Karnataka, India

D. Lodh
Department of Botany, Holy Cross College, Agartala, Tripura, India

M. A. Dar et al. (eds.), *Recent Trends in Lignocellulosic Biofuels and Bioenergy*, Clean Energy Production Technologies,
https://doi.org/10.1007/978-981-96-4636-4_3

house gases into the atmosphere than fossil fuels, they are not a practical choice at this stage of development. Better land-use planning will surely be necessary in conjunction with sustainability regulations, as these are likely to give only part of the solution.

Keywords Greenhouse gas · Land-use change · Biofuels · Biodiversity · Climate change

Abbreviations

LCA	Life cycle assessment
GHG	Greenhouse gases
SOC	Soil organic carbon
BTL	Biomass to liquid

3.1 Introductions

Energy drives industrial development, household consumption, and, ultimately, the expansion of the world economy. Potential solutions for the energy, food, and environmental problems include bioenergy (Qin et al. 2018). Biofuels, which are renewable fuels derived from biological feedstocks, exist in liquid and gaseous forms, such as hydrogen, biogas (such as methane), and bioethanol (a gasoline equivalent) (Koh and Ghazoul 2008). The most widely utilized biofuel in the world is bioethanol. About 80% of the world's energy is produced from fossil fuels, including oil, natural gas, and coal. As fuels are running out and there is a need to combat climate change, more and more people worldwide are becoming interested in sustainable energy sources such as biofuels (Koh and Ghazoul 2008). To lessen the impact of consuming fuels on the climate, governments around the world are pushing the development of biofuels (Delucchi 2010). Biofuels made from cellulosic feedstocks such as switchgrass and residual maize, as well as other conventional non-grain crops such as cassava and jatropha, result in a considerable reduction in greenhouse gas (GHG) and air pollution compared to fossil fuels such as gasoline and diesel (Qin et al. 2018). In a nutshell, biodiversity refers to the variety of life that exists at any particular time and area. In order to maintain the health and well-being of our planet for future generations, conservation initiatives often concentrate on safeguarding biodiversity through the creation or upkeep of protected areas. But more and more areas where people live, work, and grow food and fiber are coming under the conservation umbrella. This paradigm shift has been largely aided by farms because a variety of species may survive and even flourish in agricultural surroundings (Dale et al. 2010).

The environmental, social, and economic effects of biofuels are significantly influenced by the production of biomass feedstock (Gopalakrishnan et al. 2009).

However, concerns about climate change, food security, and the validity of the claims that biofuels significantly lower carbon emissions have damaged popular and political support for biofuels (Koh and Ghazoul 2008). When compared to gasoline, maize ethanol, or regular electricity, lignocellulosic feedstocks have a lot of potential to enhance conversion efficiency and emissions. However, the environmental effects of each feedstock will range slightly and may differ from one location or region to another. When all environmental costs are taken into account, some crops specifically grown for bioenergy have a higher potential for use as biofuels than crop residues (Dale et al. 2010).

Biofuels are expected to increase without depleting food-based grains, oils, and sugars or competing for land. It is less likely that biomass obtained from these sources will endanger food production or the availability of land for food crops (Qin et al. 2018). Certain qualitative generalizations can be made, despite the complexity of the technological and environmental systems influencing land use, water consumption, and climate change, as well as the challenges in developing meaningful measures. In relation to petroleum fuels, it appears that biofuels derived from crops grown using standard agricultural methods will aggravate pressures on water supply, water quality, and land use while also failing to reduce the effects of climate change (Delucchi 2010). Without affecting food production, crop residues from current croplands and prospective energy crops cultivated on marginal areas could make a substantial contribution to China's supply of biofuel feedstocks. Crop waste (30%) and energy crops (70%), when combined, may produce roughly 200 Mt. of ethanol annually or ¼ of the country's annual oil consumption by energy content (Qin et al. 2018). Several considerations must be made in this landscape approach, such as the kind and location of the plant species to be grown as feedstock for biofuels, the farming and harvesting methods used in their production, the transportation of the fuel to refineries, the kind and location of the production facilities, and the fuel's transportation to markets (Dale et al. 2010). Many nations in the Southern and Northern hemispheres are producing bioenergy from agricultural crop biomass in response to environmental concerns over the use of nonrenewable fossil fuels, including greenhouse gas emissions, air pollution, global warming, and climate change (Verdade et al. 2015).

There is a growing demand for biofuels due to factors such as rising energy prices, geopolitical concerns, and concerns about the influence of greenhouse gas emissions on climate change. Only 2% of the 1200 billion liters of gasoline consumed annually worldwide are estimated to be made from the 35 billion liters of biofuel currently anticipated to be produced. With advantages including lower greenhouse gas emissions, improved energy security, and new sources of income for farmers, it is anticipated that the proportion of biofuels in the energy supply would increase quickly. But the generation of biomass energy would severely reduce agricultural productivity in many regions of the world, as it competes with food crops for scarce land and water resources (De Fraiture et al. 2008). Land-use change, whether caused directly or indirectly by biofuel feedstock production, mediates many of the immediate environmental and socioeconomic repercussions. Compared to other industries such as food, feed, lumber, and fiber, the use of biofuels has not significantly changed land use in Asia thus far. Nonetheless, the biofuel industry could soon have a significant impact on changing how land is used (Phalan 2009).

The use of biofuels instead of fossil fuels may reduce greenhouse gas (GHG) emissions. However, the expansion of biofuel crops has required the use of damaged or marginal lands, as well as the direct or indirect uprooting of natural ecosystems. The first causes direct habitat loss, while the latter has effects more typical of agriculture (e.g., biotic and soil pollution, water eutrophication). Biofuel crops, unlike typical agricultural crops, can occasionally increase biodiversity (Verdade et al. 2015). The vast majority of biofuel produced today is grown on croplands that have traditionally been utilized for agriculture. The reclamation of reserve or abandoned croplands and the expansion of cropping into previously thought to be marginal agricultural areas are two strategies to meet the future land area requirements for biofuel production. The Conservation Reserve Program (CRP), which offers vital habitat for species that inhabit grasslands, now includes a large number of marginal areas in the United States. The economics of agricultural commodities have been significantly altered by the demand for corn ethanol when contracts expire and regions are put back into agricultural use. The loss of CRP lands has already resulted from this (Wiens et al. 2011).

Their societal value is determined by their ability to provide these needs while limiting negative social and environmental consequences (Phalan 2009). Biofuel production appears to be a sustainable means of effectively utilizing biomass and other biodegradable wastes. By converting biomass into biofuels using second-generation technologies, communities may create jobs, increase energy security, promote socioeconomic development, and provide cheap energy services. This can be done without competing with the land used for food production or creating environmental harm (Prasad and Ingle 2019). By 2020, food crops are expected to generate 170% more biofuels. Economic models suggest that food-based biofuels will cause land-use change (LUC) ranging from 67 to 365 ha 10–6 l^{-1}, leading to greater greenhouse gas emissions for decades compared to business as usual. Compared to unconverted habitat, biodiversity is diminished by more than 85% in oil palm plantations in Southeast Asia and by approximately 60% in maize and soybean fields in the United States (Fargione et al. 2010).

Invasive species connected to a range of human activities have driven many native species to extinction. They have also disturbed ecological processes, changed the timing and patterns of frequent occurrences, and changed the makeup of biological groups. As a direct result of the species and genotypes used to produce biofuels, or of the invasion of other taxa due to the use of nonnative species for biofuel production, the use of nonnative plants as feedstocks may increase the costs and risks associated with invasive species. Native species do, however, occasionally find homes among exotic species employed as feedstock. When utilizing nonnative plants with invasive features, specific management strategies must be employed to reduce the likelihood of their spread (Joly et al. 2015).

3.2 Climate Change Impacts of Biofuels

The vehicle industry has been using biofuels since its inception. For instance, when Rudolph Diesel realized that pulverized coal was insufficient, he tested his first engine with peanut oil (Michael et al. 2011). Biofuels were considered legal transportation fuels up until the 1940s, and bioethanol blends such as Agrol, Discol, and Monopolin were commonly utilized in the United States, Europe, and other nations (Michael et al. 2011). Due to growing oil prices and concerns about energy security in the late 1990s, the United States and many European nations developed initiatives to support the homegrown biofuel sector. The adoption of climate policies over the last ten years has further increased interest in biofuels, as well as strategies for reducing and mitigating greenhouse gas emissions in the transportation sector. Since then, more than 60 nations have started biofuel initiatives and established goals for mixing biofuels into their gasoline tanks.

Biofuels are high-energy materials made from biomass, including biodegradable municipal or industrial waste; forestry, agricultural, and fishery waste; and residues from these activities (Table 3.1). The type of feedstock, the conversion process, the technical characteristics of the fuel, and the intended use are some of the differentiators that can be used to identify biofuels. Biofuels are categorized in a variety of ways due to the numerous ways that they can be distinguished from one another (Bharti et al. 2021). Biofuels produced from crops used for food or animal feed are known as first-generation biofuels. Because first-generation biofuels are produced by tried-and-true methods including fermentation, distillation, and transesterification, they are frequently referred to as "conventional biofuels." Dedicated energy crops (e.g., *Miscanthus*, switchgrass, short-rotation coppice (SRC), and other lignocellulosic plants), agricultural residues, forest residues, and other waste materials (including UCO and municipal solid waste) are some of the nonfood feedstocks that are largely used in the production of second-generation biofuels. The term "third-generation biofuel" describes biodiesel made from microalgae using conventional

Table 3.1 A brief description of the feedstocks and methods used in various biofuel production processes

Generation of biofuels	Crops and foods used in production	Techniques
Conventional biofuels	Generated from the following crops: Sunflower oil, palm oil, soybean oil, wheat, corn, sugar beet, sugarcane, and sugar	Fermentation, distillation, and transesterification
Sophisticated biofuels	Produced using feedstocks other than food, such as specialty energy crops (e.g., *Jatropha*, willow, poplar, *Camelina* waste, wood waste, food waste, manure, and straw)	Fischer-Tropsch process, gasification, methanization
Advanced biofuels	Microalgae	Traditional transesterification or hydro-processing of algae oil

transesterification or hydro-treating algae oil. We sometimes refer to second- and third-generation biofuels as "advanced biofuels," even if their production processes are still in the research, development, pilot, or demonstration stages (Azapagic and Stichnothe 2011).

It is necessary to include social, economic, and environmental systems while designing, developing, and implementing sustainable renewable energy systems. Furthermore, these systems must be durable. The environmental harm that fossil fuels generate could be mitigated by locating sources of biofuel (Balajii and Niju 2019; Gunarathne et al. 2019). Additionally, the use of fossil fuels is rising, and the use of biofuels can lessen reliance on nonrenewable resources (Hassan et al. 2020; Peng et al. 2020). Disagreement exists about what defines sustainable biomass production on a national and worldwide scale (Kaloudas et al. 2021). Primarily, biofuel ought to reduce energy use and enhance energy security by utilizing fewer items derived from petroleum. Furthermore, it must protect biodiversity, soil productivity, water quality and quantity, a range of land uses, greenhouse gas emissions, and the environment. Third, it shouldn't needlessly alter the availability of food and resources while enhancing ecosystem services. Along with being economically competitive, it must also produce more energy than it consumes (Increasing feedstock production for Biofuels Economic Drivers, Environmental Implications, and the role of Research 2008; Munagala et al. 2022).

To reduce dependence on traditional diesel, two well-liked biofuels that can be used with petroleum diesel are bioethanol and biodiesel. Bioethanol is produced by fermenting and distilling sugar derived from starch or high-sugar biomass, such as grains or sugarcane. The biomass-to-liquid (BTL) process can also be used to make bioethanol or bio-oil. This process involves hydrolyzing, fermenting, and distilling lignocellulosic biomass (Table 3.2). As an alternative, vegetable oils, leftover cooking oil, and animal fats can be extracted and esterified with alcohol to create biodiesel (Sinha et al. 2013; Foteinis et al. 2020). India is reducing its dependency on imported energy by implementing renewable energy sources such as biodiesel. Since they have the ability to lower greenhouse gas emissions when used in place of conventional transportation fuels, renewable fuels such as ethanol and biodiesel are being promoted globally.

Table 3.2 Carbon emission reduction from various feedstocks during biofuel production

S. no.	Raw material	Biofuel product and coproduct	Percentage of reducing CO_2 emission
1.	Corn & straw made with wheat	Biological energy	47–50% (Cherubini and Ulgiati 2010)
2.	Raphael	Energy and biodiesel	30–50% (Gupta et al. 2022)
3.	Cane sugar	Biogas and bioethanol	Power 40–70% (Gnansounou et al. 2015)
4.	Utilized cooking oil	Biodiesel	30–55% (Bhonsle et al. 2022)
5.	Sorghum and Jatropha	Biodiesel	75–80% (Kumar et al. 2012)

3.2.1 Benefits and Drawbacks of Biofuels

When it comes to environmental, social, and economic sustainability, biofuels offer benefits and drawbacks (Azapagic and Stichnothe 2011). On the one hand, the main global drivers of biofuels are reduced greenhouse gas emissions, rural development, and energy security. Some of the drawbacks of producing more biofuels include the possibility of higher food prices, increased greenhouse gas emissions from direct and indirect land-use change (LUC) resulting from the production of biofuel feedstocks, and the possibility of ecosystem, forest, water resource, and land degradation. First-generation feedstocks, such as maize, are particularly polarizing due to concerns about exploiting agricultural land for fuel production. Harm to the ecosystem arises from increased deforestation, the removal of land with high biodiversity value to satisfy agricultural output demands, and the associated usage of pesticides, fertilizers, and freshwater. However, given the current state of the economy, with low oil prices being the key factor, several second-generation biofuels continue to be financially dubious. By using second-generation feedstocks, some of these issues may be overcome (Roy and Dutta 2013; Stephenson et al. 2010). Due to their rapid development and ability to be cultivated on nonarable soil, effluent, saline, or brackish water, third-generation (algal) biofuels may be able to circumvent the problems of food competition and land use. But the energy-dense and presently impractical method of turning microalgae into biofuels (Passell et al. 2013).

Numerous life cycle assessment (LCA) studies have evaluated the potential impact of biofuels on global warming in order to determine how they could lower life cycle greenhouse gas emissions. However, the approximations exhibit significant disparities, and their deductions occasionally lack coherence. Numerous review papers have also addressed the life cycle assessment (LCA) of biofuels; however, these have concentrated on specific aspects, such as a location, kind of biofuel, or feedstock. For instance, Shonnard et al. (2015) examined the biofuel life cycle assessment (LCA) research conducted across the North American continent. Menten et al. (2013) focused on advanced biofuels, while Roy et al. (2012) and Morales et al. (2015) focused on the life cycle assessment (LCA) studies of lignocellulosic bioethanol. While reviewing biodiesel, (Sieverding et al. 2015).

3.2.2 Drawbacks

When nitrogen fertilizer is added to the soil and organic matter decomposes, soil nitrous oxide, or N_2O, is emitted. Stronger than CO_2, N_2O has a greenhouse gas potential 265 times greater than that of CO_2 (Dar et al. 2025). For this reason, the emissions of biofuels could significantly impact the GHG balance. More fertilization of first-generation biofuel crops results in higher greenhouse gas emissions than second-generation biofuels from perennial energy crops, which are often produced without fertilizers other than during the initial crop establishment (Cherubini and Strømman 2011).

3.2.3 Land-Use Change (LUC)

The potential for natural ecosystems and agriculture to compete for land was highlighted by the growing global need for biofuels. First-generation biofuels, including maize ethanol, had a lower global warming potential (GWP) than gasoline, according to early life cycle assessment (LCA) studies on biofuels that did not include LUC (Farrell et al. 2006). But these conclusions were questioned when an attempt was made to take the LUC consequences of the growth of first-generation biofuels into consideration (Fargione et al. 2008).

3.2.4 Soil Organic Carbon (SOC)

One of the biggest carbon sinks in the terrestrial environment is soil organic carbon (SOC) (Qin et al. 2016). LUC and agricultural activities both have an impact on its equilibrium. Soil can be a source or a sink of carbon emissions, depending on a variety of soil properties and agricultural techniques. Farming causes land to mineralize, which allows them to release CO_2 into the atmosphere and remove SOC. However, cropping or the regular addition of organic manures or agricultural wastes may increase SOC (Smith 2012). When biomass breaks down, some of the carbon is absorbed by the soil. Thus, harvesting biomass lowers SOC and may result in soil degradation if it had been left to decompose (Koponen et al. 2013).

3.2.5 Biogenic Carbon

In discussions of biofuels, the carbon dioxide that is taken up from the environment during feedstock growth and released after fuel combustion is referred to as "biogenic carbon." When the amount of carbon dioxide trapped and then released is equal, "carbon neutrality" is reached. However, if there is a possible mismatch or lag between the amount of CO_2 absorbed during feedstock growth and the amount generated during the production and consumption of biofuels, carbon neutrality cannot be claimed. The short lifespans of many bioenergy products, such as annual crops and perennial grasses, lead to the assumption of carbon neutrality in life cycle assessment (LCA) standards and regulations. Because of this, the majority of LCA studies on biofuels are predicated on the idea that CO_2 uptake during feedstock growth fully offsets biogenic CO_2 emissions, which arise from burning biomass to provide energy for conversion processes as well as end-use combustion. This assumption can be contested when producing biofuel from feedstocks with harvest cycles longer than a few years, such as longer-lived lignocellulosic feedstocks from forestry, even though it makes sense for fuels from annual crops and perennial grass feedstocks (Sanchez et al. 2012).

3.3 Agriculture, Land Use, and Biodiversity

Human interaction with the natural environment has resulted in the land system (Verburg et al. 2015). Land systems offer solutions to the problems of global change and sustainable development objectives through mitigation and adaptation (Verburg et al. 2013). Tang et al. (2015) define the agricultural land system as the synthesis of the agricultural system and the land system, wherein land is the fundamental component and encompasses all agriculturally related human endeavors and outcomes. Furthermore, the agricultural land system is a dynamic and complex social structure, whose primary goal is to provide food and a means of subsistence for human security (Viana et al. 2022). Threats to global food security and the sustainable use of the agricultural land system include extreme weather, the COVID-19 pandemic, regional conflicts, and rising food and fuel prices (Global Report on Food Crises-Joint Analysis for Better Decisions; Food and Agriculture Organization (FAO); World Food Programme (WFP); International Food Policy Research Institute (IFPRI 2022). The agricultural land system is one of the main areas of study for sustainable development and global environmental change in order to meet the Sustainable Development Goals (SDGs) (Putranti and Widiastuti 2017).

3.3.1 Agriculture and Climate Change

Agriculture already contributes significantly to greenhouse gas emissions, and this contribution is likely to increase proportionately as other sectors step up their mitigation efforts. The burning of biomass and dung, the clearance of land for crops and grazing, the production of rice, the methane emissions from ruminant cattle, and the nitrous oxide emissions from fertilizer use are some of the ways that agriculture releases greenhouse gases (Dar et al. 2022). Compared to other economic sectors such as manufacturing, building, and transportation, agriculture consumes the most natural resource inputs per unit of value created. Moreover, the current incremental rate of gain in agricultural output is only a few percent each year, even with these astronomical inputs. Even if all nonagricultural fossil fuel use were discontinued, the total future greenhouse gas emissions from agriculture would still exceed the emission limit set by the Paris Agreement to avoid a 2 Celsius global temperature increase, due to land clearing, ruminants, manure, rice production, combustion, and nitrous oxide from fertilized soils (Dar et al. 2021).

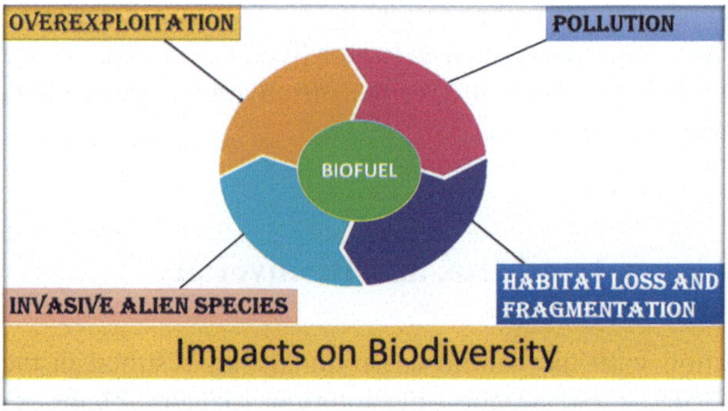

Fig. 3.1 Impacts of various factors on biodiversity

3.3.2 Agriculture and Biodiversity

A significant decline in biodiversity has resulted from the globalization of agriculture, in addition to its effects on climate (Fig. 3.1). This is due to the degradation of soils, the conversion of natural ecosystems into farms and pastures, and environmental contamination from fertilizers and pesticides. Due to increased land clearing and greater agricultural output, many communities of plants and animals will become extinct in the upcoming decades. Furthermore, the use of crop monocultures has expanded at the expense of agricultural ecosystem diversity. Orders to utilize more land for biofuels have led to the conversion of millions of acres to monoculture crops, such maize, which had not previously been produced on such area, even in wealthy nations such as the United States. Agriculture and biodiversity are intertwined, much like climate change (Table 3.3). Similar to climate change, agriculture and biodiversity are interconnected. Enhanced biodiversity benefits agriculture in several ways, including enhanced soil quality, species that fend off pests, and an increase in pollinators. For example, studies in ecology have demonstrated a strong relationship between ecological production, stability, and biodiversity (Ali et al. 2024a; Cardinale et al. 2013).

3.3.3 Nitrogen in Agriculture

The intricate relationships between agriculture and climate, along with other environmental challenges, are well shown by the usage of nitrogen in agriculture. According to Lassaletta et al. (2016), the amount of fertilizer used in agriculture worldwide has more than doubled since the 1960s. Two elements of air pollution are ammonium and nitrogen oxides; in the United States, particulate matter from these and other agricultural pollutants is estimated to be responsible for 19,000 premature deaths per year (Thakrar et al. 2020). Currently, about 8% of the heat caused by greenhouse gases is attributed to nitrous oxide, which is mostly produced by

Table 3.3 Impacts of various factors on biodiversity

Factors	Impacts on biodiversity	References
Pollution	Changes in species compositions and interactions, higher mortality rates, and nutrient loading and acidification	Bhattarai (2017); Brown et al. (2017)
Overexploitation	Genetic contamination, extinctions and decreased populations, and alien species introduced after resource depletion	Bhattarai (2017); Brown et al. (2017)
Foreign species that have invaded	Competition with and predation on native species, changes in ecosystem function, and extinctions and homogenization	Bhattarai (2017); Brown et al. (2017)
Habitat loss and fragmentation	Decrease in natural habitat, homogenization of species composition, fragmentation of landscapes, and soil degradation	Eldredge (2002)

Table 3.4 Process of emission altered by reactive nitrogen, climate forcing elements, process of warming/cooling, and their overall impacts. (Source Pinder et al. 2012)

Reactive nitrogen modifies the process	Element	Process	Environmental impacts
NOx into aerosol	Aerosol with nitrate and ammonium	NOx can promote the production of aerosols that scatter light	Chilling
Fertilizer N into the CH4 flow	CH_4	N from manure and fertilizer may interfere with CH4 uptake and emission on croplands	Warming

agricultural practices. As the amount of nitrogen consumed increases, this percentage will also rise (Ali et al. 2024b).

Will climate change lead to an increase in nitrous oxide emissions into the atmosphere? As more high-nitrogen crops are used and expand into previously unfrozen areas, the amount of nitrogen lost to the atmosphere will increase. Extreme weather conditions such as droughts and strong rains can hinder plants' capacity to absorb nitrogen, enabling more nitrogen to seep into the surrounding area (Table 3.4). More nitrogen will be mineralized by microbes during longer growing seasons and milder winters, changing it from insoluble organic forms to soluble forms that are physiologic. The amount of nitrous oxide released into the atmosphere by wasted nitrogen increases in areas where crop input rates are higher than necessary (Hao et al. 2001).

3.3.4 Agricultural Boundaries for Biodiversity

3.3.4.1 Agricultural Area

The majority of the Earth's territories were once primarily wild, but over the past three centuries, they have been predominantly occupied by humans. According to Ellis et al. (2010), half of the area free of ice in 1700 was wild, while the remaining 45% was seminatural. By 2015, the proportion of land surface directly inhabited or exploited by humans was as follows: 28% for wild ecosystems with little to no human use (i.e., 12% barren regions), 37% for grazed native ecosystems known as pastures and rangelands, and 22% for managed forests. According to Pendrill et al. (2022), the primary cause of biodiversity loss worldwide is the conversion of natural ecosystems into land used for human needs, primarily as a result of agricultural development (Newbold et al. 2015). The world's highest levels of biodiversity are found in primary vegetation and mature secondary vegetation ecosystems (Newbold et al. 2015). Certain species communities present in biomes most vulnerable to human disturbance can only be supported by primary vegetation (García-Vega and Newbold 2020). Even under the most complicated human land-use situations, many species still depend on their native environments and cannot survive in other contexts (Phalan 2018). Watson et al. claim that the bulk of terrestrial biodiversity on Earth is found in the world's intact forests, which are home to the majority of terrestrial wildlife. Many rare and endangered species that rely on the maintenance of large, uninterrupted stretches of primary forest can be found in these woods. Habitat fragmentation, which might raise additional issues even if it would not directly affect biodiversity, coexists with habitat loss (Fahrig 2017; Fahrig et al. 2019). Seventy percent of the remaining forests are vulnerable to biotic and physical stressors, such as logging, hunting, and the spread of exotic species, within one kilometer of the forest edge. Another possible source of infiltration is microclimatic changes around the boundaries (Meza-Elizalde and Armenteras-Pascual 2021).

3.3.4.2 Water Withdrawal

Irrigation is the world's largest user of freshwater, accounting for more than 70% of all human withdrawals (Siebert and Döll 2010). Seventy-five percent of the world's major rivers have been fragmented due to dam construction, which has reduced river flow for agricultural and other purposes (Pastor et al. 2014). River flow patterns are thought to be the "master variable" in biological processes related to rivers. Variations in the frequency, duration, timing, and intensity of high and low flows, as well as in the occurrence of extreme events such as drought and flooding, can all have an impact on streamflow (Poff and Zimmerman 2010).

3.3.4.3 Pesticide Pollution

Herbicides, insecticides, fungicides, nematicides, and soil fumigants are examples of bioactive and potentially dangerous pesticides. One of the main causes of the global reduction in biodiversity has been identified as the effects on individuals, entire communities, and nontarget livestock both inside and outside of fields. The variety of wild plants, ground beetles, and birds were all adversely affected by the use of fungicides and pesticides, according to a pan-European study evaluating the various facets of agricultural intensification in wheat fields (Geiger et al. 2010). Additionally, it has been shown that bacterial populations and soil fauna are harmed by herbicides, fungicides, and insecticides. It has been demonstrated that pesticides negatively affect ecosystem services and biodiversity; however, these impacts are not linear and heavily rely on the kind of substance, how it is applied, and its interactions, biogeographical characteristics, and lethal dosages, all of which differ significantly even among taxonomic groups (Pisa et al. 2015, for instance). Determining a safe critical biodiversity load for pesticides is not achievable. While 100% of pesticide inputs in high-use regions such as the EU do not accord with biodiversity conservation, their complete eradication (i.e., 0% of inputs) is not necessary for sustainable biodiversity conservation.

3.4 Feedstocks Derived from Agricultural Residues and Energy Crops

Although the development of biofuels raises issues about the environment, biodiversity, agriculture, and water resources, they could eventually replace fossil fuels. The focus of the conversation is on energy crops and agricultural wastes, which are crucial sources of biomass feedstocks. It is imperative to get a thorough comprehension of these mechanisms to facilitate the advancement of sustainable energy systems.

3.4.1 Current Biomass Feedstock Sources

Biomass feedstocks refer to organic substances suitable for use as a source of energy. Biomass feedstocks are essential in the biofuel production process, since they provide the necessary raw materials for biofuel manufacturing. Crop residues and energy crops are the two main groups into which feedstocks can be generally divided.

3.4.1.1 Crop Residues

Crop residues are the residual materials from agricultural activities that remain after the harvest of crops, including corn, wheat, sugarcane, and rice. These remnants consist of stems, foliage, outer coverings, and other inedible components of the plant (Dar et al. 2025). Crop residues are plentiful and readily accessible, rendering them a viable resource for the generation of biofuels. By employing crop wastes for the manufacturing of biofuels, farmers can make supplementary revenue from their agricultural endeavors while also mitigating waste and greenhouse gas emissions.

3.4.1.2 Energy Crops

Energy crops are only grown for the production of biofuels. The selection of these crops is based on their exceptional biomass yield and their aptitude for efficient conversion into biofuels (Ali et al. 2024b; Koçar and Civaş 2013). Energy crops include willow, poplar trees, miscanthus, and switchgrass. Energy crops have many advantages as starting points for the synthesis of biofuels, including high energy density, rapid growth rates, and the ability to flourish on unfavorable sites where food crops shouldn't be grown. Once the biomass feedstocks—such as crop residues or energy crops—are gathered, they undergo various conversion processes to yield the cellulose, oils, and sugars needed to produce biofuel. Conventional conversion techniques use thermochemical procedures, such as gasification and pyrolysis, to make biodiesel and syngas and biochemical processes, such as fermentation and anaerobic digestion, to produce ethanol and biogas (Dar et al. 2024).

3.4.2 Biofuel Production Using Crop Residues and Energy Crops

In the future, crop wastes may prove to be a useful energy source that may be employed to combat climate change and reduce greenhouse gas emissions. They are an essential instrument in reducing the consequences of climate change, since they can be utilized in place of fossil fuels. However, the extraction of crop residues from fields may adversely affect soil productivity and health. According to a study, clearing corn stover can reduce soil organic carbon (SOC) and increase the risk of soil erosion and deterioration, which will reduce soil quality and agricultural productivity. The effects of biofuel production from crop wastes and energy crops on the environment depend on a number of variables. There could be a large amount of greenhouse gas (GHG) emissions, for instance, if large grassland or woodland areas are converted into croplands to create bioenergy. This is because the conversion diminishes these ecosystems' ability to store carbon. The quantity of carbon dioxide emitted during biofuel generation from agricultural residues and energy crops can fluctuate based on the particular conversion method employed. When advanced bioethanol is produced from nonfood feedstocks, such lignocellulosic biomass, it emits fewer greenhouse gases (GHGs) than bioethanol produced from maize using standard methods.

3.4.3 Technology for Maximizing Biomass Feedstock Production

Biomass feedstock production optimization uses advanced techniques and technologies to maximize the productivity, sustainability, and efficiency of biomass crops grown for a range of uses, such as the production of bioenergy, biofuels, bioproducts, and biochemicals.

1. Precision agriculture: Precision agriculture employs modern technologies such as GPS, sensors, drones, and data analytics to enhance farming practices. By obtaining current information on crop health, weather patterns, soil conditions, and nutrient levels, farmers can maximize crop yields while minimizing inputs and making well-informed decisions. This technology can be utilized to enhance biomass feedstock production, hence improving resource efficiency and increasing overall productivity.
2. Genetic engineering and breeding: Increasing the total quantity of biomass available can be achieved by genetically modifying feedstock plants. Plant growth regulators may need to be used differently as a result of this. For example, using transgenic hybrid poplar that produces more gibberellin led to increased biomass yield and improved growth. An additional study that used genetic engineering has produced promising results in increasing biomass in a relevant crop species. Through the use of an endosperm-specific promoter, the expression of ADP-glucose pyrophosphorylase (AGP), an enzyme essential for starch production in endosperm, was increased in rice. As a result, plant biomass unexpectedly increased by 20%.
3. Agroforestry: On a single plot of land, agroforestry systems integrate crops with trees or bushes. This method maintains water resources, improves soil quality, expands the variety of revenue streams for farmers, and sequesters carbon. Through the incorporation of trees into agricultural landscapes for the purpose of producing biomass, farmers can simultaneously promote biodiversity and ecosystem services and increase overall production.
4. Biochar application: Research has shown that the addition of biochar to soil can have a beneficial impact on several soil properties and plant growth. These include improving aggregate stability and water use efficiency. A study has verified that the addition of biochar during fertilization resulted in higher levels of dissolved organic carbon and nitrogen, as well as increased soil organic matter and total nitrogen content. Furthermore, the research demonstrated that the use of biochar considerably enhanced crop productivity and nitrogen absorption.

These advanced technologies and methods provide viable options for enhancing the production of biomass feedstock by improving efficiency, productivity, and sustainability.

3.4.4 Identifying Viable Strategies to Reduce the Environmental and Agricultural Effects of Cultivating Biofuel Crops

Practices for sustainable land use: Growing biofuel crops can have a negative environmental impact, but it can be lessened by using sustainable land-use strategies, such as crop rotation, agroforestry, and intercropping. By putting these techniques into practice, you may improve soil health, encourage biodiversity, and reduce dependence on chemicals.

Water management: Minimizing the water footprint linked to biofuel crop cultivation and preserving water resources can be accomplished through the implementation of efficient water management technologies, such as rainwater harvesting and drip irrigation. Mitigating water scarcity and maintaining the long-term viability of biofuel production need effective water management.

Biodiversity conservation: Maintaining wildlife populations and preserving ecosystem services can be facilitated by promoting biodiversity conservation in and around areas where biofuel crops are grown. Maintaining biodiversity is critical to ecosystems' capacity to tolerate shocks, bounce back from them, and carry on operating sustainably over the long run. Using agricultural waste products, such as crop residues and manure, for biofuel generation or other purposes can help reduce waste output and promote the concepts of a circular economy. Biofuel production has the potential to become more environmentally helpful and sustainable by converting waste into valuable resources.

In conclusion, it is critical to apply a combination of these strategies, together with continuous research and creativity, to minimize the harm that growing biofuel crops causes to the environment and agriculture while promoting sustainable growth.

3.5 Bioenergy Potential

Bioenergy, a sustainable energy source, is obtained from living organisms, including plants and organic wastes. It is generated by the conversion of biomass into valuable energy sources, such as biofuels, thermal energy, or electricity (Dhanya et al. 2020). According to Yadav et al. (2019), bioenergy can be generated from several materials, including algae, forestry byproducts, agricultural residues, and specifically cultivated energy crops, such as switchgrass and miscanthus. Biomass is converted into bioenergy via both thermochemical and biochemical processes, such as gasification or fermentation. Bioenergy is significant because it can lower greenhouse gas emissions and lessen dependence on fossil fuels. Furthermore, the creation of bioenergy can support rural development and help manage organic waste streams by creating new job opportunities in the forestry and agriculture sectors (Dhanya et al. 2020).

3.5.1 Bioenergy Conversion Technologies

Biomass can be converted primarily through two methods: biochemical conversion and thermochemical conversion processes. Biomass can be converted into biofuels and other important goods.

- Biochemical conversion processes: Biochemical conversion techniques utilize microorganisms, including bacteria, fungi, and yeast, to decompose biomass into less complex molecules via biological reactions (Brethauer and Studer 2015).
- Anaerobic digestion: This anaerobic biological process transpires without oxygen. During this process, organic materials, including food waste, sewage sludge, and agricultural residues, are decomposed by microorganisms to generate digestate and biogas, primarily composed of carbon dioxide and methane (Manyi-Loh et al. 2019).
- Fermentation: According to Maicas (2020), fermentation is a biological process wherein microorganisms such as bacteria or yeast transform carbohydrates into organic acids or alcohol. This method is frequently employed in the generation of bioethanol from cellulosic biomass, including sugarcane or corn. When combined with gasoline, the sustainable fuel bioethanol helps reduce greenhouse gas emissions (Dev et al. 2019).
- Thermochemical conversion processes: Thermochemical conversion techniques utilize heat to break down biomass into more basic chemical molecules (Bhaskar et al. 2011).
- Pyrolysis: Pyrolysis is a heat-induced breakdown process that produces biochar, syngas, and bio-oil from biomass without needing oxygen (Imam and Capareda 2012).
- Gasification: It is the thermochemical conversion of biomass into syngas, comprising carbon dioxide, carbon monoxide, and hydrogen. Elevated temperatures and a precise quantity of oxygen or steam are employed to react with the biomass to accomplish this (Chhiti and Kemiha 2013).

3.6 Biofuels: Production and Uses

Across the globe, from bustling metropolises to sun-kissed fields, researchers are conducting a symphony of innovation, channelling their efforts toward the commercialization of sustainable biofuel production. To achieve a carbon-neutral bioeconomy, there is a labyrinth of possibilities available. Different agricultural wastes, biomass, crops, and microorganisms are identified as key sources for biofuel generation, with sustainable and renewable resources. As the world demands more and more energy and the oil wells in India run dry, scientists have started to get serious about the search for other sources of commercial energy and turning agricultural waste into solid fuel, a technology that promotes sustainable biofuel production. With global warming threatening us all and fossil supplies expected to go to zero sooner rather than later, there has been a growing interest in alternative energy sources (Shweta et al. 2024).

Choosing the right feedstock is essential for making biofuel production profitable and environmentally friendly. Common feedstock options include corn, sugarcane, soybean, switchgrass, algae, municipal solid waste, and woody biomass. Amidst this whirlwind of progress, the quest for the perfect feedstock emerges as a pivotal plotline. When evaluating feedstock options for biofuel production, it is critical to consider aspects such as raw material availability, land usage requirements, and resource utilization efficiency. Primarily used as a gasoline blending component, ethanol lowers pollutants and dependence on petroleum-based fuels. Concerns have been voiced, nevertheless, because growing food costs and competition for land usage are closely linked to the generation of ethanol from food crops (Lee and Lavoie 2013). Moreover, this process is capable of encouraging sustainability and also increasing energy security (Padhan et al. 2023).

Second-generation biofuels are produced from nonfood sources such as wood biomass, agricultural waste, and municipal solid waste. These materials promote sustainability more than food crops because they do not conflict with food production. The methods used to convert complex carbohydrates into fermentable sugars involve advanced technologies such as hydrolysis or thermochemical processes (Lee and Lavoie 2013). Pyrolysis and other advanced techniques can turn rubbish into useful substances such as bioalcohols, renewable diesel, biodiesel etc., through biochemical or thermochemical conversion processes applied on biomass (Quevedo-Amador et al. 2024). Cellulosic ethanol is thought to be capable of attaining better yields and reducing greenhouse gas emissions compared to first-generation ethanol. Sophisticated and advanced conversion technologies, such as hydrolysis and thermochemical techniques, break down complex carbohydrates into fermentable sugars. Moreover, since second-generation biofuels often employ feedstock that is seen as more sustainable than crops designed for human consumption, these materials are not in competition with food production (Lee and Lavoie 2013).

To enhance biofuel production efficiency, researchers are giving more concentration on searching and identifying new algae strains for faster biomass collection and lipid accumulation. Genetic engineering is used to boost the lipid content of algal cells, and by applying different techniques, such as putting stress on algae, researchers are trying to promote enhanced lipid accumulation (Sarker and Ghosh 2024). Microalgae biofuels offer sustainable energy sources by fixing CO_2 efficiently, reducing carbon emissions with faster growth rates. To replace fossil fuels, integration of advanced technologies has been launched, such as nanomaterials. In microalgal cells, regulating biosynthesis pathways through chemical modulation and altering chloroplast DNA through genetic engineering accelerate enhanced microalgal biofuel production by improving lipid accumulation (Tan et al. 2024). Cyanobacteria, which are photosynthetic microorganisms, are promising biocatalysts for sustainable biofuel and chemical production. Genetic engineering enhances cyanobacteria to synthesize high-value and customized biofuel (Singh et al. 2024).

Nanotechnology emphasizes sustainable biofuel production from algae, by adding nano-catalysts to reduce biofuel processing costs and by reducing carbon footprint. Nanoparticles such as iron oxide promote microalgae growth; magnesium oxide increases lipid production; magnesium sulfate enhances chlorophyll production in algae; and metal nanoparticles aid in biohydrogen production from microalgae (Kumar et al. 2023). Biofuels represent hope for a future in which people can sustain themselves. They can be used for a lot of things, such as providing power for transportation or making electricity. Industrial processes and decentralized energy systems are also possible, as they are very flexible. By using these resources, we can move toward an economy that does not rely heavily on carbon emissions and is therefore better than before from the environmental point of view. If we use this moment well, it will lead us to realize that these kinds of economies bring both wealth and longevity, because what they offer can never run out. However, if we ignore them now, our planet might start dying soon no matter how much money people have made so far. Biofuels should be appreciated more than ever before, as they work quietly behind the scenes to save us. Biofuels promise a major way of reaching sustainable energy targets as well as shifting toward a low-carbon economy by reducing carbon emissions through their production. The adoption of biofuels in energy schemes can curb global warming, improve energy self-sufficiency, and create a prosperous world that is ecologically friendly economically.

3.7 Reasons to Promote Biofuels

By using biofuels made from living organisms, it is possible to establish an alternative fuel and reduce the pollution contributing to global warming. Marketing and promotion of biofuels is crucial for various reasons (Irimescu et al. 2017). Through a comprehensive analysis of various existing reports, literatures, and empirical evidences, this paper critiques the multifaceted reasons for promoting biofuels, including their economic benefits, environmental sustainability, and superior energy security. Biofuels offer a promising avenue to mitigate the results of weather alternate, as they replace nonrenewable fossil fuels, which considerably lowers carbon dioxide emissions. In order to ensure their sustainability, it is essential to consider the environmental outcomes of increasing feedstock for biofuels, along with changes in water intake and land use. By outlining the benefits of biofuels, we can underscore their essential function in transitioning closer to a more sustainable and resilient electricity destiny. Switching to biofuels for transportation can minimize greenhouse gases from this sector, which significantly contribute to global warming. With growing concerns about climate change, unpredictable energy prices, and peaking fossil fuel reserves, it is important to quickly find clean, renewable energy options. This creates many business opportunities by popularizing biodiesel while cutting overreliance on foreign petroleum, thereby making us more energy secure. There is a need for us to shift rapidly toward sustainable sources of power in dealing with these problems. In contrast to conventional hydrocarbons, biofuels possess an important advantage in that they have positive effects on the atmosphere. Lately, they have emerged as an attractive alternative to fossil fuels, whose raw material is oil from the ground, due to the fact that they can address some of the most pressing challenges, such as climate change or energy insecurity. Despite the debates about the carbon neutrality of biofuels, they still offer a viable solution to lower global CO_2 discharge levels. The development of biofuels may also create sustainable methods of land management, which would help improve carbon sequestration and biodiversity conservation, such as afforestation, agroforestry, and soil conservation. Biofuels provide solutions for business improvement, electricity protection, and climate change by fostering worldwide cooperation, poverty reduction, and sustainable fuel production (Mathews 2008). At present, transportation significantly contributes to air pollution, especially in towns where vehicle emissions are a notable source of poor air quality and health complications for the general public (Ramos et al. 2022). There are mainly three generations of sustainable biofuel alternatives. It is inevitable for offsetting climatic change, fuel security provision, CO_2 emissions cut down and advancement of rural areas. They can improve the economy by reducing oil imports, which, if replaced with petrol, are more harmful (Zhang 2023).

It should be noted that biofuels help reduce carbon emissions from air and land transport, showing their importance in achieving global climate targets. Besides, biofuel byproducts have economic benefits as well as environmental advantages. Since it can be renewed, this form of energy decreases reliance on imported petroleum products, thereby the geopolitical exposure associated with importing natural fuels. Reducing carbon dioxide by shifting to biofuels is the only way to mitigate climate change. This calls for coherence in laws and conventions over time and across nations while at the same time advocating for a certain degree of shared competence in governance decision-making, coupled with a higher degree of societal readiness (Cavelius et al. 2023). Biofuels can replace traditional gasoline and diesel fuels in cities, resulting in noticeable improvements in air quality and a decrease in the prevalence of respiratory illnesses and other health issues. Biofuels provide a versatile, adaptable, and flexible way to cope with changing energy needs while reducing environmental effects by utilizing a variety of biomass resources. The diversity of feedstock sources improves the resilience of biofuel production systems and decreases competition with food crops. Government involvement and education are crucial for promoting the adoption of biofuels, and success depends on appropriate assistance and advantageous economic circumstances (Mukonza 2017).

Biofuel production has a complex value chain made up of distribution, processing, and agriculture, among others. This creates jobs at every step of production, thereby boosting economies as well. Transgenic crops are necessary for managing environmental degradation while meeting food as well as energy needs for sustainable development (Sexton et al. 2009). Additionally, nanotechnology offer bright possibilities toward improving competitiveness levels while reducing costs involved in producing biofuel (Zuccaro 2023). The development of biofuels encourages regional self-sufficiency and decentralizes energy production by utilizing easily available biomass resources. This lessens dependency on external energy sources and centralized energy infrastructure. Farmers can improve soil health, boost biodiversity, and earn extra revenue while helping combat climate change by cultivating specific energy crops or incorporating bioenergy crops into already existing agricultural systems.

The acceptance of biofuels is correlated with reaching sustainable development objectives for global improvement (Bonaiuto et al. 2024). Vehicles powered by biofuels fight global warming, thus helping to combat climate change. To compete with oil in the transportation sector, biofuels provide a sustainable substitute for fossil fuels (Zhang 2023). Biofuels are environmentally friendly, clean-burning alternatives to petroleum diesel, which produce less pollution. Biofuels reduce greenhouse gas emissions by 56–86%; thus, it can promote energy conservation, air quality, and safety benefits. Numerous plant products and crops can be used to produce biofuels. The process of pre-treatment is essential for turning any lignin into biofuel. The processes for producing bio-oil are pyrolysis, hydrothermal processing, and aqueous phase modification (Sherief et al. 2022). Biofuels decrease carbon footprint, substitute gasoline and diesel by using biomass, and present sustainable energy sources as an alternative, helping to combat climate change as well as contribute to sustainable development. Second-generation biofuels demonstrate a higher net energy gain compared to their first-generation counterparts. Second generation biofuels show energy yield, than first generation. Extracting biohydrogen from algae in biomass could offer a renewable energy option for third-generation biofuels. Biodiesel production uses the popular and efficient transesterification process (Saleh et al. 2022). Biofuels such as ethanol and butanol impact flame propagation in engines. Studies on engine combustion monitor variations in the performance of gasoline, ethanol, and butanol. In direct injection engines, the spread of flames and the manufacturing of soot are impacted by biofuels (Irimescu et al. 2017). Overall, selling biofuels aligns with worldwide efforts to transition toward cleaner, renewable power sources, ensuring a more sustainable and environmentally friendly future.

3.8 Future Prospects and Challenges of Biofuel Production

Concerns about the environment and the depletion of gasoline derived from petroleum have prompted research into biofuel as a potential substitute fuel source. The manufacturing capacity and engine compatibility of biofuel, however, cannot completely replace those of petroleum-based fuels. Energy is the main factor driving global, social, and economic developments. Presently, about 80% of the world's fuel requirements are met by petroleum and related fuels. According to the 2014 International Energy Report (Joshi et al. 2017), by 2040, there will be a 37% rise in the world's energy use. In the transportation sector, biofuels are thought to be the most feasible approach to reduce carbon dioxide emissions. Plant oils such as rapeseed, soybean, sunflower, and palm oils as well as other inedible oils such as mahua, neem, karanja, and jatropha are the feedstocks used to make biodiesel. Biodiesel can also be produced from refined cooking oil leftovers and animal fats such as calf tallow. Third-generation biofuels are thought to be derived from algae and other sources. in jeopardy due to the severe environmental harm it causes and the increased. Wastes from palm mills pollute significant amounts of water, which in turn affects aquatic biodiversity. Insecticides, rodenticides, fertilizers, herbicides, and palm oil mill effluents (POME) are a few other possible contaminants. However, it is asserted that the development of biofuel in developing nations takes arable land away from food crops, raising food costs for the comparatively poor.

Recently, a number of nongovernmental organizations in developing nations, environmental organizations, and even scientists have started to challenge the logic behind the current biofuel development. This may lead to carbon emissions that are equivalent to or higher than what the industry is able to collect. Many technical and economic arguments stem from analyses of the environmental and economic costs of biofuels relative to their useable production. While certain experts have periodically questioned the fundamental assumptions of these calculations, they nevertheless voice concerns about the environmental suitability of particular biofuels. The most productive blend of biodiesel (oilseeds and oil palm) could produce bioethanol annually if all of the world's arable land were planted with it. Annual crops such as corn, cane, and wheat, as well as woody perennials such as miscanthus and poplar, could only replace up to 35% of the current carbon emissions from fossil fuels. Moreover, even this low target would still require the removal of 80–120 million hectares, or more, in order to mitigate the possible harmful impacts of nonrenewable carbon emissions on the environment, such as climate change. Alternative strategies for lowering carbon emissions include growing forestry on nonarable land and using non-carbon energy sources such as nuclear, solar, wind, hydro, or tidal power generation. To put it simply, first-generation biofuels can replace fossil fuels on their own. Many people are depending on the improvements in plant breeding and processing methods, especially with regard to lignocellulosic biomass, in order to produce second-generation biofuels that are superior. Using cutting-edge breeding techniques such as transgenesis (genetic engineering), TILLING (Targeting Induced Local Lesions in Genomes), and DNA marker-assisted selection to enhance current crops or create new ones especially for biofuel production is an effective strategy. Lastly, it might be able to alter the oil yield and fatty acid content of oil-producing crops to improve the suitability of plant oils for direct fuel usage and circumvent the costly methyl ester production process. It will undoubtedly be helpful to combine several genetic engineering techniques to maximize the production of biofuels. Global cooperation is therefore crucial to integrating biofuels into our current energy system and overcoming the energy problem. This will also help to advance farming techniques and the technology used to produce biofuels (Kour et al. 2019).

References

Ali S, Dar MA, Liaqat F et al (2024a) Optimization of biomethane production from lignocellulosic biomass by a developed microbial consortium. Process Saf Environ Prot 184:1106–1118

Ali S, Rani A, Dar MA et al (2024b) Recent advances in the characterization and valorization of lignin to value-added products: challenges and future perspectives. Biomass 4:947–977. https://doi.org/10.3390/biomass4030053

Azapagic A, Stichnothe H (2011) Assessing sustainability of biofuels. In: Azapagic A, Perdan S (eds) Sustainable development in practice: case studies for engineers and scientists. John Wiley & Sons

Balajii M, Niju S (2019) Biochar-derived heterogeneous catalysts for biodiesel production. Environ Chem Lett 17(4):1447–1469. https://doi.org/10.1007/s10311-019-00885-x

Bharti MK, Chalia S, Thakur P (2021) Nanoferrites heterogeneous catalysts for biodiesel production from soybean and canola oil: a review. Environ Chem Lett 19:3727–3746

Bhaskar T, Bhavya B, Singh R, Naik DV, Kumar A, Goyal HB (2011) Thermochemical conversion of biomass to biofuels. In: Biofuels. Academic Press, pp 51–77

Bhattarai U (2017) Impacts of climate change on biodiversity and ecosystem services: direction for future research. Hydro Nepal J Water Energy Environ 20:41–48. https://doi.org/10.3126/hn.v20i0.16488

Bhonsle AK, Singh J, Trivedi J, Atray N (2022) Comparative LCA studies of biodiesel produced from used cooking oil using conventional and novel room temperature processes. Bioresour Technol Rep 18

Bonaiuto M, Mosca O, Milani A, Ariccio S, Dessi F, Fornara F (2024) Beliefs about technological and contextual features drive biofuels' social acceptance. Renew Sust Energy Rev 189:113867

Brethauer S, Studer MH (2015) Biochemical conversion processes of lignocellulosic biomass to fuels and chemicals—a review. Chimia 69(10):572

Brown ME, Carr ER, Grace KL, Wiebe K, Funk CC, Attavanich W, Backlund P, Buja L (2017) Do markets and trade help or hurt the global food system adapt to climate change? Food Policy 68:154–159. https://doi.org/10.1016/j.foodpol.2017.02.004

Cardinale BJ, Gross K, Fritschie K, Flombaum P, Fox JW, Rixen C, van Ruijven J, Reich PB, Scherer-Lorenzen M, Wilsey BJ (2013) Biodiversity simultaneously enhances the production and stability of community biomass, but the effects are independent. Ecology 94(8):1697–1707. https://doi.org/10.1890/12-1334.1

Cavelius P, Engelhart-Straub S, Mehlmer N, Lercher J, Awad D, Brügg T (2023) The potential of biofuels from first to fourth generation. PLoS Biol 21(3):1–20. https://doi.org/10.1371/journal.pbio.3002063

Cherubini F, Strømman AH (2011) Life cycle assessment of bioenergy systems: state of the art and future challenges. Bioresour Technol 102(2):437–451. https://doi.org/10.1016/j.biortech.2010.08.010

Cherubini F, Ulgiati S (2010) Crop residues as raw materials for biorefinery systems—a LCA case study. Appl Energy 87(1):47–57. https://doi.org/10.1016/j.apenergy.2009.08.024

Chhiti Y, Kemiha M (2013) Thermal conversion of biomass, pyrolysis and gasification. Int J Eng Sci (IJES) 2(3):75–85

Climate Change and Land: An IPCC Special Report on Climate Change, Desertification, Land Degradation, Sustainable Land Management, Food Security, and Greenhouse Gas Fluxes in Terrestrial Ecosystems (2019)

Dale VH, Kline KL, Wiens J, Fargione J (2010) Biofuels: implications for land use and biodiversity. Ecological Society of America, Washington, DC

Dar MA, Dhole NP, Xie RR et al (2021) Valorization potential of a novel bacterial strain, *Bacillus altitudinis* RSP75, towards lignocellulose bioconversion: an assessment of symbiotic bacteria from the stored grain pest, *Tribolium castaneum*. Microorganisms 9:1952

Dar MA, Syed R, Pawar KD, Dhole NP, Xie R, Pandit RS, Sun J (2022) Evaluation and characterization of the cellulolytic bacterium, *Bacillus pumilus* SL8 isolated from the gut of oriental leafworm, *Spodoptera litura*: an assessment of its potential value for lignocellulose bioconversion. Environ Technol Innov 27:102459

Dar MA, Xie R, Zabed HM et al (2024) Termite microbial symbiosis as a model for innovative design of lignocellulosic future biorefinery: current paradigms and future perspectives. Biomass 4:180–201. https://doi.org/10.3390/biomass4010009

Dar MA, Xie R, Zabed HM, Idrees A, Waghmare PR, Irfan M, Dhanavade MJ, Shahnawaz M, Zhu D, Sun J (2025) Can termites serve as models for future biorefinery? In: Zhu D, Dar MA, Shahnawaz M (eds) Biofuels and sustainability: life-cycle assessments, systems biology, policies and emerging technologies. Elsevier Science, pp 323–346. https://doi.org/10.1016/B978-0-443-21433-2.00017-7

De Fraiture C, Giordano M, Liao Y (2008) Biofuels and implications for agricultural water use: blue impacts of green energy. Water Policy 10(S1):67–81

Delucchi MA (2010) Impacts of biofuels on climate change, water use, and land use. Ann N Y Acad Sci 1195(1):28–45

Dev B, Zaky AS, Jayabalan R (2019) Bioethanol fermentation: the path forward for eco-friendly and sustainable development. In: Technologies for Value Addition in food products and processes. Apple Academic Press, pp 233–270

Dhanya BS, Mishra A, Chandel AK, Verma ML (2020) Development of sustainable approaches for converting the organic waste to bioenergy. Sci Total Environ 723:138109

Eldredge N (2002) Life on earth: an encyclopedia of biodiversity, ecology, and evolution. N Eldredge & Califormia (eds)

Ellis EC, Klein Goldewijk K, Siebert S, Lightman D, Ramankutty N (2010) Anthropogenic transformation of the biomes, 1700 to 2000. Glob Ecol Biogeogr 19(5):589–606. https://doi.org/10.1111/j.1466-8238.2010.00540.x

Environmental Protection Agency, Office of Transportation and Air Quality (2010) Renewable Fuel Standard Program (RFS2) Regulatory Impact Analysis

Fahrig L (2017) Ecological responses to habitat fragmentation per se. Annu Rev Ecol Evol Syst 48(1):1–23. https://doi.org/10.1146/annurev-ecolsys-110316-022612

Fahrig L, Arroyo-Rodríguez V, Bennett JR, Boucher-Lalonde V, Cazetta E, Currie DJ, Eigenbrod F, Ford AT, Harrison SP, Jaeger JAG, Koper N, Martin AE, Martin JL, Metzger JP, Morrison P, Rhodes JR, Saunders DA, Simberloff D, Smith AC et al (2019) Is habitat fragmentation bad for biodiversity? Biol Conserv 230:179–186. https://doi.org/10.1016/j.biocon.2018.12.026

Fargione J, Hill J, Tilman D, Polasky S, Hawthorne P (2008) Land clearing and the biofuel carbon debt. Science 319(5867):1235–1238. https://doi.org/10.1126/science.1152747

Fargione JE, Plevin RJ, Hill JD (2010) The ecological impact of biofuels. Annu Rev Ecol Evol Syst 41:351–377

Farrell AE, Plevin RJ, Turner BT, Jones AD, O'Hare M, Kammen DM (2006) Ethanol can contribute to energy and environmental goals. Science 311(5760):506–508. https://doi.org/10.1126/science.1121416

Food and Agriculture Organization of the United Nations (n.d.) Biofuels and the sustainability challenge: a global assessment of sustainability issues, trends and policies for biofuels and related feedstocks

Foteinis S, Chatzisymeon E, Litinas A, Tsoutsos T (2020) Used-cooking-oil biodiesel: life cycle assessment and comparison with first- and third-generation biofuel. Renew Energy 153:588–600

García-Vega D, Newbold T (2020) Assessing the effects of land use on biodiversity in the world's drylands and Mediterranean environments. Biodivers Conserv 29(2):393–408. https://doi.org/10.1007/s10531-019-01888-4

Geiger F, Bengtsson J, Berendse F, Weisser WW, Emmerson M, Morales MB, Ceryngier P, Liira J, Tscharntke T, Winqvist C, Eggers S, Bommarco R, Pärt T, Bretagnolle V, Plantegenest M, Clement LW, Dennis C, Palmer C, Oñate JJ et al (2010) Persistent negative effects of pesticides on biodiversity and biological control potential on European farmland. Basic Appl Ecol 11(2):97–105. https://doi.org/10.1016/j.baae.2009.12.001

Global Report on Food Crises-Joint Analysis for Better Decisions; Food and Agriculture Organization (FAO); World Food Programme (WFP); International Food Policy Research Institute (IFPRI) (2022)

Gnansounou E, Vaskan P, Pachón ER (2015) Comparative techno-economic assessment and LCA of selected integrated sugarcane-based biorefineries. Bioresour Technol 196:364–375. https://doi.org/10.1016/j.biortech.2015.07.072

Gopalakrishnan G, Negri MC, Wang M, Wu M, Snyder SW, Lafreniere L (2009) Biofuels, land, and water: a systems approach to sustainability. Environ Sci Technol 43(15):6094–6100

Gunarathne V, Ashiq A, Ramanayaka S (2019) Biochar from municipal solid waste for resource recovery and pollution remediation. Environ Chem Lett 17:1225–1235

Gupta R, McRoberts R, Yu Z, Smith C, Sloan W, You S (2022) Life cycle assessment of biodiesel production from rapeseed oil: influence of process parameters and scale. Bioresour Technol 360(127532):127532. https://doi.org/10.1016/j.biortech.2022.127532

Hao X, Chang C, Carefoot JM, Janzen HH, Ellert BH (2001) Nutrient cycling in agroecosystems. 60(1/3):1–8. https://doi.org/10.1023/a:1012603732435

Hassan NS, Jalil AA, Hitam C (2020) Biofuels and renewable chemicals production by catalytic pyrolysis of cellulose: a review. Environ Chem Lett 18:1625–1648

Imam T, Capareda S (2012) Characterization of bio-oil, syn-gas and bio-char from switchgrass pyrolysis at various temperatures. J Anal Appl Pyrolysis 93:170–177

Increasing feedstock production for biofuels: Economic drivers, environmental implications, and the role of research (2008) The Biomass Research and Development Board, US

Irimescu A, Merola SS, Di Iorio S, Vaglieco BM (2017) Biofuel effect on flame propagation and soot formation in a DISI engine. IOP Conf Ser Mater Sci Eng 252:012092

Joly CA, Verdade LM, Huntley BJ, Dale VH, Mace G, Muok B, Ravindranath NH (2015) Biofuel impacts on biodiversity and ecosystem services. In: Scientific Committee on Problems of the Environment (SCOPE), rapid assessment process on bioenergy and sustainability, pp 555–580

Joshi G, Pandey JK, Rana S, Rawat DS (2017) Challenges and opportunities for the application of biofuel. Renew Sust Energ Rev 79:850–866

Kaloudas D, Pavlova N, Penchovsky R (2021) Lignocellulose, algal biomass, biofuels and biohydrogen: a review. Environ Chem Lett 19:2809–2824

Koçar G, Civaş N (2013) An overview of biofuels from energy crops: current status and future prospects. Renew Sust Energ Rev 28:900–916

Koh LP, Ghazoul J (2008) Biofuels, biodiversity, and people: understanding the conflicts and finding opportunities. Biol Conserv 141:2450–2460

Koponen K, Soimakallio S, Tsupari E, Thun R, Antikainen R (2013) GHG emission performance of various liquid transportation biofuels in Finland in accordance with the EU sustainability criteria. Appl Energy 102:440–448

Kour D, Rana KL, Yadav N, Yadav AN, Rastegari AA, Singh C, Saxena AK (2019) Technologies for biofuel production: current development, challenges, and future prospects. Prospects of Renewable Bioprocessing in Future Energy Systems, pp 1–50
Kumar S, Singh J, Nanoti SM, Garg MO (2012) A comprehensive life cycle assessment (LCA) of Jatropha biodiesel production in India. Bioresour Technol 110:723–729
Kumar S, Tirlangi S, Kumar A, Imran M, Pillai HP, Koshariya AK, Sathish T, Ubaidullah M, Ayub R, Minnam Reddy VR, AlAbdulaal TH, Algarni H, Maiz F, Shkir M (2023) A review on the contribution of nanotechnology for biofuel production from algal biomass: a bridge to the reduction of carbon footprint. Sustain Energy Technol Assess 60:103498
Lassaletta L, Billen G, Garnier J, Bouwman L, Velazquez E, Mueller ND, Gerber JS (2016) Nitrogen use in the global food system: past trends and future trajectories of agronomic performance, pollution, trade, and dietary demand. Environ Res Lett 11:095007
Lee R, Lavoie JM (2013) From first- to third-generation biofuels: challenges of producing a commodity from a biomass of increasing complexity. Anim Front 3:6–11
Maicas S (2020) The role of yeasts in fermentation processes. Microorganisms 8:1142
Manyi-Loh CE, Mamphweli SN, Meyer EL, Okoh AI (2019) Microbial anaerobic digestion: process dynamics and implications from the renewable energy, environmental, and agronomy perspectives. Int J Environ Sci Technol 16:3913–3934
Mathews JA (2008) Biofuels, climate change and industrial development: can the tropical south build 2000 biorefineries in the next decade? Biofuels Bioprod Biorefin 2:103–125
Menten F, Chèze B, Patouillard L, Bouvart F (2013) A review of LCA greenhouse gas emissions results for advanced biofuels: the use of meta-regression analysis. Renew Sust Energ Rev 26:108–134
Meza-Elizalde MC, Armenteras-Pascual D (2021) Edge influence on the microclimate and vegetation of fragments of a north Amazonian forest. Forest Ecol Manag 498:119546
Michael K, Steffi N, Peter D (2011) The past, present, and future of biofuels-biobutanol as promising alternative. In: Biofuel Production-Recent Developments and Prospects, pp 451–486
Morales M, Quintero J, Conejeros R, Aroca G (2015) Life cycle assessment of lignocellulosic bioethanol: environmental impacts and energy balance. Renew Sust Energ Rev 42:1349–1361
Mukonza C (2017) Knowledge, attitudes and perceptions of stakeholders on biofuels as an enabler in a south African bio-based economy. J Energy South Afr 28:107–118
Munagala M, Shastri Y, Nagarajan S, Ranade V (2022) Production of bio-CNG from sugarcane bagasse: commercialization potential assessment in Indian context. Ind Crop Prod 188
Newbold T, Hudson LN, Hill SLL et al (2015) Global effects of land use on local terrestrial biodiversity. Nature 520(7545):45–50. https://doi.org/10.1038/nature14324
Padhan SR, Jat SL, Mishra P et al (2023) Corn for biofuel: status, prospects and implications, IntechOpen
Passell H, Dhaliwal H, Reno M et al (2013) Algae biodiesel life cycle assessment using current commercial data. J Environ Manag 129:103–111. https://doi.org/10.1016/j.jenvman.2013.06.055
Pastor AV, Ludwig F, Biemans H et al (2014) Accounting for environmental flow requirements in global water assessments. Hydrol Earth Syst Sci 18(12):5041–5059. https://doi.org/10.5194/hess-18-5041-2014
Pendrill F, Gardner TA, Meyfroidt P et al (2022) Disentangling the numbers behind agriculture-driven tropical deforestation. Science 377:6611. https://doi.org/10.1126/science.abm9267
Peng L, Fu D, Chu H (2020) Biofuel production from microalgae: a review. Environ Chem Lett 18:285–297
Phalan B (2009) The social and environmental impacts of biofuels in Asia: an overview. Appl Energy 86:S21–S29
Phalan B (2018) What have we learned from the land sparing-sharing model? Sustain For 10(6):1760. https://doi.org/10.3390/su10061760
Pinder RW, Davidson EA, Goodale CL et al (2012) Climate change impacts of US reactive nitrogen. Proc Natl Acad Sci USA 109(20):7671–7675. https://doi.org/10.1073/pnas.1114243109

Pisa LW, Amaral-Rogers V, Belzunces LP et al (2015) Effects of neonicotinoids and fipronil on non-target invertebrates. Environ Sci Pollut Res Int 22(1):68–102. https://doi.org/10.1007/s11356-014-3471-x

Poff NL, Zimmerman JKH (2010) Ecological responses to altered flow regimes: a literature review to inform the science and management of environmental flows. Freshw Biol 55(1):194–205. https://doi.org/10.1111/j.1365-2427.2009.02272.x

Prasad S, Ingle AP (2019) Impacts of sustainable biofuels production from biomass. In: Sustainable bioenergy. Elsevier, pp 327–346

Putranti AP, Widiastuti (2017) Penerapan metode Numbered Heads Together (NHT) untuk meningkatkan hasil belajar kognitif IPA di SD Kristen Gunungsitoli Utara. Polyglot 12(2):128. https://doi.org/10.19166/pji.v12i2.370

Qin Z, Dunn JB, Kwon H et al (2016) Soil carbon sequestration and land use change associated with biofuel production: empirical evidence. Global Change Biol Bioenergy 8(1):66–80. https://doi.org/10.1111/gcbb.12237

Qin Z, Zhuang Q, Cai X et al (2018) Biomass and biofuels in China: toward bioenergy resource potentials and their impacts on the environment. Renew Sust Energ Rev 82:2387–2400

Quevedo-Amador RA, Escalera-Velasco BP, Ramírez Arias AM et al (2024) Application of waste biomass for the production of biofuels and catalysts: a review. Clean Technol Environ Policy

Ramos et al (2022) Addressing the energy crisis: using microbes to make biofuels. Microbial Biotechnol 15(4):1026–1030

Roy P, Dutta A (2013) Life cycle assessment of ethanol derived from sawdust. Bioresour Technol 150:407–411. https://doi.org/10.1016/j.biortech.2013.08.057

Roy P, Tokuyasu K, Orikasa T et al (2012) A review of life cycle assessment (LCA) of bioethanol from lignocellulosic biomass. Japan Agric Res Q 46(1):41–57. https://doi.org/10.6090/jarq.46.41

Saleh AM, Saleh NM, Mahdi HH (2022) Production of biofuels from biomass as an approach towards sustainable development: a short review. Ntu J Renew Energy 3(1):9–21

Sanchez ST, Woods J, Akhurst M et al (2012) Accounting for indirect land-use change in the life cycle assessment of biofuel supply chains. J R Soc Interface 9(71):1105–1119. https://doi.org/10.1098/rsif.2011.0769

Sarker M, Ghosh S (2024) Recent advancement of algal biofuel production and its potential uses. J Adv Zool 45(1):508–510

Sexton S, Zilberman D, Rajagopal D et al (2009) The role of biotechnology in a sustainable biofuel future. AgBioforum 12(1):130–140

Sherief N, Anand M, Ramachandran M et al (2022) A review on various biofuels and its applications. REST J Emerg Trends Modelling Manuf 8(1):1–9. https://doi.org/10.46632/jemm/8/1/1

Shonnard DR, Klemetsrud B, Sacramento-Rivero J et al (2015) A review of environmental life cycle assessments of liquid transportation biofuels in the pan American region. Environ Manag 56:1356–1376. https://doi.org/10.1007/s00267-015-0543-8

Shweta, Capareda SC, Kamboj BR et al (2024) Biomass resources and biofuel technologies: a focus on Indian development. Energies 17:382

Siebert S, Döll P (2010) Quantifying blue and green virtual water contents in global crop production as well as potential production losses without irrigation. J Hydrol 384(3–4):198–217. https://doi.org/10.1016/j.jhydrol.2009.07.031

Sieverding HL, Bailey LM, Hengen TJ et al (2015) Meta-analysis of soybean-based biodiesel. J Environ Qual 44(4):1038–1048. https://doi.org/10.2134/jeq2014.07.0320

Singh VK, Jha S, Rana P et al (2024) Cyanobacteria as a biocatalyst for sustainable production of biofuels and chemicals. Energies 17(408):1–25

Sinha S, Suzuki S, Kojima T (2013) Life cycle analysis and modelling (LCAM) of Jatropha as biofuel in dynamic economic environment of newly emerging economies. Rev Asian Pacific Stud 38:119–139

Smith P (2012) Soils and climate change. Curr Opin Environ Sustain 4(5):539–544. https://doi.org/10.1016/j.cosust.2012.06.005

Stephenson AL, Dupree P, Scott SA et al (2010) The environmental and economic sustainability of potential bioethanol from willow in the UK. Bioresour Technol 101(24):9612–9623. https://doi.org/10.1016/j.biortech.2010.07.104

Tan CH, Low SS, Cheah WY, Singh J, Chai WS, Tiong SK, Show PL (2024) Futuristic opportunities for pretreatment processes in biofuel production from microalgae. GCB Bioenergy 16(e13136):1–23

Tang H, Wu W, Yu Q, Xia T, Yang P, Li Z (2015) Key research priorities for agricultural land system studies. Sci Agric Sin 48:900–910

Thakrar SK, Balasubramanian S, Adams PJ, Azevedo IML, Muller NZ, Pandis SN, Polasky S, Pope CA III, Robinson AL, Apte JS, Tessum CW, Marshall JD, Hill JD (2020) Reducing mortality from air pollution in the United States by targeting specific emission sources. Environ Sci Technol Lett 7(9):639–645. https://doi.org/10.1021/acs.estlett.0c00424

Verburg PH, Erb KH, Mertz O, Espindola G (2013) Land system science: between global challenges and local realities. Curr Opin Environ Sustain 5(5):433–437. https://doi.org/10.1016/j.cosust.2013.08.001

Verburg PH, Crossman N, Ellis EC, Heinimann A, Hostert P, Mertz O, Nagendra H, Sikor T, Erb KH, Golubiewski N, Grau R, Grove M, Konaté S, Meyfroidt P, Parker DC, Chowdhury RR, Shibata H, Thomson A, Zhen L (2015) Land system science and sustainable development of the earth system: a global land project perspective. Anthropocene 12:29–41. https://doi.org/10.1016/j.ancene.2015.09.004

Verdade LM, Pina CI, Rosalino LM (2015) Biofuels and biodiversity: challenges and opportunities. Environ Dev 15:64–78

Viana CM, Freire D, Abrantes P, Rocha J, Pereira P (2022) Agricultural land systems importance for supporting food security and sustainable development goals: a systematic review. Sci Total Environ 806(Pt 3):150718. https://doi.org/10.1016/j.scitotenv.2021.150718

Wiens J, Fargione J, Hill J (2011) Biofuels and biodiversity. Ecol Appl 21(4):1085–1095

Yadav P, Priyanka P, Kumar D, Yadav A, Yadav K (2019) Bioenergy crops: recent advances and future outlook. In: Rastegari A, Yadav A, Gupta A (eds) Prospects of renewable bioprocessing in future energy systems. Biofuel and biorefinery technologies, vol 10. Springer, Cham

Zhang F (2023) Biofuels for transportation. J Phys Conf Ser 2608(012049):1–7. https://doi.org/10.1088/1742-6596/2608/1/012049

Zuccaro G (2023) Why we should support biofuel production. Fuels 4:261–263. https://doi.org/10.3390/fuels4020016

Chapter 4
Extremophilic Microorganisms and Their Metabolic Remodulation to Explore Unconventional Cell Factories for Sustainable Biofuel and Bioenergy Production

Hossain M. Zabed, Jamila A. Tuly, Mudasir A. Dar, S. M. Hasan Mahmud, Junhua Yun, Guoyang Zhang, and Xianghui Qi

Abstract The increasing demand for energy security and environmental sustainability has intensified efforts to transition from fossil fuels to renewable energy sources such as biofuels and bioenergy. The production of biofuels involves complex upstream and midstream processes, in which microorganisms play a pivotal role. However, conventional microbial hosts used in these processes face several challenges, including contamination risks, high energy and freshwater demands, complex sterilization protocols, and significant resource-intensive operations. These limitations hinder the cost-effectiveness and scalability of biofuel production, making it difficult to compete with fossil fuel markets. To address these challenges, next-generation industrial biotechnology (NGIB) has been emerged, which focuses on the use of extremophilic microorganisms that are capable of thriving in extreme environmental conditions. Extremophiles possess unique metabolic pathways and enzymes that enable efficient biomass conversion under harsh conditions, such as high salinity, extreme pH, or high temperatures, without the need for energy-intensive sterilization. Advances in genetic and metabolic engineering, such as CRISPR/Cas9 technologies, have further enhanced the bioenergy potential of extremophiles, optimizing their metabolic traits for higher biofuel yields and resource efficiency. This chapter explores the technological interventions in

H. M. Zabed (✉) · J. A. Tuly · J. Yun · G. Zhang · X. Qi
School of Life Sciences, Guangzhou University, Guangzhou, Guangdong Province, China
e-mail: zabed@gzhu.edu.cn

M. A. Dar
Biofuels Institute, School of Environment and Safety Engineering, Jiangsu University, Zhenjiang, Jiangsu, China

S. M. H. Mahmud
Department of Computer Science, American International University-Bangladesh (AIUB), Dhaka, Bangladesh

M. A. Dar et al. (eds.), *Recent Trends in Lignocellulosic Biofuels and Bioenergy*, Clean Energy Production Technologies,
https://doi.org/10.1007/978-981-96-4636-4_4

bioengineering extremophilic microorganisms, highlighting their transformative potential in revolutionizing biofuel and bioenergy production.

Keywords Extremophiles · Unconventional chassis · Metabolic engineering · Bioprocessing · Bioenergy · Biofuels

4.1 Introduction

The growing need to achieve energy security and environmental sustainability has intensified global efforts to minimize dependence on fossil fuels. This urgency arises from rising global temperatures, concern about climate change, and the need to reduce greenhouse gas (GHG) emissions (El-Araby 2024). In this context, bioenergy and biofuels from renewable biomass resources have emerged as promising alternatives to traditional fossil fuels (Zabed et al. 2023; Elshobary et al. 2021). By reducing environmental degradation and eliminating social inequalities, bioenergy offers a viable path to a sustainable and equitable energy future (Zhao et al. 2022b). Biofuels, as a renewable energy source, are also ideal for combating climate change, which help reduce GHG emissions and therefore play a crucial role in curbing global warming (Cherwoo et al. 2023). Despite their potential, biofuels and bioenergy face economic challenges, particularly in competition with petroleum-based products in times of low oil prices and the emergence of alternative energy sources. Still, the transition to bioenergy is essential for achieving global sustainability goals (Rial 2024). By using innovative technologies, improving rural infrastructure, and promoting capacity building, bioenergy systems can be scaled to revolutionize global energy networks (Souza et al. 2017). With strategic investments and sustainable practices, biofuels and bioenergy can play a critical role in addressing the energy and climate crises of the twenty-first century (Asase et al. 2024). In this context, investments in research and development, coupled with targeted policy interventions, can further increase the competitiveness of biofuels and bioenergy while ensuring their environmental and social benefits (Rial 2024; Malode et al. 2021).

The production of biofuels and bioenergy involves a multistep process in which microorganisms play a central role in both the upstream and midstream steps. In the upstream phase, biomass, such as lignocellulosic materials, is processed to release fermentable substrates. This biomass processing mainly involves pretreatment and enzymatic saccharification to break down complex biomass into simpler molecules such as sugars, which serve as feedstock for the subsequent stages. This step, especially the saccharification phase, relies exclusively on microbial enzymes (Zabed et al. 2017a, 2024b). Enzyme production by microorganisms is critical for effective biomass hydrolysis, making this step a cornerstone of biofuel and bioenergy production (Bangaru et al. 2022). In addition, currently highly demanding biological pretreatment using microorganisms or even enzymes, has received significant attention to replace high-cost and high-energy-consuming physicochemical processes (Zabed et al. 2019). In the midstream phase, the hydrolyzed substrates undergo either fermentation or anaerobic digestion, processes that are solely mediated by microorganisms (Kumar and Kumar 2017). In fermentation, single microbial strains or microbial consortia convert sugars into biofuels such as ethanol, butanol, or biohydrogen (Zabed et al. 2017b). Similarly, in anaerobic digestion, microbial communities break down organic matter in the absence of oxygen, producing biogas—a mixture of methane and carbon dioxide—that serves as a renewable energy source (Zabed et al. 2020).

One of the major challenges in current biofuel and bioenergy production processes is the use of conventional or model strains, which poses several techno-economic challenges and makes it difficult for biofuel and bioenergy production to compete with fossil fuel market. For example, traditional microbial hosts used in the midstream process (e.g., *Escherichia coli*, *Bacillus* spp., and yeasts) are highly susceptible to microbial contamination and require strict sterilization procedures (Ye et al. 2023). This sterilization involves complex, energy-intensive processes using high-temperature or high-pressure steam to ensure microbial purity, which significantly increases production costs. In addition, sterilization temperatures in fermentation media with mixed substrates can trigger undesirable Maillard reactions, further complicating process optimization (Zhang et al. 2024). Second, traditional systems require complex operating protocols, including batch or fed-batch fermentation, which limits production efficiency and scalability (Ye et al. 2018). Third, fermentation processes using traditional microbial chassis require large amounts of fresh water and energy, making these systems resource-intensive and less sustainable (Jiang et al. 2021; Zabed et al. 2024a). Fourth, traditional methods rely on costly stainless-steel bioreactors and disinfection systems, which increases the overall investment required (Junker et al. 2006). Addressing these challenges is crucial to making the production of biofuels and bioenergy cost-competitive and environmentally sustainable (Pan et al. 2022). In this context, next-generation industrial biotechnology (NGIB), which uses extremophilic microorganisms instead of traditional hosts, is emerging as a viable alternative to overcome these obstacles (Chen and Jiang 2018).

Extremophilic microorganisms, known for thriving in extreme environmental conditions, offer promising opportunities for the further development of biofuels and bioenergy production (Aslam and Rashid 2022; Passarini et al. 2022). These microorganisms possess unique enzymes and metabolic pathways that enable efficient biomass conversion, even under harsh conditions that hinder conventional microbes. Extremophiles have evolved remarkable adaptations, such as specialized proteins and lipid membranes for stability at extreme temperatures, buffer systems for tolerance to strongly acidic or alkaline pH levels, and mechanisms for regulating osmotic pressure in saline environments (Ye et al. 2023). These properties make them durable and ideal for industrial applications. For example, halophilic bacteria, such as *Halomonas bluephagenesis*, have shown exceptional potential as an industrial chassis by growing efficiently in saline, non-sterile environments, thereby reducing contamination risks and energy requirements (Zhang et al. 2018; Ma et al. 2020). Furthermore, advances in metabolic and genetic engineering, including CRISPR/Cas9 technologies, have unlocked new possibilities to optimize the metabolic characteristics of extremophiles (Chen et al. 2018; Fokum et al. 2019). These innovations can enhance biofuel yields, lower production costs, and simplify industrial processes, positioning extremophilic microorganisms as transformative tools in the shift toward sustainable bioenergy production. Compared to traditional microbial hosts, extremophiles offer superior efficiency, stability, and scalability, making them invaluable in the quest for renewable energy solutions (Rawat et al. 2024).

4.2 Extremophilic Microorganisms: Diversity and Potential

Extremophilic microorganisms are remarkable life forms that thrive in environments that are hostile to most other organisms. They include bacteria, archaea, and certain eukaryotes, which have evolved to withstand extreme conditions, such as extremely high temperatures (thermophiles), extreme pH levels (acidophiles or alkaliphiles), and elevated salinity (halophiles) (Ye et al. 2023). Their survival in such challenging environments is made possible by distinctive physiological and biochemical adaptations, such as specialized enzymes, resilient membrane structures, and modified metabolic processes. These adaptations not only ensure their survival but also make extremophiles valuable resources for various scientific and industrial purposes.

Categories such as thermophiles, halophiles, acidophiles, and alkaliphiles possess unique traits that enable them to thrive in specific habitats (Fig. 4.1). These specialized traits not only deepen our understanding of the resilience of life but also offer significant biotechnological potential. Enzymes from extremophiles, commonly referred to as "extremozymes," exhibit exceptional stability and efficiency under extreme conditions that would inactivate conventional enzymes (Atif et al. 2024). These properties make them invaluable for industrial applications such as high-temperature manufacturing, biofuel production, and environmental cleanup, emphasizing the economic and ecological importance of extremophiles.

Extremophiles produce extremozymes, specialized enzymes that function under extreme temperatures, pH levels, salinity, or pressure, making them indispensable for industrial biotechnology. In biofuel production, extremozymes facilitate the efficient breakdown of lignocellulosic biomass, enhancing bioethanol and biodiesel yields. In the pharmaceutical sector, enzymes derived from extremophiles assist in drug synthesis, improving reaction specificity and reducing the need for harsh chemical processes. Similarly, extremozymes have transformed food manufacturing by enhancing processes such as dairy fermentation, baking, and brewing, improving product quality and reducing energy consumption.

Beyond traditional industries, extremophiles contribute significantly to environmental and technological innovations. Their role in bioremediation is critical, as they degrade pollutants and toxins in contaminated sites, including high-salinity or

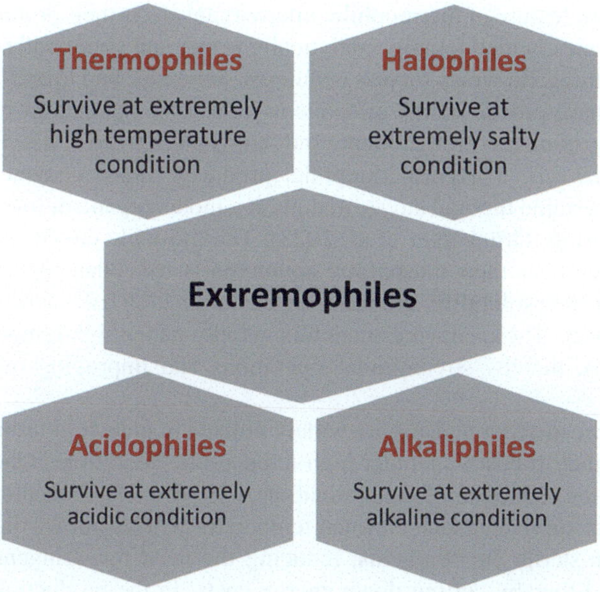

Fig. 4.1 Four major types of extremophilic microorganisms

acidic environments, where conventional microbes fail. In agriculture, extremophilic enzymes improve soil fertility and promote sustainable farming by enhancing nutrient availability and reducing dependence on chemical inputs. Moreover, extremophiles are integral to nanotechnology advancements, enabling the synthesis of novel biomaterials with customizable properties. They also offer promising solutions for space exploration, including bioremediation in extraterrestrial environments. Collectively, the versatile applications of extremophiles highlight their pivotal role in driving sustainable industrial practices, environmental protection, and technological progress.

4.2.1 Thermophiles

Thermophilic microorganisms are a unique group of extremophiles that thrive in high-temperature environments, typically ranging from 50 to 80 °C. A specialized subgroup known as hyperthermophiles can tolerate even more extreme conditions, with optimal growth temperatures exceeding 80 °C and, in certain cases, surviving temperatures up to 120 °C (Atif et al. 2024). These microorganisms are found in diverse habitats, including volcanic regions, hot springs, geothermal vents, and industrial high-temperature settings. Their ability to thrive under such extreme conditions makes them highly valuable for various industrial processes, particularly biofuel and bioenergy production.

The hallmark feature of thermophilic microorganisms is their production of thermostable enzymes, which remain catalytically active and structurally stable at elevated temperatures. Enzymes such as cellulases, amylases, and lipases derived from thermophiles have proven highly effective in industrial applications requiring high heat, such as biomass pretreatment and enzymatic hydrolysis. For example, *Pyrococcus furiosus*, a hyperthermophile, produces robust enzymes capable of withstanding extreme thermal stress, making it a model organism for studying metabolic engineering (Grünberger et al. 2023). Thermostable enzymes offer significant advantages over their mesophilic counterparts by enhancing reaction rates, improving substrate solubility, and reducing contamination risks during biofuel production processes. These enzymes maintain activity under high-temperature industrial conditions, thereby streamlining operations and improving overall process efficiency.

Thermophilic microorganisms are widely utilized in industrial fermentation processes due to their robustness under high-temperature conditions. One of their key advantages is the reduction of contamination risks, as most conventional microorganisms cannot survive at such elevated temperatures. This allows thermophiles to operate in non-sterile environments, reducing the need for stringent sterilization procedures and thereby cutting down energy costs. In the production of biofuels, thermophiles demonstrate exceptional efficiency in the breakdown and conversion of lignocellulosic biomass into fermentable substrates. Biomass pretreatment and saccharification processes, which often require high temperatures, are particularly well suited for thermophilic enzymatic systems. The use of thermophilic fermentation further minimizes cooling energy requirements in tropical and subtropical regions, where ambient temperatures are naturally higher. Engineered thermophiles have expanded the scope of biofuel production through synthetic biology and metabolic engineering. For example, recombinant *Thermoanaerobacterium saccharolyticum* has been utilized for producing n-butanol (Bhandiwad et al. 2014). Likewise, engineered *Kluyveromyces marxianus* was successfully employed for ethanol production (Löbs et al. 2017).

In conclusion, thermophilic microorganisms represent a cornerstone of modern biofuel and bioenergy production due to their unique ability to function at elevated temperatures. Their thermostable enzymes, robust metabolic pathways, and structural resilience make them ideal candidates for industrial applications requiring high efficiency, low contamination risk, and reduced energy consumption. Advances in synthetic biology and metabolic engineering have further expanded the capabilities of thermophiles, enabling them to serve as versatile platforms for next-generation biofuel production. By harnessing the remarkable traits of thermophilic microorganisms, industries can achieve cost-effective, sustainable, and scalable solutions to meet global energy demands.

4.2.2 *Halophiles*

Halophilic microorganisms, often referred to as "salt-loving" organisms, are a unique group of extremophiles that require elevated salt concentrations for optimal growth. These microorganisms demonstrate exceptional metabolic versatility and robustness, enabling them to thrive in a wide range of salinity as well as alkaline conditions. Their remarkable adaptability has made halophiles increasingly valuable for industrial biotechnology applications, including the production of bioenergy and biofuels (Fig. 4.2).

Among the halophiles, species within the *Halomonas* genus, have emerged as highly promising candidates for NGIB due to their ability to grow under extreme conditions that inhibit the growth of contaminating organisms (Chen et al. 2018). Likewise, *Halomonas* spp. have been considered to be promising candidates for bioenergy and bioproducts due to these unique physiological traits and capacity for cost-effective bioproduction processes (Park et al. 2023). Notably, *Halomonas* spp. can flourish in environments with high concentrations of NaCl and alkaline pH levels. This unique trait minimizes the need for sterile operating conditions, reducing process complexity and associated costs. For example, large-scale fermentations using *Halomonas* spp. in seawater-based artificial media have been successfully maintained for extended periods—up to 2 months—without any contamination,

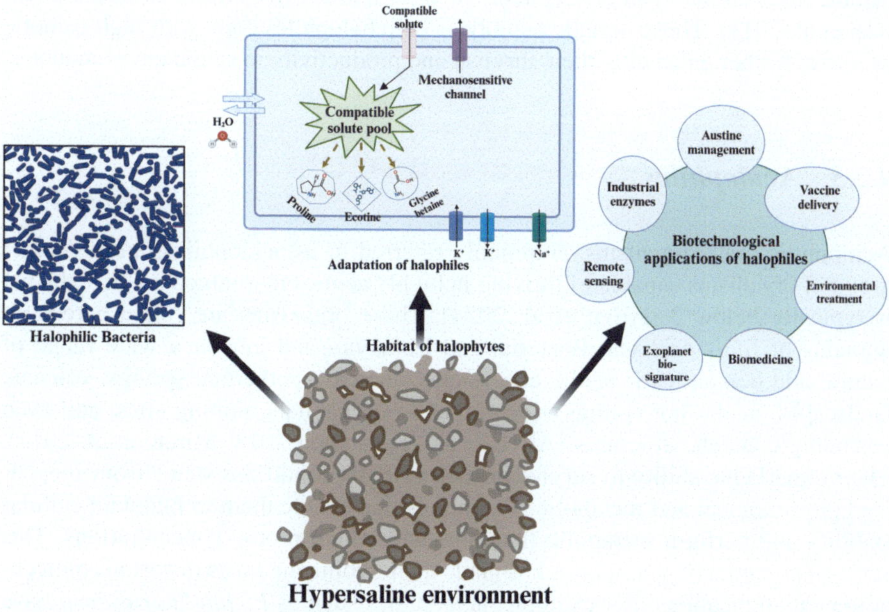

Fig. 4.2 Overview of habitats, adaptive mechanisms, and biotechnological applications of halophiles (redrwan with modification from Dutta and Bandopadhyay (2022))

showcasing their resilience and suitability for continuous, open-culture systems (Ye et al. 2018).

Within the species of halophiles, *H. bluephagenesis* has gained particular attention as a robust chassis organism for industrial-scale biomanufacturing. For example, this haloalkaliphilic bacterium grows rapidly under high-salt (e.g., 60 g/L NaCl) and alkaline (pH ~9) conditions, achieving significant biomass accumulation (Tan et al. 2011). Without the need for genetic modifications, this *H. bluephagenesis* strain provided a yield of polyhydroxybutyrate (PHB) as target product up to 80 wt% of the cell dry weight under non-sterile fermentation conditions. Another notable example of *Halomonas* species is *H. campaniensis*, a moderate halophile that can produce PHB under non-sterile conditions using inexpensive waste-based mixed substrates (Yue et al. 2014). These examples underscore the broad applicability of *Halomonas* spp. for producing renewable chemicals, biofuels, and biopolymers in economically viable, low-cost fermentation systems.

The growing interest in *Halomonas* spp. has led to significant advancements in understanding their genomes. Modern sequencing technologies have enabled the analysis of halophile genomes, providing insights into their stress resistance mechanisms and pathways for metabolic engineering. For instance, the genome of *H. bluephagenesis* TD01 has been sequenced, revealing critical information about its genetic makeup and potential for rational strain engineering. The total genome length of this strain is approximately 4.1 Mb with a GC content of 53.23% (Cai et al. 2011). Genes encoding essential components for osmotic regulation, such as betaine and ectoine synthesis systems (*betABI* and *ectABC*), have been identified (Cai et al. 2011). These organic osmolytes help halophiles cope with high osmotic pressure, further enhancing their survival and productivity in saline environments.

4.2.3 Acidophiles

Acidophilic microorganisms, commonly referred to as acidophiles, are a distinct group of organisms capable of thriving in highly acidic environments where the pH is typically below 3 (Aliyu et al. 2024). These organisms are found across all domains of life—Archaea, Bacteria, and Eukarya—and inhabit a wide range of natural and human-made acidic ecosystems, such as geothermal springs, volcanic landscapes, acidic hot springs, sulfuric acid-rich habitats, mining sites, and even corroding concrete structures (Aliyu et al. 2024; Oren 2018; Mirete et al. 2017). Their remarkable ability to survive in such extreme conditions stems from specialized physiological and metabolic adaptations that enable them to maintain cellular stability and perform metabolic functions under high proton concentrations. This survival potential of acidophiles has made them promising hosts in various biotechnological applications. For example, acidophiles such as *Pichia kudriavzevii* have been engineered to produce commercially valuable organic acids under acidic conditions (Xi et al. 2023). Likewise, acidophiles are highly suited for lignocellulosic biomass hydrolysis in biofuel production. Their acid-tolerant enzymes ensure efficient breakdown of complex plant polymers into fermentable sugars under acidic conditions (Long et al. 2024; Ashokkumar et al. 2022).

Acidophiles are highly diverse and perform various ecological roles depending on their metabolic strategies. Their categorization is based on their energy and nutrient sources, such as chemolithotrophic, heterotrophic, photoautotrophic, and thermoacidophilic acidophiles. Chemolithotrophic acidophiles oxidize inorganic compounds such as sulfur and iron to derive energy. Examples include *Acidithiobacillus ferrooxidans*, a chemolithotrophic bacterium known for its applications in biomining and metal recovery (Kondrat'eva et al. 2012). Heterotrophic acidophiles rely on organic compounds for energy and play a crucial role in organic matter degradation in acidic habitats (Johnson and Roberto 1997; González et al. 2024). Photoautotrophic acidophiles include eukaryotes like *Dunaliella acidophila* and *Chlamydomonas acidophila* that perform photosynthesis at extremely low pH levels, as low as 0–1 (Mirete et al. 2017). Thermoacidophilic archaea, such as *Sulfolobus* spp., *Acidanus* spp., and *Picrophilus oshimae*, thrive in environments with both high temperatures and low pH levels (Schocke et al. 2019; Razia et al. 2023). For instance, *Sulfolobus* and *Acidanus* species grow at pH levels of 1–5 and temperatures reaching 96 °C, while *Picrophilus oshimae* can withstand pH levels below 0 and temperatures up to 65 °C (Oren 2018).

The ability of acidophilic microorganisms to survive in extremely low pH environments is due to an array of cellular, molecular, and biochemical adaptations. These specialized mechanisms protect their cellular machinery from acid-induced damage while maintaining intracellular homeostasis. Firstly, many acidophiles sustain an exceptionally steep proton gradient across their cytoplasmic membranes, often spanning up to five orders of magnitude (Atif et al. 2024). This gradient prevents the excessive influx of protons, which could otherwise disrupt the delicate balance of intracellular pH. Secondly, unlike neutrophiles and alkaliphiles, acidophiles frequently maintain a positive charge inside their cells. This reversed membrane potential counteracts the inward flow of protons, providing additional stability to the intracellular environment (Baker-Austin and Dopson 2007). Thirdly, acidophiles are equipped with highly proton-impermeable cell membranes. This crucial property allows them to maintain a near-neutral intracellular pH despite the acidic conditions of their surroundings (Baker-Austin and Dopson 2007). Fourthly, acidophiles actively expel excess protons through specialized transport systems such as H^+ ATPases and antiporters (Atif et al. 2024; Rao et al. 2022). These systems efficiently pump protons out of the cell, further aiding in pH regulation. Fifthly, these microorganisms produce buffering amino acids and chaperone proteins that shield cellular components from acid-induced damage (Baker-Austin and Dopson 2007). These molecules play a critical role in stabilizing intracellular pH and safeguarding essential enzymes and cellular structures. Finally, acidophiles synthesize extracellular enzymes that remain functional and exhibit optimal activity in highly acidic conditions. For instance, carboxylesterase enzymes derived from *Ferroplasma acidiphilum* are highly active at pH 2, making them particularly suited for processes like lignocellulosic biomass hydrolysis in acidic biorefineries (Ye et al. 2023).

These adaptations collectively enable acidophilic microorganisms to not only survive but also thrive in environments that are inhospitable to most forms of life.

4.2.4 Alkaliphiles

Alkaliphilic microorganisms, or alkaliphiles, are a unique group of extremophiles that thrive in environments with high pH values, typically above pH 9, with an optimum growth pH at around 10 (Atif et al. 2024). Alkaliphilic microorganisms are predominantly found in alkaline soda lakes, such as those in the Great Rift Valley in East Africa, as well as in soda deserts and soils undergoing ammonification. Interestingly, some alkaliphiles have also been isolated from non-alkaline environments, indicating the presence of micro-niches that support their survival (Kulkarni et al. 2019; Mamo and Mattiasson 2016). Alkaliphiles are further categorized into two types based on their growth conditions: obligate alkaliphiles, which can only grow at pH levels above 9, and facultative alkaliphiles, which prefer alkaline conditions but can also survive in near-neutral pH environments (Preiss et al. 2015).This diversity of alkaliphiles presents significant biotechnological potential, particularly in the production of industrially valuable enzymes, antibiotics, exopolysaccharides, and polyhydroxyalkanoates (Kulkarni et al. 2019). These microorganisms produce enzymes that exhibit exceptional stability and functionality in high-pH environments, making them ideal for industrial applications such as detergent formulations, biorefining, and waste management (Tuly et al. 2022).

To survive under extreme alkaline conditions, alkaliphilic microorganisms have evolved specialized adaptations to maintain cellular integrity and pH homeostasis. These include pH regulation, protective outer layer, stabilized membrane proteins, and adapted membrane properties (Mamo and Mattiasson 2016). Alkaliphiles utilize secondary cation/proton antiporters to regulate proton uptake and maintain a balanced cytoplasmic pH, preventing the detrimental effects of the highly alkaline external environment (Krulwich et al. 1997). Their cell walls often contain acidic compounds such as teichuronic acid and teichuronopeptides, which create a protective barrier to shield the cell from extreme pH (Aono et al. 1999). Some alkaliphiles produce cytochrome C on their outer membranes to stabilize proton flow and ensure membrane integrity under high-pH conditions (Atif et al. 2024). In alkaliphiles like *Bacillus* spp., proton-coupled ATP synthases operate efficiently despite low proton-motive force, highlighting their evolutionary adaptations to optimize energy production (Krulwich et al. 1998). These adaptations enable alkaliphiles to flourish in environments that are otherwise inhospitable, making them invaluable for various industrial and environmental applications.

4.2.5 *Other Extremophilic Microorganisms*

In addition to the large extremophilic microorganisms mentioned above, certain other microorganisms exhibit remarkable resistance to extreme environments, such as toxic substances, including both inorganic (Cu, Zn, Hg, Cr, etc.), and organic compounds (various alcohols, fatty acids, and aldehydes) (Ye et al. 2023). Among them, microbes that can thrive under heavy metal conditions are often explored for their potential in bioremediation due to their ability to metabolize toxic organic compounds (Giovanella et al. 2020). Acetogenic bacteria represent another unique group capable of converting gaseous substrates, such as carbon monoxide-rich syngas, into valuable products like ethanol and acetate (Molitor et al. 2016). This process, often employed in integrated biorefineries, is gaining attention as a sustainable method for utilizing industrial emissions. Another group of extremophiles with promising applications is methanotrophs. These bacteria are capable of oxidizing methane, a potent greenhouse gas, and play a significant role in mitigating its impact on global warming. These microbes are utilized in bioreactors designed to reduce methane emissions from landfills and coal mines. Additionally, methanotrophs are being explored for their potential to degrade other toxic substances, demonstrating their versatility in environmental bioengineering (Jiang et al. 2010). Some microorganisms are also highly resistant to radiation, called radioresistant microbes, which are known for their ability to survive under radiation levels that are lethal to most conventional life forms. Their resilience is often attributed to protective cellular mechanisms, such as the accumulation of specialized proteins, pigments, and other molecules that mitigate oxidative stress caused by reactive oxygen species during radiation exposure (Gabani and Singh 2013). For example, certain enzymes involved in DNA repair in these organisms play a crucial role in maintaining cellular integrity under such extreme conditions (Ye et al. 2023).

4.3 Tools for Engineering Extremophiles

4.3.1 *Vectors and Expression Systems for Extremophiles*

Genetic engineering in extremophilic microorganisms has seen significant progress, driven by the development of specialized plasmids, vectors, and expression systems. For instance, plasmids and vectors have been instrumental in manipulating genes in organisms such as *Acidithiobacillus ferrooxidans* (Jung et al. 2022). Early genetic studies of *A. ferrooxidans* initially borrowed techniques established for *Escherichia coli*, but the physiological differences between the two organisms necessitated new methods. The discovery of native plasmids like pTF1 and pTF-FC2 in the 1980s led to self-transmissible and mobilizable plasmids, allowing the creation of chimeric constructs capable of maintaining and expressing *A. ferrooxidans* genes in *E. coli* (Rawlings and Woods 1985). This breakthrough laid the foundation for the development of conjugation-based gene transfer systems. One such system involved the mobilizable shuttle plasmid pJRD215, which shared similarities with the RSF1010 plasmid (Peng et al. 1994). This vector enabled effective replication and expression of exogenous genes in *A. ferrooxidans*. Optimization of conjugation techniques, including adjusting growth media and mating conditions, enhanced plasmid transfer efficiency, as methods like electroporation proved less reliable due to poor reproducibility. Selectable markers like kanamycin and streptomycin facilitated these advancements, while antibiotics such as ampicillin were less effective due to rapid degradation or limited resistance gene expression in the host cells (Jung et al. 2022).

In parallel, significant progress has been made in engineering other extremophilic microorganisms, such as *Halomonas* spp. The development of a T7-like expression system tailored for *Halomonas bluephagenesis* addressed the need for highly inducible and efficient promoters (Ma et al. 2020). Traditional T7 RNA polymerase (RNAP) systems performed sub-optimally in these organisms, prompting genome screening efforts that identified alternative T7-like systems such as K1F, VP4, and MmP1. These systems, integrated into the genome and controlled by lacI-regulated P_{tac} promoters, minimized expression leakage prior to induction and achieved remarkable performance metrics (Chen et al. 2018). This T7-like system has found applications in industrial-scale processes. For instance, it enabled the overexpression of the *phaCAB* operon, critical for polyhydroxyalkanoate (PHA) production. In a 7 L bioreactor setup, PHA accumulation reached an impressive 92% of the cell dry weight, demonstrating the system's potential for biomanufacturing (Ma et al. 2020). These advancements collectively underscore the importance of tailored genetic tools and expression systems in harnessing the biotechnological potential of extremophilic microorganisms for diverse applications, including environmental remediation and industrial-scale bioproduction. Another plasmid named "Standard European Vector Architecture (SEVA)" plasmid system has been developed for *Halomonas* spp., featuring three core components: a selective marker (typically an antibiotic resistance gene), a replication origin, and a cargo region (Silva-Rocha et al. 2013). With the addition of an *oriT* sequence for conjugation, SEVA plasmids serve as versatile tools for genetic engineering in organisms like *Halomonas* spp. (Chen et al. 2018).

4.3.2 CRISPR/Cas9-Based Tools

The CRISPR/Cas9 system has transformed microbial genome editing by providing precise and effective methods for targeted genetic modifications (Fig. 4.3). This technology is increasingly being applied to extremophilic microorganisms to optimize their metabolic processes for industrial and biotechnological purposes. Additionally, CRISPR/Cas9 has demonstrated significant efficiency in introducing large genes, which expand the biosynthetic capabilities of extremophilic microbes. Beyond its precision, CRISPR/Cas9 simplifies and accelerates genome editing,

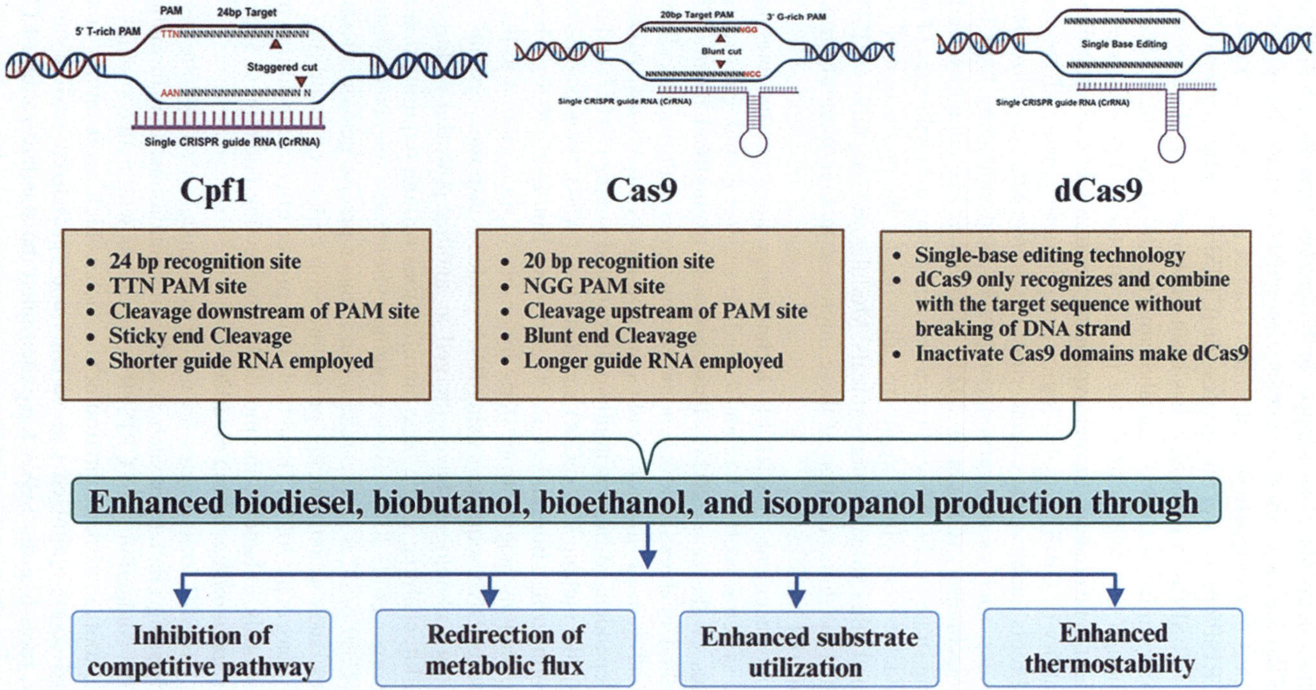

Fig. 4.3 CRISPR/Cas-based tools for engineering extremophiles for the purpose of producing biofuels (redrawn from Garg et al. (2023))

enabling scientists to develop new metabolic pathways and adapt extremophiles for various applications, including biofuel production and the synthesis of bioplastics and high-value chemicals. For instance, in *Halomonas* species, CRISPR interference (CRISPRi) has been used to enhance biopolymer yields by repressing key metabolic genes that compete for substrates (Tao et al. 2017). For transcriptional regulation, CRISPR tools offer both repression (CRISPRi) and activation (CRISPRa) capabilities. CRISPRi, in particular, has been instrumental in silencing metabolic pathways to optimize microbial metabolism (Gurdo et al. 2022). Meanwhile, CRISPRa, though less widely used due to its complexity, has shown promise in activating silent gene clusters, further expanding the utility of extremophiles.

The versatility of CRISPR technologies extends further, with advancements in RNA-guided tools enabling the engineering of extremophiles to utilize unconventional substrates, such as one-carbon compounds, and enhancing their utility in industrial biotechnology (Garg et al. 2023). Innovations like counter-selection and gene silencing strategies have been pivotal in overcoming the challenges of working with non-model extremophilic organisms, allowing for the production of biofuels, bioplastics, and other valuable biochemicals. Modular and adaptable CRISPR tools have facilitated the domestication of these microbes, making them integral to sustainable industrial workflows.

Another advancement in CRISPR technologies involves the integration of recombinase systems, such as Cre and λ-Red, to support homologous recombination processes (Volke et al. 2023). This approach, exemplified by the CRAGE-CRISPR method, enables scarless genome editing across multiple loci, allowing for both loss- and gain-of-function studies in non-model organisms. Additionally, template-independent editing via nonhomologous end joining (NHEJ) and modified Cas proteins, like nicking Cas (nCas) and dead Cas (dCas), broadens the scope of microbial engineering (Bertrand et al. 2019; Finger-Bou et al. 2020). These tools facilitate applications ranging from steric hindrance of transcription to the precise insertion of large DNA sequences using RNA-guided transposase systems.

Base-editing technologies have emerged as another powerful addition to the CRISPR toolkit, enabling single-base modifications without requiring donor DNA templates (Komor et al. 2016). By leveraging cytidine and adenosine deaminases fused with nCas9 or dCas9, researchers can achieve precise genetic alterations in extremophiles, such as introducing premature stop codons or modifying biosynthetic genes (Garg et al. 2023). These tools have proven effective in editing multiple loci simultaneously, enhancing the metabolic engineering potential of species like *Rhodobacter sphaeroides* (Luo et al. 2020).

Adapting CRISPR technologies for use in non-model extremophilic organisms requires addressing challenges such as culturing conditions, transformation efficiency, and the compatibility of CRISPR components with native systems. Efforts to tailor these tools to extreme environments, such as high temperatures or salinity, have included codon-optimized Cas variants and specialized auxiliary enzymes. Additionally, tools like the SEVA have facilitated the standardization of CRISPR systems, ensuring their compatibility with diverse microbial hosts. The integration of CRISPR/Cas systems in extremophilic microbial engineering underscores their transformative potential. By enabling precise genetic modifications, these tools open new avenues for sustainable biomanufacturing, leveraging the unique properties of extremophiles to address challenges in industrial biotechnology.

4.3.3 Omics-Based Tools

The term "omics" encompasses advanced methodologies such as genomics, transcriptomics, proteomics, and metabolomics, which have revolutionized the discovery of enzymes with exceptional efficiency and versatility. By combining omics with bioinformatics and computational techniques, researchers have developed methods like mutagenesis and genetic shuffling to improve enzyme stability and robustness. One significant challenge in utilizing extremophiles for various biotechnological applications is the difficulty of culturing many of these microorganisms, which limits access to their unique enzymatic capabilities. Omics-based approaches address this limitation by enabling the identification of enzymes and metabolic pathways directly from unculturable microbes (Atif et al. 2024). Over the last two decades, genome sequencing has been instrumental in enhancing our understanding of how extremophiles adapt to harsh environments. These insights have revealed mechanisms related to enzyme structure and function, metabolic processes, and substrate transformation.

Genomic studies have been instrumental in elucidating the adaptive mechanisms of extremophiles. For instance, genomic analysis of *Comamonas* species has uncovered multiple pathways for lignin degradation, highlighting its potential for industrial applications (Yancey et al. 2014). Advances in genomics have been further bolstered by evolutionary techniques such as directed evolution and synthetic biology, which, when combined with computational modeling, have accelerated the discovery of industrially relevant enzymes. Genomic data from thermophilic organisms, including *Thermotoga*, *Thermus*, and *Pyrococcus*, have revealed numerous enzymes suited for commercial applications, with over 120 thermophilic species cataloged in public databases (Atif et al. 2024).

Transcriptomics, the study of RNA expression, provides detailed information on gene regulation and expression patterns, enabling researchers to understand how extremophiles adapt to challenging conditions. For example, transcriptomic analysis of *Deinococcus radiodurans* under heat stress revealed significant transcriptional reprogramming, identifying key functional genes, including DR_0127, DR_1083, and DR_1325, as crucial contributors to heat tolerance (Xue et al. 2019). This study highlights the adaptive mechanisms of this extremophilic strain to maintain cellular homeostasis under high-temperature conditions, providing new insights into its heat stress response. Similarly, analyses of termite gut systems have identified enzymes and regulatory networks involved in lignin breakdown, offering valuable insights for biomass conversion in industrial processes (Geng et al. 2018). Transcriptional engineering, which modifies gene expression to enhance enzyme activity and stability, has shown promise in improving enzymes such as mutant α-amylases from halophilic and thermotolerant *Bacillus* strains for industrial applications (Bibra et al. 2017).

Proteomics focuses on protein expression and function, providing a deeper understanding of the survival strategies of extremophiles in extreme conditions. For example, proteomic studies on *Bacillus ligniniphilus* have identified enzymes involved in lignin degradation, such as oxidoreductases and peroxiredoxins (Zhu et al. 2017), while analyses of *Oleispira antarctica* have highlighted proteins associated with hydrocarbon metabolism (Gregson et al. 2020). These findings have significant implications for biotechnological applications, as they enable the optimization of enzyme properties—such as thermal stability, substrate specificity, and solvent tolerance—through recombinant protein engineering and modification.

The integration of genomics, transcriptomics, and proteomics has provided a comprehensive framework for understanding the metabolic pathways and regulatory networks of extremophiles. By combining insights from these omics disciplines, researchers can identify novel metabolic targets and develop optimized microbial strains for industrial use. For example, multi-omics approaches have been used to enhance enzyme production, improve microbial strain performance, and overcome the challenges associated with unculturable microorganisms (Zhu et al. 2020). These advancements hold the potential to drive sustainable bioprocesses, such as biofuel production and biorefinery operations.

Omics technologies have transformed our approach to studying extremophilic microorganisms by unveiling their unique adaptive mechanisms and metabolic capabilities. The integration of these methodologies with computational tools and synthetic biology has facilitated the discovery and optimization of enzymes for industrial applications. Through these advancements, extremophiles are becoming increasingly important for sustainable and efficient biotechnological solutions, paving the way for innovations in fields such as bioremediation, biofuels, and biopolymer production.

4.3.4 AI-Based Tools

The integration of artificial intelligence (AI) into the engineering of extremophilic microorganisms is revolutionizing the field of synthetic biology, enabling the development of efficient and cost-effective solutions for biomanufacturing. Traditional methods of pathway optimization, often reliant on static strategies and high-throughput construct designs, face challenges such as high costs and prolonged research cycles. AI-based tools, including machine learning (ML) and deep learning (DL) models, are addressing these limitations by offering dynamic, data-driven approaches to optimize multigene regulatory networks and fine-tune metabolic pathways. For instance, AI has facilitated the rational flux-tuning of *Halomonas bluephagenesis* for the co-production of bioplastic polyhydroxybutyrate (PHB) and ectoine, significantly reducing the time and resources needed for optimization (Ma et al. 2020). By leveraging machine learning algorithms to analyze vast datasets, researchers can rapidly identify optimal pathways and improve strain performance.

Deep learning, in particular, is emerging as a transformative tool in the genomic analysis of extremophiles like *Halomonas* species (Chen et al. 2022a). Unlike traditional ML models that rely on small-scale mathematical frameworks with limited predictive capacity, DL employs highly complex neural network architectures containing hundreds of thousands of parameters. These sophisticated models are capable of decoding intricate biological mechanisms and predicting the functions of bioproduct synthesis with unprecedented accuracy. For example, DL is increasingly being used to analyze genomic and proteomic data to uncover the underlying mechanisms of bioproduct synthesis in *Halomonas* spp., enabling researchers to design strains with enhanced production capabilities (Zhao et al. 2024; Chen et al. 2022a). As the field advances, AI-based approaches will play a critical role in accelerating the discovery and engineering of extremophiles, paving the way for innovative applications in industrial biotechnology.

4.4 Engineering Extremophilic Microorganisms

4.4.1 Engineering Halophiles

Recent progress in synthetic biology and genetic engineering has expanded the potential of extremophiles, particularly halophilic species, as platforms for sustainable and efficient biomanufacturing, including the generation of biofuels and bioenergy. Various genome engineering tools optimized for halophilic strains such as CRISPR/Cas9, RNA interference (RNAi), and expression systems like the SEVA plasmid library enable precise modifications of their genomes (Qin et al. 2018; Wang et al. 2022; Chen et al. 2018). A key breakthrough for *Halomonas* spp. has been the development of T7-like expression systems, which mimic the highly efficient T7 system in *Escherichia coli*. These systems employ an RNA polymerase controlled by a tac promoter and a target gene driven by a T7-like promoter, regulated by *lac* operon elements and inducible with IPTG (Zhao et al. 2017). Such innovations ensure the controlled, high-level expression of genes in *Halomonas* spp., supporting their scalability for industrial applications.

Robust promoters like the porin promoter (P_{porin}) integrate target genes directly into the genome, eliminating the need for plasmid maintenance and antibiotics. Introducing ribosome-binding sites (RBS) has further improved translational efficiency. These approaches allow for gene expression modulation, deletion of nonessential genes, and metabolic pathway rewiring, which enhance the production of target products. For example, ***H. bluephagenesis*** has been engineered using advanced techniques such as CRISPR interference (CRISPRi) and RNA interference (RNAi) technologies to significantly improve target product yields, cell morphology, and the compositional modifications of the product (Tao et al. 2017; Wang et al. 2022).

Beyond genetic modifications, engineering strategies focused on regulatory elements and physiological processes further optimize extremophile performance. Promoter and terminator engineering, along with T7-like induction systems, enables precise control of gene expression, enhancing yield and metabolic efficiency (Shen et al. 2018; Zhao et al. 2017; Xu et al. 2022a). Metabolic engineering approaches, such as NADH/NAD+ balancing and oxygen uptake enhancement, have been instrumental in improving redox balance and metabolic activity, increasing bioproduct accumulation (Ling et al. 2018). Surface and morphological modifications, including the deletion of outer membrane-related genes, facilitate downstream processing, improve self-flocculation, and simplify recovery (Xu et al. 2022b). Furthermore, innovative coculture systems and conjugation methods harness the synergy between different extremophile strains, paving the way for higher yields of industrially relevant chemicals like cadaverine (Fu et al. 2014; Zhao et al. 2022a). In addition to optimizing metabolic pathways, genome reduction in extremophiles has proven to be effective. For instance, deleting large nonessential gene clusters, such as the flagellar operon in *Halomonas* spp., has reduced energy expenditure, enhancing metabolic output and bioproduct synthesis (Xu et al. 2022b). Innovative metabolic strategies, such as "stimulus-response-based flux-tuning" (SR-FT), further enhance the synthesis of complex biomolecules (Ye et al. 2020). For example, *Halomonas bluephagenesis* has been engineered using SR-FT to fine-tune metabolic fluxes and increase ectoine production from glucose. Coupled with machine learning tools for pathway optimization, these approaches reduce development cycles and enhance production efficiencies (Ma et al. 2020).

4.4.2 Engineering Thermophiles

The engineering of thermophilic microorganisms for the production of bio-based fuels and chemicals leverages their unique physiological and metabolic traits, which are well suited to high-temperature environments. One major advantage of utilizing thermophiles is their ability to facilitate contamination-free bioprocessing, as their extreme growth conditions naturally inhibit the proliferation of contaminants. Due to their relatively small genomes, thermophiles were among the first organisms to be fully sequenced, paving the way for the application of systems biology to explore their unique features (Chen and Jiang 2018). This genomic information, combined with the availability of genetic manipulation tools, has enabled the transition from basic discovery to practical applications in metabolic engineering.

Recent advancements in genetic tools have enabled these microorganisms to serve as effective hosts for metabolic engineering, offering a high-temperature alternative to traditional mesophilic platforms. The genetic modification of extreme thermophiles reveals substantial potential for industrial biofuel synthesis, especially in converting lignocellulosic substrates to bioethanol. While thermophilic bacteria exhibit efficiency, adaptability, and lower production costs, the commercial-scale use of these microbes faces several challenges. For instance, high temperatures enhance cellulose breakdown and enzyme stability but inhibit the growth of ethanol-producing microbes. However, their application in metabolic engineering is still in its infancy. Advances in genome editing, particularly using CRISPR/Cas9, have enabled the modification of thermo-resistant proteins, paving the way for enhanced production of the target products (Mandal et al. 2024). For example, a secretory protein expression system has been developed in *Geobacillus thermoglucosidasius* (Bartosiak-Jentys et al. 2013). More recently, a novel heterologous gene expression system, driven by a silica-inducible promoter, has been established in *Thermus thermophilus*, expanding the potential for biotechnological applications (Fujino et al. 2020).

Thermotolerance, which is a multigene trait of thermophiles, has been reported to be linked to the accumulation of heat shock proteins (HSPs) and trehalose (Garg et al. 2023). Prominent thermophilic microorganisms have demonstrated significant progress through genetic engineering. For example, *Clostridium bescii* can ferment diverse cellulosic biomass into fuels. Genetic modifications, such as the overexpression of substrate-binding proteins like tapirins, have improved lignocellulose degradation (Lee et al. 2018). Furthermore, the introduction of heterologous pathways, such as the *AdhE* gene, has increased ethanol production from switchgrass (Williams-Rhaesa et al. 2018). Another thermophile, *Thermococcus kodakarensis*, has shown potential in hydrogen gas production from renewable carbon sources, and genetic engineering in this strain has further enhanced hydrogen yields (Cao et al. 2022). Similarly, *Pyrococcus furiosus* has been metabolically engineered to produce butanol (Keller et al. 2015). Despite low yields, the introduction of heterologous enzymes and novel pathways has improved its efficiency, demonstrating its potential as a metabolic engineering host.

Overall, engineering thermophilic microorganisms offers a promising pathway for advancing biofuel production. While significant progress has been made, further innovations in genome editing and metabolic engineering are needed to overcome existing challenges and fully harness the potential of these extremophiles in industrial biotechnology.

4.4.3 Engineering Acidophiles

Engineering acidophilic microorganisms has gained significant attention due to their remarkable ability to thrive in extremely acidic environments, presenting opportunities for industrial and biotechnological advancements. These microorganisms naturally exhibit robust mechanisms to withstand high proton concentrations, such as specialized membrane structures and efficient proton efflux systems. Genetic engineering efforts have focused on enhancing these inherent traits while introducing novel functionalities for industrial applications. The integration of CRISPR/Cas9 technology and other advanced genome-editing tools has enabled precise genetic modifications in acidophiles, facilitating metabolic rewiring and optimizing pathways for the production of biofuels and value-added products (Chen et al. 2022b; Mandal et al. 2024). Additionally, the development of plasmid-based and chromosomal expression systems, tailored to acidophilic hosts, allows stable and efficient gene expression under low pH conditions (Farkas et al. 2013; Inaba et al. 2018). Another critical aspect of acidophile engineering involves the optimization of regulatory elements, such as promoters and terminators, to improve metabolic efficiency (Öztürk et al. 2017; Peng et al. 2012). For instance, engineered promoter systems have been employed to drive the high-level expression of genes encoding acid-stable enzymes, which are of significant interest in industries such as bioenergy (Öztürk et al. 2017).

Among the acidophiles, acidophilic chemolithoautotrophic microorganisms, such as those belonging to the *Acidithiobacillus* genus, have been extensively studied and genetically modified to enhance their metabolic capabilities and environmental adaptability (Jung et al. 2022). Various *Acidithiobacillus* species, including *A. thiooxidans*, *A. caldus*, and *A. ferrooxidans*, have undergone gene knockout, recombinant expression, overexpression, and complementation techniques to target specific genes linked to their sulfur-oxidizing and iron-oxidizing functions. For instance, in *A. thiooxidans*, genes like *cueO*, which encodes a putative multicopper oxidase, and *pelD*, associated with c-di-GMP signaling, have been knocked out to study their roles (Wen et al. 2014; Díaz et al. 2018). In *A. caldus*, genetic modifications have focused on a wide range of genes, such as *sor* (sulfur oxygenase reductase), *arsB* (arsenic resistance), and *tetH* (tetrathionate hydrolase) (Chen et al. 2012; van Zyl et al. 2008). Techniques such as overexpression and recombinant expression of genes like *eGFP* (enhanced green fluorescent protein) and arsenic resistance operons have provided insights into protein localization and resistance mechanisms of *A. caldus* in extreme environments (Zhang et al. 2014). For iron- and sulfur-oxidizing microorganisms like *A. ferrooxidans*, genetic modifications have targeted genes essential for energy generation and stress response. Examples include knocking out the *qs-I* operon, which encodes a LuxI/R-like quorum-sensing system, and the *tetH* gene (Gao et al. 2020; Yu et al. 2014). Collectively, these genetic engineering efforts demonstrate the potential of *Acidithiobacillus* spp. to serve as robust models for studying extremophile biology and developing biotechnological applications.

4.4.4 Engineering Alkaliphiles

Engineering alkaliphilic microorganisms has emerged as a promising area of research due to their natural ability to thrive in high pH environments, making them valuable for diverse industrial applications (Mamo and Mattiasson 2016). These microorganisms possess unique adaptations, such as specialized ion transport systems and modified cell walls, to maintain intracellular pH homeostasis under alkaline conditions (Kulkarni et al. 2019). Genetic engineering approaches have leveraged these traits to enhance the metabolic capabilities of alkaliphiles, enabling the efficient production of biotechnologically important compounds. The application of advanced genetic tools, such as CRISPR/Cas systems, has significantly improved the genetic manipulation of alkaliphilic microorganisms, allowing precise editing of their genomes (Khot et al. 2023; Huang et al. 2022). These technologies facilitate the optimization of metabolic pathways, leading to increased production yields of target metabolites, including those used as biofuels. Additionally, tailored promoter systems and regulatory elements have been developed to ensure stable and high-level gene expression in alkaline environments, further enhancing the efficiency of bioprocesses (Ojala et al. 2011; Horikoshi 2021).

4.5 Conclusion

In conclusion, extremophiles and their associated extremozymes represent a transformative force across diverse industries, from environmental remediation to sustainable manufacturing and advanced biotechnologies. These organisms thrive in extreme conditions, offering robust and versatile biocatalysts that enable processes where traditional systems fail. The integration of synthetic biology, systems biology, and advanced genetic engineering has significantly expanded the scope of extremophile applications, with organisms like *H. bluephagenesis* emerging as industrial chassis for low-cost and contamination-resistant biomanufacturing. Developments in CRISPR/Cas technologies, omics-based tools, and dynamic regulatory systems have enhanced metabolic engineering and strain optimization, unlocking new possibilities for high-value product generation.

Despite these advancements, several challenges persist, including the scalability of cultivation under extreme conditions, genetic manipulation inefficiencies, and contamination risks in non-sterile processes. Overcoming these hurdles requires innovative approaches, such as the development of phage-resistant strains, advanced bioreactor materials, and machine learning-driven metabolic flux optimization. Furthermore, exploring unculturable extremophiles and their enzymatic diversity remains an untapped opportunity, offering a vast potential for novel biotechnological applications.

Looking ahead, extremophiles hold promise for addressing global challenges, particularly bioenergy and biofuel production. The adoption of integrated frameworks like the "design-build-test-learn" (DBTL) approach, combined with advancements in molecular biology, will pave the way for the rational engineering of extremophiles. These efforts are crucial not only for improving current industrial practices but also for driving innovation in emerging fields such as space exploration and nanotechnology. As research continues to bridge the gap between laboratory-scale discoveries and industrial-scale applications, extremophiles are poised to remain at the forefront of biotechnological innovation, fostering a sustainable and resilient future.

References

Aliyu GO, Ezugworie FN, Onwosi CO, Nnamchi CI, Ekwealor CC, Igbokwe VC, Sani RK (2024) Multi-stress adaptive lifestyle of acidophiles enhances their robustness for biotechnological and environmental applications. Sci Total Environ 954:176190. https://doi.org/10.1016/j.scitotenv.2024.176190

Aono R, Ito M, Machida T (1999) Contribution of the cell wall component teichuronopeptide to pH homeostasis and alkaliphily in the alkaliphile Bacillus lentus C-125. J Bacteriol 181(21):6600–6606. https://doi.org/10.1128/jb.181.21.6600-6606.1999

Asase RV, Okechukwu QN, Ivantsova MN (2024) Biofuels: present and future. Environ Dev Sustain. https://doi.org/10.1007/s10668-024-04992-w

Ashokkumar V, Venkatkarthick R, Jayashree S, Chuetor S, Dharmaraj S, Kumar G, Chen W-H, Ngamcharussrivichai C (2022) Recent advances in lignocellulosic biomass for biofuels and value-added bioproducts—a critical review. Bioresour Technol 344:126195. https://doi.org/10.1016/j.biortech.2021.126195

Aslam M, Rashid N (2022) Bioenergy production in extremophiles. In: Kuddus M (ed) Microbial extremozymes. Academic Press, pp 231–246. https://doi.org/10.1016/B978-0-12-822945-3.00014-2

Atif F, Maqsood N, Ali W, Ali W, Irfan M (2024) Extremophiles and their enzymatic diversity and biotechnological potential. Syst Microbiol Biomanuf 4(3):833–849. https://doi.org/10.1007/s43393-024-00275-7

Baker-Austin C, Dopson M (2007) Life in acid: pH homeostasis in acidophiles. Trends Microbiol 15(4):165–171. https://doi.org/10.1016/j.tim.2007.02.005

Bangaru A, Sree KA, Kruthiventi C, Banala M, Shreya V, Vineetha Y, Shalini A, Mishra B, Yadavalli R, Chandrasekhar K, Reddy CN (2022) Role of enzymes in biofuel production: recent developments and challenges. In: Chowdhary P, Khanna N, Pandit S, Kumar R (eds) Bio-clean energy technologies: volume 1. Springer Nature Singapore, Singapore, pp 81–112. https://doi.org/10.1007/978-981-16-8090-8_4

Bartosiak-Jentys J, Hussein AH, Lewis CJ, Leak DJ (2013) Modular system for assessment of glycosyl hydrolase secretion in Geobacillus thermoglucosidasius. Microbiology 159(Pt_7):1267–1275. https://doi.org/10.1099/mic.0.066332-0

Bertrand C, Thibessard A, Bruand C, Lecointe F, Leblond P (2019) Bacterial NHEJ: a never ending story. Mol Microbiol 111(5):1139–1151. https://doi.org/10.1111/mmi.14218

Bhandiwad A, Shaw AJ, Guss A, Guseva A, Bahl H, Lynd LR (2014) Metabolic engineering of Thermoanaerobacterium saccharolyticum for n-butanol production. Metab Eng 21:17–25. https://doi.org/10.1016/j.ymben.2013.10.012

Bibra M, Navanietha Krishnaraj R, Sani RK (2017) An overview on extremophilic chitinases. In: Sani RK, Krishnaraj RN (eds) Extremophilic enzymatic processing of lignocellulosic feedstocks to bioenergy. Springer International Publishing, Cham, pp 225–247. https://doi.org/10.1007/978-3-319-54684-1_12

Cai L, Tan D, Aibaidula G, Dong X-R, Chen J-C, Tian W-D, Chen G-Q (2011) Comparative genomics study of polyhydroxyalkanoates (PHA) and ectoine relevant genes from Halomonas sp. TD01 revealed extensive horizontal gene transfer events and co-evolutionary relationships. Microb Cell Factories 10:1–15. https://doi.org/10.1186/1475-2859-10-88

Cao Y, Liu H, Liu W, Guo J, Xian M (2022) Debottlenecking the biological hydrogen production pathway of dark fermentation: insight into the impact of strain improvement. Microb Cell Factories 21(1):166. https://doi.org/10.1186/s12934-022-01893-3

Chen G-Q, Jiang X-R (2018) Next generation industrial biotechnology based on extremophilic bacteria. Curr Opin Biotechnol 50:94–100. https://doi.org/10.1016/j.copbio.2017.11.016

Chen L, Ren Y, Lin J, Liu X, Pang X, Lin J (2012) Acidithiobacillus caldus sulfur oxidation model based on transcriptome analysis between the wild type and sulfur oxygenase reductase defective mutant. https://doi.org/10.1371/journal.pone.0039470

Chen X, Yu L, Qiao G, Chen G-Q (2018) Reprogramming Halomonas for industrial production of chemicals. J Ind Microbiol Biotechnol 45(7):545–554. https://doi.org/10.1007/s10295-018-2055-z

Chen G-Q, Zhang X, Liu X, Huang W, Xie Z, Han J, Xu T, Mitra R, Zhou C, Zhang J (2022a) Halomonas spp., as chassis for low-cost production of chemicals. Appl Microbiol Biotechnol 106(21):6977–6992. https://doi.org/10.1042/EBC20200159

Chen J, Liu Y, Diep P, Mahadevan R (2022b) Genetic engineering of extremely acidophilic Acidithiobacillus species for biomining: Progress and perspectives. J Hazard Mater 438:129456. https://doi.org/10.1016/j.jhazmat.2022.129456

Cherwoo L, Gupta I, Flora G, Verma R, Kapil M, Arya SK, Ravindran B, Khoo KS, Bhatia SK, Chang SW, Ngamcharussrivichai C, Ashokkumar V (2023) Biofuels an alternative to traditional fossil fuels: a comprehensive review. Sustain Energy Technol Assess 60:103503. https://doi.org/10.1016/j.seta.2023.103503

Díaz M, Castro M, Copaja S, Guiliani N (2018) Biofilm formation by the acidophile bacterium Acidithiobacillus thiooxidans involves c-di-GMP pathway and Pel exopolysaccharide. Genes 9(2):113. https://doi.org/10.3390/genes9020113

Dutta B, Bandopadhyay R (2022) Biotechnological potentials of halophilic microorganisms and their impact on mankind. Beni Suef Univ J Basic Appl Sci 11(1):75. https://doi.org/10.1186/s43088-022-00252-w

El-Araby R (2024) Biofuel production: exploring renewable energy solutions for a greener future. Biotechnol Biofuels Bioprod 17(1):129. https://doi.org/10.1186/s13068-024-02571-9

Elshobary ME, Zabed HM, Yun J, Zhang G, Qi X (2021) Recent insights into microalgae-assisted microbial fuel cells for generating sustainable bioelectricity. Int J Hydrog Energy 46(4):3135–3159. https://doi.org/10.1016/j.ijhydene.2020.06.251

Farkas JA, Picking JW, Santangelo TJ (2013) Genetic techniques for the archaea. Annu Rev Genet 47(1):539–561. https://doi.org/10.1146/annurev-genet-111212-133225

Finger-Bou M, Orsi E, van der Oost J, Staals RHJ (2020) CRISPR with a happy ending: non-templated DNA repair for prokaryotic genome engineering. Biotechnol J 15(7):1900404. https://doi.org/10.1002/biot.201900404

Fokum E, Zabed HM, Guo Q, Yun JH, Yang MM, Pang H, An YF, Li W, Qi XH (2019) Metabolic engineering of bacterial strains using CRISPR/Cas9 systems for biosynthesis of value-added products. Food Biosci 28:125–132. https://doi.org/10.1016/j.fbio.2019.01.003

Fu X-Z, Tan D, Aibaidula G, Wu Q, Chen J-C, Chen G-Q (2014) Development of Halomonas TD01 as a host for open production of chemicals. Metab Eng 23:78–91. https://doi.org/10.1016/j.ymben.2014.02.006

Fujino Y, Goda S, Suematsu Y, Doi K (2020) Development of a new gene expression vector for Thermus thermophilus using a silica-inducible promoter. Microb Cell Factories 19(1):126. https://doi.org/10.1186/s12934-020-01385-2

Gabani P, Singh OV (2013) Radiation-resistant extremophiles and their potential in biotechnology and therapeutics. Appl Microbiol Biotechnol 97(3):993–1004. https://doi.org/10.1007/s00253-012-4642-7

Gao X-Y, Liu X-J, Fu C-A, Gu X-F, Lin J-Q, Liu X-M, Pang X, Lin J-Q, Chen L-X (2020) Novel strategy for improvement of the bioleaching efficiency of Acidithiobacillus ferrooxidans based on the AfeI/R quorum sensing system. Fortschr Mineral 10(3):222. https://doi.org/10.3390/min10030222

Garg D, Samota MK, Kontis N, Patel N, Bala S, Rosado AS (2023) Revolutionizing biofuel generation: unleashing the power of CRISPR-Cas mediated gene editing of extremophiles. Microbiol Res 274:127443. https://doi.org/10.1016/j.micres.2023.127443

Geng A, Cheng Y, Wang Y, Zhu D, Le Y, Wu J, Xie R, Yuan JS, Sun J (2018) Transcriptome analysis of the digestive system of a wood-feeding termite (Coptotermes formosanus) revealed a unique mechanism for effective biomass degradation. Biotechnol Biofuels 11:24. https://doi.org/10.1186/s13068-018-1015-1

Giovanella P, Vieira GA, Otero IVR, Pellizzer EP, de Jesus FB, Sette LD (2020) Metal and organic pollutants bioremediation by extremophile microorganisms. J Hazard Mater 382:121024. https://doi.org/10.1016/j.jhazmat.2019.121024

González E, Vera F, Scott F, Guerrero C, Bolívar JM, Aroca G, Muñoz JÁ, Ladero M, Santos VE (2024) Acidophilic heterotrophs: basic aspects and technological applications. Front Microbiol 15:1374800. https://doi.org/10.3389/fmicb.2024.1374800

Gregson BH, Metodieva G, Metodiev MV, Golyshin PN, McKew BA (2020) Protein expression in the obligate hydrocarbon-degrading psychrophile Oleispira Antarctica RB-8 during alkane degradation and cold tolerance. Environ Microbiol 22(5):1870–1883. https://doi.org/10.1111/1462-2920.14956

Grünberger F, Schmid G, El Ahmad Z, Fenk M, Vogl K, Reichelt R, Hausner W, Urlaub H, Lenz C, Grohmann D (2023) Uncovering the temporal dynamics and regulatory networks of thermal stress response in a hyperthermophile using transcriptomics and proteomics. MBio 14(6):e02174–e02123. https://doi.org/10.1128/mbio.02174-23

Gurdo N, Volke DC, Nikel PI (2022) Merging automation and fundamental discovery into the design–build–test–learn cycle of nontraditional microbes. Trends Biotechnol 40(10):1148–1159. https://doi.org/10.1016/j.tibtech.2022.03.004

Horikoshi K (2021) Alkaliphiles. Routledge, London. https://doi.org/10.1201/9781315137155

Huang S, Xue Y, Zhou C, Ma Y (2022) An efficient CRISPR/Cas9-based genome editing system for alkaliphilic Bacillus sp. N16-5 and application in engineering xylose utilization for d-lactic acid production. Microb Biotechnol 15(11):2730–2743. https://doi.org/10.1111/1751-7915.14131

Inaba Y, Banerjee I, Kernan T, Banta S (2018) Transposase-mediated chromosomal integration of exogenous genes in Acidithiobacillus ferrooxidans. Appl Environ Microbiol 84(21):e01381–e01318. https://doi.org/10.1128/AEM.01381-18

Jiang H, Chen Y, Jiang P, Zhang C, Smith TJ, Murrell JC, Xing X-H (2010) Methanotrophs: multifunctional bacteria with promising applications in environmental bioengineering. Biochem Eng J 49(3):277–288. https://doi.org/10.1016/j.bej.2010.01.003

Jiang X-R, Yan X, Yu L-P, Liu X-Y, Chen G-Q (2021) Hyperproduction of 3-hydroxypropionate by Halomonas bluephagenesis. Nat Commun 12(1):1–13. https://doi.org/10.1038/s41467-021-21632-3

Johnson DB, Roberto FF (1997) Heterotrophic Acidophiles and their roles in the bioleaching of sulfide minerals. In: Rawlings DE (ed) Biomining: theory, microbes and industrial processes. Springer, Berlin Heidelberg, Berlin, Heidelberg, pp 259–279. https://doi.org/10.1007/978-3-662-06111-4_13

Jung H, Inaba Y, Banta S (2022) Genetic engineering of the acidophilic chemolithoautotroph Acidithiobacillus ferrooxidans. Trends Biotechnol 40(6):677–692. https://doi.org/10.1016/j.tibtech.2021.10.004

Junker B, Lester M, Leporati J, Schmitt J, Kovatch M, Borysewicz S, Maciejak W, Seeley A, Hesse M, Connors N, Brix T, Creveling E, Salmon P (2006) Sustainable reduction of bioreactor contamination in an industrial fermentation pilot plant. J Biosci Bioeng 102(4):251–268. https://doi.org/10.1263/jbb.102.251

Keller MW, Lipscomb GL, Loder AJ, Schut GJ, Kelly RM, Adams MW (2015) A hybrid synthetic pathway for butanol production by a hyperthermophilic microbe. Metab Eng 27:101–106. https://doi.org/10.1016/j.ymben.2014.11.004

Khot V, Strous M, Dong X, Kiesser AK (2023) Viral diversity and dynamics and CRISPR-Cas-mediated immunity in a robust alkaliphilic cyanobacterial consortium. Microbiol Spectr 11(6):e02217–e02223. https://doi.org/10.1128/spectrum.02217-23

Komor AC, Kim YB, Packer MS, Zuris JA, Liu DR (2016) Programmable editing of a target base in genomic DNA without double-stranded DNA cleavage. Nature 533(7603):420–424. https://doi.org/10.1038/nature17946

Kondrat'eva TF, Pivovarova TA, Tsaplina IA, Fomchenko NV, Zhuravleva AE, Murav'ev MI, Melamud VS, Bulayev AG (2012) Diversity of the communities of acidophilic chemolithotrophic microorganisms in natural and technogenic ecosystems. Microbiology 81(1):1–24. https://doi.org/10.1134/S0026261712010080

Krulwich TA, Ito M, Gilmour R, Guffanti AA (1997) Mechanisms of cytoplasmic pH regulation in alkaliphilic strains of Bacillus. Extremophiles 1:163–170. https://doi.org/10.1007/s007920050030

Krulwich TA, Gilmour R, Hicks DB, Guffanti AA, Ito M (1998) Energetics of alkaliphilic Bacillus species: physiology and molecules. Adv Microb Physiol 40:401–438. https://doi.org/10.1016/S0065-2911(08)60136-8

Kulkarni S, Dhakar K, Joshi A (2019) Alkaliphiles: diversity and bioprospection. In: Das S, Dash HR (eds) Microbial diversity in the genomic era. Academic Press, pp 239–263. https://doi.org/10.1016/B978-0-12-814849-5.00015-0

Kumar R, Kumar P (2017) Future microbial applications for bioenergy production: a perspective. Front Microbiol 8:450. https://doi.org/10.3389/fmicb.2017.00450

Lee LL, Blumer-Schuette SE, Izquierdo JA, Zurawski JV, Loder AJ, Conway JM, Elkins JG, Podar M, Clum A, Jones PC (2018) Genus-wide assessment of lignocellulose utilization in the extremely thermophilic genus Caldicellulosiruptor by genomic, pangenomic, and metagenomic analyses. Appl Environ Microbiol 84(9):e02694–e02617. https://doi.org/10.1128/AEM.02694-17

Ling C, Qiao G-Q, Shuai B-W, Olavarria K, Yin J, Xiang R-J, Song K-N, Shen Y-H, Guo Y, Chen G-Q (2018) Engineering NADH/NAD+ ratio in Halomonas bluephagenesis for enhanced production of polyhydroxyalkanoates (PHA). Metab Eng 49:275–286. https://doi.org/10.1016/j.ymben.2018.09.007

Löbs A-K, Engel R, Schwartz C, Flores A, Wheeldon I (2017) CRISPR–Cas9-enabled genetic disruptions for understanding ethanol and ethyl acetate biosynthesis in Kluyveromyces marxianus. Biotechnol Biofuels 10:1–14. https://doi.org/10.1186/s13068-017-0854-5

Long B, Zhang F, Dai SY, Foston M, Tang YJ, Yuan JS (2024) Engineering strategies to optimize lignocellulosic biorefineries. Nat Rev Bioeng. https://doi.org/10.1038/s44222-024-00247-5

Luo Y, Ge M, Wang B, Sun C, Wang J, Dong Y, Xi JJ (2020) CRISPR/Cas9-deaminase enables robust base editing in Rhodobacter sphaeroides 2.4. 1. Microb Cell Factories 19:1–14. https://doi.org/10.1186/s12934-020-01345-w

Ma H, Zhao Y, Huang W, Zhang L, Wu F, Ye J, Chen G-Q (2020) Rational flux-tuning of Halomonas bluephagenesis for co-production of bioplastic PHB and ectoine. Nat Commun 11(1):3313. https://doi.org/10.1038/s41467-020-17223-3

Malode SJ, Prabhu KK, Mascarenhas RJ, Shetti NP, Aminabhavi TM (2021) Recent advances and viability in biofuel production. Energy Convers Manag 10:100070. https://doi.org/10.1016/j.ecmx.2020.100070

Mamo G, Mattiasson B (2016) Alkaliphilic microorganisms in biotechnology. In: Rampelotto P (ed) Biotechnology of extremophiles: grand challenges in biology and biotechnology, vol 1. Springer, Cham, pp 243–272. https://doi.org/10.1007/978-3-319-13521-2_8

Mandal S, Ganguli K, Roy A (2024) Applications of genome editing in extremophiles. In: Shah MP, Dey S (eds) Trends in biotechnology of polyextremophiles. Springer, Cham, pp 63–76. https://doi.org/10.1007/978-3-031-55032-4_3

Mirete S, Morgante V, González-Pastor JE (2017) Acidophiles: diversity and mechanisms of adaptation to acidic environments. In: Stan-Lotter H, Fendrihan S (eds) Adaption of microbial life to environmental extremes: novel research results and application. Springer International Publishing, Cham, pp 227–251. https://doi.org/10.1007/978-3-319-48327-6_9

Molitor B, Richter H, Martin ME, Jensen RO, Juminaga A, Mihalcea C, Angenent LT (2016) Carbon recovery by fermentation of CO-rich off gases–turning steel mills into biorefineries. Bioresour Technol 215:386–396. https://doi.org/10.1016/j.biortech.2016.03.094

Ojala DS, Beck DA, Kalyuzhnaya MG (2011) Genetic systems for moderately halo(alkali)philic bacteria of the genus Methylomicrobium. Methods Enzymol 495:99–118. https://doi.org/10.1016/B978-0-12-386905-0.00007-3

Oren A (2018) Acidophiles. In: eLS. pp 1–14. https://doi.org/10.1002/9780470015902.a0000336.pub2

Öztürk S, Ergün BG, Çalık P (2017) Double promoter expression systems for recombinant protein production by industrial microorganisms. Appl Microbiol Biotechnol 101:7459–7475. https://doi.org/10.1007/s00253-017-8487-y

Pan S, Zabed HM, Wei Y, Qi X (2022) Technoeconomic and environmental perspectives of biofuel production from sugarcane bagasse: current status, challenges and future outlook. Ind Crop Prod 188:115684. https://doi.org/10.1016/j.indcrop.2022.115684

Park H, Toogood HS, Chen GQ, Scrutton NS (2023) Co-production of biofuel, bioplastics and biochemicals during extended fermentation of Halomonas bluephagenesis. Microb Biotechnol 16(2):307–321. https://doi.org/10.1111/1751-7915.14158

Passarini MRZ, Duarte AWF, Rosa LH, de Oliveira VM, Ottoni JR (2022) Extremofuels: production of biofuels by extremophile microbes as an alternative to avoid climate change effects. In: Kumar A, Singh J, Ferreira LFR (eds) Microbiome under changing climate. Woodhead Publishing, pp 237–256. https://doi.org/10.1016/B978-0-323-90571-8.00010-9

Peng J-B, Yan W-M, Bao X-Z (1994) Plasmid and transposon transfer to Thiobacillus ferrooxidans. J Bacteriol 176(10):2892–2897. https://doi.org/10.1128/jb.176.10.2892-2897.1994

Peng N, Deng L, Mei Y, Jiang D, Hu Y, Awayez M, Liang Y, She Q (2012) A synthetic arabinose-inducible promoter confers high levels of recombinant protein expression in hyperthermophilic archaeon Sulfolobus islandicus. Appl Environ Microbiol 78(16):5630–5637. https://doi.org/10.1128/AEM.00855-12

Preiss L, Hicks DB, Suzuki S, Meier T, Krulwich TA (2015) Alkaliphilic bacteria with impact on industrial applications, concepts of early life forms, and bioenergetics of ATP synthesis. Front Bioeng Biotechnol 3:75. https://doi.org/10.3389/fbioe.2015.00075

Qin Q, Ling C, Zhao Y, Yang T, Yin J, Guo Y, Chen GQ (2018) CRISPR/Cas9 editing genome of extremophile Halomonas spp. Metab Eng 47:219–229. https://doi.org/10.1016/j.ymben.2018.03.018

Rao AS, Nair A, Nivetha K, More VS, Anantharaju KS, More SS (2022) Molecular adaptations in proteins and enzymes produced by extremophilic microorganisms. In: Arora NK, Agnihotri S, Mishra J (eds) Extremozymes and their industrial applications. Academic Press, pp 205–230. https://doi.org/10.1016/B978-0-323-90274-8.00002-2

Rawat M, Chauhan M, Pandey A (2024) Extremophiles and their expanding biotechnological applications. Arch Microbiol 206(6):247. https://doi.org/10.1007/s00203-024-03981-x

Rawlings DE, Woods DR (1985) Mobilization of Thiobacillus ferrooxidans plasmids among Escherichia coli strains. Appl Environ Microbiol 49(5):1323–1325. https://doi.org/10.1128/aem.49.5.1323-1325.1985

Razia S, Hadibarata T, Lau SY (2023) Acidophilic microorganisms in remediation of contaminants present in extremely acidic conditions. Bioprocess Biosyst Eng 46(3):341–358. https://doi.org/10.1007/s00449-022-02844-3

Rial RC (2024) Biofuels versus climate change: exploring potentials and challenges in the energy transition. Renew Sust Energy Rev 196:114369. https://doi.org/10.1016/j.rser.2024.114369

Schocke L, Bräsen C, Siebers B (2019) Thermoacidophilic Sulfolobus species as source for extremozymes and as novel archaeal platform organisms. Curr Opin Biotechnol 59:71–77. https://doi.org/10.1016/j.copbio.2019.02.012

Shen R, Yin J, Ye J-W, Xiang R-J, Ning Z-Y, Huang W-Z, Chen G-Q (2018) Promoter engineering for enhanced P (3HB-co-4HB) production by Halomonas bluephagenesis. ACS Synth Biol 7(8):1897–1906. https://doi.org/10.1021/acssynbio.8b00102

Silva-Rocha R, Martínez-García E, Calles B, Chavarría M, Arce-Rodríguez A, de Las HA, Paez-Espino AD, Durante-Rodríguez G, Kim J, Nikel PI (2013) The Sandard European Vctor Achitecture (SEVA): a coherent platform for the analysis and deployment of complex prokaryotic phenotypes. Nucleic Acids Res 41(D1):D666–D675. https://doi.org/10.1093/nar/gks1119

Souza GM, Ballester MVR, de Brito Cruz CH, Chum H, Dale B, Dale VH, Fernandes ECM, Foust T, Karp A, Lynd L, Maciel Filho R, Milanez A, Nigro F, Osseweijer P, Verdade LM, Victoria RL, Van der Wielen L (2017) The role of bioenergy in a climate-changing world. Environ Dev 23:57–64. https://doi.org/10.1016/j.envdev.2017.02.008

Tan D, Xue Y-S, Aibaidula G, Chen G-Q (2011) Unsterile and continuous production of polyhydroxybutyrate by Halomonas TD01. Bioresour Technol 102(17):8130–8136. https://doi.org/10.1016/j.biortech.2011.05.068

Tao W, Lv L, Chen G-Q (2017) Engineering Halomonas species TD01 for enhanced polyhydroxyalkanoates synthesis via CRISPRi. Microb Cell Factories 16(1):48. https://doi.org/10.1186/s12934-017-0655-3

Tuly JA, Ma H, Zabed HM, Dong Y, Chen G, Guo L, Betchem G, Igbokwe CJ (2022) Exploring magnetic field treatment into solid-state fermentation of organic waste for improving structural and physiological properties of keratin peptides. Food Biosci 49:101872. https://doi.org/10.1016/j.fbio.2022.101872

van Zyl LJ, van Munster JM, Rawlings DE (2008) Construction of arsB and tetH mutants of the sulfur-oxidizing bacterium Acidithiobacillus caldus by marker exchange. Appl Environ Microbiol 74(18):5686–5694. https://doi.org/10.1128/AEM.01235-08

Volke DC, Orsi E, Nikel PI (2023) Emergent CRISPR–Cas-based technologies for engineering nonmodel bacteria. Curr Opin Microbiol 75:102353. https://doi.org/10.1016/j.mib.2023.102353

Wang L-J, Jiang X-R, Hou J, Wang C-H, Chen G-Q (2022) Engineering Halomonas bluephagenesis via small regulatory RNAs. Metab Eng 73:58–69. https://doi.org/10.1016/j.ymben.2022.06.005

Wen Q, Liu X, Wang H, Lin J (2014) A versatile and efficient markerless gene disruption system for Aidithiobacillus thiooxidans: application for characterizing a copper tolerance related multicopper oxidase gene. Environ Microbiol 16(11):3499–3514. https://doi.org/10.1111/1462-2920.12494

Williams-Rhaesa AM, Rubinstein GM, Scott IM, Lipscomb GL, Poole FL, Kelly RM, Adams MW (2018) Engineering redox-balanced ethanol production in the cellulolytic and extremely thermophilic bacterium, Caldicellulosiruptor besciiMetab Eng Commun 7:e00073. https://doi.org/10.1016/j.mec.2018.e00073

Xi Y, Xu H, Zhan T, Qin Y, Fan F, Zhang X (2023) Metabolic engineering of the acid-tolerant yeast Pichia kudriavzevii for efficient L-malic acid production at low pH. Metab Eng 75:170–180. https://doi.org/10.1016/j.ymben.2022.12.007

Xu M, Chang Y, Zhang Y, Wang W, Hong J, Zhao J, Lu X, Tan D (2022a) Development and application of transcription terminators for polyhydroxylkanoates production in halophilic Halomonas bluephagenesis TD01. Front Microbiol 13:941306. https://doi.org/10.3389/fmicb.2022.941306

Xu T, Chen J, Mitra R, Lin L, Xie Z, Chen G-Q, Xiang H, Han J (2022b) Deficiency of exopolysaccharides and O-antigen makes Halomonas bluephagenesis self-flocculating and amenable to electrotransformation. Commun Biol 5(1):623. https://doi.org/10.1038/s42003-022-03570-y

Xue D, Liu W, Chen Y, Liu Y, Han J, Geng X, Li J, Jiang S, Zhou Z, Zhang W, Chen M, Lin M, Ongena M, Wang J (2019) RNA-Seq-based comparative transcriptome analysis highlights new features of the heat-stress response in the Extremophilic bacterium Deinococcus radiodurans. Int J Mol Sci 20:22. https://doi.org/10.3390/ijms20225603

Yancey PH, Gerringer ME, Drazen JC, Rowden AA, Jamieson A (2014) Marine fish may be biochemically constrained from inhabiting the deepest ocean depths. Proc Natl Acad Sci 111(12):4461–4465. https://doi.org/10.1073/pnas.1322003111

Ye J, Huang W, Wang D, Chen F, Yin J, Li T, Zhang H, Chen G-Q (2018) Pilot scale-up of poly(3-hydroxybutyrate-co-4-hydroxybutyrate) production by Halomonas bluephagenesis via cell growth adapted optimization process. Biotechnol J 13(5):1800074. https://doi.org/10.1002/biot.201800074

Ye J, Hu D, Yin J, Huang W, Xiang R, Zhang L, Wang X, Han J, Chen G-Q (2020) Stimulus response-based fine-tuning of polyhydroxyalkanoate pathway in Halomonas. Metab Eng 57:85–95. https://doi.org/10.1016/j.ymben.2019.10.007

Ye J-W, Lin Y-N, Yi X-Q, Yu Z-X, Liu X, Chen G-Q (2023) Synthetic biology of extremophiles: a new wave of biomanufacturing. Trends Biotechnol 41(3):342–357. https://doi.org/10.1016/j.tibtech.2022.11.010

Yu Y, Liu X, Wang H, Li X, Lin J (2014) Construction and characterization of tetH overexpression and knockout strains of Acidithiobacillus ferrooxidans. J Bacteriol 196(12):2255–2264. https://doi.org/10.1128/JB.01472-13

Yue H, Ling C, Yang T, Chen X, Chen Y, Deng H, Wu Q, Chen J, Chen G-Q (2014) A seawater-based open and continuous process for polyhydroxyalkanoates production by recombinant Halomonas campaniensis LS21 grown in mixed substrates. Biotechnol Biofuels 7:1–12. https://doi.org/10.1186/1754-6834-7-108

Zabed H, Sahu JN, Suely A (2017a) Bioethanol production from lignocellulosic biomass: An overview of pretreatment, hydrolysis, and fermentation. In: Mondal P, Daiai AK (eds) Sustainable utilization of natural resources. CRC Press, Boca Raton, FL, pp 145–186. https://doi.org/10.1201/9781315153292-7

Zabed H, Sahu JN, Suely A, Boyce AN, Faruq G (2017b) Bioethanol production from renewable sources: current perspectives and technological progress. Renew Sustain Energy Rev 71:475–501. https://doi.org/10.1016/j.rser.2016.12.076

Zabed HM, Akter S, Yun JH, Zhang GY, Awad FN, Qi XH, Sahu JN (2019) Recent advances in biological pretreatment of microalgae and lignocellulosic biomass for biofuel production. Renew Sustain Energy Rev 105:105–128. https://doi.org/10.1016/j.rser.2019.01.048

Zabed HM, Akter S, Yun JH, Zhang GY, Zhang YF, Qi XH (2020) Biogas from microalgae: technologies, challenges and opportunities. Renew Sustain Energy Rev 117:109503. https://doi.org/10.1016/j.rser.2019.109503

Zabed HM, Akter S, Yun J, Zhang G, Zhao M, Mofijur M, Awasthi MK, Kalam MA, Ragauskas A, Qi X (2023) Towards the sustainable conversion of corn Stover into bioenergy and bioproducts through biochemical route: technical, economic and strategic perspectives. J Clean Prod 400:136699. https://doi.org/10.1016/j.jclepro.2023.136699

Zabed HM, Akter S, Tian Y, Dar MA, Yun J, Zhao M, Ragauskas AJ, Li J, Qi X (2024a) Assessing microbial systems and process configurations for improved ethanol production from sugary stovers by integrating soluble sugars and holocellulose. Ind Crop Prod 212:118269. https://doi.org/10.1016/j.indcrop.2024.118269

Zabed HM, Akter S, Yun J, Elshobary ME, Haldar SS, Zhao M, Chowdhury FI, Li J, Qi X (2024b) Tailoring whole slurry bioprocessing for sugary stovers to augment sugar production by integrating soluble and insoluble carbohydrates. J Clean Prod 450:141844. https://doi.org/10.1016/j.jclepro.2024.141844

Zhang M-J, Jiang C-Y, You X-Y, Liu S-J (2014) Construction and application of an expression vector from the new plasmid pLAtc1 of Acidithiobacillus caldus. Appl Microbiol Biotechnol 98:4083–4094. https://doi.org/10.1007/s00253-014-5507-z

Zhang X, Lin Y, Chen G-Q (2018) Halophiles as chassis for bioproduction. Adv Biosyst 2(11):1800088. https://doi.org/10.1002/adbi.201800088

Zhang J, Yan X, Park H, Scrutton NS, Chen T, Chen G-Q (2024) Nonsterile microbial production of chemicals based on Halomonas spp. Curr Opin Biotechnol 85:103064. https://doi.org/10.1016/j.copbio.2023.103064

Zhao H, Zhang HM, Chen X, Li T, Wu Q, Ouyang Q, Chen G-Q (2017) Novel T7-like expression systems used for Halomonas. Metab Eng 39:128–140. https://doi.org/10.1016/j.ymben.2016.11.007

Zhao C, Zheng T, Feng Y, Wang X, Zhang L, Hu Q, Chen J, Wu F, Chen G-Q (2022a) Engineered Halomonas spp. for production of l-lysine and cadaverine. Bioresour Technol 349:126865. https://doi.org/10.1016/j.biortech.2022.126865

Zhao L, Sun Z-F, Zhang C-C, Nan J, Ren N-Q, Lee D-J, Chen C (2022b) Advances in pretreatment of lignocellulosic biomass for bioenergy production: challenges and perspectives. Bioresour Technol 343:126123. https://doi.org/10.1016/j.biortech.2021.126123

Zhao C, Guan Y, Yan S, Li J (2024) Exploring the promoter generation and prediction of Halomonas spp. based on GAN and multi-model fusion methods. Int J Mol Sci 25(23):13137. https://doi.org/10.3390/ijms252313137

Zhu D, Zhang P, Xie C, Zhang W, Sun J, Qian W-J, Yang B (2017) Biodegradation of alkaline lignin by Bcillus ligniniphilus L1. Biotechnol Biofuels 10(1):44. https://doi.org/10.1186/s13068-017-0735-y

Zhu D, Adebisi WA, Ahmad F, Sethupathy S, Danso B, Sun J (2020) Recent development of extremophilic bacteria and their application in biorefinery. Front Bioeng Biotechnol 8:483. https://doi.org/10.3389/fbioe.2020.00483

Chapter 5
Inspiration from Natural Biomass Utilization System for a Sustainable Lignocellulosic Refinery

Naiem Harun Nadaf, Ali Abdulmawjood Mohammed, and Salama Harun Nadaf

Abstract A lignocellulose refinery is a sustainable way to produce biofuels and other value-added chemicals. The pretreatment of lignocellulose biomass (LCB) at a large scale is one of the challenging tasks when considering its large availability in nature as a raw material for energy and beneficial products. Eventually, the processing of LCB became a major area of research and development in recent times, and there is always scope for developing novel, nontoxic pretreatment methods. In nature, some animals, insects, and microorganisms demonstrate energy-efficient and eco-friendly models of LCB utilization. Studies on such natural systems bring insights for the development of inspired LCB biorefineries. Therefore, any proposed nature-inspired refinery not just mimics natural systems in lab models but also attempts to minimize integral limitations of indigenous natural and physicochemical systems. The general attitude of designing the LCB refinery is to improve the productivity of processes in an eco-friendly and sustainable manner. This chapter recognizes the relationship between LCB refinery and nature-inspired systems by understanding their mechanisms, features, and prospects for sustainable development.

Keywords Chemicals · Enzymes · Fuel · Lignocellulosic biomass · Nature-inspired · Refinery

---

N. H. Nadaf (✉)
Department of Microbiology, Shivaji University, Kolhapur, Maharashtra, India

A. A. Mohammed
College of Science, Mustansiriyah University, Baghdad, Iraq

S. H. Nadaf
Department of Biotechnology, Vivekananda College, Kolhapur, Maharashtra, India

M. A. Dar et al. (eds.), *Recent Trends in Lignocellulosic Biofuels and Bioenergy*, Clean Energy Production Technologies,
https://doi.org/10.1007/978-981-96-4636-4_5

5.1 Introduction

Energy has become a most essential need of humans in today's world, which is increasing exponentially with the increased population. The accomplishment of this huge energy demand requires massive energy resources and feedstock (Panwar et al. 2011). The excessive usage of conventional petrochemicals leads to its shortage. That also results in greenhouse gas emissions and other pollution-related environmental crises (Sharma et al. 2022a, b; Nigam and Singh 2011). Hence, sustainable technology is required to generate more clean energy to overcome the upcoming energy calamity and minimize carbon and greenhouse emissions (Dahman et al. 2019; Azhar et al. 2017). Some conventional and renewable energy resources were already known for their energy efficiency and environmental safety. Biofuel production from LCB is one of the current approaches among all clean energy generation strategies. Lignocellulose biomass mainly comprises cellulose, hemicellulose, lignin, and other components. Though LCB is the most abundant biopolymer on Earth, it is also known for its structural complexity, making it difficult to degrade. The underutilization of this source will lead to the wastage of resources and feedstocks (Sharma et al. 2022a, b; Nigam and Singh 2011; Dahman et al. 2019; Abo et al. 2019; Panwar et al. 2011; Arumugam et al. 2021).

The conversion of LCB into valuable energy products through physicochemical and biological processes is called a lignocellulosic refinery (Banu et al. 2021). The physicochemical pretreatment methods are incompetent for overall lignocellulose degradation (Kumar et al. 2009; Aslanzadeh et al. 2014). Biological pretreatments play a promising role and are preferred to minimize the limitation of physicochemical techniques. Biological pretreatments are cost-effective and environment-friendly and require less energy consumption for LCB transformation (Wu et al. 2022; Akhtar et al. 2016; Saritha et al. 2012). The biorefineries of LCB mainly comprise bacteria, fungi, etc. Some limitations of these processes include contamination, mutation, and a slow rate of lignin degradation (Niu et al. 2023; González-Bautista et al. 2020; Sasaki and Yamanaka 2020). These limitations of biological systems led to the discovery of improved versions of ubiquitous natural systems to understand efficient lignocellulose degradation. This is the point where the inception of nature-inspired technology in LCB refineries was initiated. Mimicking natural LCB degradation systems for the development of LCB refineries is the basic concept of nature-inspired LCB refinery. Nature-inspired processes help to transmute mechanisms and techniques of biological systems into technological interventions (Vincent et al. 2006). Some examples of such naturally occurring LCB-degrading systems include woodcutter insects, ruminant animals, fungal cellulose degradation, etc. However, the main difficulty in executing natural degradative systems as LCB refineries is their complexity (Chen and Davaritouchaee 2023). Therefore, to design a sustainable nature-inspired lignocellulose refinery, it is necessary to understand the ethnic system of LCB degradation. This chapter provides an overview of the development of nature-inspired refineries, their impact on the environment, and the biotransformation of LCB.

5.2 Understanding the Mechanism of LCB Pretreatments

5.2.1 Physicochemical Methods

For biofuel production a major part of LCB is derived from agricultural raw materials, making it helpful in agricultural waste management. The refinery of LCB includes different stages such as hydrolysis, production/fermentation, and downstreaming (Haldar et al. 2022; Hasanov et al. 2020). Hydrolysis of lignocellulose was done chemically by using diluted acids; alkaline, ionic, elastic, and organic solvents, and ozonolysis (Jatoi et al. 2021). For the acid pretreatments diluted acids like sulfuric acids, hydrochloric acid, and nitric acid, are used. The acid treatment causes the separation of cellulose fraction from lignin by breaking glycosidic linkages (Nargotra et al. 2016; Mankar et al. 2021). On the other hand, for lignin solubilization, alkali bases such as calcium hydroxide, ammonium hydroxide, sodium hydroxides, etc. are used (Sharma et al. 2022a, b). Methods including carbon dioxide explosion and ammonium fiber explosion are also used in physicochemical LCB pretreatment. These methods result in the dissociation of integrated LCB structural components and increased surface area. This ultimately facilitates the closer proximity of cellulase enzymes to cellulose without its crystallization (Vassilev et al. 2022; Srivastava et al. 2015; He et al. 2017). However, these techniques have some drawbacks, including high cost, corrosion of equipment, and synthesis of enzyme inhibitors. Alkali treatment can reduce some limitations of acid treatments, but it is costly and time-consuming (Vaidya et al. 2022). Methods like pyrolysis and comminution require a high energy-consuming process either in the form of heat or mechanical strength. Therefore, physicochemical pretreatments involved in the conversion of LCB have problems in terms of pollution, higher energy input, partial degradation of lignocellulose, less productivity, etc. (Vu et al. 2020; Kumar et al. 2009; Aslanzadeh et al. 2014). Thus, to design a sustainable LCB refinery, pretreatments should be cost-effective and eco-friendly. The naturally occurring LCB-utilizing processes can remain a model for such sustainable technology (Vincent et al. 2006). Moreover, in nature-inspired LCB treatment systems, there is always scope for newer developments by simulating natural mechanisms to increase productivity. Improvising these systems by integrating biological and physicochemical approaches enhances the chances of developing a comprehensive LCB refinery (Chen and Davaritouchaee 2023; Sun et al. 2013).

5.2.2 Biological Methods

In general, the conversion of LCB into fuel and other products was achieved through two platforms. The first is a thermochemical platform, where lignocellulose is converted into a mixture of liquids and gasses at higher pressure and temperature. Downstreaming of this mixture resulted in the separation of different chemicals, some of which can be utilized directly as fuel, while others can be used as value-added products. The second platform is based on biochemical processing, where cellulose is first separated from the lignocellulose complex, which then serves as a substrate for fermentative ethanol production (Bayer et al. 1998). Lignocellulose is an example of a well-engineered complex polymer, where lignin can resist the enzymatic degradation of cellulose and hemicellulose (Fig. 5.1). Furthermore, the hydrophobicity of this molecule aids in reducing the difficulty level of its complete degradation. Therefore, separating cellulose from lignin and hemicelluloses will be necessary. This can be achieved by altering or breaking down this structure using different pretreatment methods (Chen et al. 2010). The overall success of biofuel production using nature-inspired refineries depends on the maximum accessibility of cellulose to microorganisms and their enzymes. There is always scope to design compatible biochemical processes to separate lignin from cellulose and hemicellulose. Here, ligninolytic enzymes acted as a promising option. These enzymes can free the lignin's integrated structure, making cellulose available for further saccharification and fermentation (Ali et al. 2024a). Furthermore, due to its biological origin, there is always scope for the progression of these methods using industrial biotechnology, molecular biology, microbiology, biochemistry, strain improvement studies, etc. These could also encourage scientists and engineers to develop nature-inspired refineries for LCB (Dar et al. 2025a, b; Chen et al. 2010).

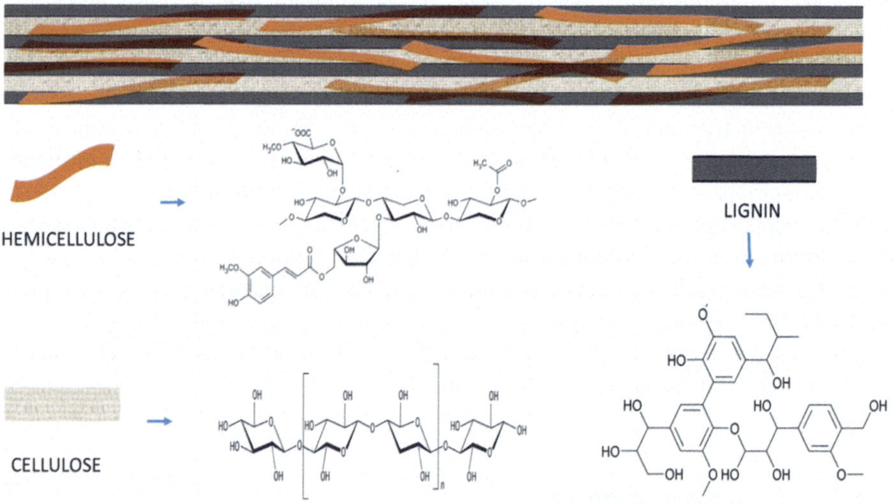

Fig. 5.1 Presentation of the organized structure of lignocellulose with its subunits

5.3 Evaluation of Enzymatic Potential in LCB Refinery Pretreatments

Reactive oxygen species (ROS) are critical in the oxidative delignification processes. ROS are nothing but groups of free radicals or ions with an oxygen molecule. ROS have a short life, but they can react with almost all kinds of molecules in plant cells. Superoxide's peroxides and hydroxyl radicals are examples of ROS, where oxygen molecules contain one or more unpaired electrons. ROS can be produced in living cells through various types of reactions, leading to oxidative denaturation or alteration of various chemicals (Glaze et al. 1987). In various biological systems, ROS-assisted enzyme reactions can result in the delignification of LCB (Davaritouchaee et al. 2019). Lignin peroxidases, manganese peroxidase, peroxidases, and laccases are some ROS requiring ligninolytic enzymes (Martínez 2002). The structural complexity of the LCB molecule favors ligninolytic enzymes acting on it, as the heterogeneous structure of lignocellulose and electron-rich lignin molecules are cross-linked with various functional groups. ROS molecules such as peroxide, hydroxyl peroxide, superoxide, and sulfate can ideally be reacted with these functional groups of LCB polymer (Chen and Davaritouchaee 2023). Further degradation of cellulose is then achieved by other enzymes, namely, cellobiose oxidoreductase, cellobiose dehydrogenase, glucose oxidase, glyoxal oxidase, etc. (Wong 2009). Cellulases and xylanases are involved in the conversion of lignocellulose into fine biochemicals such as sugars, organic acids, amino acids, and partially hydrolyzed cellulose (Sharma et al. 2019). Endoglucanase and exoglucanase with β-glucosidase are also involved in cellulose degradation, where endoglucanase acted on β-1,4 bonds of cellulose and converted them into uneven long or short chains of glucose. At the same time, exoglucanase acted on these chains and converted them into disaccharides and monomeric sugars. The hemicellulose part of LCB was catabolized by xylanase, which converts hemicellulose into sugars utilized in fuel production refineries. Natural LCB-utilizing systems such as the rumen and gut microbiota of insects, animals, nematodes, and aquatic zooplanktonic animals were investigated as sources of these enzymes. The gut systems of these animals and insects harbor commensals, including bacteria, fungi, and molds, which are capable of LCB biotransformation with their fascinating enzyme systems (Kumar et al. 2020; Haq et al. 2021; Kumar et al. 2019; Yadav 2017).

5.4 Examples of Nature-Inspired Systems for the LCB Refinery

5.4.1 Termite Symbiotic Systems

Among wood-cutting insects, termites are the predominant lignocellulose utilizers observed across the globe in varied climatic conditions, but the majority of them are found in the tropics (Brune 2014). Due to their multifunctional role in environmental ecology and biogeochemical cycles, termites are among the most studied lignocellulose-metabolizing insects. For food, termites—mainly worker termites—rely on cellulose and hence utilize LCB abundantly (Eggleton 2011).

To understand the biochemistry of lignocellulose degradation in termites, it is essential to study its gut physiology (Dar et al. 2024). The digestive system of termites is mainly divided into two sections: the interior section, where ingested wooden material is broken down into small pieces, and the posterior gut section, where assimilation and uptake of nutrients are carried out (Scharf and Tartar 2008). The processing of ingested lignocellulosic biomass starts with the action of the mandible—a part of the buccal cavity—where the mechanical chopping of biomass converts it into small particles. This ground material then reaches the foregut, midgut, and hindgut for its further metabolism. Indigenous enzymes and physicochemical conditions found in each gut compartment acted sequentially on this complex LCB structure for its complete metabolism (Ke and Chen 2013) (Fig. 5.2). The termite's gut system is an ideal habitat for metabolically diverse microbial groups. These microbial species are scattered based on the physiological properties of each part of the gut (Dar et al. 2024; Köhler et al. 2012). A maximum number of microbial species was observed in the hindgut compared to the foregut and midgut (Brune 2014).

The termite gut can inhabit up to 10^{11} mL^{-1} bacterial cells. Among all bacteria, mostly lignocellulolytic bacteria were studied by different research groups (Table 5.1). *Bacteroidetes*, *Actinobacteria*, *Elusimicrobia*, and *Proteobacteria* are the main phyla of endosymbiotic bacteria found in termites (Dar et al. 2022; Brune 2014; Brune and Dietrich 2015; Köhler et al. 2012; Noda et al. 2005; Strassert et al. 2012). Numerous bacterial species remain unstudied due to their uncultivable or fastidious nature. Therefore, the symbiotic interaction among bacteria and termites

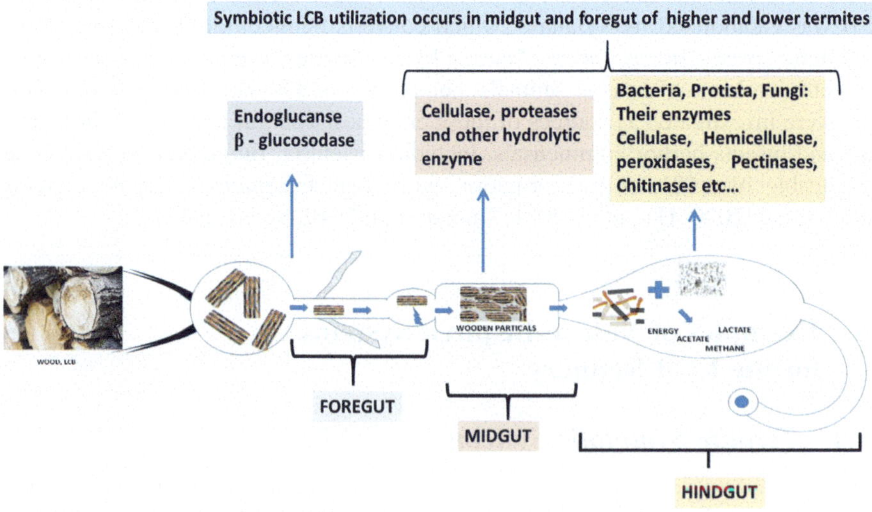

Fig. 5.2 LCB utilization and cellulose digestion in higher and lower termites

is partially explored. In such conditions, genome-based analysis became useful for understanding the symbiotic interaction between uncultivable bacteria and termites (Dar et al. 2024). The termite shares a complex relationship with fungi. Some species of fungi and yeast have been observed in the termite gut, and these termite species are mainly found in Asian and African continents (Bignell and Eggleton 2000). The symbiotic association between termites of the *Macrotermitinae* subfamily and *Termitomyces* spp. is a well-known example of fungal association with termites, where termites harbor *Termitomyces* spp. and fungi build their combs on termites' lignin-rich feces (Dar et al. 2025b; Berchtold and König 1995). The partially hydrolyzed lignocellulose-containing fungal comb is again consumed by the worker termites. This feeding activity of termites can have ingested fungal spores as well as partially hydrolyzed lignocellulose. Fungal species belonging to the genera *Ascomycota*, *Spiromatix*, and *Byssochlamys* are commonly observed in termite species of the genera *Microcerotermes*, *Nasutitermes*, and *Termes* (Vikram et al. 2021). Termites can also host symbiotic protists with their cellulolytic activity. Without this association, termites would fail to degrade cellulosic biomass and could be eradicated from the ecosystem due to starvation. In this association, symbiotic protists secrete their cellulases to metabolize cellulose for energy. Representative examples of these species include *Holomastigotoides*, *Pseudotrichonympha*, and *Spirotrichonympha* from *Coptotermes* spp., *Cryptotermes* spp., *Schedorhinotermes* spp., and *Hodotermopsis* spp. (Gile 2024; Ohkuma et al. 2000, 2007; Nalepa 2020); *Metadevescovina cuspidata* from the host *Incisitermes minor* (Todaka et al. 2006; Tartar et al. 2009); genera *Devescovina*, *Foaina*, and *Oxymonas* from the host *Neotermes* spp.; and *Staurojoenina* from the host *Epicalotermes* spp. (Dolan et al. 2004).

Table 5.1 Examples of isolated microorganisms from termites used in biofuel and valuable chemical production

Sr. no.	Isolated microorganism	Source	Substrate used	Products	References
1	*Clostridium termitidis* strain CT1112	*Nasutitermes lujae*	Cellobiose	Acetate, lactate, and formate	Ramachandran et al. (2008)
2	*Vanrija humicola* SSA-1544	*Coptotermes formosanus*	Saw dust	Ethanol	Dar et al. (2022)
3	*Bacillus* sp. BMP01	*Cryptotermes brevis*	Wheat straw	Sugars	Tsegaye et al. (2018)
4	*Bacillus* and *Clostridium* sp.	*Odontotermes formosanus*	Humic acid	—	Mathew et al. (2013)
	Mixed culture	*Nasutitermes ephratae*	Wheat straw	Volatile fatty acids	Lazuka et al. (2018)
5	*Meyerozyma guilliermondii* SSA-1543 T	*Coptotermes formosanus*	Saw dust	Ethanol	Ali et al. (2019)
6	Mixed culture	*Microcerotermes parvus*	Wheat straw	Volatile fatty acids	Auer et al. (2017)

Despite being a century-old research area, termite biology has recently evolved into a multidisciplinary area that will pave the way for innovations and future breakthroughs in LCB bioconversions. Termites' gut physiology explores multiple possibilities for advanced biofuel production (Table 5.1). The symbiotic microorganisms in some termite species reveal a novel genetic makeup for catalytic LCB utilization. The termites' gut system with its microbes can become an ideal nature-inspired system to build an LCB refinery (Dar et al. 2024).

5.4.2 White and Brown Rot Fungi

White rot fungi are one of those enzymatically distinct groups of fungi that can decompose every molecule of LCB. This fungal group is well known for the production of extracellular enzymes like cellulases, hemicellulose, pectinase, and xylanase. In addition to that, this fungal group is also recognized as a source of a wide variety of oxidative ligninolytic enzymes such as lignin peroxidase, manganese peroxidase, laccase, etc. Such versatile enzyme systems can completely mineralize lignin into H_2O_2 and CO_2. This fungus has a selective and consistent pattern of LCB degradation. Based on its type, in the case of selective rot, the lignocellulose structure randomly degraded lignin and hemicellulose at a higher rate than cellulose. In case of consistent rot, each component of LCB is degraded uniformly (Xiao et al. 2011; Asgher et al. 2008). The metabolic versatility of lignin degradation in white rot fungi will make them a potential tool for LCB pretreatment, wastewater treatment, etc. However, the only limitation of its implementation is the slower processing rate (Witomski et al. 2014).

Substrate selectivity is a remarkable property of white rot fungi, where most members of this group can do delignification only and specifically attack lignin but not cellulose and hemicellulose. Therefore, they have become an ideal choice for biofuel production (Fackler et al. 2007). White rot fungi can degrade lignin through three modes of action: first, by oxidative cleavage; second, by modifying side chains; and third, by oxidative cleaving of aromatic rings (Ali et al. 2024b). Due to all such properties, white rot fungi have also been studied extensively for LCB refinery. Some representative examples of widely studied white rot fungi are *Phanerochaete chrysosporium*, *Pleurotus eryngii*, *Phlebia radiata*, *Pleurotus pulmonarius*, *Pleurotus sapidus*, and *Phlebia tremellosa* (Giatti et al. 2004; Linke et al. 2005; Hatakka et al. 2003).

Brown rot fungi are prominently seen in coniferous forest regions to cause the rot of plants. The plants turn brown due to their high lignin content. Brown rot fungi are known for LCB degradation, but in contrast to white rot fungi, they are efficient in carbohydrate mineralization. Therefore, brown rot fungi can completely degrade cellulose after dissociation from lignin. This fungal group is well known for its ability to weaken wood strength through its cellulolytic activity rather than its ligninolytic activity because it lacks genes for synthesizing lignin-degrading enzymes (Witomski et al. 2014). Brown rot fungi are also known for the rapid degradation of cellulose in the initial stages of plant growth, which is surprisingly observed in a nonenzymatic way. Based on the pore size observed in such a complex, it is proposed that it is a small diffusible agent rather than an enzyme that causes the formation of these pores (Ferraz et al. 2000). However, the ability to degrade a large amount of cellulose created a mild acidic environment and was responsible for the generation of free radicals through the Fenton reaction. Brown rot fungal metabolites and enzymes like hydroquinones and cellobiose dehydrogenase mediate the Fenton reaction seen in LCB degradation. In this reaction, peroxide ions are generated to reduce Fe^{3+} to Fe^{2}. This generated oxidant helps in cellulose and hemicellulose degradation (Kajisa et al. 2004).

5.4.3 Ruminant Animals

The rumen is a typical digestive organ that evolved in ruminant animals and nourishes diverse microorganisms. These organisms remain obligatory to the rumen environment for their growth and help host animals in food digestion, as such animals lack genes to synthesize enzymes for LCB degradation, especially cellulases (Hess et al. 2011; Mizrahi et al. 2021; Wang and McAllister 2002). The diversity of rumen microorganisms depends on various factors including diet, anatomical features, feeding strategies, and climatic and ecological conditions. The microbial communities of the rumen are continuously undergoing succession and adapting throughout the life span of the animals (Jami et al. 2013; Rey et al. 2014). These microorganisms convert plants' LCB into energy intermediates such as volatile fatty acids, sugars, amino acids, etc. to gain 70% of daily energy (Flint et al. 2008). Among all microbial populations, the potent inhabitants of the rumen include archaebacteria, bacteria, fungi, and protozoa. These microorganisms can help their host in cellulose digestion and maintain the host's physiology, immunity, and gene expression (Flint et al. 2008; Lin et al. 2019; Malmuthuge et al. 2019). Archaebacteria belonging to methanogenic groups are more abundantly found in the rumen than other groups (Mizrahi et al. 2021). The archaebacteria like *Methanobrevibacter are* primarily involved in methanogenesis. Among bacterial genera, *Bacteroidetes* and *Firmicutes* are found in the rumen at higher numbers. However, archaebacteria can't degrade lignocellulose but can help in this process by producing methane. Methane can serve as an electron source for microorganisms in the anaerobic degradation of lignocellulose. Protozoa may contribute to half of the rumen microbial biomass but do not have any active role in lignocellulose degradation. However, their role in the horizontal gene transfer of bacteria and fungi is noted (Mizrahi and Jami 2018; Henderson et al. 2015; Mizrahi et al. 2021; Newbold et al. 2015; Williams et al. 2020). Most ciliate protozoans maintained rumen microbial (protein) equilibrium by predating on them. Among fungi, the anaerobic group is effective in lignocellulose degradation, but their number is relatively less (Gruninger et al. 2014). Some rumen bacteria are very efficient in lignocellulose degradation, while others are

Table 5.2 Examples of isolated microorganisms from the rumen used in biofuel and valuable chemical production

Sr. No.	Isolated microorganism	Source	Substrate used	Products	References
1	Mixed microorganism	Cattle	Napier grass	Sugars and hydrogen	Chang et al. (2010)
2	*C. Xylanolyticum* Ru15	Cattle	Napier grass	Ethanol volatile fatty acids	Ho et al. (2011)
3	Mixed rumen inoculum	Cattle	Corn Stover	Volatile fatty acids	Li et al. (2017)
4	Mixed rumen inoculum	Cattle	Rapeseed	Methane and ethanol	Baba et al. (2017)
5	Mixed culture	Bull	Wheat straw	Hydrogen, propionate, acetate	Ozbayram et al. (2017)
6	Mixed culture	Sheep	Wheat straw	Methane	Ozbayram et al. (2017)
7	Mixed rumen inoculum	Deer	Sugars	Hydrogen, ethanol, acetate, propionate	Rieu-Lesme et al. (1995)

involved in the further utilization of the produced metabolic products (Mizrahi et al. 2021).

Therefore, it can be seen that the rumen can serve as an ideal prototype for the LCB degradation and refinery. The microbial consortium observed in the rumen will work in an organized manner and produce different metabolites that could be utilized further as value-added products and biofuels. The versatile activity of ruminant microorganisms can be utilized in the biological pretreatment of LCB (Table 5.2).

5.5 Advancements and Future Prospects of Nature-Inspired Pretreatment

All the naturally occurring systems of LCB utilization discussed above can be taken as models for sustainable refinery in biofuel generation. Mimicking these systems into energy-efficient refineries is an approach behind the nature-inspired processes. As biological systems evolved to degrade the LCB complex in an energy-efficient manner, nature-inspired refineries have emerged as cost-effective and sustainable technologies for biofuel generation. Some good examples of such implemented nature-inspired treatment processes have already been studied for LCB utilization (Vincent et al. 2006), such as oxidative enzyme-based lignin degradation, mainly considered for nature-inspired LCB refineries. However, to develop such in vitro biological LCB transformation systems, the natural mechanisms of LCB degradation must be thoroughly understood. This helps to improvise the preexisting methods to upgrade the newer technologies with novel approaches and better efficiencies.

Table 5.3 Physicochemical process-assisted microbial pretreatment systems

Physicochemical method	Microorganism	Substrate	Reference
Oxidative treatment	*Pleurotus ostreatus*	Rice hull	Yu et al. (2009)
Oxidative treatment	*Cupriavidus basilensis* B-8	Rice straw	Zhang et al. (2018)
Oxidative treatment	*Bacillus* sp. B-3	Rice straw	Si et al. (2019)
Acid treatment	*Trametes velutina* D1014	*Populus tomentosa*	Wang et al. (2013)
Acid treatment	*Cupriavidus basilensis* B-8	Rice straw	Yan et al. (2017)
Acid treatment	*Acinetobacter* sp. B-2	Rice straw	Si et al. (2019)
Alkali treatment	*Sphingobacterium* sp. LD-1	Rice straw	Dai et al. (2015)
Alkali treatment	*Acinetobacter* sp. B-2	Rice straw	Si et al. (2019)
Alkali treatment	*Irpex lacteus*	Corn stalks	Yu et al. (2010)
Organosolv pretreatment	*Ceriporiopsis subvermispora*	*Pinus radiata* woods chips	Munoz et al. (2007)
Organosolv treatment	*Gloeophyllum trabeum*	*Pinus radiata*	Fissore et al. (2010)
Organosolv treatment	*Ceriporiopsis subvermispora*	Sugarcane straw	Saad et al. (2008)

The nature-inspired refinery facilitates and encourages the design of reactors for the enzyme system rather than for whole organisms (Zhang et al. 2022). Moreover, the synergistic approach of combining physicochemical and biological pretreatments can also result in the LCB refinery (Table 5.3), where ROS is generated by chemical reaction for oxidative lignin degradation (Fig. 5.3).

5.6 Conclusion

The lignocellulose refinery is an emerging concept for the production of valuable products. These products include fine chemicals and energy. For the development of sustainable and eco-friendly refinery processes, there is a need for novel, efficient, and low-cost treatment methods. The physicochemical pretreatment for refineries is costly and can create pollution problems. Hence, nature-inspired biological pretreatments for LCB refineries are mostly preferred. These pretreatments seem to be cost-effective and help minimize pollution problems by producing clean energy from lignocellulosic waste. Nature-inspired refinery is nothing but translating biological systems into LCB pretreatments. The basic approach of these systems is based on chemical pretreatment principles but differs in its working mechanism. These nature-inspired pretreatments can modify LCB using an ROS-mediated system, but the radical atom (ROS) synthesis can be achieved in an eco-friendly and

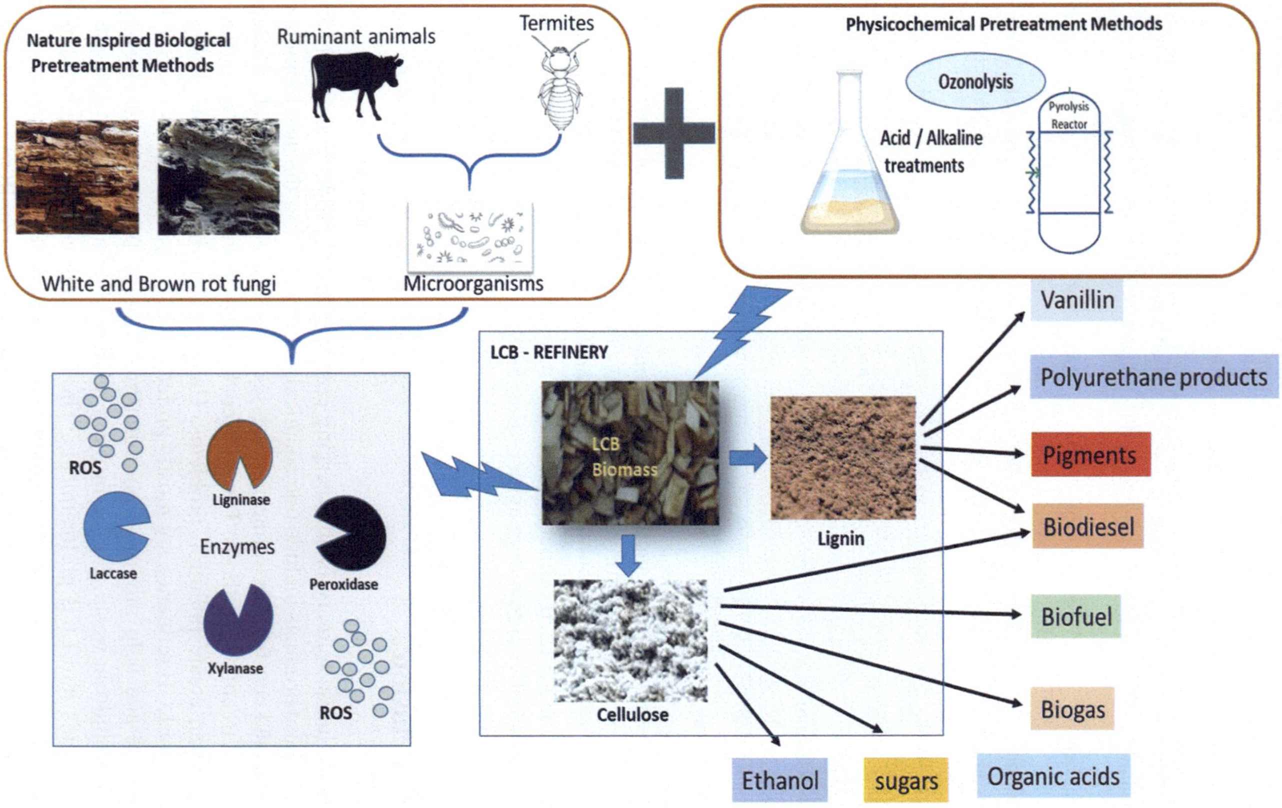

Fig. 5.3 Nature-inspired systems to develop a sustainable lignocellulose refinery

relatively simpler manner. This is followed by the conversion of cellulose and hemicellulose by the glycosidic enzyme system. The development of such sustainable and more efficient LCB pretreatment is still in the developing stage and requires higher output in the form of research. For that, it is necessary to recognize naturally occurring energy-efficient LCB-utilizing systems, such as termites, the rumen, ligninolytic microorganisms, and their enzymes. Hence, the nature-inspired refinery for the utilization of LCB is not only a mimicking approach but also a strategy to develop sustainable energy as well as value-added products at higher concentrations without polluting the environment.

References

Abo BO, Gao M, Wang Y, Wu C, Ma H, Wang Q (2019) Lignocellulosic biomass for bioethanol: an overview on pretreatment, hydrolysis and fermentation processes. Rev Environ Health 34(1):57–68. https://doi.org/10.1515/reveh-2018-0054

Akhtar N, Gupta K, Goyal D, Goyal A (2016) Recent advances in pretreatment technologies for efficient hydrolysis of lignocellulosic biomass. Environ Prog Sustain Energy 35(2):489–511. https://doi.org/10.1002/ep.12257

Ali SS, Al-Tohamy R, Sun J, Wu J, Huizi L (2019) Screening and construction of a novel microbial consortium SSA-6 enriched from the gut symbionts of wood-feeding termite, Coptotermes formosanus and its biomass-based biorefineries. Fuel 236: 1128–1145. https://doi.org/10.1016/j.fuel.2018.08.117

Ali S, Rani A, Dar MA et al (2024a) Recent advances in the characterization and valorization of lignin to value-added products: challenges and future perspectives. Biomass 4:947–977. https://doi.org/10.3390/biomass4030053

Ali S, Dar MA, Liaqat F et al (2024b) Optimization of biomethane production from lignocellulosic biomass by a developed microbial consortium. Process Saf Environ Prot 184:1106–1118

Arumugam A, Malolan VV, Ponnusami V (2021) Contemporary pretreatment strategies for bioethanol production from corncobs: a comprehensive review. Waste Biomass Valorization 12:577–612. https://doi.org/10.1007/s12649-020-00983-w

Asgher M, Bhatti HN, Ashraf M, Legge RL (2008) Recent developments in biodegradation of industrial pollutants by white rot fungi and their enzyme system. Biodegradation 19:771–783. https://doi.org/10.1007/s10532-008-9185-3

Aslanzadeh S, Ishola MM, Richards T, Taherzadeh MJ (2014) An overview of existing individual unit operations. Bioref 1:3–6. https://doi.org/10.1016/B978-0-444-59498-3.000014

Auer L, Lazuka A, Sillam-Dussès D, Miambi E, O'Donohue M, Hernandez-Raquet G (2017) Uncovering the potential of termite gut microbiome for lignocellulose bioconversion in anaerobic batch bioreactors. Front Microbiol 8:1–14. https://doi.org/10.3389/fmicb.2017.02623

Azhar SHM, Abdulla R, Jambo SA, Marbawi H, Gansau JA, Faik AAM, Rodrigues KF (2017) Yeasts in sustainable bioethanol production: a review. Biochem Biophys Rep 10:52–61. https://doi.org/10.1016/j.bbrep.2017.03.003

Baba Y, Matsuki Y, Mori Y, Suyama Y, Tada C, Fukuda Y et al (2017) Pretreatment of lignocellulosic biomass by cattle rumen fluid for methane production: bacterial flora and enzyme activity analysis. J Biosci Bioeng 123:489–496. https://doi.org/10.1016/j.jbiosc.2016.11.008

Banu JR, Kavitha S, Tyagi VK, Gunasekaran M, Karthikeyan OP, Kumar G (2021) Lignocellulosic biomass based biorefinery: a successful platform towards circular bioeconomy. Fuel 302:121086. https://doi.org/10.1016/j.fuel.2021.121086

Bayer EA, Chanzy H, Lamed R, Shoham Y (1998) Cellulose, cellulases and cellulosomes. Curr Opin Struct Biol 8(5):548–557. https://doi.org/10.1016/S0959-440X(98)80143-7

Berchtold M, König H (1995) Phylogenetic position of the two uncultivated trichomonads Pentatrichomonoides scroa Kirby and Metadevescovina extranea Kirby from the hindgut of the termite Mastotermes darwiniensis Froggatt. Syst Appl Microbiol 18(4):567–573. https://doi.org/10.1016/S0723-2020(11)80417-0

Bignell DE, Eggleton P (2000) Termites in ecosystems. In: Abe T, Bignell DE, Higashi M (eds) Termites: evolution, sociality, symbioses, ecology. Springer, Dordrecht, pp 1–363. https://doi.org/10.1007/978-94-017-3223-9_17

Brune A (2014) Symbiotic digestion of lignocellulose in termite guts. Nat Rev Microbiol 12(3):168–180. https://doi.org/10.1038/nrmicro3182

Brune A, Dietrich C (2015) The gut microbiota of termites: digesting the diversity in the light of ecology and evolution. Ann Rev Microbiol 69:145–166. https://doi.org/10.1146/annurev-micro-092412-155715

Chang JJ, Lin JJ, Ho CY, Chin WC, Huang CC (2010) Establishment of rumen-mimic bacterial consortia: a functional union for bio-hydrogen production from cellulosic bioresource. Int J Hydrog Energy 35:13399–13406. https://doi.org/10.1016/j.ijhydene.2009.11.119

Chen S, Davaritouchaee M (2023) Nature-inspired pretreatment of lignocellulose–perspective and development. Bioresour Technol 369:12845. https://doi.org/10.1016/j.biortech.2022.128456

Chen S, Zhang X, Singh D, Yu H, Yang X (2010) Biological pretreatment of lignocellulosics: potential, progress and challenges. Biofuels (London) 1(1):177–199. https://doi.org/10.4155/bfs.09.13

Dahman Y, Syed K, Begum S, Roy P, Mohtasebi B (2019) Biofuels: their characteristics and analysis. In: Biomass, biopolymer-based materials, and bioenergy. Woodhead Publishing, pp 277–325. https://doi.org/10.1016/B978-0-08-102426-3.00014-X

Dai Y, Si M, Chen Y, Zhang N, Zhou M, Liao Q, Shi D, Liu Y (2015) Combination of biological pretreatment with NaOH/urea pretreatment at cold temperature to enhance enzymatic hydrolysis of rice straw. Bioresour Technol 198:725–731. https://doi.org/10.1016/j.biortech.2015.09.091

Dar MA, Xie R, Pandit RS, Danso B, Dong C, Sun J (2022) Exploring the region-wise diversity and functions of symbiotic bacteria in the gut system of wood-feeding termite, Coptotermes formosanus, toward the degradation of cellulose, hemicellulose, and organic dyes. Insect Sci 29(5):1414–1432. https://doi.org/10.1111/1744-7917.13012

Dar MA, Xie R, Zabed HM, Ali S, Zhu D, Sun J (2024) Termite microbial symbiosis as a model for innovative design of lignocellulosic future biorefinery: current paradigms and future perspectives. Biomass 4(1):180–201. https://doi.org/10.3390/biomass4010009

Dar MA, Xie R, Zabed HM, Idrees A, Waghmare PR, Irfan M, Dhanavade MJ, Shahnawaz M, Zhu D, Sun J (2025a) Can termites serve as models for future biorefinery? In: Zhu D, Dar MA, Shahnawaz M (eds) Biofuels and sustainability: life-cycle assessments, systems biology, policies and emerging technologies. Elsevier Science, pp 323–346. https://doi.org/10.1016/B978-0-443-21433-2.00017-7

Dar MA, Ali S, Xie R, Zabed HM, Idrees A, Sun J (2025b) Non-termite insects as biomimetic models to advance the production of second and third generation biofuels. In: Zhu D, Dar MA, Shahnawaz M (eds) Biofuels and sustainability: life-cycle assessments, systems biology, policies and emerging technologies. Elsevier Science, pp 347–364. https://doi.org/10.1016/B978-0-443-21433-2.00002-5

Davaritouchaee M, Hiscox WC, Martinez-Fernandez J, Fu X, Mancini RJ, Chen S (2019) Effect of reactive oxygen species on biomass structure in different oxidative processes. Ind Crops Prod 137:484–494. https://doi.org/10.1016/j.indcrop.2019.05.063

Dolan MF, Wier AM, Melnitsky H, Whiteside JH, Margulis L (2004) Cysts and symbionts of Staurojoenina assimilis Kirby from Neotermes. Eur J Protist 40(4):257–264. https://doi.org/10.1016/j.ejop.2004.01.005

Eggleton P (2011) An introduction to termites: biology, taxonomy and functional morphology. In: Bignell D, Roisin Y, Lo N (eds) Biology of termites: a modern synthesis, 2nd edn. Springer, Dordrecht, The Netherlands; Singapore, pp 1–26. https://doi.org/10.1007/978-90-481-3977-4_1

Fackler K, Gradinger C, Schmutzer M, Tavzes C, Burgert I, Schwanninger M, Hinterstoisser B, Watanabe T, Messner K (2007) Biotechnological wood modification with selective white-rot fungi and its molecular mechanisms. Food Technol Biotechnol 45(3):269–276

Ferraz A, Mendonça R, da Silva FT (2000) Organosolv delignification of white- and brown-rotted *Eucalyptus grandis* hardwood. J Chem Technol Biotechnol 75:18–24. https://doi.org/10.1002/(SICI)1097-4660(200001)75:1<18::AID-JCTB169>3.0.CO;2-Z

Fissore A, Carrasco L, Reyes P, Rodríguez J, Freer J, Mendonça RT (2010) Evaluation of a combined brown rot decay–chemical delignification process as a pretreatment for bioethanol production from *Pinus radiata* wood chips. J Ind Microbiol Biotechnol 37:893–900. https://doi.org/10.1007/s10295-010-0736-3

Flint HJ, Bayer EA, Rincon MT, Lamed R, White BA (2008) Polysaccharide utilization by gut bacteria: potential for new insights from genomic analysis. Nat Rev Microbiol 6(2):121–131. https://doi.org/10.1038/nrmicro1817

Giatti MDSC, Kirst Tychanowicz G, Farani De Souza D, Peralta RM (2004) Production of laccase isoforms by *Pleurotus pulmonarius* in response to presence of phenolic and aromatic compounds. J Basic Microbiol 44(2):129–136. https://doi.org/10.1002/jobm.200310365

Gile GH (2024) Protist symbionts of termites: diversity, distribution, and coevolution. Biol Rev 99(2):622–652. https://doi.org/10.1111/brv.13038

González-Bautista E, Alarcón-Gutiérrez E, Dupuy N, Gaime-Perraud I, Ziarelli F, Foli L, Farnet-Da-Silva AM (2020) Preparation of a sugarcane bagasse-based substrate for second-generation ethanol: effect of pasteurisation conditions on dephenolisation. Renew Energy 157:859–866. https://doi.org/10.1016/j.renene.2020.05.116

Gruninger RJ, Puniya AK, Callaghan TM, Edwards JE, Youssef N, Dagar SS, Fliegerova K, Griffith GW, Forster R, Tsang A, McAllister T (2014) Anaerobic fungi (phylum Neocallimastigomycota): advances in understanding their taxonomy, life cycle, ecology, role, and biotechnological potential. FEMS Microbiol Ecol 90(1):1–17. https://doi.org/10.1111/1574-6941.12383

Glaze WH, Kang JW, Chapin DH (1987) The chemistry of water treatment processes involving ozone, hydrogen peroxide and ultraviolet radiation. Ozone Sci Eng 9(4):335–352. https://doi.org/10.1080/01919518708552148

Haldar D, Dey P, Patel AK, Dong CD, Singhania RR (2022) A critical review on the effect of lignin redeposition on biomass in controlling the process of enzymatic hydrolysis. Bioenergy Res 15:863–874. https://doi.org/10.1007/s12155-021-10374-1

Haq IU, Qaisar K, Nawaz A, Akram F, Mukhtar H, Xu Y, Mumtaz MW, Rashid U, Ghani WA, Choong TS (2021) Advances in valorization of lignocellulosic biomass towards energy generation. Catalysts 11(3):309. https://doi.org/10.3390/catal11030309

Hasanov I, Raud M, Kikas T (2020) The role of ionic liquids in the lignin separation from lignocellulosic biomass. Energies 13:4864. https://doi.org/10.3390/en13184864

Hatakka A, Lundell T, Hofrichter M, Maijala P (2003) Manganese peroxidase and its role in the degradation of wood lignin. In: Mansfield SD, Saddler JN (eds) Applications of enzymes to Lignocellulosics. ACS symposium series 855. American Chemical Society, WA, USA, pp 230–243. https://doi.org/10.1021/bk-2003-0855.ch014

He Z, Wang Z, Zhao Z, Yi S, Mu J, Wang X (2017) Influence of ultrasound pretreatment on wood physiochemical structure. Ultrason Sonochem 34:136–141. https://doi.org/10.1016/j.ultsonch.2016.05.035

Henderson G, Cox F, Ganesh S, Jonker A, Young W, Janssen PH (2015) Rumen microbial community composition varies with diet and host, but a core microbiome is found across a wide geographical range. Sci Rep 5(1):14567. https://doi.org/10.1038/srep14567

Hess M, Sczyrba A, Egan R, Kim TW, Chokhawala H, Schroth G, Luo S, Clark DS, Chen F, Zhang T, Mackie RI (2011) Metagenomic discovery of biomass-degrading genes and genomes from cow rumen. Science 331(6016):463–467. https://doi.org/10.1126/science.1200387

Ho CY, Chang JJ, Lin JJ, Chin TY, Mathew GM, Huang CC (2011) Establishment of functional rumen bacterial consortia (FRBC) for simultaneous biohydrogen and bioethanol production from lignocellulose. Int J Hydrog Energy 36:12168–12176. https://doi.org/10.1016/j.ijhydene.2011.06.125

Jami E, Israel A, Kotser A, Mizrahi I (2013) Exploring the bovine rumen bacterial community from birth to adulthood. ISME J 7(6):1069–1079. https://doi.org/10.1038/ismej.2013.2

Jatoi AS, Abbasi SA, Hashmi Z, Shah AK, Alam MS, Bhatti ZA, Maitlo G, Hussain S, Khandro GA, Usto MA, Iqbal A (2021) Recent trends and future perspectives of lignocellulose biomass for biofuel production: a comprehensive review. Biomass Convers Biorefinery 11:1–13. https://doi.org/10.1007/s13399-021-01853-8

Kajisa T, Yoshida M, Igarashi K, Katayama A, Nishino T, Samejima M (2004) Characterization and molecular cloning of cellobiose dehydrogenase from the brown-rot fungus *Coniophora puteana*. J Biosci Bioeng 98(1):57–63. https://doi.org/10.1016/S1389-1723(04)70242-X

Ke J, Chen S (2013) In: Sun J, Ding S, Peterson JD (eds) In biological conversion of biomass for fuels and chemicals: explorations from natural utilization systems. The Royal Society of Chemistry, pp 177–194. https://doi.org/10.1039/9781849734738

Köhler T, Dietrich C, Scheffrahn RH, Brune A (2012) High-resolution analysis of gut environment and bacterial microbiota reveals functional compartmentation of the gut in wood-feeding higher termites (*Nasutitermes* spp.). Appl Environ Microbiol 78(13):4691–4701. https://doi.org/10.1128/AEM.00683-12

Kumar P, Barrett DM, Delwiche MJ, Stroeve P (2009) Methods for pretreatment of lignocellulosic biomass for efficient hydrolysis and biofuel production. Ind Eng Chem Res 48(8):3713–3729. https://doi.org/10.1021/ie801542g

Kumar V, Patel SK, Gupta RK, Otari SV, Gao H, Lee JK, Zhang L (2019) Enhanced saccharification and fermentation of rice straw by reducing the concentration of phenolic compounds using an immobilized enzyme cocktail. Biotechnol J 14(6):1800468. https://doi.org/10.1002/biot.201800468

Kumar B, Bhardwaj N, Agrawal K, Chaturvedi V, Verma P (2020) Current perspective on pretreatment technologies using lignocellulosic biomass: An emerging biorefinery concept. Fuel Process Technol 199:106244. https://doi.org/10.1016/j.fuproc.2019.106244

Lazuka A, Auer L, O'Donohue M, Hernandez-Raquet G (2018) Anaerobic lignocellulolytic microbial consortium derived from termite gut: enrichment, lignocellulose degradation and community dynamics. Biotechnol Biofuels 11:1–14. https://doi.org/10.1186/s13068-018-1282-x

Li F, Zhang P, Zhang G, Tang X, Wang S, Jin S (2017) Enhancement of corn Stover hydrolysis with rumen fluid pretreatment at different solid contents: effect, structural changes and enzymes participation. Int Biodeterior Biodegrad 119:405–412. https://doi.org/10.1016/j.ibiod.2016.10.038

Lin L, Xie F, Sun D, Liu J, Zhu W, Mao S (2019) Ruminal microbiome-host crosstalk stimulates the development of the ruminal epithelium in a lamb model. Microbiome 7(1):83. https://doi.org/10.1186/s40168-019-0701-y

Linke D, Bouws H, Peters T, Nimtz M, Berger RG, Zorn H (2005) Laccases of *Pleurotus sapidus*: characterization and cloning. J Agric Food Chem 53(24):9498–9505. https://doi.org/10.1021/jf052012f

Malmuthuge N, Liang G, Guan LL (2019) Regulation of rumen development in neonatal ruminants through microbial metagenomes and host transcriptomes. Genome Biol 20(1):172. https://doi.org/10.1186/s13059-019-1786-0

Mankar AR, Pandey A, Modak A, Pant KK (2021) Pretreatment of lignocellulosic biomass: a review on recent advances. Bioresour Technol 334:125235. https://doi.org/10.1016/j.biortech.2021.125235

Martínez AT (2002) Molecular biology and structure-function of lignin-degrading heme peroxidases. Enzym Microb Technol 30:425–444. https://doi.org/10.1016/S0141-0229(01)00521-X

Mathew GM, Mathew DC, Lo SC, Alexios GM, Yang JC, Sashikumar JM et al (2013) Synergistic collaboration of gut symbionts in *Odontotermes formosanus* for lignocellulosic degradation and bio-hydrogen production. Bioresour Technol 145:337–344. https://doi.org/10.1016/j.biortech.2012.12.055

Mizrahi I, Jami E (2018) Review: the compositional variation of the rumen microbiome and its effect on host performance and methane emission. Animal 12(s2):s220–s232. https://doi.org/10.1017/S1751731118001957

Mizrahi I, Wallace RJ, Moraïs S (2021) The rumen microbiome: balancing food security and environmental impacts. Nat Rev Microbiol 19(9):553–566. https://doi.org/10.1038/s41579-021-00543-6

Munoz C, Mendonça R, Baeza J, Berlin A, Saddler J, Freer J (2007) Bioethanol production from bio-organosolv pulps of *Pinus radiata* and *Acacia dealbata*. J Chem Technol Biotechnol 82:767–774. https://doi.org/10.1002/jctb.1737

Nalepa CA (2020) Origin of mutualism between termites and flagellated gut protists: transition from horizontal to vertical transmission. Front Ecol Evol 6(8):14. https://doi.org/10.3389/fevo.2020.00014

Nargotra P, Vaid S, Bajaj BK (2016) Cellulase production from *Bacillus subtilis* SV1 and its application potential for saccharification of ionic liquid pretreated pine needle biomass under one pot consolidated bioprocess. Fermentation 2(4):19. https://doi.org/10.3390/fermentation2040019

Newbold CJ, De La Fuente G, Belanche A, Ramos-Morales E, McEwan NR (2015) The role of ciliate protozoa in the rumen. Front Microbiol 6:1313. https://doi.org/10.3389/fmicb.2015.01313

Nigam PS, Singh A (2011) Production of liquid biofuels from renewable resources. Prog Energy Combust Sci 37:52–68. https://doi.org/10.1016/j.pecs.2010.01.003

Niu D, An W, Yu C, Zhu P, Li C, Yin D, Zhi J, Jiang X, Ren J (2023) Pre-pasteurization enhances the fermentation of wheat straw by *Irpex lacteus*: chemical composition, enzymatic hydrolysis, and microbial community. Ind Crop Prod 15(202):116962. https://doi.org/10.1016/j.indcrop.2023.116962

Noda S, Iida T, Kitade O, Nakajima H, Kudo T, Ohkuma M (2005) Endosymbiotic *Bacteroidales* bacteria of the flagellated protist *Pseudotrichonympha grassii* in the gut of the termite *Coptotermes formosanus*. Appl Environ Microbiol 71(12):8811–8817. https://doi.org/10.1128/AEM.71.12.8811-8817.2005

Ohkuma M, Ohtoko K, Iida T, Tokura M, Moriya S, Usami R, Horikoshi K, Kudo T (2000) Phylogenetic identification of hypermastigotes, *Pseudotrichonympha*, *Spirotrichonympha*, *Holomastigotoides*, and parabasalian symbionts in the hindgut of termites. J Eukar Microbiol 47(3):249–259. https://doi.org/10.1111/j.1550-7408.2000.tb00044.x

Ohkuma M, Sato T, Noda S, Ui S, Kudo T, Hongoh Y (2007) The candidate phylum 'termite group 1' of bacteria: phylogenetic diversity, distribution, and endosymbiont members of various gut flagellated protists. FEMS Microbiol Ecol 60(3):467–476. https://doi.org/10.1111/j.1574-6941.2007.00311.x

Ozbayram EG, Kleinsteuber S, Nikolausz M, Ince B, Ince O (2017) Effect of bioaugmentation by cellulolytic bacteria enriched from sheep rumen on methane production from wheat straw. Anaerobe 46:122–130. https://doi.org/10.1016/j.anaerobe.2017.03.013

Panwar NL, Kaushik SC, Kothari S (2011) Role of renewable energy sources in environmental protection: a review. Renew Sust Energ Rev 15(3):1513–1524. https://doi.org/10.1016/j.rser.2010.11.037

Ramachandran U, Wrana N, Cicek N, Sparling R, Levin DB (2008) Hydrogen production and end-product synthesis patterns by *clostridium termitidis* strain CT1112 in batch fermentation cultures with cellobiose or α-cellulose. Int J Hydrog Energy 33:7006–7012. https://doi.org/10.1016/j.ijhydene.2008.09.022

Rey M, Enjalbert F, Combes S, Cauquil L, Bouchez O, Monteils V (2014) Establishment of ruminal bacterial community in dairy calves from birth to weaning is sequential. J Appl Microbiol 116(2):245–257. https://doi.org/10.1111/jam.12405

Rieu-Lesme F, Fonty G, Doré J (1995) Isolation and characterization of a new hydrogen-utilizing bacterium from the rumen. FEMS Microbiol Lett 125:77–82. https://doi.org/10.1111/j.1574-6968.1995.tb07338.x

Saad MBW, Oliveira LRM, Cândido RG, Quintana G, Rocha GJM, Gonçalves AR (2008) Preliminary studies on fungal treatment of sugarcane straw for organosolv pulping. Enzym Microb Technol 43:220–225. https://doi.org/10.1016/j.enzmictec.2008.03.006

Saritha M, Arora A, Lata (2012) Biological pretreatment of lignocellulosic substrates for enhanced delignification and enzymatic digestibility. Ind J Microbiol 52:122–130. https://doi.org/10.1007/s12088-011-0199-x

Sasaki C, Yamanaka S (2020) Novel sterilization method combining food preservative use and low temperature steaming for treatment of lignocellulosic biomass with white rot fungi. Ind Crop Prod 155:112765. https://doi.org/10.1016/j.indcrop.2020.112765

Scharf ME, Tartar A (2008) Termite digestomes as sources for novel lignocellulases. Biofuels Bioprod Biorefin 2(6):540–552. https://doi.org/10.1002/bbb.107

Sharma HK, Xu C, Qin W (2019) Biological pretreatment of lignocellulosic biomass for biofuels and bioproducts: an overview. Waste Biomass Valorization 10:235–251. https://doi.org/10.1007/s12649-017-0059-y

Sharma V, Tsai ML, Nargotra P, Chen CW, Sun PP, Singhania RR, Patel AK, Dong CD (2022a) Journey of lignin from a roadblock to bridge for lignocellulose biorefineries: a comprehensive review. Sci Total Environ 863:160560. https://doi.org/10.1016/j.scitotenv.2022.160560

Sharma S, Tsai ML, Sharma V, Sun PP, Nargotra P, Bajaj BK, Chen CW, Dong CD (2022b) Environment friendly pretreatment approaches for the bioconversion of lignocellulosic biomass into biofuels and value-added products. Environments 10:6. https://doi.org/10.3390/environments10010006

Si M, Liu D, Liu M, Yan X, Gao C, Chai L, Shi Y (2019) Complementary effect of combined bacterial-chemical pretreatment to promote enzymatic digestibility of lignocellulose biomass. Bioresour Technol 272:275–280. https://doi.org/10.1016/j.biortech.2018.10.036

Srivastava N, Rawat R, Singh Oberoi H, Ramteke PW (2015) A review on fuel ethanol production from lignocellulosic biomass. Int J Green Energy 12(9):949–960. https://doi.org/10.1080/15435075.2014.890104

Strassert JF, Köhler T, Wienemann TH, Ikeda-Ohtsubo W, Faivre N, Franckenberg S, Plarre R, Radek R, Brune A (2012) *Candidatus Ancillula trichonymphae*, a novel lineage of endosymbiotic Actinobacteria in termite gut flagellates of the genus *Trichonympha*. Environ Microbiol 14(12):3259–3270. https://doi.org/10.1111/1462-2920.12012

Sun J, Ding SY, Doran-Peterson J (2013) Biomass and its biorefinery: novel approaches from nature-inspired strategies and technology. Biological conversion of biomass for fuels and chemicals: exploration from natural utilization systems. Pp 1–13. https://doi.org/10.1039/9781849734738-00001

Tartar A, Wheeler MM, Zhou X, Coy MR, Boucias DG, Scharf ME (2009) Parallel metatranscriptome analyses of host and symbiont gene expression in the gut of the termite *Reticulitermes flavipes*. Biotechnol Biofuels 2:1–9. https://doi.org/10.1186/1754-6834-2-25

Todaka N, Moriya S, Saita K, Hondo T, Kiuchi I, Takasu H, Ohkuma M, Piero C, Hayashizaki Y, Kudo T (2006) Environmental cDNA analysis of the genes involved in lignocellulose digestion in the symbiotic protist community of *Reticulitermes speratus*. FEMS Microbiol Ecol 59(3):592–599. https://doi.org/10.1111/j.1574-6941.2006.00237.x

Tsegaye B, Balomajumder C, Roy P (2018) Biodegradation of wheat straw by *Ochrobactrum oryzae* BMP03 and *Bacillus* sp. BMP01 bacteria to enhance biofuel production by increasing total reducing sugars yield. Environ Sci Pollut Res 25:30585–30596. https://doi.org/10.1007/s11356-018-3056-1

Vaidya AA, Murton KD, Smith DA, Dedual G (2022) A review on organosolv pretreatment of softwood with a focus on enzymatic hydrolysis of cellulose. Biomass Convers Biorefinery 11:5427–5442. https://doi.org/10.1007/s13399-022-02373-9

Vassilev SV, Baxter D, Andersen LK, Vassileva CG, Morgan TJ (2022) An overview of the organic and inorganic phase composition of biomass. Fuel 94:1–33. https://doi.org/10.1016/j.fuel.2011.09.030

Vikram S, Arneodo JD, Calcagno J, Ortiz M, Mon ML, Etcheverry C, Cowan DA, Talia P (2021) Diversity structure of the microbial communities in the guts of four neotropical termite species. Peer J 9:e10959. https://doi.org/10.7717/peerj.10959

Vincent JF, Bogatyreva OA, Bogatyrev NR, Bowyer A, Pahl AK (2006) Biomimetics: its practice and theory. J R Soc Interface 3(9):471–482. https://doi.org/10.1098/rsif.2006.0127

Vu HP, Nguyen LN, Vu MT, Johir MA, McLaughlan R, Nghiem LD (2020) A comprehensive review on the framework to valorize lignocellulosic biomass as biorefinery feedstocks. Sci Total Environ 743:140630. https://doi.org/10.1016/j.scitotenv.2020.140630

Wang Y, McAllister TA (2002) Rumen microbes, enzymes, and feed digestion—a review. Asian Australas J Anim Sci 15(11):1659–1676. https://doi.org/10.5713/ajas.2002.1659

Wang W, Yuan T, Cui B, Dai Y (2013) Investigating lignin and hemicellulose in white rot fungus-pretreated wood that affect enzymatic hydrolysis. Bioresour Technol 134:381–385. https://doi.org/10.1016/j.biortech.2013.02.042

Williams CL, Thomas BJ, McEwan NR, Rees Stevens P, Creevey CJ, Huws SA (2020) Rumen protozoa play a significant role in fungal predation and plant carbohydrate breakdown. Front Microbiol 11:720. https://doi.org/10.3389/fmicb.2020.00720

Witomski P, Olek W, Bonarski JT (2014) Effects of white and brown rot decay on changes of wood ultrastructure. Bioresources 9(4):7363–7371

Wong DW (2009) Structure and action mechanism of ligninolytic enzymes. Appl Biochem Biotechnol 157:174–209. https://doi.org/10.1007/s12010-008-8279-z

Wu Z, Peng K, Zhang Y, Wang M, Yong C, Chen L, Qu P, Huang H, Sun E, Pan M (2022) Lignocellulose dissociation with biological pretreatment towards the biochemical platform: a review. Mater Today Biol 16:100445. https://doi.org/10.1016/j.mtbio.2022.100445

Xiao P, Mori T, Kamei I, Kondo R (2011) A novel metabolic pathway for biodegradation of DDT by the white rot fungi, *Phlebia lindtneri* and *Phlebia brevispora*. Biodegradation 22:859–867. https://doi.org/10.1007/s10532-010-9443-z

Yadav SK (2017) Technological advances and applications of hydrolytic enzymes for valorization of lignocellulosic biomass. Bioresour Technol 245:1727–1739. https://doi.org/10.1016/j.biortech.2017.05.066

Yan X, Wang Z, Zhang K, Si M, Liu M, Chai L, Shi Y. (2017) Bacteria-enhanced dilute acid pretreatment of lignocellulosic biomass. Bioresour Technol 245:419–425. https://doi.org/10.1016/j.biortech.2017.08.037

Yu J, Zhang J, He J, Liu Z, Yu Z (2009) Combinations of mild physical or chemical pretreatment with biological pretreatment for enzymatic hydrolysis of rice hull. Bioresour Technol 100:903–908. https://doi.org/10.1016/j.biortech.2008.07.025

Yu H, Du W, Zhang J, Ma F, Zhang X, Zhong W (2010) Fungal treatment of cornstalks enhances the delignification and xylan loss during mild alkaline pretreatment and enzymatic digestibility of glucan. Bioresour Technol 101:6728–6734. https://doi.org/10.1016/j.biortech.2010.03.119

Zhang K, Si M, Liu D, Zhuo S, Liu M, Liu H, Yan X, Shi Y (2018) A bionic system with Fenton reaction and bacteria as a model for bioprocessing lignocellulosic biomass. Biotechnol Biofuels 11:146. https://doi.org/10.1186/s13068-018-1154-1

Zhang J, Li K, Liu S, Huang S, Xu C (2022) Alkaline hydrogen peroxide pretreatment combined with bio-additives to boost high-solids enzymatic hydrolysis of sugarcane bagasse for succinic acid processing. Bioresour Technol 345:126550. https://doi.org/10.1016/j.biortech.2021.126550

Chapter 6
Bioengineering the Rumen Microbiota as an Advanced Biocatalyst for Renewable Fuels

Maryam Gheibipour, Mahta Yousefipour, Sahand Jorfi, Zahra Eslamifar, Maryam Panahizade, Fateme Gheibipour, Niloofar Rahimi babadi, and Arezoo Gheibipour

Abstract The world's fossil energy resources are decreasing, which has led to an increasing desire to produce fuels from organic waste to create a competitive, resource-efficient, and low-carbon sustainable economy. In natural ecosystems like rumen microbiome (RM), different activities such as hydrolysis, acetogenesis, acidogenesis, and methanogenesis of lignocellulosic biomass (LB) have been observed. These processes are carried out through physical penetration and syner-

M. Gheibipour (✉)
Environmental Technologies Research Center, Ahvaz Jundishapur University of Medical Sciences, Ahvaz, Iran

Center for Climate Change and Health Research (CCCHR), Dezful University of Medical Sciences, Dezful, Iran

Department of Animal Science, Faculty of Agriculture, University of Birjand, Birjand, Iran
e-mail: gheibipour.m@ajums.ac.ir; gheibipour.mariya94@birjand.ac.ir

M. Yousefipour
Center for Climate Change and Health Research (CCCHR), Dezful University of Medical Sciences, Dezful, Iran

Department of Educational Sciences, Farhangian University, Tehran, Iran

S. Jorfi
Environmental Technologies Research Center, Ahvaz Jundishapur University of Medical Sciences, Ahvaz, Iran

Department of Environmental Health Engineering, School of Health, Ahvaz Jundishapur University of Medical Sciences, Ahvaz, Iran

Z. Eslamifar
Department of Laboratory Sciences, School of Allied Medical Sciences, Dezful University of Medical Sciences, Dezful, Iran

M. A. Dar et al. (eds.), *Recent Trends in Lignocellulosic Biofuels and Bioenergy*, Clean Energy Production Technologies,
https://doi.org/10.1007/978-981-96-4636-4_6

gistic enzyme activities that microbial communities (fungi, protozoa, archaea, bacteria, and viruses) are responsible for them. Here, microbial fuel cells act as bioreactors and microbes as biocatalysts that can directly transform chemical energy stored in LB into electrical energy. Therefore, microbiome engineering as a practical technique can enhance host performance through the selection and manipulation of specific microbial communities. It is done by leveraging scientific fundamental programs and designing quantitative methods to form microbiomes with effective functions. It promises sustainable energy production from renewable biomass. The purpose of this chapter is to review the functions, enzymatic potentials of RM and its applications for the efficient bioconversion of LB to biofuels (BFs). Finally, the current status of RM applications, diverse metabolic and genomic approaches for the development of RM bioreactors for efficient bioconversion of LBs to renewable fuels, and the challenges ahead for the practical application of bioenergy have been highlighted.

Keywords Biocatalysts · Bioengineering · Biofuels · Lignocellulose biomass · Renewable resources · Rumen microbiota

6.1 Biofuels and Bioenergy

In the twenty-first century, fossil fuels have become an essential need for human survival, serving numerous requirements such as generating electricity, heating, transportation, initiating businesses, and manufacturing of various products. The increasing demand for fossil fuels has led to a rise in greenhouse gas emissions (Naseema Rasheed et al. 2023). As the primary sources of energy, fossil fuels account for 80% of the energy consumed, with the transportation sector alone utilizing approximately 58%. The excessive reliance on fossil fuels significantly contributes to the emission of harmful gases, air pollution, climate change, and the destruction of biodiversity and natural glaciers (Shweta et al. 2024). Consequently, the quest for alternative natural or renewable resources that can fulfill human needs has become increasingly challenging (Das et al. 2021). Recently, it has been

M. Panahizade
Department of Animal Science, Faculty of Animal Science and Food Industry, Ramin Khuzestan University of Agriculture and Natural Resources, Molasani, Iran

F. Gheibipour
Department of Humanities, Payam Noor Andimeshk University, Andimeshk, Iran

N. Rahimi babadi
Department of Midwifery, Faculty of Medical sciences, Karaj Branch, Islamic Azad University, Karaj, Iran

A. Gheibipour
Department of Physics, Faculty of Sciences, Urmia University, Urmia, Iran

demonstrated that biomass derived from plants, animals, and microorganisms is the most effective for producing a new generation of bio-based chemicals, such as BFs (Bhaskar et al. 2024). These organic matters are affordable, renewable, and abundantly available resources, that are composed of complex and sugar-rich polymers such as cellulose, hemicellulose, and lignin. The depolymerization and degradation of this biomass have traditionally required costly enzymatic and chemical processes, which increase overall expense of decomposition methods (Zhang et al. 2012). To address this issue, certain enzymes produced by bacteria or fungi have been proposed. These complex microbial multiprotein enzymes possess the ability to degrade or hydrolyze biomass effectively (Bhaskar et al. 2024). The depletion of readily available fossil energy resources, climate changes, and the political imperative to achieve energy self-sufficiency and sustainable solutions have prompted governments to pursue the strategic and intelligent use of renewable resources to foster a robust bioeconomy (Aragao 2013). Thus, bioengineering (BE) can play a pivotal role in achieving sustainable energy supply approaches. It integrates various sciences, including engineering, mathematics, and life sciences, to identify and address different challenges in biology, medicine, healthcare, and the environment (Aragao 2013). The BE is a relatively novel discipline that combines numerous contemporary engineering fields, such as chemical, electrical, and mechanical engineering, along with several scientific specialties, including biomedical, biomolecular, biomechanical, biotechnological, and biochemical engineering. Although these areas focus on different interests and aspects, all strive to improve life (Xue et al. 2024; Cameron et al. 2024; Kosky et al. 2021; Peng et al. 2010). This chapter aims to review RM, and investigate their microbial products for the bioconversion of LB into renewable fuels. It also highlights advances in metabolic and genetic engineering that elucidate the structure and functions of the RM to exploit their hidden biotechnological potential. In addition, future perspectives on the integrated bioprocessing approach to reduce the cost of BFs production, exploration of new microorganisms and enzymes from rumen for the production of BFs, valuable chemicals and development of rumen biomimetic systems in an anaerobic digester are discussed.

6.2 Lignocellulose: Potential Sources for Biofuel Production

The "bio" refers to any living being, while "fuel" recourse to energy-releasing something. The BFs is a type of nonfossil fuel derived from the organic biomass of renewable resources, including plants, animals, microbes, and organic waste. The energy generated from BFs is referred to as bioenergy (Rai et al. 2022). The BFs are categorized into three primary groups: (i) liquid (e.g., ethanol and biodiesel), (ii) solid (such as charcoal, wood pellets, and fuelwood), and (iii) gas (such as biogas). Unlike fossil fuels, BFs are renewable because they are produced by converting photosynthetic solar energy into chemical energy. Based on the source of production (i.e., biomass raw materials) and the technology used, BFs are further classified into four generations, ranging from first to fourth generation (Alalwan et al. 2019). Here, four generations of BFs will be briefly reviewed:

1. First-generation BFs including biodiesel, bioethanol, and biogas are primarily derived from edible plant materials, such as grains and oilseeds. These BFs are produced from sources like sugar, starch, vegetable oils, and fats (Hirani et al. 2018).
2. Second-generation BFs are comparatively advanced fuels derived from various nonfood biomass sources, such as lignocellulose. The LBs are produced from nonedible plants or nonedible parts of plants, such as vegetable grasses, forest residues, and agricultural waste. These materials are abundant in the natural ecosystem and can serve as feedstock for BFs production. Ethanol, butanol, and mixed alcohols are examples of lignocellulosic BFs (Rai et al. 2022).
3. Third-generation BFs are derived from photosynthetic microbes, such as microalgae and other autotrophic organisms. In this process, carbon dioxide, light, and various nutrients are utilized in the synthesis of biomass, which is subsequently used in BFs production (Rai et al. 2022; Dragone et al. 2010).
4. Fourth-generation BFs: This technology is not yet common and is still in the developmental stage, having emerged only a few years ago. It utilizes genetically altered photosynthetic microbes, such as cyanobacteria, algae, and fungi, as well as feedstocks. These microbes possess the ability to convert atmospheric CO_2 into BFs (Alalwan et al. 2019).

The three most abundant components of a plant cell wall, celluloses, hemicelluloses, and lignin are complex polysaccharides. These compounds account for a significant portion of fodder, and their hydrolysis presents industrial challenges, particularly with lignocellulosic materials. The LB produced through photosynthesis on Earth exceeds 100 billion tons annually (Rai et al. 2022). However, only a small fraction is utilized by human activities, resulting in significant resource waste. Therefore, the bioconversion of LB does not compete with food and feed (Takizawa et al. 2020). The primary components of lignocellulosic feedstock are cellulose and hemicelluloses, which together account for about two-thirds of the dry mass of plants. These components are converted into sugars through thermochemical and biological processes (Rai et al. 2022; Sawatdeenarunat et al. 2015). Lignocellulosic feedstocks are categorized into three types: (1) herbaceous and woody energy crops, (2) agricultural residues, and (3) forest residues. The cellulose present in the cell wall must be broken down to form sugars, which can then be further converted into ethanol or fuels such as biodiesel or butanol. However, it is important to note that the morphological characteristics of plant cell walls hinder the production of cellulosic BFs (Rai et al. 2022). The products of the biofuel process include biodiesel, biobutanol, bioethanol, biohydrogen, bio-methanol, bio-oil, and biogas. Using corn straw and nonagricultural plants to produce bioethanol helps reduce greenhouse gas emissions in both the agricultural and energy sectors (Shweta et al. 2024). Nowdays the efficient and inexpensive production of bioethanol using various grains and non-crop plants has gained global popularity as a gasoline substitute due to its carbon neutrality (Kratzel et al. 2020). In some countries, biodiesel is produced using oils from soybeans, corn, canola, linseed, jatropha, castor, and animal fat, among others (Shweta et al. 2024).

6.3 Pretreatment of Lignocellulose Biomass

The chemical composition of agriculture residues consists of soluble components (e.g., sugar, amino acids, and organic acids), insoluble components (e.g., lignin, hemicellulose, and cellulose), and other components such as protein, fats, oil waxes, pigment, resins, and mineral (Rai et al. 2020). This biomass can be used to improve quality of soil, supply animal feed, and produce BFs. Due to complex structure of LB, its use for BFs production is limited. The LB is contained 40–60% cellulose, 20–40% hemicellulose, and 10–24% lignin. Therefore, pretreatment of LBs is a crucial step in BFs production, which occurs through different physical, chemical, physicochemical, and biological pretreatments (Bhujbal et al. 2022). It is implemented during the processes of BFs and bioenergy production and primarily consists of two methods:

(i) Physical and chemical pretreatment: These methods are widely employed in BFs production to enhance the quality of the substrate and facilitate further digestion. Physical techniques include the use of heat, pressure, steam, hot water, and ultrasonics, while chemical methods encompass oxidation, ozonation, and acid or base pretreatment. These techniques are typically used in combination to achieve optimal results (Wagner et al. 2018).
(ii) Physiochemical pretreatment: it includes liquid hot water, ammonia-fiber explosion, supercritical CO_2, and steam explosion (Chettri et al. 2022).
(iii) Biological pretreatment: this method is primarily aimed to break down and disrupt lignin to expose cellulose structures and makes them more susceptible to microbial digestion. In this process, microorganisms play a crucial role, and valuable by-products are also produced (Wagner et al. 2018).

Biological treatment employs microorganisms to effectively degrade LBs using hydrolyzing enzymes such as peroxidases, manganese peroxidases, ligninase and laccases. For example, white-rot fungi was produced some of these enzymes to remove lignin (Adegboye et al. 2021). As reported previously, sulfuric acid could decline 89.5% of hemicellulose solubilization (Dionísio et al. 2021). These pretreatments are costly and have environmental hazards, while biological pretreatment is cheap and environmentally friendly (Bhujbal et al. 2022).

6.4 Biodegradation of Lignocellulose in the Rumen

Ruminant animals, such as cattle, sheep, goats, buffalo, camels, deer, rams, giraffes, okapis, chevrotains, and pronghorns, possess a large anaerobic chamber known as the rumen, which houses a diverse array of microorganisms. In the rumen, forage degradation is an essential process that supports the growth and performance of ruminants. However, research into RM has revealed valuable insights into the degradation of plant cells through microbial activity. Rumen microorganisms are involved in colonizing and breaking down fiber into progressively smaller pieces (Osorio-Doblado et al. 2023). The RM is one of the most fascinating and complex ecosystems known to science (Singh et al. 2022; Juengel et al. 2014). It comprises numerous, intricate, and diverse communities of interacting microorganisms, including bacteria, protozoa, fungi, and archaea. This microbiome acts as an ecosystem, consisting of bacteria (up to 10^{11} viable cells per gram), ciliate protozoa (10^4–10^6 viable cells per gram), anaerobic fungi (10^2–10^4 viable cells per gram), viruses, and archaea, with numbers exceeding ten times the total number of cells in the host body (Marappan et al. 2024). The competition between microbiome components and pathogenic microorganisms for food particles leads to decomposition, transformation, and protective processes. Microbe-host interactions are crucial for the normal development and physiology of ruminants (Choudhury et al. 2015). Rumen is an anaerobic environment with optimal conditions for microorganism growth, that has a pH of 5.5–6.9 and a temperature between 38 and 41 °C. Rumen optimal condition is maintained with the assistance of its microbiomes, that facilitates ideal growth and transforms feed particles into desirable products (Marappan et al. 2024). In the rumen, bacterial conversion of complex polysaccharides occurs, breaking them down lignocellulosic molecules into oligosaccharides and subsequently into hydrogen and volatile fatty acids (VFAs) that are considered the end products of fiber fermentation in the rumen, including acetate, propionate, and butyrate. Additionally, small amounts of valeric, caproic, isobutyric, and isovaleric acids serve as significant energy sources for host animal (He et al. 2024). Zhang et al. (2016) were treated the rice straw with rumen fluid for methane production. Recent studies have indicated that pretreatment of rice straw with RM reduces the volatile solids, while increasing the biomethane was without any physicochemical pretreatment (Candia-García et al. 2018; Thongbunrod and Chaiprasert 2021). Similarly, Nguyen et al. (2020) have found that RM could produce four times more VFAs than anaerobic sludge.

6.5 Rumen as a Source of Potential Biocatalysts

Biocatalysts are natural substances that increase the catalyze of chemical reactions, and include a group of proteins called enzymes or a combination of enzymes and cells (whole cell). These biomolecules selectively act on special substrates and can use to produce desired metabolites such as organic acids, antimicrobials, and other chemicals. Biocatalysts have diverse biotechnological applications and advantages due to the non toxicity, high rate of reaction, and biodegradability. They are used in different industrial processes to the synthesis of food and feed additives, herbicides, medicines, biofuels, and etc. (Patil and Rathod 2021). Rumen, as vat of biocatalysts, engages in a close and intensive interaction with ruminants, residing in the rumen to facilitate digestion and ferment biomass into nutrients. The end products of this fermentation are subsequently utilized by the host, enabling ruminants to convert plant materials that are indigestible to humans into high quality animal protein, such as meat and milk. Many agricultural by-products and food residues have been proposed for ruminant feeding due to the interesting activities of RM, which can reduce competition between humans and animals for grains (Liu et al. 2021). The long development of the digestive gut under significant biomass pressure has led to the rumen being recognized as one of the most efficient bioreactors. The RM consists of representatives from all kingdoms of life, including protozoa, fungi, methanogens, bacteriophages, ciliates, and bacteria, with the latter group being the most predominant (Aragao, 2013). The rumen contains a diverse microbial biomass including ciliate protozoa (up to 50% of total microorganisms), anaerobic fungi (8–20%), archaea (0.3–4%), and anaerobic bacteria (approximately 26%), which degrade LB by producing different enzymes (Table 6.1, Tapio et al. 2017). As previously reported, tannase-producing bacteria isolated from rumen of deer could enhance degradability of lignocellulosic forages (conocarpus and eucalyptus leaves), and improve fermentative parameters, in vitro. These studies reported that tannase enzyme, as a biocatalyst, was effective in processing the tannin-rich leaves of tropical trees, and had potential to use in practical ruminant feeding (Mohammadabadi et al. 2021, 2023; Gheibipour 2017). Furthermore, biocatalysts such as cellulase, tannase, amylase, and protease were identified in the rumen of deer, goat, camel, and wild ram, that also suppressed pathogenic strains by producing antimicrobial biomolecules (Ghiasi et al. 2024; Gheibipour et al. 2022, 2023, 2024a, b). Gheibipour et al. (2024c) were proved the positive effect of tannase biocatalyst on the gas production and in vitro digestibility of animal diets containing pomegranate peel.

6.5.1 Bacteria

The most diverse group of microorganisms, bacteria, have a prevalence of 10^{10}–10^{11} cells/mL in rumen fluid; however, solid-adherent bacteria constitute 70–80% of its microflora. The rumen bacteria diversity is estimated around 7000 species, that approximately 30% remain unrecognized. So far, more than 2500 species of them have been identified that anaerobic bacteria make up a complex and diverse collection of them (Tapio et al. 2017). Each of these bacteria competes with the others through different substrate fermentation and biochemical/ecological strategies. Due to variable diets, rumen bacteria may degrade various substrates including fiber, starch, protein, and sugars. Hence, their metabolic processes change rapidly in different conditions (Rai et al. 2020). Ghiasi et al. (2024) observed that the bacterial genera producing the cellulase biocatalyst, which were isolated from the rumen, varied according to the animal species and the type of forages they consumed. Also, the most abundant ruminal bacterial phyla include *Proteobacteria*, *Bacteroidetes*, and *Firmicutes* (Lan and Yang 2019), while the most identified bacteria belonged to *Ruminococcus*, *Fibrobacter*, *Prevotella*, *Selenomonas*, *Streptococcus*, and *Ruminobacter* genera (Russell and Mantovani 2002). Cellulolytic bacteria (Table 6.1) enable ruminants to digest fibrous residues (Marappan et al. 2024). Some of them, such as fibrolytic bacteria, particularly cellulose-degrading species of *Ruminococcus* and *Eubacterium* (*Firmicutes*) are effective H_2 producers

Table 6.1 Microbial groups identified in the rumen with different activities

Bacteria	
Acetogens	*Ruminococcaceae acetogens*, *Acetitomaculum ruminis*, *Eubacterium limosum*, *Lachnospiraceae*
Acid utilizers	*Desulfovibrio desulfuricans*, *Wolinella succinogenes*, *Megasphaera elsdeni*, *Veillonella gazogene*, *Succiniclasticum ruminis*, *Micrococcus lactolytica*, *Oxalobacter formigenes*, *Desulfotomaculum ruminis*
Cellulolytic	*Fibrobacter succinogenes*, *Klebsiella oxytoca*, *Pectobacterium carotovorum*, *Butyrivibrio fibrisolvens*, *Ornithinibacillus bavariensis*, *Ruminococcus flavefaciens*, *Ruminococcus albus*, *Clostridium cellobioparum*, *Enterobacter* sp., *Clostridium longisporum*, *Clostridium lochheadii*, *Eubacterium cellulosolvens*, *Escherichia coli*, *Raoultella planticola*, *Escherichia fergusonii*, *Enterobacter cloacae*, *Enterobacter asburiae*, *Klebsiella pneumoniae*, *Citrobacter murliniae*, *Citrobacter braakii*, *Bacillus subtilis*, *Bacillus circulans*, *Citrobacter* sp., *Pseudomonas* sp., *Klebsiella* sp., *Klebsiella pneumonia*, *Klebsiella variicola*, *Lelliottia amnigena*, *Bordetella* sp., *Bacillus stratosphericus*, *Paenibacillus dendritiformis*
Hemicellulolytic	*Eubacterium xylanophilum*, *Prevotella ruminicola*, *Eubacterium uniformis*
Lipolytic	*Anaerovibrio lipolytica*, *Paenibacillus dendritiformis*, *Bacillus circulans*, *Citrobacter* sp., *Pseudomonas* sp., *Klebsiella* sp., *Enterobacter* sp., *Klebsiella pneumonia*, *Klebsiella variicola*, *Lelliottia amnigena*, *Bordetella* sp., *Bacillus stratosphericus*
Pectinolytic	*Lachnospira multiparus*, *Treponema saccharophilum*
Proteolytic	*Prevotella ruminicola*, *Citrobacter braakii*, *Ruminobacter amylophilus*, *Clostridium bifermentans*, *Ornithinibacillus bavariensis*, *Enterobacter cloacae*, *Enterobacter asburiae*, *Raoultella ornithinolytica*, *Klebsiella pneumoniae*, *Bacillus stratosphericus*, *Klebsiella oxytoca*, *Escherichia fergusonii*, *Escherichia coli*, *Enterobacter cloacae*, *Pectobacterium carotovorum*, *Escherichia coli*, *Citrobacter murliniae*, *Bacillus subtilis*, *Bacillus circulans*, *Citrobacter* sp., *Pseudomonas* sp., *Klebsiella* sp., *Enterobacter* sp., *Klebsiella pneumonia*, *Klebsiella variicola*, *Lelliottia amnigena*, *Bordetella* sp., *Paenibacillus dendritiformis*
Amylolytic	*Streptococcus bovis*, *Prevotella ruminicola*, *Klebsiella pneumoniae*, *Enterobacter cloacae*, *Escherichia coli*, *Bacillus subtilis*, *Bacillus circulans*, *Citrobacter* sp., *Pseudomonas* sp., *Klebsiella* sp., *Enterobacter* sp., *Klebsiella pneumonia*, *Klebsiella variicola*, *Lelliottia amnigena*, *Bordetella* sp., *Bacillus stratosphericus*, *Paenibacillus dendritiformis*, *Ruminobacter amylophilus*,
Saccharolytic	*Succinivibrio dextrinosolvens*, *Pseudomonas* sp., *Klebsiella* sp., *Succinivibrio amylolytica*, *Selenomonas ruminantium*, *Lactobacillus acidophilus*, *Lactobacillus fermentum*, *Lactobacillus plantarum*, *Lactobacillus brevis*, *Lactobacillus helveticus*, *Bifidobacterium globosum*, *Bifidobacterium longum*, *Bifidobacterium thermophilum*, *Bifidobacterium ruminale*, *Bifidobacterium ruminantium*, *Bacillus circulans*, *Citrobacter* sp., *lactobacillus casei*, *Enterobacter* sp., *Klebsiella pneumonia*, *Klebsiella variicola*, *Lelliottia amnigena*, *Bordetella* sp., *Bacillus stratosphericus*, *Paenibacillus dendritiformis*

(continued)

Table 6.1 (continued)

Tanninolytic	*Streptococcus caprinus*, *Citrobacter murliniae*, *Ornithinibacillus bavariensis*, *Citrobacter braakii*, *Bacillus subtilis*, *Eubacterium oxidoreducens*, *Escherichia coli*, *Raoultella planticola*, *Escherichia fergusonii*, *Enterobacter cloacae*, *Enterobacter asburiae*, *Raoultella ornithinolytica*, *Klebsiella pneumoniae*, *Klebsiella oxytoca*, *Klebsiella michiganensis*, *Pectobacterium carotovorum*, *Acinetobacter baumannii*
Ureolytic	*Megasphaera elsdenii*, *Pseudomonas*, *Haemophilus*, *Neisseria*, *Streptococcus*, *Actinomyces*, *Bacillus*, *Succinivibrio dextrinosolvens*, *Treponema* sp., *Ruminococcus bromii*, *Butyrivibrio* sp., *Bifidobacterium* sp., *Prevotella ruminicola*, *Peptostreptococcus productus*
Bacteriophages	
	Methanobacterium phage Ψ M10, *Methanothermobacter phage Ψ M100*, *Methanobacterium phage Ψ M100*, *Methanobacterium phage ΨM2*, *Methanobacterium phage Ψ M1*
Fungi	
	Neocallimastix patriciarum, *Cyllamyces icaris*, *Orpinomyces joynii*, *Orpinomyces intercalaris*, *Piromyces communis*, *Piromyces mae*, *Piromyces minutus*, *Piromyces dumbonicus*, *Piromyces citronii*, *Piromyces polycephalus*, *Anaeromyces mucronatus*, *Anaeromyces elegans*, *Caecomyces communis*, *Caecomyces equi*, *Cyllamyces aberensis*, *Neocallimastix frontalis*, *Neocallimastix hurleyensis*, *Neocallimastix variabilis*, *Piromyces spiralis*, *Piromyces rhizinflatus*, *Caecomyces sympodialis*
Methanogens	
	Methanobrevibacter ruminantium, *Methanosarcina barkeri*, *Methanobacterium formicicum*, *Methanobacterium bryantii*, *Methanomicrobium mobile*, *Methanobrevibacter smithii*, *Methanoculleus olentangyi*
Protozoa	
	Polyplastron multivesiculatum, *Entodinium bovis*, *Entodinium bursa*, *Entodinium chatterjee*, *Entodinium parvum*, *Entodinium longinucleatum*, *Entodinium dubardi*, *Epidinium caudatum*, *Entodinium bubalum*, *Isotricha prostoma*, *Isotricha intestinalis*, *Diplodinium indicum*, *Oligoisotricha bubali*, *Eremoplastron asiaticus*, *Eremoplastron bubalus*, *Entodinium caudatum*, *Diplodinium dentatum*, *Entodinium exiguum*, *Dasytricha ruminantium*

Source: Ghiasi et al. (2024); Srivastava et al. (2021); Gheibipour et al. (2022, 2023, 2024a); Gheibipour (2017); Jin et al. (2016); Gagen et al. (2015a, b); Choudhury et al. (2015); Wozny et al. (1977)

(Li et al. 2022). Although, the cellulolytic *Fibrobacter* do not possess the ability to produce H_2, Bacteroidetes are net H_2 utilizers. The analysis of bacterial community revealed that structural differences in the sheep RM affect methane emissions (Tapio et al. 2017). The methanogens embedded in bacterial biofilms on feed particles exhibit a unique interaction. These two microbial groups can interact through inter-species H_2 transfer, which significantly impacts fermentation, VFAs profile, energy conservation, and methanogenesis in the rumen environment (Lan and Yang 2019). The valuable initiatives, Global Rumen Census and the Rumen Microbial Genomics Network's Hungate 1000 project, have enhanced findings about the role of ruminal bacteria in the digestion of BM (Tapio et al. 2017). Notably, *Ruminococcus*, *Butyrivibrio*, and *Fibrobacter* are recognized as key lignocellulose decomposers. Also, *Ruminococcus flavefaciens* produces a complex cellulosome containing up to 14 enzymes, which plays a crucial role in lignocellulose degradation (Christopherson and Suen 2013). Others have demonstrated that the tannase biocatalyst, isolated from the rumen of deer, wild rams, camels, and goats, can degrade tannic acid into gallic acid, while simultaneously exhibiting antibacterial activity (Gheibipour et al. 2022, 2023, 2024a).

6.5.2 Fungi

Anaerobic fungi are one of the most important eukaryotes in the RM, which are considered as flagellated protozoa. But, they were later categorized as aquatic fungi (Chytridiomycetes). Despite their low population ($\sim 10^6$ zoospores/g rumen content), they are the first and most important strictly anaerobic organisms to colonize fibrous substrates and degrade plant cell in the rumen. Rumen fungi secrete diverse polysaccharide-degrading biocatalysts to decompose pectin, cellulose, and hemicellulose. They encode crucial carbohydrate esterases to degrade cellulose and hemicellulose from their linkage with lignin (Cann et al. 2025). So far, real-time PCR, a molecular biological technique, has helped distinguish more than 18 species of anaerobic fungi in the rumen (Lan and Yang 2019). As previously reported, ruminal anaerobic fungi decompose the cell walls of plant fibers (Orpin 1977). Numerous studies demonstrate that anaerobic fungi are among the most effective microorganisms for degrading lignocellulose in the rumen. So far, six genera of rumen fungi, including *Caecomyces*, *Neocallimastix*, *Oorpinomyces*, *Anaeromyces*, *Piromyces*, and *Cyllamyces* were identified, which are involved in the decomposition of biomass. Compared to others, *Neocallimastix* and *Cyllamyces* have shown high efficiency in the lignocellulosic biomass degradation (Liang et al. 2020; Lan and Yang 2019; Newbold and Ramos-Morales 2020). Furthermore, the multienzyme cellulosome consists of several glycosyl hydrolases, such as endoglucanase, exoglucanase, β-glucosidase, and xylanase secrete by ruminal fungi to enhance lignocellulose degradation efficiency. Also, rumen fungi have a significant advantage over other microorganisms in lignocellulose degradation due to their ability to attach to plant cell walls and produce effective enzymes (Table 6.1). They can generate H_2, CO_2, formate, and acetate as metabolic by-products. The function of anaerobic fungi is most effective when combined with hydrogen-utilizing bacteria and methanogenic archaea that closely associated with fungal hyphae (Liang et al. 2020; Lan and Yang 2019; Newbold and Ramos-Morales 2020; Tapio et al. 2017).

6.5.3 Protozoa

Protozoa are motile eukaryotes that make up almost 50% of the RM, in sizes 10–100 times that of bacteria. These organisms are include two groups Entodiniomorphida and Holotricha (Cann et al. 2025). They ingest starch, thus likely possess amylase biocatalyst for producing glucose, which play an essential role in total VFAs production. These organisms also gobble bacteria and fungi, therefor, their cellulases and hemicellulases activities probably originate from the ingested microorganisms (Cann et al. 2025). Protozoa are difficult to grow in the laboratory, so less is known about their physiology and ecology (Rai et al. 2020). Despite the variety of genera and species, the dominant genus in the rumen is *Entodinium* (Choudhury et al. 2015). Although protozoa are not methane producers, but they play an important role in this pathway by (1) producing significant amounts of hydrogen gas in their hydrogenosomes, (2) hosting symbiotic methanogens, and (3) protecting methanogens from toxicity oxygen (Shinkai et al. 2024; Tapio et al. 2017; Lan and Yang 2019). Compared to entodiniomorphids, holotrichs have a more significant impact on methanogenesis due to more active hydrogenosomes and the ability to synthesize glycogen in the presence of excess carbohydrates (Lan and Yang 2019). Rumen metagenomic analyses have proved the various glycoside hydrolases in protozoa, indicating their capacity to degrade complex carbohydrates (Table 6.1). Furthermore, horizontal transfer of genes from bacteria and archaea has been observed in the rumen protozoan genome, which indicates their close interaction with other species and high fibrolytic ability (Shinkai et al. 2024; Tapio et al. 2017; Lan and Yang 2019). The presence or absence of protozoa affects the structure and diversity of methanogens, and the elimination of protozoa can significantly reduce methanogenesis. Methanogens are closely associated with these protozoa, either on their surfaces or as endosymbionts, to enhance hydrogen availability (Ungerfeld and Pitta 2024). Studies have reported a methane reduction of 9–37% through the elimination of protozoa (Shinkai et al. 2024; Tapio et al. 2017; Lan and Yang 2019).

6.5.4 Methanogen Archaea

Extremophile archaea are primitive single-celled microorganisms that exist in extreme ecosystems. Archaea, particularly methanogens, have a relative abundance in the rumen measured at 5%, based on total bacterial and archaeal 16S rRNA genes (Bath et al. 2013). Methanogens can be either free-living or dependent on protozoa or fungi to enhance their access to H_2 (Janssen and Kirs 2008). Archaea are exclusively anaerobic methanogens with diverse metabolic functions (Table 6.1) that use hydrogen and carbon dioxide produced by other microbes for methane synthesis. Rumen archaea have a high affinity for hydrogen, which drastically lowers the H_2 pressure, triggering the spontaneous conversion of NADH to NAD (Morgavi et al.

2010). Therefore, methanogenesis is a reduction-equivalent sink in the rumen, and without it, the feed catabolism slows or stops due to of NADH accumulation. Methanogens have a vital role to ensure optimal rumen function and often live on and within protozoa or on fungi. Methane gas is an important component in the carbon cycle, and strategies to reduce methane seem necessary because its production causes a loss of carbon and dietary energy up to 12% (McAllister et al. 2015). Analysis of the archaeal 16S rRNA gene sequences in the dataset is revealed that most of the rumen methanogen archaea (92.3%) are classified into three main genera: *Methanobrevibacter* (61.6%), the uncultured Rumen Cluster C (RCC) (15.8%), and *Methanomicrobium* (14.9%) (Jeyamalar 2010). The genus *Methanobrevibacter* is a primary hydrogenotrophic methanogen in the rumen. In addition, *Methanimicrococcus*, *Methanosphaera*, and *Methanobacterium* are key hydrogenotrophic genera among ruminal methanogen members. The *Methanosarcinales*, *Methanosphaera*, and *Methanomassiliicoccaceae*, methylotrophic archaea, are less numerous. While they are undesirable in the biohydrogen-producer reactor several pretreatments are implemented to inhibit methanogens archaea (Bundhoo et al. 2015).

6.5.5 Bacteriophage

Bacteriophage (Table 6.1) present typically at >10^9 particles per mL and are obligate pathogens for bacteria (Choudhury et al. 2015). They are viruses that exclusively prey on bacteria, infect bacteria, snatch bacterial replication machinery, lysis their cell, and release cell contents into the environment. Despite their unknown role, they play a crucial role in daily changes in the bacterial communities and also in nutrient cycling such as nitrogen (Rai et al. 2020). Bacteriophage plays a role in bacterial mass turnover and makes protein available to ruminant host (Choudhury et al. 2015).

Recently, RM engineering has been applied in the livestock industry to improve animal health and performance. In most of the studies, microbial feed additives, including enzymes, probiotics and prebiotics, are added to the diet to manipulate the rumen microflora (Kholif and Patra 2024). Also, the rumen enzymes such as tannase, cellulase, hemicellulase, phytase, amylase, xylanase, beta-glucanase, beta-mannanase, protease, and lysozyme are incorporated in animal nutrition to enhance feed digestion (Kholif and Patra 2024; Saini and Mishra 2024; Ojha et al. 2019). In some cases, tannase feed additive reduced the diarrhea in the weaning lambs, which likely due to its antibacterial properties and its ability to inhibit pathogens that cause diarrhea (Gheibipour et al. 2024b). Furthermore, exogenous enzymes, as alternatives to antibiotics, can enhance growth performance and nutrient digestion in poultry, swine, ruminants, as well as fish, fur-bearing animals, and pets (Anadón et al. 2019). Enzymes such as phytase, xylanases, β glucanases, pectinases, amylases, and proteases can influence the intestinal microbiome by increasing the availability of digested substrates, detoxifying anti-nutritional factors, and producing

Table 6.2 Biofuels produced using lignocellulose and rumen microbiota

Microbial inoculum	Plant biomass	Volume (mL/g)	Reference
Biohydrogen			
Rumen bacteria	Napier grass	6.95	Ho et al. (2011)
RF	Wastepaper	42	Botta et al. (2016)
Rumen *Escherichia coli*	Corn straw	4.71	Pang et al. (2017)
RF non-methanogens	Crude paper	8.2	Botta et al. (2020)
RF bacterial consortia	Napier grass	0.33	Chang et al. (2010)
Biomethane			
RF	Rice straw	82.23	Thongbunrod and Chaiprasert (2021)
RF	Wheat straw	106	Li et al. (2017)
RC	Rice straw	>287	Candia García et al. (2018)
RF	Potato peel slurry	1318.83	Sema et al. (2020)
RF	Seaweed	108.54	Zou et al. (2018)
RF	*Brassica napus*	20.77	Baba et al. (2019)
Bioethanol			
RF cellulase	Wheat straw	0.46 g/g	Patel et al. (2020)
Buffalo RF—*Pichia kudriavzevii* yeast	Lignocellulosic biomass	9.1 g/L	Avchar et al. (2021)
Buffalo RF—*Candida glabrata* yeast	Lignocellulosic biomass	9 g/L	Avchar et al. (2021)
VFAs			
Cow RF	Wheat straw	2.92 ± 0.45 g C L	Lazuka et al. (2015)
Beef cattle RF	Corn Stover	13.271 mg/L	Li et al. (2017)
Cow RF	Maize silage	438 mg VFA/g	Nguyen et al. (2020)
Beef cattle RF	Rice straw	0.37 g/g VS	Liang et al. (2021)

RF rumen fluid, *RC* rumen content

oligosaccharides (Anadón et al. 2019; Femia et al. 2010). These additives likely modify the gut microbiota through production of (i) short-chain fatty acids (SCFAs), which lower luminal pH, provide energy sources, regulate metabolism, and modulate immune responses, (ii) antimicrobial compounds that inhibit pathogens, and (iii) prebiotic derivatives that reduce pathogen adhesion to host cells (Samal and Behura 2015). However, the diverse hydrolytic abilities observed in the rumen confirm the potential of RM in processing of LB and production of BFs and other value-added products.

6.6 Rumen Microbiota in Biofuel Production

The terrestrial plants have a cell wall rich in energy as form of cellulose, hemicellulose, pectin, and lignin, which can convert to second-generation BFs. Most LB is converted to BFs by the action of specific enzymes that depolymerize and catalyze polysaccharides (Srivastava et al. 2021). The RM is an interactive resource of carbohydrate-active enzymes (CAZymes) that divide into six important classes, i.e., glycosyltransferase, glycosyl hydrolase, carbohydrate esterases, polysaccharide lyases, and auxiliary activities. These enzymes have a crucial role in carbohydrate metabolism, including the hydrolysis, formation, and modification of glycosidic bonds in LB, converting complex feed ingredients into easily absorbable monomers (Chettri et al. 2022). The CAZymes are effective in BFs production by converting this recalcitrant polysaccharide into fermentable carbohydrate monomers and offer as a sustainable alternative to solve the challenge of conventional chemical catalysts (Saini et al. 2022). Here, anaerobic fungi and bacteria are promising candidates for lignocellulosic BFs production through CAZymes with natural diversity. As some studies have reported (Table 6.2), RM can be an efficient reactor for the degradation of LB and production of BFs (Chettri et al. 2022). In rumen, hydrogen is generated via anaerobic digestion of organic biomass by different microbial species for instance methanogenic archaea (Bundhoo et al. 2015). Herein, endoglucanases (EGN) are essential for initiation and wide degradation of cellulose substrates through randomized cleavage of 1,4-glycosidic bonds. This process lead to produce different end-products of glucose chains to use in downstream cellulolytic activities. After that, cellobiohydrolases (CHL) bind to the end-products or cleavage sites created by endoglucanases (Cann et al. 2025). They also can attach to the free ends in the cellulosic microfibrils, then slid sequentially along the glucose chain and separate the cellubiose. Accumulation of cellobiose suppress EGN and CHL, while the some of cellodextrinases can cleave cellobiose and cellodextrins to glucose units (Cann et al. 2025). Therefore, enzymes are biocatalysts for the release of fermentable sugars from plants LB to produce BFs. Zhang et al. (2016) reported that pretreatment of rice straw using rumen fluid was enhanced methane production by 82.6%. Recent studies have indicated that pretreatment of rice straw with RM reduced the volatile solids (41.23–82.23%) and increased biomethane (287–310.25 mL/g VS) without physicochemical pretreatment (Candia-García et al. 2018; Thongbunrod and Chaiprasert 2021). Nguyen et al. (2020) also found that RM could produce four times more VFAs than anaerobic sludge. As a biological agent, rumen fluid has been used to treat wheat straw for producing VFAs and biogas. After 93-day anaerobic fermentation, the efficiency of cellulose, hemicellulose, and lignin degradation was reported 97.6%, 95.8%, and 42.4%, respectively, and the VFA production was equal to 0.484 g COD/g VS (Xing et al. 2020)

In a study, microbial analysis revealed that *Ruminococcus*, as hydrogen-producing genus, was dominant at 35 °C, and the *Thermicanus* was the dominant genus at 55 °C, as methanol producer (Botta et al. 2020). A recent study was isolated *Escherichia coli* from cow rumen for producing hydrogen and ethanol using corn straw. They observed that the degradation ratio of cellulose and hemicellulose was equal to 14.30% and 11.39%, respectively, which produced 4.71 mL/g H2 and 0.36 g/L ethanol (Pang et al. 2017). Another BF is biogas, which can be produced using LB and RM. Some authors have used rumen fluid as an inoculum for anaerobic biodegradation of lignocellulose residues to biogas (Bhujbal et al. 2022). A study by Li et al. (2018) inoculated rumen fluid into a continuous anaerobic digestion tank of paper mill, which resulted in 182.7 mL/g methane production at 10% (w/w) of substrate. Using RM, the acetic acid was the major SCFA during fermentation of LB. It is a main precursor for the synthesis of vinyl acetate monomer, ethanol, and terephthalic acid (Nguyen et al. 2020).

6.7 Metabolic Pathways in Rumen Microbiota

The diversity of RM activities leads to the production of SCFAs, microbial proteins, ammonia, and other metabolites (Rey et al. 2014). Approximately 50–85% of SCFAs are directly absorbed through rumen epithelium, serving as the primary energy source for ruminants. These SCFAs also influence the expression of host genes related to energy metabolism, enabling adaptation to the extreme plateau environment through coevolution with the genome (Zhang et al. 2016). Furthermore, rumen metabolites play an important role in regulating host physiological functions, including epithelial cell growths, insulin secretion, and immune response (Sha et al. 2024; Zened et al. 2024). The SCFAs can stimulate the activation of the sympathetic nervous system, enhance the body's energy expenditure, and inhibit fat synthesis. These findings collectively demonstrate a significant connection between gastrointestinal microbiota and host function, indicating that both host genetic structure and microbiota influence the host's metabolic phenotype (Sha et al. 2024). The RM, metabolites, mRNA, and serum metabolites are involved in regulating the fermentation metabolic function and immune levels of Tibetan sheep at different ages (Sha et al. 2024). During rumen fermentation, methanogens archaea mainly utilize by-products to synthesize methane (Thorpe 2009; Patra et al. 2017). There is a fascinating interaction between different microorganisms, as that soluble polysaccharides released by fibrolytic bacteria are mostly absorbed by non-cellulolytic types (Aragao 2013). These bacteria ferment sugars and polysaccharides into primary products such as acetate, propionate, butyrate, carbon dioxide, and hydrogen, which are subsequently utilized as nutrients by the host. Therefore, rumen metabolic pathways have the potential to enhance our understanding of the biodegradation and transformation of biomass into valuable products such as biofuels, as shown in Fig. 6.1 (Aragao 2013). Also, the abundance of *Megasphaera elsdenii* in some microbiomes increases the energy availability for the host. In addition, acrylate metabolic pathway genes were increased in high-yielding cows, but methanogenic archaea and methanogenesis pathway were less prevalent (Bielak 2019). A study by Rico et al. (2021) stated that *Alistipes* had correlations with butyric acid production and *Eubacterium nodatum* and *Clostridiales bacterium* with pentanoic acid manufacture. Additionally, VFAs can provide energy and modulate host intestinal barrier function (Liu et al. 2021).

The electrons derived from hydrogen and formate may also reduce conversion of CO_2 to methane. Some methanogen species oxidize methyl groups, including methylamines and methanol, into CO_2 to generate electrons, which are then utilized to reduce methyl groups to methane (Kelly et al. 2022). The *Methanosarcinales*

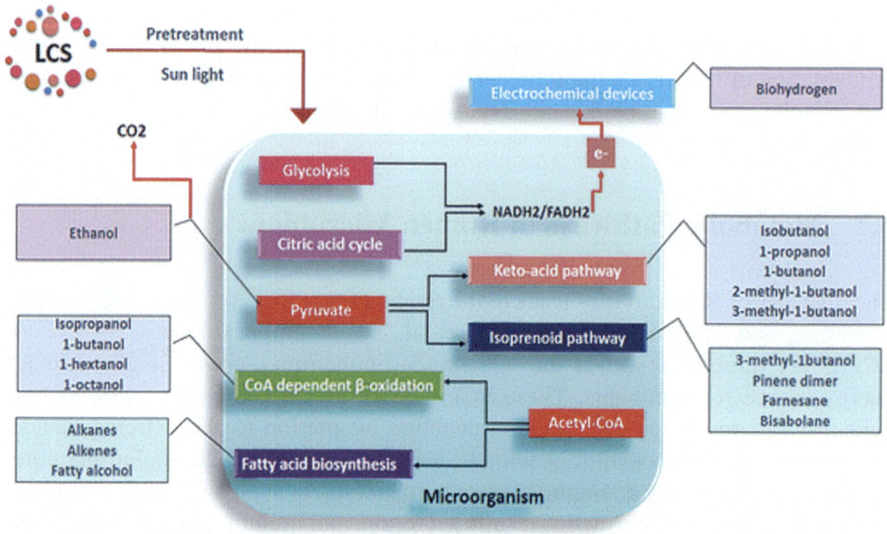

Fig. 6.1 Microbial metabolic pathways for biofuel productions

group can even grow using acetate, effectively producing both CO_2 and methane. The hydrogenotrophic pathway is the most significant route for methane production, accounting for nearly 82% of ruminal methane synthesis, which results from the reduction of CO_2 with H_2 (Janssen and Kirs 2008). These observations highlight the crucial role of specific metabolic pathways in the rumen microbiota and their importance in acquiring nutrients from feed. However, the causal relationships and physiological mechanisms underlying host-microbiota interactions remain unclear.

6.8 Bioengineering Strategies Employed for Rumen Microbiome

Through millions of years of coevolution, microorganisms have adapted to form microbial ecosystems on Earth, known as microbiomes. These communities can persistently exist in or on hosts, including air, soil, oceans, plants, humans, and animals. The microorganisms within these microbiomes engage in dynamic interactions with one another and their hosts, creating synergistic communications that determine the composition and abundance of the microbiome (Chaparro et al. 2012). Synergistic microbial relationships not only influence the characteristics of the microbiome when resident but also have profound effects on host physiology and environmental conditions (Foo et al. 2017). The characteristics of microbiome ecosystems are shaped by their core microbiome. The unique features of a microbiome result from complex interactions among microbial populations, between the microbes and their hosts, and the metabolites they produce. These interactions

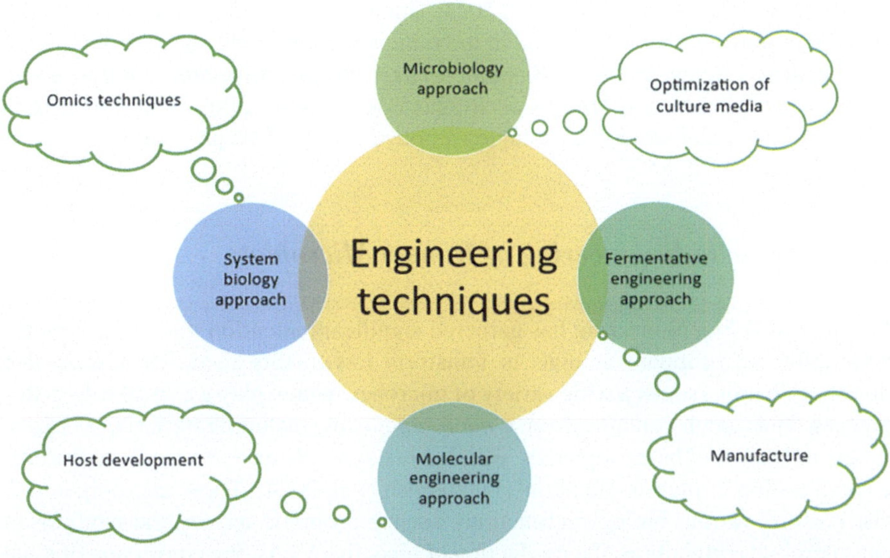

Fig. 6.2 Different engineering approaches for enhanced microbial production of biofuels

illustrate how microbiome composition can impact ecosystems. Consequently, there is a growing interest in engineering microbiomes to modify ecosystems, as illustrated in Fig. 6.2 (Waldor et al. 2015; Round and Mazmanian 2009). Bio-compounds can serve as substitutes for alternatives to fuels, which are currently produced primarily from crude oil resources. Remarkable processes occur naturally in anaerobic environments, such as the pre-stomach of herbivores, where biomass is converted into various compounds important for the biorefinery industry (Beltrán-Ramírez et al. 2019). The discovery of anaerobic microorganisms for industrial biotechnology applications facilities effective and flexible processes with lower performance and operational costs, making them viable for development in both emerging and established economies. Despite the increasing interest in anaerobic microorganisms for industrial biotechnology, there is still limited information regarding their performance compared to what is known about their aerobic counterparts (Aragao 2013).

Bacterial decomposition of LB is emerging as a solution to these challenges (Chukwuma et al. 2021). Bacterial cellulases, hemicellulases, and ligninases break down lignocellulose by converting polysaccharides into simple sugars and lignin into smaller molecules, respectively (Kumari et al. 2024). Certain genera, including *Bacillus*, *Cellulomonas*, *Clostridium*, *Pseudomonas*, and *Streptomyces*, are more effective at degrading lignocellulose than others due to their extensive enzymatic repertoire (Chukwuma et al. 2021). Under aerobic degradation, sugars are oxidized more efficiency, resulting in higher biofuel yields, while, anaerobic method is preferred for biogas production leading to formation of methane (Chandra et al. 2012). Due to their effective hydrolysis and acidification capabilities, rumen microorganisms and their enzymes can produce carboxylates, biohydrogen, biomethane, and bioethanol (Gharechahi et al. 2015, 2023; Bhujbal et al. 2022; Rabee et al. 2022) stated that fibrolytic bacteria found in the rumen of camels and sheep, belonging to the genera *Selenomonas*, *Ruminococcus*, *Succiniclasticum*, *Butyrivibrio*, *Saccharofermentans*, and *Streptococcus*, can convert lignocellulosic residues, such as barley and wheat straw, into soluble carbohydrates for BFs production.

6.8.1 Genetic Engineering of Rumen Microbiota

Recently, RM bioengineering has garnered significant attention due to the remarkable ability of ruminant animals to transform low-quality feed. Their anaerobic digestive chambers host a wide variety of microbes, which play a critical role in this process. The rumen contains diverse microorganisms, including bacteria, protozoa, fungi, archaea, and bacteriophages, which work synergistically to facilitate the efficient digestion of ingested materials (Marappan et al. 2024). These microorganisms also perform various biological functions, such as probiotic activity, the synthesis of antimicrobial metabolites, the production of bioactive VFAs, the conversion of plant biomass, and the generation of BFs. Consequently, researchers are actively seeking new classes of these microorganisms for potential commercial and industrial applications (Mohammadi et al. 2024). Host-dependent species are found in the digestive tracts of ruminants (Seshadri et al. 2018), allowing certain rumen microbes to synthesize polysaccharides, aromatics, and proteins, which primarily consist of lignocellulose. Consequently, these microbes are ideal candidates for converting low-cost and sustainable biomass—such as agricultural, forestry, and municipal waste—into value-added products (Kumari et al. 2024). The model microorganisms for BFs production must possess the ability to decompose LB and produce high yields of desirable end products on an industrial scale. These strains require several attributes to efficiently utilize elevated levels of LB sugars. They should achieve a high growth rate for cell mass in a short period and facilitate the recovery of BFs from LB-derived hydrolysates. However, these hydrolysates may contain inhibitory compounds, such as aldehydes, aromatic compounds, and acetate. Furthermore, the microorganisms must be capable of utilizing various sugars, including hexoses, pentoses, and disaccharides, which play a crucial role in BFs production. Identifying such microbes can be accomplished through screening or by combining relevant genes (Adegboye et al. 2021). As reported by Kumari et al. (2024), the efficiency of BFs production through bacterial degradation is noteworthy. The metabolism of long-chain fatty acids derived from lignocellulosic waste by *Candidatus Weimeria bifida* holds significant potential for developing that are compatible with current petroleum infrastructure (Tolonen 2024). The strain *F. natans* synthesizes the protein fraction of organic matter to metabolize isobutyrate and n-butyrate, which are utilized in the production of plastics, artificial fibers, herbicides, and isovalerate, which is used for flavoring and perfumes (Zaplana et al. 2024). *Lachnospiraceae acetogens* compete with methanogens in the cow rumen, demonstrating a method to reduce bovine methane production (Lan and Yang 2019; Zaplana et al. 2024). Additionally, lignocellulosic fermentation produces hydrogen that can be used by methanogens to fix carbon dioxide. It has also been indicated that *Lachnospiraceae* can activate the transformation of cellulose into methane through methanogenic consortia (Dai et al. 2016). The combination of *Lachnospiraceae* with other microorganisms to create synthetic consortia has been widely employed to enhance the rate of product formation (Zaplana et al. 2024).

Engineered consortia comprising biomass-fermenting *Lachnospiraceae* and methanogens or acetogens have the potential for the direct transformation of lignocellulose into methane and other valuable biochemicals (Tolonen 2024). Utilizing genetic engineering approaches, genes associated with the ethanol-producing pathway have been expressed in certain microbes that exhibit deficiencies in metabolic pathways for specific BFs. Consequently, these microbes were transformed from non-BF-producing strains into BF-generating strains. This engineering approach is advantageous for BFs production, as it allows for the exploitation of various substrates (Ramamurthy et al. 2021). A novel biomimetic bioprocess utilizing dry pulverization of lignocellulose and continuous flow of rumen bacteria at a stable pH produced 69.6 ± 12.2 mmol of VFAs from 20 g of rice straw within 24 h of digestion. This research indicates that the pulverization of lignocellulose is essential for mimicking rumen fermentation and increasing the surface area of the feedstock. Additionally, the digestion rate can be enhanced by employing efficient cellulolytic bacteria (Agematu et al. 2017). The bioengineering of *Saccharomyces cerevisiae* and *Escherichia coli* as industrial strains has been previously reported (Pasotti et al. 2017; Mendez-Perez et al. 2017). In the presence of glucose, most industrial microorganisms lack the ability to metabolize other sugars due to carbon catabolite repression, which limits their utilization of biomass such as lignocellulose (Adegboye et al. 2021). One study addressed this challenge in *E. coli* strains through genetic engineering, controlling catabolite repression via a point mutation in a transcriptional activator for catabolic operons (Flores et al. 2019). Conversely, the introduction of vectors expressing genes that facilitate the utilization of alternative sugars has reduced glucose repression and improved co-fermentation (Bilal et al. 2018). This approach has been validated by engineering *E. coli* and yeast strains for the co-metabolism of various sugar combinations. Given the presence of toxic compounds in LB hydrolysates, including weak organic acids, furan derivatives, and phenols, it is crucial to engineer robust strains with high tolerance to these inhibitors to enhance BFs production. The accumulation of end products and by-products can negatively impact microbial growth (Adegboye et al. 2021). To achieve this, rational engineering can be performed through the direct manipulation of known genetic components, such as transcriptional regulators, transporters, and key pathway enzymes. In contrast, evolutionary engineering can employ indirect manipulation through adaptation or mutagenesis, leading to the evolution of strains. High-throughput screening is essential for selecting strains with elevated tolerance levels (Alper et al. 2006). Atsumi et al. (2008) conducted five rounds of transferred cultures of *E. coli* to enrich for isobutanol resistance until tolerant strains were obtained. Ultimately, the mutant bacterial strain exhibited tolerance to 2.0% (w/v) isobutanol, while the activity of wild-type *E. coli* was inhibited at a concentration of 1.5% (w/v) isobutanol (Atsumi et al. 2008).

6.8.2 Metabolic Engineering for Pathway Improvements

The most LB is converted into BFs through the action of specific enzymes that depolymerize and catalyze polysaccharides. In the rumen, a highly specialized ecosystem, microorganisms convert biomass into various forms of essential nutrients for the host (Srivastava et al. 2021). Consequently, the relevance of certain industrial microorganisms for biorefinery and bioremediation can be observed in such environments. The model strain for BFs production must be capable of utilizing large amounts of substrate, transporting sugars through rapid and deregulated pathways, withstanding inhibitory agents and end products, and exhibiting higher metabolic fluxes to enhance fermentation yields (Adegboye et al. 2021). These approaches could be highly effective in technologically producing BFs from LB by engineering microorganisms. Such methods target specific metabolic pathways in a cost-effective manner. Metabolic engineering is a sophisticated and contemporary process that involves the manipulation of microbial cells, leading to advancements in biotechnological development (Adegboye et al. 2021). This approach has been successfully employed to modify the biosynthetic machinery in various native and recombinant hosts, enabling the production of advanced intermediates and end-use products. Engineered enzymes designed to synthesize specific products have also been effectively utilized to develop new metabolic pathways and optimize existing ones in organisms (Simeonidis and Price 2015; Pirie et al. 2013). Research has shown that genetically engineering strains, combined with optimizing fermentation conditions, significantly enhances BFs production. While, this poss a considerable challenge in process optimization to maximize yields across various methods, the integration of process engineering with fermentation technology, enzyme engineering, and metabolic engineering has greatly benefited the industry (Lee et al. 2008). Multi-parallel metabolic pathways in various microorganisms have significantly enhanced the production of key metabolites essential for BFs synthesis. Additionally, engineering these pathways and evolving enzymes from various metabolic routes serves to enhance both the quantity and diversity of bioactive compounds that lower organisms can biosynthesize. The development of metabolic networks that can produce advanced BFs, such as alkanes, alkenes, and aromatics, will require extensive reprogramming of the metabolism of the microbes used (Adegboye et al. 2021; Lee et al. 2008). Overall, lignocellulose is decomposed by various enzymes, including cellulases, hemicellulases, pectinases, ligninases, and polysaccharide oxygenases. The expansion of the cell wall occurs due to a reduction in the weight and thickness of the matrix polymers, facilitated by wall-loosening substances (Adegboye et al. 2021; Liao et al. 2016). Additionally, other studies have indicated that bioengineering electron metabolism in the bacterium *Clostridium thermocellum* enhances bioethanol production (Lo et al. 2017). As illustrated in Table 6.3, metabolic bioengineering has been employed to augment the potential for biofuel production by microorganisms.

Table 6.3 Biofuel production by metabolic engineered strains

Microorganism	Pathway	Substrate	Product
E. Coli	Glycerol	Glucose, xylose	3-Hydroxypropic acid
Klebsiella pneumoniae	Butanediol biosynthetic	Glucose, galactose	2,3-Butanediol
Synechocystis sp.	Ehrlich	Glucose	Isobutanol
Caldicellulosiruptor	Glycolytic	Lignocellulose	Hydrogen
Thermoanaerobacterium saccharolyticum	*n-Butanol*	Xylose	N-butanol
Clostridium thermocellum and Thermoanaerobacterium saccharolyticum	Embden-Meyerhof	Cellulose	Ethanol

Source: Adegboye et al. (2021)

6.8.3 Role of Synthetic Biology for Biofuel Production

Recent advances in metabolic engineering have transformed the synthesis of various cell wall polymers, resulting in increased polysaccharide availability, a reduced inhibitory effect of polymer-derived products, and an enhanced hexose/pentose ratio in biomass (Fig. 6.3; Joshi et al. 2022). Significant efforts have been made to enhance cellulose biosynthesis and remodeling in the lignocellulose to facilitate cost-effective BFs production. In certain plants, the regulation of gene expression has resulted in substantial changes in cellulose content, crystallinity, and overall biomass, as well as in the activity of sucrose synthase, cellulose synthase, glycosyl hydrolases, and glycosyltransferases (Lao et al. 2014). New synthetic metabolic pathways enhance the production of BFs, including isoprene-based products. Some studies have observed an increased flux of exogenous isoprenol or prenol, resulting in the production of dimethylallyl pyrophosphate and isopentenyl pyrophosphate (Lund et al. 2019; Chatzivasileiou et al. 2019). These findings can be utilized to improve hydrogen (H_2) production. Furthermore, *Escherichia coli* and *Saccharomyces cerevisiae* have been genetically modified to express cell surface hydrolytic enzymes, which is expected to enhance product yields in the short term and promote their application in biorefineries (Ramamurthy et al. 2021). *Escherichia coli* and *Saccharomyces cerevisiae* are also genetically engineered because of their cell-surfaces hydrolytic enzymes. This approach to enhance product yield in the short term and facilitate their use in the biorefinery (Joshi et al. 2022).

6.8.4 Designing Novel Consortia

The ideal microbial strains for BFs production must effectively utilize both pentose and hexose sugars derived from LB feedstock. Furthermore, these strains should also be capable of tolerating the inhibitory compounds generated during the

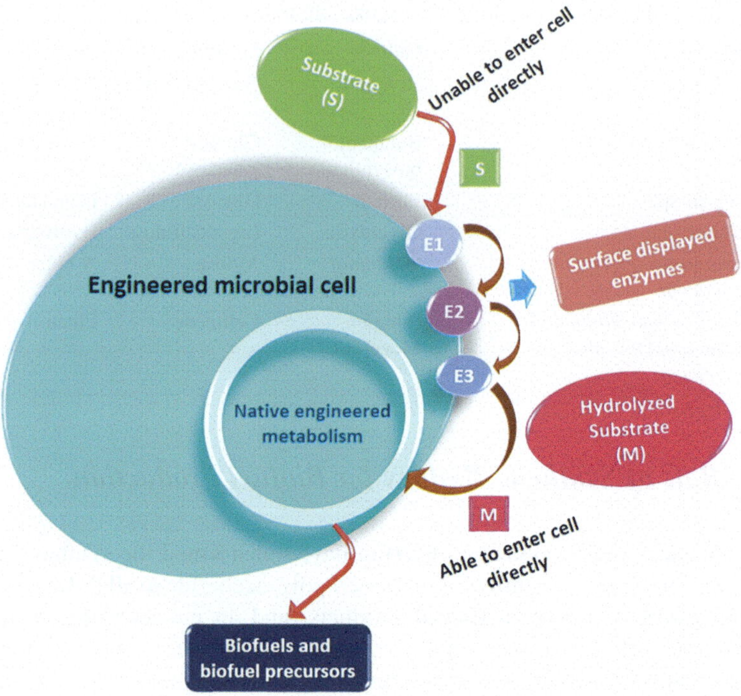

Fig. 6.3 A model for metabolically engineered microbial cell

pretreatment stages. However, most organisms involved in fermentation are unable to utilize pentose sugars, and, conversely, fermenters are often inhibited by the end products of the fermentation process (Adegboye et al. 2021).

6.8.5 Development of Microbial Chassis Systems

Ruminal microbial consortia have shown considerable potential for biohydrogen production, owing to the physiological synergy among species and their capacity to return to their original state after disturbances. This characteristic makes them promising candidates for a range of industrial applications, including energy production (Ramamurthy et al. 2021). Researchers have utilized buffalo rumen waste and sludge for biohydrogen production from buffalo sludge and low-protein cheese whey (scotta). The rumen inoculum demonstrated greater efficiency to be more efficient than sludge during biogas production, yielding 120.8 ml H2 per gram of volatile solids (g VS−1) compared to 65.4 ml H2 g VS−1 for sludge, with a biohydrogen content of 48.1% (v/v) and a maximum peak at 64.6. Additionally, it was also reported that the majority of bacterial genera belonged to the phylum *Firmicutes*, particularly *Proteobacteria*, *Bacteroidetes*, *Clostridia*, and *Tenericutes* (Chiariotti and Crisà 2018). In samples inoculated with buffalo rumen inoculum, *Clostridia* sp. produced the highest volume of biohydrogen, while the family XIII, C. *Incertae sedis*, was the most abundant. Additionally, *Euryarchaeota* and *Methanobrevibacter* were the most frequently observed archaea and genera. Several studies have reported biohydrogen production using various inoculants and substrates, including bovine rumen fluid and cellulose (Nissila et al. 2010), rumen fluid pretreated with $HClO_4$ and cellulose (Ratti et al. 2014), as well as rumen fluid pretreated with HClO4 and office paper (Botta et al. 2016). Gándara-Arteaga et al. (2024) found that the maximum biohydrogen production using ruminal consortia from bovines and lignocellulose occurred after 16 h of incubation. The diverse microbial communities in the rumen can also be utilized to produce methane from plant biomass and animal waste, such as maize silage, bovine manure, and chicken manure. Analysis of RNA-Seq data revealed that cellulases are highly transcribed during biogas fermentation (Güllert et al. 2016). Compared to *Bacteroidetes*, cellulases were four times more frequently associated with *Firmicutes*, although these enzymes were evenly distributed in elephant fecal samples. A study reported an increase in methane yield through the co-inoculation of rumen content culture with anaerobic methanogen seeds derived from sludge (Ozbayram et al. 2018). *Lachnospiraceae* have been identified as effective biocatalysts for coversion of biomass into hydrogen (Bu et al. 2021), with certain strains producing 2–3 moles of hydrogen per mole of glucose equivalent. In addition to its application as a fuel, hydrogen generated from LB digestion serves as a reductant for the fixation of CO and CO_2 by acetogens through gas fermentation. This process presents an ideal method for transforming industrial carbon emissions into valuable biochemical products (Liew et al. 2016).

6.9 Integrating Rumen Microbial Systems with Biorefinery

Fibrolytic enzymes are employed for the degradation and utilization of LB as a carbon source for the production of BFs and bio-derivatives (Gharechahi et al. 2023). However, the high cost of enzyme production, coupled with issues related to their stability and catalytic efficiency under stress conditions, poses a significant challenge. Most conventional commercial enzymes demonstrate low viability, while the gastrointestinal tracts of herbivores may provide a promising source of highly efficient enzymes (Christopherson and Suen 2013). Ruminal enzymes are classified into six major categories: glycosyl hydrolases, glycosyltransferases, polysaccharide lyases, carbohydrate esterases, and auxiliary activities. They perform a variety of functions, including the hydrolysis, formation, and modification of glycosidic bonds, making them essential for carbohydrate metabolism (Chettri et al. 2020). These enzymes posses significant industrial value and are vital to the concept of eco-friendly development. Due to their sustainable nature, BFs represent an important energy source. The BFs are derived from gaseous, liquid, and solid products generated through the thermochemical or biochemical processing of agricultural,

Fig. 6.4 Schematic design of biotechnological applications of rumen microbiota

forestry, animal, and vegetable waste, as well as organic domestic, industrial, and urban waste (Sungur 2024).

The RM rapidly converts lignocellulose into SCFAs, which can be absorbed by the ruminant host (Fig. 6.4). Due to its high diversity, functional abundance, and ease of cultivation, this mixed microbiome serves as a unique carboxylate platform for producing BFs and chemical precursors. Unlike other anaerobic reactor microbiomes, the RM naturally produces high concentrations of SCFAs while generating only modest amounts of methane, owing to the absence of proton-declining acetogens and aceticlastic methanogens (Weimer and Kohn 2016). In a study, anaerobic digester sludge, bison rumen waste, and ruminal cattle inoculum were cultivated on a cellulose substrate to produce acetic acid in anaerobic reactors (Rico et al. 2021). The highest concentration of propionic acid was obtained from bovine rumen reactors. Furthermore, ruminal microorganisms efficiently hydrolyze various biomass sources to achieve a significant yield of VFAs (Liang et al. 2020), which can subsequently be transformed into medium-chain VFAs. Ruminal microbes produce a diverse array of enzymes, including cellulases, hemicellulases, and esterases, which play a critical role in biomass degradation (Astuti et al. 2021; Hinsu et al. 2021). This degradation process converts biomass into mono- or oligosaccharides, which are then rapidly transformed into pyruvate, ultimately leading to the production of acetate, propionate, and butyrate through various metabolic pathways (Won et al. 2020). The VFAs produced supply approximately 70% of the energy requirements of ruminants. In another study, the highest concentrations of butyric and pentanoic acids were synthesized using bison rumen (Rico et al. 2021).

6.10 Challenges and Future Prospects

Nowadays, applying the principles of rumen fermentation to industrial anaerobic digestion systems is highly appealing. This process involves rumen microorganisms, and integrates aspects of animal physiology, biological environmental engineering, and lignocellulosic waste management. Anaerobic reactors are designed to enhance the generation efficiency of VFAs and biogas production (Rico et al. 2021). Typical operating conditions in the rumen include semisolid materials (12–18% solids), the selection of feed particles and its retention time, semicontinuous flow, and a temperature range of 38–40 °C. Various reactor designs, such as continuous flow, two-phase anaerobic, anaerobic sequential batch, and anaerobic membrane types, have been implemented based on these rumen characteristics (Liang et al. 2020). The RM produce different biogas through four major phases: hydrolysis, acidogenesis, acetogenesis, and methanogenesis. The first three phases are facilitated by various bacterial groups that inhabit the rumen; however, the fourth phase exclusively carried out by methanogenic archaea (Angelidaki et al. 2011). The RM effectively decomposes LB into soluble carbohydrates, which are subsequently converted into transformed VFAs for energy synthesis (Bhujbal et al. 2022) by the ruminal microbiota, which includes bacteria, fungi, protozoa, and archaea (Gharechahi et al. 2023; Liang et al. 2024). These microbes can serve as biological pretreatment or bioaugmentation agents, significantly amends the efficiency of LB digestion and CH_4 output (Liang et al. 2020).

6.11 Conclusions

In recent years, the significance and role of microbiomes in the physiology of hosts and their ecosystems have been extensively documented. The growing understanding of microbiomes and microbiome engineering has underscored their critical roles in disease treatment, enhancing agricultural productivity, and reducing environmental pollutants. However, the field of microbiome engineering remains in its infancy and holds substantial potential for further development. This approach can be employed to study microbial populations over time, thereby advancing our knowledge of microbiomes and facilitating progress in microbiome engineering. Although ruminal microbes and enzymes can transform biomass waste into biogas, bioethanol, organic acids, and other valuable chemicals, in-depth studies on the structural complexity of the ruminal microbiota and their various mechanisms are essential. This includes investigating uncultured ruminal microbes in vitro, their functions, and the interactions between ruminal microbes and their enzymes that work synergistically to decompose ingested biomass into volatile fatty acids (VFAs), sugars, and proteins. Consequently, despite being a relatively new field for microbiota modification, it offers promising solutions for human health and environmental sustainability, as well as addressing ecological challenges and developing innovative tools and technologies. Overall, continued research in understanding and managing the microbiome, along with microbiome engineering, will be a major strategy to enhance human health and agriculture.

Acknowledgments The authors sincerely thank Dezful University of Medical Sciences (Dezful City, Khuzestan Province, Iran) for their financial support in publishing this chapter.

Conflict of Interest The authors declare that this chapter was written without any commercial or financial relationships that could be percieved as a potential conflict of interest.

References

Adegboye MF, Ojuederie OB, Talia PM, Babalola OO (2021) Bioprospecting of microbial strains for biofuel production: metabolic engineering, applications, and challenges. Biotechnol Biofuels 14:5

Agematu H, Takahashi T, Hamano Y (2017) Continuous volatile fatty acid production from lignocellulosic biomass by a novel rumen-mimetic bioprocess. J Biosci Bioeng 124:528–533

Alalwan HA, Alminshid AH, Aljaafari HA (2019) Promising evolution of biofuel generations: subject review. Renew Energy Focus 28:127–139

Alper H, Moxley J, Nevoigt E, Fink GR, Stephanopoulos G (2006) Engineering yeast transcription machinery for improved ethanol tolerance and production. Science 314:1565–1568

Anadón A, Ares I, Martínez-Larrañaga MR, Martínez MA (2019) Enzymes in feed and animal health. Nutraceutical Vet Med 2019:303–313

Angelidaki I, Karakashev D, Batstone DJ, Plugge CM, Stams AJ (2011) Biomethanation and its potential. Methods Enzymol 494:327–335. https://doi.org/10.1016/B978-0-12-385112-3.00016-0

Aragao R (2013) Exploring anaerobic bacteria for industrial biotechnology—diversity studies, screening, and biorefinery applications

Astuti T, Jamarun N, Yanti G (2021) Effect fermentation of sugarcane shoots with *Phanerochaete chrysosporium* on the activity of laccase enzymes, lignin peroxidase, and manganese peroxidase. IOP Conf Ser Earth Environ Sci 709:012065

Atsumi S, Hanai T, Liao JC (2008) Non-fermentative pathways for synthesis of branched-chain higher alcohols as biofuels. Nature 451:86–89

Avchar R, Lanjekar V, Kshirsagar P, Dhakephalkar PK, Dagar SS, Baghela A (2021) Buffalo rumen harbours diverse thermotolerant yeasts capable of producing second-generation bioethanol from lignocellulosic biomass. Renew Energy 173:795–807

Baba Y, Matsuki YU, Takizawa S, Suyama Y, Tada C, Fukuda Y, Nakai Y (2019) Pretreatment of lignocellulosic biomass with cattle rumen fluid for methane production: fate of added rumen microbes and indigenous microbes of methane seed sludge. Microbes Environ 34:421–428

Bath C, Morrison M, Ross EM, Hayes BJ, Cocks BG (2013) The symbiotic rumen microbiome and cattle performance: a brief review. Anim Prod Sci 53:876–881

Beltrán-Ramírez F, Orona-Tamayo D, Cornejo-Corona I, González-Cervantes JLN, de Jesús Esparza-Claudio J, Quintana-Rodríguez E (2019) Agro-industrial waste revalorization: the growing biorefinery. In: Biomass for bioenergy—recent trends and future challenges, pp 83–102

Bhaskar A, Dubey S, Singhania RR (2024) Bacterial cellulosomes: production and applications. In: Biomass hydrolyzing enzymes. CRC Press, pp 26–47

Bhujbal SK, Ghosh P, Vijay VK, Rathour R, Kumar M, Singh L, Kapley A (2022) Biotechnological potential of rumen microbiota for sustainable bioconversion of lignocellulosic waste to biofuels and value-added products. Sci Total Environ 814:152773

Bielak A (2019) Relation between metabolic state, microbial community structure and methane production in dairy cows. Doctoral dissertation, Leipzig, Universität Leipzig

Bilal M, Iqbal HM, Hu H, Wang W, Zhang X (2018) Metabolic engineering and enzyme-mediated processing: a biotechnological venture towards biofuel production—a review. Renew Sust Energ Rev 82:436–447

Botta LS, Ratti RP, Sakamoto IK, Ramos LR, Silva EL, Varesche MBA (2016) Bioconversion of waste office paper to hydrogen using pretreated rumen fluid inoculum. Bioprocess Biosyst Eng 39:1887–1897

Botta LS, Delforno TP, Rabelo CABS, Silva EL, Varesche MBA (2020) Microbial community analyses by high-throughput sequencing of rumen microorganisms fermenting office paper in mesophilic and thermophilic lysimeters. Process Saf Environ Prot 136:182–193

Bu J, Wei HL, Wang YT, Cheng JR, Zhu MJ (2021) Biochar boosts dark fermentative H2 production from sugarcane bagasse by selective enrichment/colonization of functional bacteria and enhancing extracellular electron transfer. Water Res 202:117440

Bundhoo MZ, Mohee R, Hassan MA (2015) Effects of pre-treatment technologies on dark fermentative biohydrogen production: a review. J Environ Manag 157:20–48

Cameron O, Neves JF, Gentleman E (2024) Listen to your gut: key concepts for bioengineering advanced models of the intestine. Adv Sci 11:2302165

Candia-García C, Delgadillo-Mirquez L, Hernandez M (2018) Biodegradation of rice straw under anaerobic digestion. Environ Technol Innov 10:215–222

Cann I, Cheng Y, Alhawsawi MAB, Moran M, Li Y, Gong T, Zhu W, Mackie RI (2025) Rumen-targeted mining of enzymes for bioenergy production. Annu Rev Anim Biosci 13(1):343–369. https://doi.org/10.1146/annurev-animal-021022-030040

Chandra R, Takeuchi H, Hasegawa T (2012) Methane production from lignocellulosic agricultural crop wastes: a review in context to second generation of biofuel production. Renew Sust Energ Rev 16:1462–1476

Chang JJ, Lin JJ, Ho CY, Chin WC, Huang CC (2010) Establishment of rumen-mimic bacterial consortia: a functional union for bio-hydrogen production from cellulosic bioresource. Int J Hydrog Energy 35:13399–13406

Chaparro JM, Sheflin AM, Manter DK, Vivanco JM (2012) Manipulating the soil microbiome to increase soil health and plant fertility. Biol Fertil Soils 48:489–499

Chatzivasileiou AO, Ward V, Edgar SM, Stephanopoulos G (2019) Two-step pathway for isoprenoid synthesis. Proc Natl Acad Sci USA 116:506–511

Chettri D, Verma AK, Verma AK (2020) Innovations in CAZyme gene diversity and its modification for biorefinery applications. Biotechnol Rep 28:e00525

Chettri D, Nad S, Konar U, Verma AK (2022) CAZyme from gut microbiome for efficient lignocellulose degradation and biofuel production. Front Chem Eng 4:1054242

Chiariotti A, Crisà A (2018) Bio-hydrogen production from buffalo waste with rumen inoculum and metagenomic characterization of bacterial and archaeal community. Front Sustain Food Syst 2:13

Choudhury PK, Salem AZM, Jena R, Kumar S, Singh R, Puniya AK (2015) Rumen microbiology: an overview. In: *rumen microbiology: from evolution to revolution*, pp. 3–16

Christopherson MR, Suen G (2013) Nature's bioreactor: the rumen as a model for biofuel production. Biofuels 4:511–521

Chukwuma OB, Rafatullah M, Tajarudin HA, Ismail N (2021) A review on bacterial contribution to lignocellulose breakdown into useful bio-products. Int J Environ Res Public Health 18:6001

Dai Y, Yan Z, Jia L, Zhang S, Gao L, Wei X, Liu X (2016) The composition, localization, and function of low-temperature-adapted microbial communities involved in methanogenic degradations of cellulose and chitin from Qinghai–Tibetan Plateau wetland soils. J Appl Microbiol 121:163–176

Das BK, Hassan R, Tushar MSH, Zaman F, Hasan M, Das P (2021) Techno-economic and environmental assessment of a hybrid renewable energy system using multi-objective genetic algorithm: a case study for remote Island in Bangladesh. Energy Convers Manag 230:113823

Dionísio SR, Santoro DCJ, Bonan CIDG, Soares LB, Biazi LE, Rabelo SC, Ienczak JL (2021) Second-generation ethanol process for integral use of hemicellulosic and cellulosic hydrolysates from diluted sulfuric acid pretreatment of sugarcane bagasse. Fuel 304:121290

Dragone G, Fernandes B, Vicente AA, Teixeira JA (2010) Third-generation biofuels from microalgae. Curr Res Technol Educ Top Appl Microbiol Microb Biotechnol 2:1355–1366

Femia AP, Salvadori M, Broekaert WF, François IE, Delcour JA, Courtin CM, Caderni G (2010) Arabinoxylan-oligosaccharides (AXOS) reduce preneoplastic lesions in the colon of rats treated with 1, 2-dimethylhydrazine (DMH). Eur J Nutr 49:127–132

Flores AD, Ayla EZ, Nielsen DR, Wang X (2019) Engineering a synthetic, catabolically orthogonal coculture system for enhanced conversion of lignocellulose-derived sugars to ethanol. ACS Synth Biol 8:1089–1099

Foo JL, Ling H, Lee YS, Chang MW (2017) Microbiome engineering: current applications and its future. Biotechnol J 12:1600099

Gagen EJ, Denman SE, McSweeney CS (2015a) Acetogenesis as an alternative to methanogenesis in the rumen. In: Livestock production and climate change. CABI, Wallingford, pp 292–303

Gagen EJ, Padmanabha J, Denman SE, McSweeney CS (2015b) Hydrogenotrophic culture enrichment reveals rumen Lachnospiraceae and Ruminococcaceae acetogens and hydrogen-responsive Bacteroidetes from pasture-fed cattle. FEMS Microbiol Lett 362:fnv104

Gándara-Arteaga VM, Guatemala-Morales GM, Martínez-Gómez ÁDJ, Toriz G, Pelayo-Ortiz C, Corona-González RI (2024) Production of hydrogen with ruminal microbiota: finding culture conditions for high yields. Fermentation 10:274

Gharechahi J, Zahiri HS, Noghabi KA, Salekdeh GH (2015) In-depth diversity analysis of the bacterial community resident in the camel rumen. Syst Appl Microbiol 38:67–76

Gharechahi J, Vahidi MF, Sharifi G, Ariaeenejad S, Ding XZ, Han JL, Salekdeh GH (2023) Lignocellulose degradation by rumen bacterial communities: new insights from metagenome analyses. Environ Res 229:115925

Gheibipour M (2017) Isolation of tannin-degrading bacteria from the gastrointestinal tract of deer and using for improving the nutritional value of tannin-rich feeds. [Dissertation]. Khuzestan Ramin Agriculture and Natural Resource University, Molasani, Iran

Gheibipour M, Ghiasi SE, Bashtani M, Torbati MBM, Motamedi H (2022) The potential of tannin-degrading bacteria isolated from rumen of Iranian Urial ram as silage additives. Bioresour Technol Rep 18:101024. https://doi.org/10.1016/j.biteb.2022.101024

Gheibipour M, Ghiasi SE, Bashtani M, Montazer Torbati MB, Motamedi H (2023) Tannase-producing bacteria isolated from the rumen of Fallow deer (Dama dama): livestock potential feed additives. Microbial Microorganism 12:27–40

Gheibipour M, Ghiasi SE, Bashtani M, Torbati MBM, Motamedi H (2024a) Screening the rumen of Balochi camel (Camelus dromedarius) and Cashmere goat (Capra hircus) to isolate enzyme-producing bacteria as potential additives for animal feed. Indian J Microbiol 64:1–11. https://doi.org/10.1007/s12088-024-01197-7

Gheibipour M, Ghiasi SE, Bashtani M, Torbati MBM (2024b) The effect of probiotic tannase bacterial additive on diarrhea score in weaning lambs fed with pomegranate peel. The First International And The Sixth National Conference On Biodiversity And Its Impact On Agriculture And Environment, Urmia University, Urmia, Iran. https://civilica.com/doc/2096937

Gheibipour M, Ghiasi SE, Bashtani M, Torbati MBM (2024c) The effect of tannase-producing supplementation with probiotic potential on ruminal fermentation of diets containing pomegranate peel in vitro. The First International And The Sixth National Conference On Biodiversity And Its Impact On Agriculture And Environment, Urmia, Urmia University, Iran. https://civilica.com/doc/2096936.

Ghiasi SE, Gheibipour M, Motamedi H, Dar MA (2024) Screening and evaluation of the ruminal cellulolytic bacteria and their potential application as probiotics. Iran J Microbiol 16(3):389–400. https://doi.org/10.18502/ijm.v16i3.15796

Güllert S, Fischer MA, Turaev D, Noebauer B, Ilmberger N, Wemheuer B, Alawi M, Rattei T, Daniel R, Schmitz RA, Grundhoff A, Streit WR (2016) Deep metagenome and metatranscriptome analyses of microbial communities affiliated with an industrial biogas fermenter, a cow rumen, and elephant feces reveal major differences in carbohydrate hydrolysis strategies. Biotechnol Biofuels 9:121. https://doi.org/10.1186/s13068-016-0534-x

He S, Zhao S, Wang Z, Dai S, Mao H, Wu D (2024) Impact of Seasonal Variation in Pasture on Rumen Microbial Community and Volatile Fatty Acids in Grazing Yaks: Insights from High-Altitude Environments. Microorganism 12(8):1701. https://doi.org/10.3390/microorganisms12081701

Hinsu AT, Tulsani NJ, Panchal KJ, Pandit RJ, Jyotsana B, Dafale NA, Patil NV, Purohit HJ, Joshi CG, Jakhesara SJ (2021) Characterizing rumen microbiota and CAZyme profile of Indian dromedary camel (Camelus dromedarius) in response to different roughages. Sci Rep 11(1):9400. https://doi.org/10.1038/s41598-021-88943-9

Hirani AH, Javed N, Asif M, Basu SK, Kumar A (2018) A review on first- and second-generation biofuel productions. In A Kumar, S Ogita, Y-Y Yau (eds.), Biofuels: Greenhouse Gas Mitigation and Global Warming (pp. 141–154). Springer

Ho C-Y, Chang J-J, Lin J-J, Chin T-Y, Mathew GM, Huang C-C (2011) Establishment of functional rumen bacterial consortia (FRBC) for simultaneous biohydrogen and bioethanol production from lignocellulose. Int J Hydrogen Energy 36(19):12168–12176. https://doi.org/10.1016/j.ijhydene.2011.06.125

Janssen PH, Kirs M (2008) Structure of the archaeal community of the rumen. Appl Environ Microbiol 74(12): 3619–3625. https://doi.org/10.1128/AEM.02812-07

Jeyamalar J (2010) Investigation of rumen methanogens in New Zealand livestock: a thesis presented in partial fulfillment of the requirements for the degree of Doctor of Philosophy in Animal Science at Massey University, Palmerston North, New Zealand (Doctoral dissertation, Massey University). http://hdl.handle.net/10179/2501

Jin D, Zhao S, Wang P, Zheng N, Bu D, Beckers Y, Wang J (2016) Insights into abundant rumen ureolytic bacterial community using rumen simulation system. Front Microbiol 7:1006. https://doi.org/10.3389/fmicb.2016.01006

Joshi A, Verma KK, Rajput VD, Minkina T, Arora J (2022) Recent advances in metabolic engineering of microorganisms for advancing lignocellulose-derived biofuels. Bioengineered 13(4):8135–8163. https://doi.org/10.1080/21655979.2022.2051856

Juengel JL, Miyamoto A, Price C, Reynolds LP, Smith MF, Webb R (2014) Reproduction in Domestic Ruminants VIII. In Proceedings of the Ninth International Symposium on Reproduction in Domestic Ruminants. England: Context 493–506. 101651512

Kelly WJ, Mackie RI, Attwood GT, Janssen PH, McAllister TA, Leahy SC (2022) Hydrogen and formate production and utilisation in the rumen and the human colon. Anim Microbiome 4(1):22. https://doi.org/10.1186/s42523-022-00174-z

Kholif AE, Patra AK (2024) Dietary applications of exogenous enzymes to improve nutrient utilization and performance in ruminants. In MS Mahesh, VK Yata (eds.), Feed Additives and Supplements for Ruminants (pp. 1–28). Springer Nature Singapore, Singapore. https://doi.org/10.1007/978-981-97-0794-2_1

Kosky P, Balmer R, Keat W, Wise G (2021) Bioengineering. In: Exploring engineering, 5th edn. Academic Press, pp 363–382. ISBN 9780128150733

Kratzel A, Todt D, V'kovski P, Steiner S, Gultom M, Thao TTN, Ebert N, Holwerda M, Steinmann J, Niemeyer D, Dijkman R, Kampf G, Drosten C, Steinmann E, Thiel V, Pfaender S (2020) Inactivation of Severe Acute Respiratory Syndrome Coronavirus 2 by WHO-Recommended Hand Rub Formulations and Alcohols. Emerg Infect Dis 26(7):1592–1595. https://doi.org/10.3201/eid2607.200915

Kumari A, Kumar M, Bhoi B (2024) Role of Bacterial Degradation in Lignocellulosic Biomass for Biofuel Production. In AL Srivastav, AK Bhardwaj, M Kumar (eds.), Valorization of Biomass Wastes for Environmental Sustainability: Green Practices for the Rural Circular Economy (pp. 303–315). Springer Nature Switzerland, Cham

Lan W, Yang C (2019) Ruminal methane production: Associated microorganisms and the potential of applying hydrogen-utilizing bacteria for mitigation. Sci Total Environ 654:1270–1283. https://doi.org/10.1016/j.scitotenv.2018.11.180

Lao J, Oikawa A, Bromley JR, McInerney P, Suttangkakul A, Smith-Moritz AM, Plahar H, Chiu TY, González Fernández-Niño SM, Ebert B, Yang F, Christiansen KM, Hansen SF, Stonebloom

S, Adams PD, Ronald PC, Hillson NJ, Hadi MZ, Vega-Sánchez ME, Loqué D, Scheller HV, Heazlewood JL (2014) The plant glycosyltransferase clone collection for functional genomics. Plant J 79(3):517–529. https://doi.org/10.1111/tpj.2014.79.issue-3. https://doi.org/10.1111/tpj.12577

Lazuka A, Auer L, Bozonnet S, Morgavi DP, O'Donohue M, Hernandez-Raquet G (2015) Efficient anaerobic transformation of raw wheat straw by a robust cow rumen-derived microbial consortium. Bioresour Technol 196:241–249. https://doi.org/10.1016/j.biortech.2015.07.084

Lee SK, Chou H, Ham TS, Lee TS, Keasling JD (2008) Metabolic engineering of microorganisms for biofuels production: from bugs to synthetic biology to fuels. Curr Opin Biotechnol 19(6):556–563. https://doi.org/10.1016/j.copbio.2008.10.014

Li QS, Wang R, Ma ZY, Zhang XM, Jiao JZ, Zhang ZG, Ungerfeld EM, Yi KL, Zhang BZ, Long L, Long Y, Tao Y, Huang T, Greening C, Tan ZL, Wang M (2022) Dietary selection of metabolically distinct microorganisms drives hydrogen metabolism in ruminants. ISME J 16(11):2535–2546. https://doi.org/10.1038/s41396-022-01294-9

Li N, Yang F, Xiao H, Zhang J, Ping Q (2018) Effect of feedstock concentration on biogas production by inoculating rumen microorganisms in biomass solid waste. Appl Biochem Biotechnol 184(4):1219–1231. https://doi.org/10.1007/s12010-017-2615-0

Li K, Zhu H, Zhang Y, Zhang H (2017) Characterization of the microbial communities in rumen fluid inoculated reactors for the biogas digestion of wheat straw. Sustainability 9(2):243. https://doi.org/10.3390/su9020243

Liang J, Nabi M, Zhang P, Zhang G, Cai Y, Wang Q, Zhou Z, Ding Y (2020) Promising biological conversion of lignocellulosic biomass to renewable energy with rumen microorganisms: a comprehensive review. Renew Sustain Energy Rev 134:110335. https://doi.org/10.1016/j.rser.2020.110335

Liang J, Zhang R, Chang J, Chen L, Nabi M, Zhang H, Zhang G, Zhang P (2024) Rumen microbes, enzymes, metabolisms, and application in lignocellulosic waste conversion - A comprehensive review. Biotechnol Adv 71:108308. https://doi.org/10.1016/j.biotechadv.2024.108308

Liang J, Zhang H, Zhang P, Zhang G, Cai Y, Wang Q, Zhou Z, Ding Y, Zubair M (2021) Effect of substrate load on anaerobic fermentation of rice straw with rumen liquid as inoculum: Hydrolysis and acidogenesis efficiency, enzymatic activities and rumen bacterial community structure. Waste Manag 124:235–243. https://doi.org/10.1016/j.wasman.2021.02.017

Liao JC, Mi L, Pontrelli S, Luo S (2016) Fuelling the future: microbial engineering for the production of sustainable biofuels. Nat Rev Microbiol 14(5):288–304. https://doi.org/10.1038/nrmicro.2016.32

Liew F, Martin ME, Tappel RC, Heijstra BD, Mihalcea C, Köpke M (2016) Gas Fermentation—A Flexible Platform for Commercial Scale Production of Low-Carbon-Fuels and Chemicals from Waste and Renewable Feedstocks. Front Microbiol 7. https://doi.org/10.3389/fmicb.2016.00694

Liu J, Wu X, Zhang Z, Fan H, Zou J, Jin J (2021) Novel insights into microbial communities of dairy manure composting revealed by metagenomic analysis. Sci Total Environ 756:143682

Lo J, Olson DG, Murphy SJ, Tian L, Hon S, Lanahan A, Guss AM, Lynd LR (2017) Engineering electron metabolism to increase ethanol production in Clostridium thermocellum. Metab Eng 39:71–79. https://doi.org/10.1016/j.ymben.2016.10.018

Lund S, Hall R, Williams GJ (2019) An Artificial Pathway for Isoprenoid Biosynthesis Decoupled from Native Hemiterpene Metabolism. ACS Synth Biol 8(2):232–238. https://doi.org/10.1021/acssynbio.8b00383

Marappan G, Rajendran D, Tellis JM, Heena H, Soren NM (2024) Rumen microbes: exploring its potential for productivity and commercial use. Indian J Anim Sci 94(2):95–100

McAllister TA, Meale SJ, Valle E, Guan LL, Zhou M, Kelly WJ, Janssen PH (2015) Ruminant nutrition symposium: use of genomics and transcriptomics to identify strategies to lower ruminal methanogenesis. J Anim Sci 93(4):1431–1449

Mendez-Perez D, Alonso-Gutierrez J, Hu Q, Molinas M, Baidoo EE, Wang G, Lee TS (2017) Production of jet fuel precursor monoterpenoids from engineered *Escherichia coli*. Biotechnol Bioeng 114(8):1703–1712

Mohammadi M, Alian M, Dale B, Ubanwa B, Balan V (2024) Multifaceted application of AFEX-pretreated biomass in producing second-generation biofuels, ruminant animal feed, and value-added bioproducts. Biotechnol Adv 108341
Mohammadabadi T, Gheibipour M, Motamedi H, Chaji M, Abbas BA (2021) Isolation and identification of tannin-degrading bacteria from deer gut and potency for improving nutritional value of tannin rich plants. Journal Homepage: http://www.ivj.ir/, https://doi.org/10.22055/IVJ.2021.257994.2322
Mohammadabadi T, Gheibipour M, Motamedi H, Chaji M, Abbas BA, Ali HJ (2023) Effect of processing with tannin degrading bacteria isolated from gut of European fallow deer on digestibility and fermentation of conocarpus and eucalyptus leaves. AIP Conf Proc 2977(1):040022. AIP Publishing, https://doi.org/10.1063/5.0186664
Morgavi DP, Forano E, Martin C, Newbold CJ (2010) Microbial ecosystem and methanogenesis in ruminants. Anim 4(7):1024–1036
Naseema Rasheed R, Pourbakhtiar A, Mehdizadeh Allaf M, Baharlooeian M, Rafiei N, Alishah Aratboni H, Winck FV (2023) Microalgal co-cultivation: recent methods, trends in omic-studies, applications, and future challenges. Front Bioeng Biotechnol 11:1193424
Newbold CJ, Ramos-Morales E (2020) Review: ruminal microbiome and microbial metabolome: effects of diet and ruminant host. Anim 14(S1):S78–S86
Nguyen AQ, Nguyen LN, Johir MAH, Ngo HH, Chaves AV, Nghiem LD (2020) Derivation of volatile fatty acid from crop residues digestion using a rumen membrane bioreactor: a feasibility study. Bioresour Technol 312:123571. https://doi.org/10.1016/j.biortech.2020.123571
Nissila ME, Tähti HP, Rintala JA, Puhakka JA (2010) Thermophilic hydrogen production from cellulose with rumen fluid enrichment cultures: effects of different heat treatments. Int J Hydrog Energy 36:1482–1490
Ojha BK, Singh PK, Shrivastava N (2019) Enzymes in the animal feed industry. In Enzymes in Food Biotechnology (pp. 93–109). Elsevier. https://doi.org/10.1016/B978-0-12-813280-7.00007-4
Orpin CG (1977) Invasion of plant tissue in the rumen by the flagellate *Neocallimastix frontalis*. Microbiology 98(2):423–430
Osorio-Doblado AM, Feldmann KP, Lourenco JM, Stewart RL, Smith WB, Tedeschi LO, Callaway TR (2023) Forages and pastures symposium: forage biodegradation: advances in ruminal microbial ecology. J Anim Sci 101:skad178
Ozbayram EG, Akyol Ç, Ince B, Karakoç C, Ince O (2018) Rumen bacteria at work: bioaugmentation strategies to enhance biogas production from cow manure. J Appl Microbiol 124(2):491–502
Pang J, Liu ZY, Hao M, Zhang YF, Qi QS (2017) An isolated cellulolytic *Escherichia coli* from bovine rumen produces ethanol and hydrogen from corn straw. Biotechnol Biofuels 10:1–10
Pasotti L, Zucca S, Casanova M, Micoli G, Cusella De Angelis MG, Magni P (2017) Fermentation of lactose to ethanol in cheese whey permeate and concentrated permeate by engineered *Escherichia coli*. BMC Biotechnol 17:1–12
Patel M, Patel HM, Dave S (2020) Determination of bioethanol production potential from lignocellulosic biomass using novel Cel-5m isolated from cow rumen metagenome. Int J Biol Macromol 153:1099–1106
Patil SS, Rathod VK (2021) Intensification of extraction of biomolecules using three-phase partitioning. In Three Phase Partitioning (pp. 285-322). Elsevier. https://doi.org/10.1016/B978-0-12-824418-0.00009-6
Patra A, Park T, Kim M, Yu Z (2017) Rumen methanogens and mitigation of methane emission by anti-methanogenic compounds and substances. J Anim Sci Biotechnol 8:1–18
Peng Y, Zhang Y, Wang L (2010) Artificial intelligence in biomedical engineering and informatics: an introduction and review. Artif Intell Med 48(2-3):71–73. https://doi.org/10.1016/j.artmed.2009.07.007
Pirie CM, De Mey M, Prather KLJ, Ajikumar PK (2013) Integrating the protein and metabolic engineering toolkits for next-generation chemical biosynthesis. ACS Chem Biol 8(4):662–672

Rabee AE, Sayed Alahl AA, Lamara M, Ishaq SL (2022) Fibrolytic rumen bacteria of camel and sheep and their applications in the bioconversion of barley straw to soluble sugars for biofuel production. PLoS One 17:e0262304
Rai R, Basotra N, Kaur B, Di Falco M, Tsang A, Chadha BS (2020) Exoproteome profile reveals thermophilic fungus Crassicarpon thermophilum (strain 6GKB; syn. Corynascus thermophilus) as a rich source of cellobiose dehydrogenase for enhanced saccharification of bagasse. Biomass Bioenergy 132:105438
Rai AK, Al Makishah NH, Wen Z, Gupta G, Pandit S, Prasad R (2022) Recent developments in lignocellulosic biofuels, a renewable source of bioenergy. Fermentation 8:161
Ramamurthy PC, Singh S, Kapoor D, Parihar P, Samuel J, Prasad R, Singh J (2021) Microbial biotechnological approaches: renewable bioprocessing for the future energy systems. Microb Cell Factories 20:1–11
Ratti RP, Botta LS, Sakamoto IK, Silva EL, Amancio MB (2014) Production of H2 from cellulose by rumen microorganisms: effects of inocula pre-treatment and enzymatic hydrolysis. Biotechnol Lett 36:537–546
Rey M, Enjalbert F, Combes S, Cauquil L, Bouchez O, Monteils V (2014) Establishment of ruminal bacterial community in dairy calves from birth to weaning is sequential. J Appl Microbiol 116:245–257
Rico JL, Reardon KF, Susan K (2021) Inoculum microbiome composition impacts fatty acid product profile from cellulosic feedstock. Bioresour Technol 323:124532
Round JL, Mazmanian SK (2009) The gut microbiota shapes intestinal immune responses during health and disease. Nat Rev Immunol 9:313–323
Russell JB, Mantovani HC (2002) The bacteriocins of ruminal bacteria and their potential as an alternative to antibiotics. J Mol Microbiol Biotechnol 4:347–355
Saini P, Mishra P (2024) Microbial bioprospecting for biorefinery application: an overview of a sustainable and renewable source of energy. Bioprospect Microbial Resour Agric Environ Biochem Ind:109–136
Saini JK, Kaur A, Mathur A (2022) Strategies to enhance enzymatic hydrolysis of lignocellulosic biomass for biorefinery applications: a review. Bioresour Technol 360:127517
Samal L, Behura NC (2015) Prebiotics: an emerging nutritional approach for improving gut health of livestock and poultry. Asian J Anim Vet Adv 10:724–739
Sawatdeenarunat C, Surendra KC, Takara D, Oechsner H, Khanal SK (2015) Anaerobic digestion of lignocellulosic biomass: challenges and opportunities. Bioresour Technol 178:178–186
Sema YA, Urgessa OE, Ayele AK (2020) Evaluating rumen fluid-inoculated lignocellulosic substrates for biogas production. East Afr J Sci 14:99–110
Seshadri R, Leahy SC, Attwood GT, Teh KH, Lambie SC, Cookson AL, Kelly WJ (2018) Cultivation and sequencing of rumen microbiome members from the Hungate1000 collection. Nat Biotechnol 36:359–367
Sha Y, Liu X, Li X, Wang Z, Shao P, Jiao T, Zhao S (2024) Succession of rumen microbiota and metabolites across different reproductive periods in different sheep breeds and their impact on the growth and development of offspring lambs. Microbiome 12:172
Shinkai T, Takizawa S, Fujimori M, Mitsumori M (2024) The role of rumen microbiota in enteric methane mitigation for sustainable ruminant production. Anim Biosci 37:360
Shweta CSC, Kamboj BR, Malik K, Singh K, Bhisnoi DK, Arya S (2024) Biomass resources and biofuel technologies: a focus on Indian development. Energies 17:382
Singh K, Byun C, Bux F (2022) Ecological restoration of degraded ecosystems in India: science and practices. Ecol Eng 182:106708. https://doi.org/10.1016/j.ecoleng.2022.106708
Simeonidis E, Price ND (2015) Genome-scale modeling for metabolic engineering. J Ind Microbiol Biotechnol 42:327–338
Srivastava S, Dafale NA, Jakhesara SJ, Joshi CG, Patil NV, Purohit HJ (2021) Unraveling the camel rumen microbiome through metaculturomics approach for agriculture waste hydrolytic potential. Arch Microbiol 203:107–123

Sungur Ş (2024) Biofuels. In: Handbook of emerging materials for sustainable energy. Elsevier, pp 399–417
Takizawa S, Asano R, Fukuda Y, Feng M, Baba Y, Abe K, Nakai Y (2020) Change of endoglucanase activity and rumen microbial community during biodegradation of cellulose using rumen microbiota. Front Microbiol 11:603818
Tapio I, Snelling TJ, Strozzi F, Wallace RJ (2017) The ruminal microbiome associated with methane emissions from ruminant livestock. J Anim Sci Biotechnol 8:1–11
Thongbunrod N, Chaiprasert P (2021) Efficacy and metagenomic analysis of the stabilized anaerobic lignocellulolytic microbial consortium from Bubalus bubalis rumen with rice straw enrichment for methane production. Bioenergy Res 14:870–890
Thorpe A (2009) Enteric fermentation and ruminant eructation: the role (and control?) of methane in the climate change debate. Clim Chang 93:407–431
Tolonen AC (2024) Engineering of Lachnospiraceae fermentation for health and the bioeconomy (Doctoral dissertation, Universite Paris Saclay)
Ungerfeld EM, Pitta D (2024) Biological consequences of the inhibition of rumen methanogenesis. Anim:101170
Wagner AO, Lackner N, Mutschlechner M, Prem EM, Markt R, Illmer P (2018) Biological pretreatment strategies for second-generation lignocellulosic resources to enhance biogas production. Energies 11:1797
Waldor MK, Tyson G, Borenstein E, Ochman H, Moeller A, Finlay BB, Smith H (2015) Where next for microbiome research? PLoS Biol 13:e1002050
Weimer PJ, Kohn RA (2016) Impacts of ruminal microorganisms on the production of fuels: how can we intercede from the outside? Appl Microbiol Biotechnol 100:3389–3398
Won SM, Chen S, Lee SY, Lee KE, Park KW, Yoon JH (2020) Lactobacillus sakei ADM14 induces anti-obesity effects and changes in gut microbiome in high-fat diet-induced obese mice. Nutrients 12:3703
Wozny MA, Bryant MP, Holdeman LT, Moore W (1977) Urease assay and urease-producing species of anaerobes in the bovine rumen and human feces. Appl Environ Microbiol 33:1097–1104
Xing BS, Han Y, Wang XC, Wen J, Cao S, Zhang K, Yuan H (2020) Persistent action of cow rumen microorganisms in enhancing biodegradation of wheat straw by rumen fermentation. Sci Total Environ 715:136529
Xue X, Liu Y, Fu J (2024) Bioengineering embryo models. Nat Rev Bioeng 3:1–19
Zaplana T, Miele S, Tolonen AC (2024) Lachnospiraceae are emerging industrial biocatalysts and biotherapeutics. Front Bioeng Biotechnol 11:1324396
Zened A, Julien C, Cauquil L, Pascal G, Canlet C, Tremblay-Franco M, Combes S (2024) Milk replacer feeding once or twice a day did not change the ruminal metabolomic profile and the microbial diversity of dairy calves from birth to weaning. J Dairy Sci 107:5574
Zhang H, Pang H, Shi J, Fu T, Liao B (2012) Investigation of liquefied wood residues based on cellulose, hemicellulose, and lignin. J Appl Polym Sci 123:850–856
Zhang R, Zhu W, Mao S (2016) High-concentrate feeding upregulates the expression of inflammation-related genes in the ruminal epithelium of dairy cattle. J Anim Sci Biotechnol 7:1–13
Zou Y, Xu X, Li L, Yang F, Zhang S (2018) Enhancing methane production from U. lactuca using combined anaerobically digested sludge (ADS) and rumen fluid pre-treatment and the effect on the solubilization of microbial community structures. Bioresour Technol 254:83–90

Chapter 7
Termite-Based Biorefinery as an Innovative Model for Biohydrogen and Biomethane Productions

Mudasir A. Dar, Rongrong Xie, Shehbaz Ali, Md Muzammel Hossain, Farhan Ahmad, Hossain M. Zabed, Mohd Shahnawaz, Daochen Zhu, Sameh S. Ali, and Jianzhong Sun

Abstract Lignocellulosic biomass is a renewable resource with significant potential for sustainable energy production. Termites, as natural bioreactors, offer a unique model for efficient lignocellulose degradation, driven by their gut microbiota composed of bacteria, archaea, and eukaryotic symbionts. This microbial symbiosis enables termites to ferment lignocellulose into energy resources including biohydrogen and methane. The symbiotic mechanisms, underpinned by the gut's physicochemical conditions, facilitate anaerobic fermentation and carbon recycling. Termite gut symbionts demonstrate unparalleled efficiency in converting cellulosic substrates into biohydrogen and biomethane, with rates surpassing those of other natural systems. Here we explore the lignocellulolytic capabilities of termite gut systems, emphasizing their role as biomimetic models for biohydrogen and biomethane production from lignocellulosic biomass. The chapter identifies critical research gaps in understanding symbiont functions and fermentation dynamics while highlighting termites' potential for sustainable biomass utilization. By leveraging termite-inspired biotechnologies, we can address global challenges in renewable energy and waste management, paving the way for innovative solutions in lignocellulosic biorefineries.

Keywords Termites · Gut microbes · Symbiosis · Fermentation · Lignocellulosic biomass · Biomimetics

---

M. A. Dar · R. Xie · S. Ali · Md M. Hossain · F. Ahmad · D. Zhu · S. S. Ali · J. Sun (✉)
Biofuels Institute, School of Environment and Safety Engineering, Jiangsu University, Zhenjiang, Jiangsu, China
e-mail: jzsun1002@ujs.edu.cn

H. M. Zabed
School of Life Science, Guangzhou University, Guangzhou, Guangdong Province, China

M. Shahnawaz
Department of Botany, Govt. Degree College Drass, A Constituent College of University of Ladakh, Ladakh, India

M. A. Dar et al. (eds.), *Recent Trends in Lignocellulosic Biofuels and Bioenergy*, Clean Energy Production Technologies,
https://doi.org/10.1007/978-981-96-4636-4_7

7.1 Introduction

The global energy landscape is undergoing a rapid transformation driven by a variety of factors including technological innovation, shifts in energy demand, environmental concerns, and the pursuit of sustainability (Dar et al. 2022a, 2024a). Climate change impacts necessitate swift development of mitigation and adaptation strategies to limit global warming, with many countries exploring carbon-neutral pathways in line with the Paris Agreement's goals (Yacout et al. 2024; Ali et al. 2024a). The Intergovernmental Panel on Climate Change (IPCC) emphasizes that achieving a 1.5 °C limit requires integrating carbon dioxide (CO_2) removal and adopting negative emission technologies (Anderson et al. 2015). In support of these efforts, the United Nations established 17 Sustainable Development Goals (SDGs), with two—SDG7, focused on ensuring universal access to affordable, sustainable energy, and SDG13, targeted to climate action—directly addressing climate change mitigation. Achieving these objectives necessitates the exploration of innovative and renewable energy resources to ensure sustainability. Among such resources, lignocellulose stands out as renewable and abundant biomass that can be harvested year-round (Dar et al. 2024b). This biomass is produced in substantial quantities annually through photosynthesis and encompasses municipal solid waste (MSW), agricultural and forestry residues, and dedicated energy crops (Dar et al. 2018a, 2021a). These lignocellulosic resources not only offer vast availability but also hold significant potential to mitigate greenhouse gas emissions (Sadalage et al. 2020). Through the concerted action of cellulases and hemicellulases, lignocellulose can be enzymatically hydrolyzed into simple sugars, which are subsequently fermented and converted into hydrogen or biomethane through anaerobic routes (Bhatia et al. 2022). This enzymatic bioconversion offers a promising and practical approach to generating novel renewable energy sources (Dar et al. 2022a). For example, biofuels derived from lignocellulose via enzymatic hydrolysis and fermentation are recognized as a viable alternative to conventional fossil fuels (Xie et al. 2024). With an estimated global production of 3–5 gigatonnes annually, lignocellulosic biomass has the potential to provide approximately 50–85 petajoules of energy per year, accounting for 10–20% of the world's current energy demands (Cragg et al. 2015; Dar et al. 2015). Beyond energy production, lignocellulose bioconversion contributes to net energy savings, reduces industrial CO_2 emissions, and promotes carbon sequestration in soils via perennial energy crops (Zabed et al. 2023). While bioconversion to energy and commodity chemicals holds promise, the targeted hydrolysis of lignocellulose into platform chemicals appears more efficient in terms of atom economy and mass utilization (Mujtaba et al. 2023). Thus, lignocellulose bioconversion is integral to sustainable development, facilitating the production of renewable energy while safeguarding environmental health.

Anaerobic digestion is a biological process that occurs both in the gut and other natural environments, wherein microbial communities decompose organic matter in the absence of oxygen into methane (Kumar et al. 2024; Ali et al. 2024b). Within the gut, a vast population of anaerobic bacteria performs critical functions, including anaerobic digestion and hydrogen (H_2) production. Biohydrogen and biogas (biomethane) are considered among the most energy-efficient and environmentally friendly biofuels (Sahrin et al. 2022). Production of these energy resources can be achieved through various advanced anaerobic digestion technologies such as dark fermentation and photo fermentation of biomass (Chong et al. 2022). The production of biomethane involves the breakdown of organic wastes by microorganisms in an oxygen-free environment, resulting in biogas that typically contains 45–85% methane and 25–50% CO_2. This biogas can be upgraded through different methods to pure methane that may serve multiple applications, including electricity generation, heating, or as a vehicle fuel. Biomethane has a carbon footprint that's eight times lower than natural gas (Ali et al. 2024b). On the other hand, biohydrogenation emerges as a promising strategy to address future energy demands, attributed to its sustainability, regenerative potential, cost-effectiveness, and minimal environmental impact in comparison to traditional methods (Ram et al. 2024). Biological hydrogen production techniques, although characterized by prolonged reaction durations and lower yields (4 mol of H_2 per mol of glucose) (Akhlaghi and Najafpour-Darzi 2020), exhibit superior H_2 production selectivity. These methods deliver sustainable, environmentally friendly, and economically viable advantages (Kazmi et al. 2024). Research has shown that enhancing bioprocessing efficiency and H_2 production can be most effectively achieved through the integration of microorganisms. Moreover, the production of hydrogen via biological processes using renewable resources such as biomass and organic wastes represents a significant advancement (Goria et al. 2024). Biohydrogen production offers a promising alternative for achieving a sustainable H_2 supply with minimal pollution and high efficiency. It has been demonstrated that microorganisms and their intrinsic biochemical machinery are essential to the bioconversion of lignocellulosic biomass to H_2 (Dar et al. 2021a; Li et al. 2024). Initially, anaerobic microbes that thrive in carbohydrate-rich gut environments of wood-feeding insects like termites undergo pretreatment, during which they hydrolyze substrates into simpler sugars. Subsequently, these microorganisms engage in anaerobic digestion, yielding byproducts such as alcohols, CH_4, H_2, volatile fatty acids (VFAs), and CO_2 (Ali et al. 2024a, b, c).

In nature, lignocellulosic biomass is efficiently utilized by various natural biomass utilization systems (NBUS), including insects, termites, snails, and ruminants (Show et al. 2022; Dar et al. 2018b, 2021b, 2025a). Among these, termites exhibit the most advanced and efficient lignocellulolytic mechanisms, primarily due to their symbiotic association with gut microbiota comprising bacteria, protists, and fungi (Brune 2007; Dar et al. 2022b). Termites rely on these microbial communities to digest complex lignocellulosic materials, showcasing exceptional proficiency in producing lignocellulose-degrading enzymes (Dar et al. 2025b; Peng et al. 2021). Advances in biotechnology have led to a significant expansion of knowledge regarding lignocellulose decomposition by termites, revealing promising applications for biorefinery processes (Scharf 2015). In this context, here we discuss the practical implications of termite gut microbial systems, focusing on the lignocellulolytic capabilities of their microbial symbionts. Additionally, we explore the potential of these systems as a model for developing biomimetic strategies in lignocellulosic biorefineries, paving the way for innovative and sustainable solutions to energy and material challenges. The chapter aims to provide a thorough literature review to elucidate the current knowledge landscape, pinpoint critical research gaps, and highlight potential limitations in Bio-H_2 production from lignocellulose.

7.2 Termites as Ecosystem Engineers

Termites (*Isoptera*) predominantly inhabit regions between 45° N and 45° S latitudes, with a significant abundance in warm ecosystems (Gomati et al. 2011). These highly social insects form large colonies comprising up to several million individuals. According to the Termite database of the University of Brazil (accessed on 27 October 2024), termites are classified into 3212 species, spanning 393 genera and 13 families (Hellemans et al. 2024). These families include Archeorhinotermitidae, Archotermopsidae, Cratomastotermitidae, Heterotermitidae, Hodotermitidae, Kalotermitidae, Mastotermitidae, Rhinotermitidae, Serritermitidae, Stolotermitidae, Stylotermitidae, Termitidae, and Termopsidae. Termites are highly adaptable insects that thrive in diverse environments, including soil, trees, and wooden structures (Dar et al. 2022b; Gomati et al. 2011). As one of the most ecologically and evolutionarily successful insect groups, termites are distributed across all continents except Antarctica, with their highest prevalence in tropical and subtropical regions (Dar et al. 2024b). They exhibit a unique social structure, with distinct castes performing specialized roles. The worker caste, while essential for colony sustenance, is particularly associated with extensive decomposition of forestry resources and wooden structures (Xie et al. 2024). Termites can be categorized into three main feeding groups based on their dietary habits: soil-feeding (humivorous) termites that decompose decayed organic soil material, xylophagous termites that feed on wood (both decomposed and living), and fungus-feeding termites that maintain a symbiotic relationship with fungi (Eggleton 2000).

In tropical regions, termites pose substantial threats to forestry by attacking living trees and undermining afforestation efforts. Species such as *Odontotermes* and *Microtermes*, collectively known as fungus-growing termites, are particularly destructive, initiating damage in the upper 15–25 cm of soil (Xie et al. 2024)). During periods of drought, termites can consume plant biomass at rates comparable to grazing livestock, intensifying competition for forage resources. Despite their reputation as pests, termites play a vital role in ecosystem functioning, providing critical services such as organic matter decomposition, nutrient recycling, and soil fertility enhancement (Myer and Forschler 2019). These contributions, which support primary productivity, have earned termites the title of "ecosystem engineers." They are particularly important in tropical savannas, forests, and arid ecosystems, where their activities facilitate the decomposition of lignocellulosic biomass and contribute to carbon cycling. From a sustainability perspective, termites are recognized as one of the most efficient natural bioreactors on Earth, with an unparalleled ability to convert lignocellulosic biomass into valuable energy substrates (Ohkuma 2003).

Termites possess a remarkable capacity for biomass degradation, significantly contributing to global carbon recycling. Among termite castes, workers are the primary feeders, processing large quantities of lignocellulose. This ability is primarily due to the metabolic efficiency of their "gut digestome," which consists of a diverse array of intestinal symbiotic microorganisms, including bacteria, archaea, yeasts, and other eukaryotic symbionts. The termite digestive tract is rich in species, enzymes, and genes that facilitate the breakdown of complex organic materials (Janusz et al. 2020). Termites are valuable sources of wood-degrading enzymes, including cellulases, xylanases, laccases, and glucosidases, which are potentially involved in the degradation of agro-industrial wastes (Bhujbal et al. 2021). Metagenomic analysis of the hindgut microbiota of higher termites revealed a diverse array of enzymes, including endoxylanases, endoglucanases, GH94 phosphorylases, glucosidases, nitrogenases, enzymes for carbon dioxide reduction, and novel enzymes used for lignocellulose-based biofuel production (Warnecke et al. 2007) and acetate production (Brune 2014).

The termite gut microbiota plays a pivotal role in the bioconversion of lignocellulose into fermentable sugars, a process with promising applications in biofuel production and biorefineries (Scharf and Tartar 2008). Over a century of research has positioned termites as model organisms for studying symbiotic processes. Their gut systems, with microbial complexity rivaling that of mammalian microbiomes, offer unique insights into host-microbe interactions and biodegradation pathways. Furthermore, termites' roles in carbon recycling and efficient lignocellulose utilization make them sophisticated models for applied research in renewable energy and sustainability. Over millions of years of evolution, termites have developed advanced lignocellulolytic mechanisms to overcome the structural complexity of lignocellulose (Dar et al. 2022b). These mechanisms, underpinned by their symbiotic gut microbiota, represent a critical natural innovation for biomass degradation. As natural bioreactors, termites are increasingly recognized for their potential to inspire bioengineering applications aimed at converting renewable biomass into biofuels and other value-added products (Ohkuma 2003). The gut systems of wood-feeding termites, in particular, offer novel opportunities for the development of nature-inspired technologies with industrial relevance. Since the early twentieth century, termites have been studied as reservoirs of novel microorganisms and potent biocatalysts. Their symbiotic lignocellulolytic systems provide a robust framework for engineering advanced biotechnologies, particularly in the context of renewable bioenergy production. Despite this, the specialized functions of these gut symbionts, particularly bacteria symbionts, remain poorly understood, highlighting the need for further research to unlock their full potential. Continued research into termite microbiomes and their biodegradation pathways promises to yield innovative solutions for sustainable biomass utilization and biofuel development (Sun and Chen 2010).

7.2.1 Termite Digestomes

The termite gut, particularly in the worker caste, is a highly specialized and compartmentalized organ adapted for efficient lignocellulose digestion. It consists of three primary regions: the foregut, midgut, and hindgut, each with distinct functions (Dar et al. 2021b, 2022a, b). The foregut and anterior gut play roles in mechanical grinding, while the midgut contributes to enzymatic hydrolysis, including cellulase production in some higher termite species (Mikaelyan et al. 2017). The hindgut, the largest and most complex region, serves as the primary site for nutrient absorption, fermentation, and lignocellulose metabolism. Considerable differences exist between lower and higher termites in gut structure and function. Lower termites have shorter, slightly acidic guts (pH 5.5–7.0) and host flagellates alongside prokaryotes for cellulose digestion. In contrast, higher termites exhibit more developed and compartmentalized hindguts, including unique structures like the "mixed segment." The hindgut is further subdivided into segments (P1–P5), with P3 being the largest in higher termites (Brune and Dietrich 2015). Physicochemical variations along the gut influence digestive physiology and microbial symbiosis. The hindgut functions as a fermentation chamber, enabling anaerobic breakdown of cellulose into hydrogen and energy by endosymbiotic prokaryotes. It harbors the highest microbial diversity within the gut, in contrast to the relatively low microbial load of the foregut and midgut. These features underscore the termite gut's role as a microhabitat optimized for lignocellulose bioconversion.

While some bacteria inhabit the foregut and midgut, the majority of intestinal microbiota resides in the hindgut, particularly in the paunch, which is the area immediately behind the enteric valve (Brune 2014). According to Köhler et al. (2012), termite gut offers a favorable environment for microbial activity, with the type and distribution of microflora being influenced by physicochemical conditions such as pH, redox potential, and temperature. Among the various physicochemical factors, pH and redox potential are crucial in determining the type of microflora present in the gut (Mikaelyan et al. 2017). The pH of the foregut and midgut is nearly neutral, whereas the paunch, colon, and rectum are slightly acidic. Furthermore, the foregut and midgut of termites are aerobic with an E_0 exceeding +100 mV (Quevedo et al. 2021). In contrast, the paunch and colon are anaerobic with E_0 values ranging from −230 to −270 mV, and the hindgut of termites shows E_0 between −120 and − 270 mV, confirming the anaerobic conditions of the gut lumen (Gomati et al. 2011).

7.2.2 Termite Microbial Symbiosis for Lignocellulose Bioconversion

Termites exhibit a unique symbiotic relationship within their gut, hosting diverse microorganisms from all three domains of life—Archaea, Bacteria, and Eukarya. This complex microbial ecosystem enables termites to efficiently degrade lignocellulosic biomass, including cellulose and lignin-derived compounds, a capability unparalleled in the animal kingdom (Dar et al. 2024a). For example, the wood-feeding termite *Coptotermes formosanus* demonstrates remarkable cellulolytic activity, producing 10^3–10^4 μmol of reducing sugars per minute per milliliter using carboxymethyl cellulose as a substrate (Brune 2014). Genomic studies have highlighted the cellulolytic potential of termites, but further research into protein functions, enzyme localization, and symbiont interactions is needed to fully elucidate their mechanisms. Recent advancements in molecular biology have revealed that termite gut microbiota performs diverse ecological functions, including carbohydrate digestion, nitrogen fixation and metabolism, oxygen consumption, and the production of hydrogen and methane (Rouland-Lefèvre 2010). These functions are critical for energy production and nutrient cycling within the gut microenvironment.

In lower termites, such as *Coptotermes* and *Reticulitermes*, flagellates are integral to cellulose digestion, while prokaryotic endosymbionts contribute to acetogenesis, methanogenesis, and fermentation (Scharf and Tartar 2008). The termite gut hosts over 200 species of bacteria, highlighting its remarkable microbial diversity (Dar et al. 2022a, b, 2024b). A list of the different groups of microorganisms present in termite gut systems are detailed in Table 7.1. Key endosymbionts, such as *Candidatus Endomicrobium trichonymphae* and *Candidatus Azobacteroides pseudotrichonymphae*, play specialized roles in fermenting carbohydrates to acetate. These bacteria exhibit substrate specificity: the former metabolizes glucose-6-phosphate, while the latter processes glucose, xylose, and hexuronates. The availability of these substrates in the termite gut is attributed to the enzymatic digestion of cellulose and hemicellulose. Research over the past two decades has significantly advanced our understanding of termite digestion, nitrogen cycling, and the intricate termite-microbe symbiosis, highlighting their role as ecological engineers. The synergistic termite-endosymbiont system provides a natural blueprint for lignocellulose bioconversion, offering insights into sustainable technologies for renewable energy production. These findings underscore termites' ecological importance and their potential as models for innovative biotechnologies aimed at addressing global challenges in bioenergy and waste management.

7.2.3 Biohydrogen Potential of Termites

The termite gut environment is particularly efficient in hydrogen production, with a turnover rate of hydrogen ranging from 9 to 33 m^3 per m^3 of hindgut, aligning with 22–26% of the termites' respiratory activity. This indicates that hydrogen (H2) acts as a central free intermediate during lignocellulose degradation, positioning the termite gut, with its high rates of reductive acetogenesis, as the smallest and most efficient natural bioreactor currently known (Sun and Zou 2014). Figure 7.1 illustrates the production of hydrogen and methane by the termites microbial symbiosis from lignocellulosic biomass. The small size of the termite gut (compared to

Table 7.1 Some of the representative symbiotic bacteria of termites are essential for the digestion of lignocellulose and the subsequent fermentation into biohydrogen and biomethane. These microorganisms participate in various fermentation pathways, leading to the generation of hydrogen, acetate, methane, and other metabolites that are highly valuable for biotechnological applications, especially in the realm of bioenergy production

Microorganism	Termite host	Biofuel type	Carbon source	Highlights	References
Bacteroides, *Prevotella*	*Nasutitermes ephratae*, *Termes bellicosus*, *Reticulitermes grassei*	Biomethane	Wheat straw	Dominant groups in termite gut support hydrogen production	Lazuka et al. (2018)
Streptomyces sp.	*Microcerotermes* species	Bioethanol	Wheat straw		Dar et al. (2024a, b, c)
Bacillus sp. BMP01, *Ochrobactrum oryzae* BMP03	*Coptotermes formosanus*, *Cryptotermes brevis*, *Odontotermes formosanus*	Hydrogen	CMC, xylan, lignin	Supports clostridium, promotes H_2 production in gut system	Tsegaye et al. (2019)
Bacterial symbionts	*Nasutitermes ephratae*, *Microcerotermes parvus*, *N. lujae*, *Termes hospes*	Biomethane	Wheat straw		Auer et al. (2017)
Actinobacteria	*M. Nervosus*, *Macrognathothermes sunteri*, *Tumulitermes pastinato*	Hydrogen, biomethane	Organic carbon		Chiri et al. (2021)
Symbionts	*Reticulitermes speratus*	Hydrogen	Wood		Konishi et al. (2020)
Treponema primitia	*Neotermes castaneus*, *Zootermopsis angusticollis*, *Cryptotermes cavifrons*	Acetate, hydrogen, formate	Wood	Lignocellulose bioconversion; Cospeciating	Graber and Breznak (2004)
Clostridium termitidis, *Clostridium beijerinckii*	*Nasutitermes* spp.	Hydrogen	Cellulose	Strict anaerobe, main biohydrogen producer	Gomez-Flores et al. (2017)
Methylocella sp., and other symbionts	*R. speratus*, *Z. nevadensis*, *Cubitermes orthognathus*	Biomethane, hydrogen	Wood		Pester and Brune (2007)
Enterococcus sp.	*Reticulitermes flavipes*	Acetate	Lignocellulose		Tholen et al. (1997)
Sporomusa termitida, *Sporomusa* sp. strain TmAO3	Termites	Hydrogen	Lignocellulose	Acetogenic; key contributor to hydrogen formation	Boga et al. (2003)

Microorganism	Termite host	Biofuel type	Carbon source	Highlights	References
Microbiota	*Cubitermes* spp.	Biomethane	Lignocellulose		Schmitt-Wagner and Brune (1999)
Sporomusa aerivorans	*Thoracotermes macrothorax*	Hydrogen	Lactate		Boga et al. (2003)
Sporotalea propionica	*T. macrothorax*	Hydrogen	Glucose		
Acetonema longum	*Pterotermes occidentis*	Hydrogen	Yeast extract, rumen fluid, resazurin	Acetogen helps in hydrogen formation	Kane and Breznak (1991)
Enterobacter cloacae KBH3	*Globitermes* sp.	Hydrogen	Glucose	Facultative anaerobe; H2 producer under microoxic conditions	Harun et al. (2012)
Pseudotrichonympha grassii	*C. formosanus*	Hydrogen	Wood cellulose		Inoue et al. (2007)
Trichonympha sphaerica	*Z. termites*	Hydrogen	Lignocellulose		Yamin (1981)
Trichomitopsis termopsidis, *Hexamastix termopsidis* and *Tricercomitus termopsidis*, *gut microbiota*	*Reticulitermes santonensis*, *Z. nevadensis*, *Cryptotermes secundus*	Hydrogen, biomethane	CO_2, Formate, lactate		Pester and Brune (2007)
Trichomitopsis termopsidis, *Trichonympha sphaerica*	*Z. termites*	Hydrogen, biomethane	Cellulose, corncob, cereal leaves		Odelson and Breznak (1985)
E. aerogenes	*Reticulitermes lucifugus*	Acetate, hydrogen	Lignocellulose	Acetogenic; important for hydrogen production	Carbonero-Pacheco et al. (2023)
Clostridium mayombei	*Nasutitermes lujae*	Acetate, hydrogen	Wood	Acetogenic bacterium; key contributor to hydrogen formation in termites	
Ruminococcus sp.	*Macrotermes bellicosus*	Acetate and hydrogen,	Cellulose	Hydrogen producer	

CMC carboxymethyl cellulose

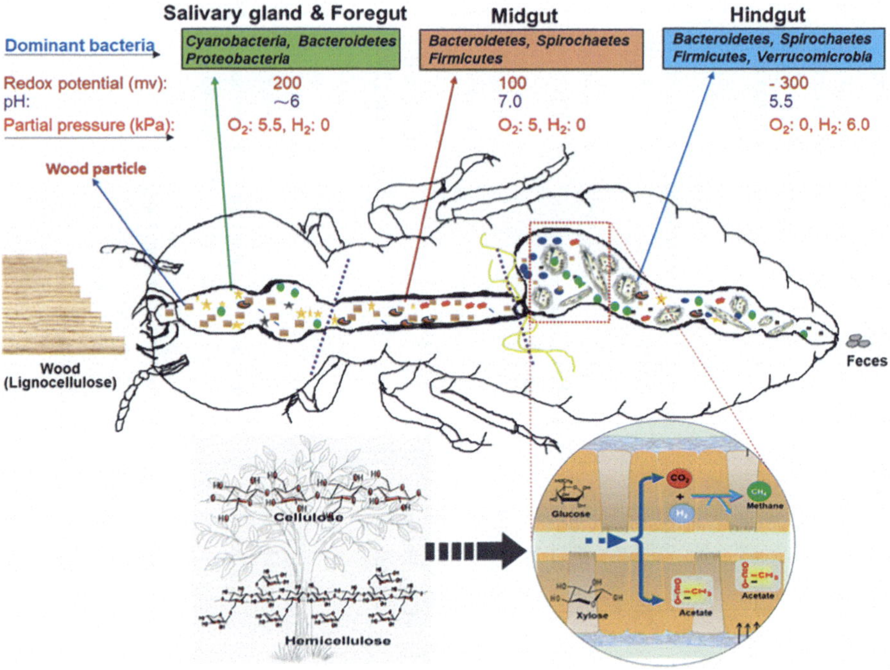

Fig. 7.1 The gut system of lower wood-feeding termites showing the complex interactions between gut microbial symbionts and host for the efficient bioconversion of lignocellulosic into hydrogen and biomethane. The hindgut is densely populated with microorganisms, including bacteria, archaea, and protists, which play crucial roles in the fermentation of fibrous plant materials

rumens) allows for a 500-fold increase in oxygen influx per unit volume, enhancing the efficiency of microbial processes. These efficient, anaerobic processes are highly conducive to biomimicry, with studies like Mathew et al. (2013) demonstrating that co-culturing termite gut microbes, such as *Bacillus* and *Clostridium* species, can replicate these processes in vitro, producing significant amounts of hydrogen (up to 4.08 mmol/mL using glucose as a substrate). Termite gut symbionts can convert a sheet of A4 paper into two liters of hydrogen, highlighting their potential for use as an inorganic energy source. Estimates suggest that subterranean termites can produce up to 3858 μmol of molecular hydrogen per gram of cellulose, showcasing the gut's exceptional efficiency in converting cellulosic substrates into biofuels including biohydrogen (Dar et al. 2024a). Additional studies have identified facultative and obligate anaerobes, such as *Clostridium* sp., *Enterobacter cloacae*, and *E. aerogenes*, which contribute to biohydrogen production through dark fermentation under microoxic conditions. *E. cloacae* strains, in particular, have shown promising hydrogen production rates under batch fermentation, further validating the potential of termite gut microbiota for large-scale bioenergy production (Ali et al. 2024a, b, c). Despite these advances, there remains much to explore regarding the roles of unculturable bacteria and other microbes in termite guts.

Further investigation into the complex fermentative patterns and competitive mechanisms of termite gut acetogens will be essential for optimizing hydrogen production and improving energy yields. Ongoing global research is focused on enhancing the fermentation processes of termite symbionts to develop more efficient, scalable biohydrogen production technologies.

7.2.4 Biomethane Potential of Termites

Anaerobic digestion, classified as a biological process, yields methane, which has been commercially proven. Anaerobic digestion involves the breakdown of organic waste matter in the absence of oxygen by microorganisms, producing biogas, mainly methane and carbon dioxide (Tomar et al. 2023). This biogas can be upgraded to biomethane, which is a clean fuel equivalent to natural gas (Stephen and Periyasamy 2018). The use of biogas from this waste contributes to reducing greenhouse gas emissions (Tawfik et al. 2022). Thus, anaerobic digestion becomes a viable process at the heart of a great concept which is the waste biorefinery.

Methane (CH4) production by termites was first documented by Cook (1932), and subsequent studies have confirmed that methane production occurs in the termite gut across nearly all species (Brune 2018). In higher termites, which are more evolutionarily advanced and predominantly found in tropical forests, methane production is facilitated by symbiotic bacteria, whereas in lower termites, it is facilitated by flagellate protozoa in association with bacterial symbionts (Brune 2018). Zimmerman et al. (1982) quantified methane emissions from individual termites and estimated global termite emissions to be approximately 150 Tg CH4 per year, accounting for about 40% of global natural methane emissions at that time. However, more recent estimates vary widely, ranging from 2 to 15 Tg CH4 per year, which represents approximately 0.5–4% of the total natural methane emissions (Saunois et al. 2020). Despite their lower contributions compared to major natural sources such as wetlands and freshwater systems, the role of termite emissions in the methane budget of local tropical ecosystems remains significant. On an individual level, each termite produces an average of about half a microgram of methane per day. Although this may appear trivial at first glance, the vast distribution and sheer number of termites globally amplify this effect significantly. Termites are found in 70% of the world's countries, inhabiting diverse ecosystems from tropical rainforests to temperate regions. Currently, there are over 3150 classified species of termites, each contributing to the collective methane emissions. Given the high species count and substantial populations, the cumulative methane production by termites becomes substantial. It is estimated that termites collectively release approximately 20 million tons of methane annually. This impressive quantity of methane emanating from termites represents a significant portion of global methane emissions. Termites are estimated to contribute between 1% and 3% of the world's total methane emissions.

Global estimates of CH4 emissions from termites are derived from both field and laboratory measurements. The common approach involves measuring CH4 emissions per individual termite (mg CH_4 termite^{-1} h^{-1}) or per unit of termite biomass (mg CH_4 g^{-1} termite h^{-1}), with biomass either measured directly or referenced from existing literature (Law et al. 2024). However, this method has the drawback of potentially altering termite emissions and behavior since termites are removed from their natural habitat. Alternatively, in situ measurements of CH_4 emissions from termite nests in the field can be conducted, expressed per mound or nest (mg CH_4 mound^{-1} h^{-1}). Although this method avoids disturbing the natural environment, accurately estimating the size of termite colonies within the nests remains challenging, making it difficult to convert these values into per-termite emissions (Jones et al. 2005). Studies often concurrently measure CH_4 and CH_4 emissions, finding a generally consistent relationship between the two, with species-specific ratios (Jamali et al. 2013). Current global CH_4 emission estimates from termites are based on a limited number of studies, indicating a need for more comprehensive data (Brune, 2018). Existing research has predominantly focused on species from Australia, Asia, and Africa (Jamali et al. 2013). Future research should focus on refining in situ measurement techniques to accurately estimate CH_4 emissions without disturbing termite colonies and expanding studies to underrepresented regions such as Alaska, Amazon, Russia boreal forest, etc. Additionally, investigating species-specific emission ratios and the impact of environmental variables on CH_4 production can enhance predictive models and global emission estimates.

The contribution of termites to methane emissions is an often-overlooked aspect of the global carbon cycle. Their digestion processes, which involve symbiotic microorganisms that break down cellulose in wood, produce methane as a byproduct. The high efficiency of these natural bioreactors, operating within the termites' guts, highlights the significance of these small organisms for renewable energy. By studying the biomethane production process in termites, we can develop biomimetic systems that replicate their highly efficient natural digestion mechanisms. Understanding the ecological role and impact of termites is crucial for comprehensive climate models and mitigation strategies. Further research into their contribution to methane emissions and potential methods to mimic their gut symbionts can provide valuable insights for future biorefinery. By closely mimicking the natural digestive process of termites, the biochemical reactors could transform agricultural and forestry waste into valuable biomethane, providing a sustainable and efficient source of renewable energy. By mimicking the termite symbiotic digestion mechanisms, we can design bioreactors that emulate the termite gut environment, optimizing conditions such as pH, temperature, and anaerobic conditions to enhance microbial activity. These engineered systems could efficiently convert agricultural and forestry waste into biomethane, providing a sustainable and renewable energy source while reducing reliance on fossil fuels and minimizing waste. To biomimic or learn from termites' biomethane production from lignocellulose, artificial bioreactors need to be designed and created that can emulate the conditions within the termite gut. The first step involves studying and isolating the specific microorganisms in termite guts responsible for lignocellulose breakdown and methane production. These microorganisms can then be cultured and introduced into bioreactors designed to replicate the gut environment, including maintaining anaerobic conditions, optimal pH, and temperature similar to the termite gut systems. Advanced biotechnological methods, such as genetic engineering, could be helpful to enhance the efficiency of these microbes. Additionally, incorporating a continuous feed system for lignocellulosic biomass and efficient gas collection mechanisms can further optimize the process.

7.2.5 Mechanism of Biohydrogen and Biomethane Production

In addition to methane, termite emissions encompass various other gases including molecular hydrogen (Inoue et al. 2007). In gut systems, the microorganisms involved in biohydrogen production include both biohydrogen producers and nonproducers that work synergistically. Biohydrogen producers can convert complex organic substrates into biohydrogen, either as single strains or in diverse microbial communities. Hydrogen producers are frequently found in gut environments (Dar et al. 2024b). Generally, *Clostridium* and *Enterobacter* are the predominant genera under anaerobic gut conditions (Kumar et al. 2018; Dar et al. 2024c). Other genera, including *Bacillus*, *Ethanoligenens*, *Klebsiella*, *Citrobacter*, and *Escherichia*, are also frequently reported as biohydrogen producers (Dzulkarnain et al. 2022). The biochemical pathways involved in dark fermentation largely overlap with those of anaerobic digestion, where diverse microbial communities synergistically degrade organic substrates to ensure a stable process (Stolze et al. 2016). These pathways are divided into four phases: hydrolysis, acidogenesis, acetogenesis, and methanogenesis. During anaerobic digestion, H2 is produced during the acidogenesis and acetogenesis phases by hydrolytic and fermentative bacteria. This hydrogen is later consumed during methanogenesis when methanogenic archaea utilize H2 and CO2 to CH_4 (Hassa et al. 2018). Therefore, inhibiting methanogenesis is necessary to redirect the pathway towards hydrogen production in the final step of dark fermentation.

The initial hydrolysis phase involves breaking down complex polysaccharide substrates (cellulose and hemicellulose) into monomer residues (sugars, glucose, xylose) through the action of extracellular hydrolytic enzymes such as cellulase, xylanases, ligninases, etc. The microbial taxa responsible for these hydrolytic activities are predominantly from the *Firmicutes* and *Bacteroidetes* phyla, specifically the *Clostridium* and *Bacteroides* genera. These phyla are abundant in termite guts due to their degradative capabilities, primarily breaking down cellulolytic materials (Abendroth et al. 2015). Members of these phyla are known for their rapid growth and ability to utilize hydrolyzed products for fermentation throughout the degradation process. They also exhibit resilience to environmental changes (Wang et al. 2018). Hydrolysis is often the rate-limiting step for recalcitrant substrates like lignocellulose, which typically requires pretreatment through munching and mincing by a host to achieve complete hydrolysis (Menzel et al. 2020). Next, during the acidogenesis phase, hydrolyzed products are metabolized into H2, CO2, and short-chain fatty acids (SCFAs) such as acetate, butyrate, formate, and propionate by acidogenic microbial communities. The predominant phyla involved in this phase are *Bacteroidetes, Firmicutes, Chloroflexi*, and *Proteobacteria* (Campbell et al. 2023). We have recently observed a tremendous diversity of these bacteria in the gut system of the wood-feeding lower termite, *C. formosanus* (Dar et al. 2022a, 2024b). Acidogenesis is typically a rapid process characterized by the accumulation of SCFAs and a subsequent decrease in gut pH. The microorganisms active in this phase include both facultative and obligate anaerobes, commonly referred to as acidogens or acid formers. Notable genera participating in this phase include *Clostridium*, *Bacteroides*, *Bifidobacterium*, *Bacillus*, and *Streptococcus* (Gonzalez-Martinez et al. 2016). During acidogenesis, carbohydrates, predominantly glucose, are converted to pyruvate through the glycolytic pathway (Saravanan et al. 2021). Under gut microenvironmental conditions, the primary H_2-yielding fermentation routes include obligate anaerobic (*Clostridium* type) and facultative anaerobic (*Enterobacteria* type) pathways (Dzulkarnain et al. 2022; Campbell et al. 2023).

In the facultative anaerobic fermentation route, pyruvate is further converted to acetyl-CoA and formate, by pyruvate formate lyase (PFL), and H_2 is produced from formate by the formate hydrogen lyases enzyme complex. The strict anaerobic fermentation route involves the oxidization of pyruvate to acetyl-CoA and reduced ferredoxin (Fd) by pyruvate ferredoxin oxidoreductase (PFOR). Hydrogen is then released from the reduced Fd by the action of hydrogenase. Additional molecules of H_2 can also be produced from the oxidization of nicotinamide adenine dinucleotide (NADH) to reduced Fd by NADH: ferredoxin oxidoreductase (NFOR), followed by the subsequent release of H_2 from the reduced Fd by hydrogenase. However, the activities of NFOR are usually inhibited under standard conditions and can only proceed when H_2 partial pressure is very low, as opposed to PFOR which is still active at standard H_2 partial pressure (Dzulkarnain et al. 2022). In addition to H_2, acetyl-CoA can also be further converted to nongaseous products including SCFA (acetate, lactate, butyrate, propionate), alcohols (ethanol, butanol), and ketones (acetone). The overall theoretical maximum yield of H_2 from the reduced Fd pathway is 4 mol of H_2 per 1 mol of glucose and 2 mol of H_2 per 1 mol of glucose from the formate pathway (Sun et al. 2014). The yield is influenced by the fermentation end products generated alongside H_2. Theoretically, based on the "Thauer limit," the maximum yield of 4 mol of H_2 can be obtained with acetate as the fermentation end product, 2 mol with butyrate or propionate, and much lower yields when alcohols are the end product (Keskin et al. 2019). Biohydrogen production from the dark fermentation of organic wastes rarely surpasses 2 moles H_2 per mole of hexose (Wang and Yin 2019). To address these challenges, various strategies have been investigated, including optimizing reactor configurations, operational conditions, inoculum and substrate types, pretreatment methods, and integrating multiple biohydrogen production systems (Audu et al. 2021).

In the acetogenesis phase, intermediate products like H_2, CO_2, and acetate can be directly utilized by methanogens for CH_4 production. Other products, such as butyrate, propionate, and valerate, require further transformation through syntrophic acetogenesis (Lim et al. 2020). During this phase, hydrolytic and fermentative bacteria that lack hydrolytic activities convert unusable substrates into acetate, CO_2, and H_2. Notably, acetogenesis is considered the rate-limiting phase. The H_2 produced during acetogenesis is subsequently converted to CH_4 by hydrogenotrophic methanogens (Venkiteshwaran et al. 2015). The oxidation of nongaseous acidogenesis products involves a reverse electron transfer process, which is thermodynamically unfavorable and requires energy input. This process entails multiple enzyme systems, including formate dehydrogenases, ferredoxin: NAD oxidoreductases, hydrogenases, reactive quinone complexes, and c-type cytochromes (Sieber et al. 2012). However, when this oxidation is coupled with methane production, energy conversion becomes more feasible due to the reduction in H_2 pressure caused by methanogenic activity (Sikora et al. 2017). The most commonly reported syntrophic acetogens in anaerobic digesters include propionate degraders from the genera *Pelotomaculum*, *Smithella*, and *Syntrophobacter*, while the oxidation of butyrate and other fatty acids is carried out by *Syntrophus* and *Syntrophomonas* (Venkiteshwaran et al. 2015). Acetogenesis is crucial for ensuring rapid and stable anaerobic digester operation by preventing methanogenic inhibition caused by high acid concentrations (Wang et al. 2018).

Methanogenesis represents the final phase in anaerobic digestion, where acetate, H_2, and CO_2 produced during the acidogenic and acetogenic phases are further converted into biogas, primarily CH_4 and CO_2 (Fig. 7.2). Methanogens, which are the primary hydrogen consumers in anaerobic environments, are typically suppressed in biohydrogen dark fermentation to maximize hydrogen yield (Kumar et al. 2024). Unlike the preceding phases dominated by fermentative bacteria, methanogenesis is exclusively driven by methanogenic archaea, which are characterized by slow growth and show sensitivity to environmental changes. Table 7.2 depicts the most important methanogens observed in the termite gut systems. Methanogenesis can proceed through three pathways based on the available substrates: acetoclastic, methylotrophic, or hydrogenotrophic methanogenesis. Acetoclastic methanogens utilize acetate to produce methane, methylotrophic methanogens process methyl-based compounds, and hydrogenotrophic methanogens use CO_2 and H_2 for methane production (Menzel et al. 2020). The genera *Methanosaeta* and *Methanosarcina* are commonly observed acetoclastic methanogens, while *Methanobacterium*, *Methanospirillum*, *Methanococcus*, and *Methanobrevibacter* are prevalent hydrogenotrophic methanogens (Castellano-Hinojosa et al. 2018). Acetoclastic methanogens, such as *Methanosaeta*, are slow growers with a high affinity for acetate, predominantly responsible for about 70% of methane production in anaerobic digesters (Gonzalez-Martinez et al. 2016). In contrast, *Methanosarcina*, which can utilize a wider range of substrates and grow faster, tend to dominate in high acetate concentrations but are outcompeted by *Methanosaeta* at low acetate concentrations due to their higher affinity for acetate (Lim et al. 2020).

Fermentation of the lignocellulose in the termite gut systems is a complex process often referred to as a "black box" due to the myriad of metabolic activities and interactions within the microbial community and the limited availability of biogas-producing microbial genomes in reference databases (Ozbayram et al. 2020). Cao et al. (2012) investigated the effects of antibiotics on H_2 production by termite gut

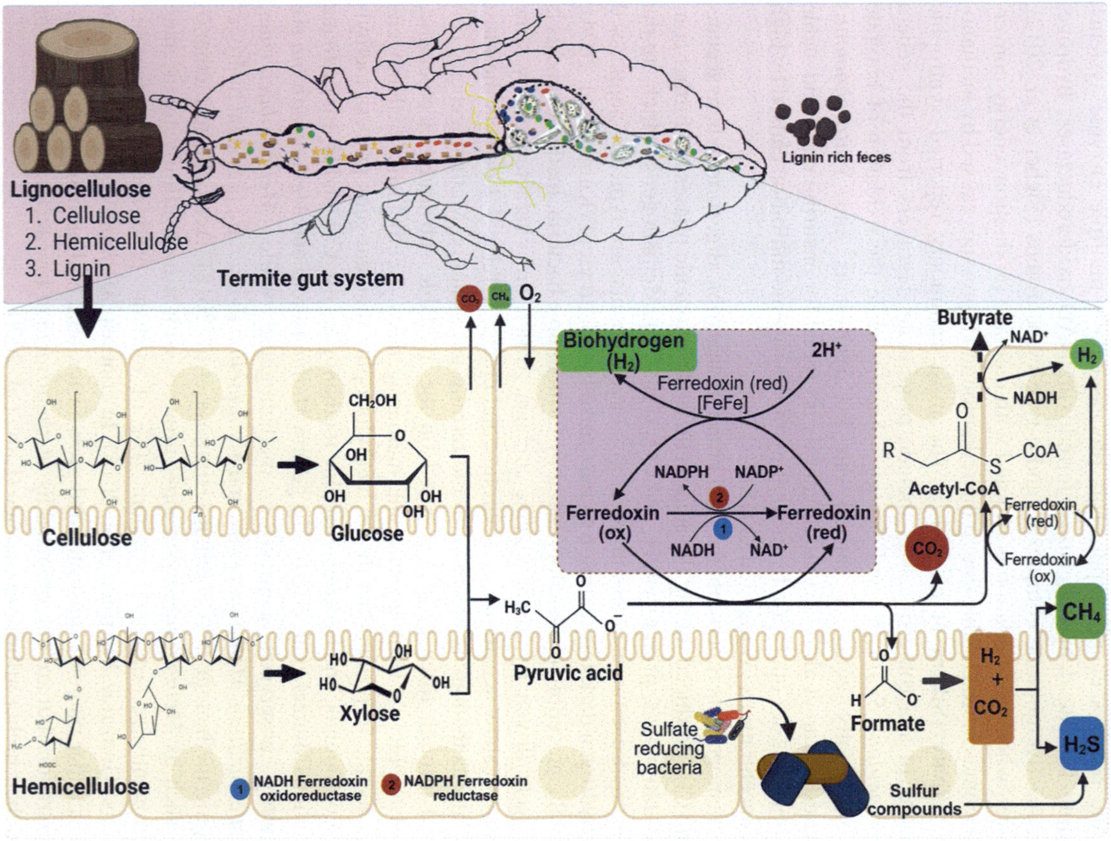

Fig. 7.2 Propose pathways for the biohydrogen and biomethane production from lignocellulosic biomass within the gut system of termites

Table 7.2 Some of the methanogenic bacteria that have been reported in the gut system of termites

Methanogen species	Termite source	Reference(s)
Methanobrevibacter smithii, Methanobrevibacter ruminantium, Methanospirillum hungatei, Methanosarcina acetivorans, Methanothermus fervidus	*Coptotermes gestroi*	Do et al. (2023)
Methylocystis parvus 610	*Mastotermes darwiniensis*	Reuß et al. (2015)
Methylocystis bryophila 613	*Neotermes castaneus*	
Methylocystis parvus 626	*Mastotermes darwiniensis*	
Methylocystis parvus 639	*Mastotermes darwiniensis*	
Methylocystis parvus 640	*Incisitermes marginipennis*	
Methanobrevibacter sp.	*Reticulitermes flavipes*	Ohkuma et al. (1999)
Methanobrevibacter arboriphilus		
Methanobacterium bryantii		
Methanobrevibacter arboriphilus Mc30	*Microcerotermes crassus*	Deevong et al. (2004)
Methanobacterium bryantii Ps21	*Pericapritermes* sp.	
Methanobacterium bryantii Tc3	*Termes comis*	
Methanobacterium ivanovii Mg38	*Macrotermes gilvus*	
Methanobrevibacter sp. *MPn19*	*Pericapritermes nitobei*	Ohkuma et al. (1999)
Methanobrevibacter arboriphilus P3Ar12	*Cubitermes orthognathus*	Friedrich et al. (2001)
Methanobrevibacter sp. Rs801, *Methanobrevibacter* sp. Rs802	*Reticulitermes speratus*	Hara et al. (2004)

symbionts. The authors revealed that antibiotics significantly reduced H_2 production, underscoring the critical role of gut symbionts in H_2 production. Antibiotic administration disrupted the termite gut microbiota, reducing symbiont diversity, which could impair cellulose digestion and hydrogen production. The termite gut systems are stable and controlled microenvironments crucial for microbial H_2 production, facilitating the decomposition of complex polymers and uninterrupted fermentation. The microbiomes within these systems are critical to their functionality, making the identification and understanding of the microbial populations and their behaviors essential for enhancing the robustness of dark fermentation processes. Thus, exploring the effects of antibiotics on H_2 production across diverse termite species could unveil species-specific symbiotic dynamics and resilience mechanisms that may further inspire bioengineered systems for sustainable H_2 production from lignocellulose.

7.3 Challenges and Recommendations

Recent advances in termite biotechnology have led to significant breakthroughs, particularly in the design of reactors that may simulate termite gut processes for LCB pretreatment, hydrolysis, and fermentation. For instance, Dröge et al. (2006) demonstrated that spirochetes isolated from the hindgut of the lower dry-wood termite *Neotermes castaneus* could ferment lignocellulose into products like acetate, formate, and ethanol. Other studies have identified similar symbiotic bacteria in lower termites that degrade cellulose into acetic acid, essential for termite respiration (Show et al. 2022). Key bacteria such as *Sporomusa termitida*, *Acetonema longum*, and *Clostridium mayombei* are integral to acetic acid production in termite guts, with some species contributing up to one-third of the acetic acid required for termite metabolism. Notably, *lactococci* and *enterococci*, known for fermenting cellobiose and xylose, may also be involved in acetic acid production, while hydrogen production has been observed in higher termites through microbial fermentation, including the hydrogen-producing bacterium *Clostridium termitidis* (Rajeswari et al. 2021). Despite the potential of termites' biochemical pathways in biohydrogen and biomethane production, harnessing this capability for biorefinery faces challenges such as efficiently replicating the complex microbial symbiosis and achieving stable process conditions. By leveraging the biohydrogen and biomethane potential of termites, we may be able to create more sustainable and efficient future biorefineries, opening new avenues for renewable energy sources and waste management solutions.

Future perspectives for enhancing dark fermentation can draw inspiration from the highly efficient biochemical pathways observed in termite gut systems. These natural systems, characterized by diverse microbial consortia from the *Firmicutes* and *Bacteroidetes* phyla, demonstrate remarkable efficiency in breaking down complex lignocellulosic materials under anaerobic conditions (Xie et al. 2023). By studying and mimicking the enzymatic mechanisms and synergistic microbial interactions within termite guts, researchers can develop tailored microbial consortia and enzyme cocktails for improved substrate hydrolysis and hydrogen production. Additionally, leveraging the resilience and adaptability of these microbes to environmental fluctuations can inform the design of robust bioprocesses. Advanced genetic engineering and synthetic biology approaches could enable the replication or enhancement of termite gut metabolic pathways, paving the way for innovative reactor designs and scalable biohydrogen production systems. Integrating these insights into pretreatment strategies and process optimization could transform dark fermentation into a more sustainable and efficient technology.

7.4 Conclusions and Future Perspectives

Termites have garnered significant attention in recent years due to their symbiotic digestion of lignocellulose (Auer et al. 2017; Lazuka et al. 2018). Research on termite biology has advanced considerably, revealing their unique ability to break down complex lignocellulosic biomass through a series of biocatalytic processes involving the hydrolysis of lignin, hemicellulose, and cellulose. This knowledge has provided critical insights for the development of innovative bioreactors designed to ferment lignocellulosic biomass into bioenergy (Pester and Brune 2007). The stepwise, enzyme-driven digestion process in termites offers a model for biomimetic lignocellulose processing systems that could be applied in biorefining technologies. Termite gut biodiversity and biochemistry remain largely unexplored, representing a promising source of novel catalytic processes and potential biocatalysts for industrial applications, including biofuels production.

Undoubtedly termites offer a promising model for the development of future biorefineries, attributed to their exceptional ability to efficiently degrade and utilize lignocellulosic biomass for both energy and nutrient cycling. Their specialized gut microbiota, which includes a diverse array of bacteria, archaea, and protozoa, plays a pivotal role in breaking down complex plant materials into simpler, more accessible compounds. By studying these symbiotic systems, researchers can glean valuable insights into the microbial processes that drive lignocellulose degradation and fermentation. This knowledge can be applied to the design of innovative, sustainable biorefining technologies that mirror the natural efficiency of termite digestion. Ultimately, such advances could contribute to the development of more sustainable, bio-based energy systems, helping to drive the transition toward a circular, low-carbon economy.

Recent advances in termite microbiology have underscored the importance of specific microbial communities in driving lignocellulose breakdown, with particular attention to the microbial symbionts responsible for the fermentation of sugars and the production of biofuels such as hydrogen and acetate. By decoding the biochemical pathways and microbial interactions occurring within the termite gut, researchers are beginning to unravel the complex mechanisms behind these processes, paving the way for their application in industrial-scale biorefining. Furthermore, the study of termite gut systems offers a promising avenue for the design of novel, biomimetic technologies capable of improving the efficiency and sustainability of lignocellulosic biomass conversion. Inspired by the termite's natural digestion processes, engineers and biotechnologists can develop innovative bioreactors and fermentation systems that mimic the termite gut's anaerobic conditions, optimizing the conversion of biomass into biofuels. Such innovations are poised to contribute significantly to the development of bio-based, low-carbon energy solutions. Ultimately, termite-inspired biorefining technologies have the potential to not only improve the efficiency of biomass conversion but also provide a foundation for a more sustainable bioeconomy. By harnessing the natural mechanisms of lignocellulose digestion in termites, we can advance the development of renewable energy systems and reduce our reliance on fossil fuels. As research continues to explore the rich biodiversity and biochemistry of termite gut microbiota, the full potential of these fascinating organisms in shaping the future of bioenergy and industrial biotechnology remains an exciting frontier.

Acknowledgments This study was funded by the National Natural Science Foundation of China (grant number 32250410285), the National Key R&D Program of China (2023YFC3403600), and the Foreign Expert Scientist Program under the Ministry of Science and Technology of China (WGXZ2023020L).

Declarations The authors declare that they have no conflict of interest.

References

Abendroth C, Vilanova C, Günther T et al (2015) Eubacteria and archaea communities in seven mesophile anaerobic digester plants in Germany. Biotechnol Biofuels 8:87. https://doi.org/10.1186/s13068-015-0271-6

Akhlaghi N, Najafpour-Darzi G (2020) A comprehensive review on biological hydrogen production. Int J Hydrog Energy 45:22492–22512. https://doi.org/10.1016/j.ijhydene.2020.06.182

Ali SS, Al-Tohamy R, Elsamahy T, Sun J (2024a) Harnessing recalcitrant lignocellulosic biomass for enhanced biohydrogen production: recent advances, challenges, and future perspective. Biotechnol Adv 72:108344. https://doi.org/10.1016/j.biotechadv.2024.108344

Ali S, Dar MA, Liaqat F et al (2024b) Optimization of biomethane production from lignocellulosic biomass by a developed microbial consortium. Process Saf Environ Prot 184:1106–1118

Ali S, Rani A, Dar MA et al (2024c) Recent advances in the characterization and valorization of lignin to value-added products: challenges and future perspectives. Biomass 4:947–977. https://doi.org/10.3390/biomass4030053

Anderson JE, Wulfhorst G, Lan W (2015) Energy analysis of the built environment—a review and outlook. Renew Sust Energ Rev 44:149–158. https://doi.org/10.1016/j.rser.2014.12.027

Audu JO et al (2021) Optimization of the operational parameters for mesophilic biohydrogen production from palm oil mill effluent using enriched mixed culture. Biomass Convers Biorefinery 13:4915. https://doi.org/10.1007/s13399-021-01488-9

Auer L, Lazuka A, Sillam-Dussès D, Miambi E, Hernandez-Raquet G (2017) Uncovering the potential of termite gut microbiome for lignocellulose bioconversion in anaerobic batch bioreactors. Front Microbiol 8:2623

Bhatia L, Sarangi PK, Singh AK, Prakash A, Shadangi KP (2022) Lignocellulosic waste biomass for biohydrogen production: future challenges and bio-economic perspectives. Biofuels Bioprod Biorefin 16:838–858. https://doi.org/10.1002/bbb.2329

Bhujbal SK, Kumar M, Vijay VK, Kumar V, Ghosh P (2021) Potential of termite gut microbiota for biomethanation of lignocellulosic wastes: current status and future perspectives. Rev Environ Sci Biotechnol 20:419–438. https://doi.org/10.1007/s11157-021-09576-y

Boga HI, Ludwig W, Brune A (2003) Sporomusa aerivorans sp. nov., an oxygen-reducing homoacetogenic bacterium from the gut of a soil-feeding termite. Int J Syst Evol Microbiol 53:1397–1404

Brune A (2007) Woodworker's digest. Nature 450:487–488

Brune A (2014) Symbiotic digestion of lignocellulose in termite guts. Nat Rev Microbiol 12:168–180

Brune A (2018) Methanogenesis in the digestive tracts of insects and other arthropods. In: Biogenesis of hydrocarbons, handbook of hydrocarbon and lipid microbiology. Springer, Cham, pp 1–32

Brune A, Dietrich C (2015) The gut microbiota of termites: digesting the diversity in the light of ecology and evolution. Ann Rev Microbiol 69:145–166. https://doi.org/10.1146/annurev-micro-092412-155715

Campbell A, Gdanetz K, Schmidt AW et al (2023) H2 generated by fermentation in the human gut microbiome influences metabolism and competitive fitness of gut butyrate producers. Microbiome 11:133. https://doi.org/10.1186/s40168-023-01565-3

Cao Y, Sun JZ, Rodriguez JM (2012) Effects of antibiotics on hydrogen production and gut symbionts in the Formosan subterranean termite *Coptotermes formosanus* (Isoptera: Rhinotermitidae). Insect Sci 19:346–354. https://doi.org/10.1111/j.1744-7917.2011.01489.x

Carbonero-Pacheco J, Aguilar J, Raya MC, Trapero A, Gaju-Ricart M, Agustí-Brisach C (2023) Diversity of cellulolytic microorganisms associated with the subterranean termite *Reticulitermes grassei*. J Fungi 9(3):294. https://doi.org/10.3390/jof9030294

Castellano-Hinojosa A et al (2018) New concepts in anaerobic digestion processes: recent advances and biological aspects. Appl Microbiol Biotechnol 102:5065–5076. https://doi.org/10.1007/s00253-018-9039-9

Chiri E, Nauer PA, Lappan R et al (2021) Termite gas emissions select for hydrogenotrophic microbial communities in termite mounds. Proc Natl Acad Sci USA 118:e2102625118

Chong CC, Cheng YW, Ng KH, Vo DVN, Lam MK, Lim JW (2022) Biohydrogen production from steam reforming of liquid biomass wastes and biomass-derived oxygenates: a review. Fuel 311:122623. https://doi.org/10.1016/j.fuel.2021.122623

Cook S (1932) The respiratory gas exchange in *Termopsis nevadensis*. Biol Bull 63:246–257

Cragg SM, Beckham GT, Bruce NC et al (2015) Lignocellulose degradation mechanisms across the tree of life. Curr Opin Chem Biol 29:108–119

Dar MA, Pawar KD, Jadhav JP, Pandit RS (2015) Isolation of cellulolytic bacteria from the gastro-intestinal tract of *Achatina fulica* (Gastropoda: Pulmonata) and their evaluation for cellulose biodegradation. Int Biodeterior Biodegrad 98:73–80

Dar MA, Pawar KD, Pandit RS (2018a) Prospecting the gut fluid of giant African land snail, *Achatina fulica* for cellulose degrading bacteria. Int Biodeterior Biodegrad 126:103–111

Dar MA, Shaikh AF, Pawar KD, Pandit RS (2018b) Exploring the gut of *Helicoverpa* armigera for cellulose degrading bacteria and evaluation of a potential strain for lignocellulosic biomass deconstruction. Process Biochem 73:142–153

Dar MA, Shaikh AF, Pawar KD, Xie RR, Sun JZ, Kandasamy S, Pandit RS (2021a) Evaluation of cellulose degrading bacteria isolated from the gut-system of cotton bollworm, *Helicoverpa armigera* and their potential values in biomass conversion. PeerJ Life Environ 9:e11254

Dar MA, Dhole NP, Xie RR et al (2021b) Valorization potential of a novel bacterial strain, *Bacillus altitudinis* RSP75, towards lignocellulose bioconversion: an assessment of symbiotic bacteria from the stored grain pest, *Tribolium castaneum*. Microorganisms 9:1952

Dar MA, Syed R, Pawar KD, Dhole NP, Xie R, Pandit RS, Sun J (2022a) Evaluation and characterization of the cellulolytic bacterium, *Bacillus pumilus* SL8 isolated from the gut of oriental leafworm, *Spodoptera litura*: an assessment of its potential value for lignocellulose bioconversion. Environ Technol Innov 27:102459

Dar MA, Xie R, Pandit RS, Danso B, Dong CC, Sun J (2022b) Exploring the region-wise diversity and functions of symbiotic bacteria in the gut-system of wood-feeding termite, *Coptotermes formosanus*, towards lignocellulose degradation. Insect Sci 29:1414–1432

Dar MA, Xie R, Luohui J et al (2024a) Elucidating the structure and composition of bacterial symbionts in the gut regions of wood-feeding termite, *Coptotermes formosanus*, and their functional profile towards lignocellulolytic systems. Front Microbiol 15:1395568. https://doi.org/10.3389/fmicb.2024.1395568

Dar MA, Xie R, Zabed HM et al (2024b) Termite microbial symbiosis as a model for innovative design of lignocellulosic future biorefinery: current paradigms and future perspectives. Biomass 4:180–201. https://doi.org/10.3390/biomass4010009

Dar MA, Xie R, Zabed HM et al (2024c) Current paradigms and future challenges in harnessing gut bacterial symbionts of insects for biodegradation of plastic wastes. Insect Sci 31:1–28. https://doi.org/10.1111/1744-7917.13417

Dar MA, Xie R, Zabed HM, Idrees A, Waghmare PR, Irfan M, Dhanavade MJ, Shahnawaz M, Zhu D, Sun J (2025a) Can termites serve as models for future biorefinery? In: Zhu D, Dar MA, Shahnawaz M (eds) Biofuels and sustainability: life-cycle assessments, systems biology, policies and emerging technologies. Elsevier Science, pp 323–346. https://doi.org/10.1016/B978-0-443-21433-2.00017-7

Dar MA, Ali S, Xie R, Zabed HM, Idrees A, Sun J (2025b) Non-termite insects as biomimetic models to advance the production of second and third generation biofuels. In: Zhu D, Dar MA, Shahnawaz M (eds) Biofuels and sustainability: life-cycle assessments, systems biology, policies and emerging technologies. Elsevier Science, pp 347–364. https://doi.org/10.1016/B978-0-443-21433-2.00002-5

Deevong P, Hattori S, Yamada A et al (2004) Isolation and detection of methanogens from the gut of higher termites. Microb Environ 19:221–226

Do TH, Dao TK, Nguyen HD, Truong NH (2023) Understanding the role of free-living bacteria in the gut of the lower termite *Coptotermes gestroi* based on metagenomic DNA analysis. Insects 14(11):832. https://doi.org/10.3390/insects14110832

Dröge S, Fröhlich J, Radek R, König H (2006) *Spirochaeta coccoides* sp. nov., a novel coccoid spirochete from the hindgut of the termite *Neotermes castaneus*. Appl Environ Microbiol 72:392–397. https://doi.org/10.1128/AEM.72.1.392-397.2006

Dzulkarnain ELN, Audu JO, Wan Dagang WRZ, Abdul-Wahab MF (2022) Microbiomes of biohydrogen production from dark fermentation of industrial wastes: current trends, advanced tools and future outlook. Bioresour Bioprocess 9(1):16. https://doi.org/10.1186/s40643-022-00504-8

Eggleton P (2000) Global patterns of termite diversity. In: Termites: evolution, sociality, symbioses, ecology. Springer, Dordrecht, pp 25–51

Friedrich MW, Schmitt-Wagner D, Lueders T, Brune A (2001) Axial differences in community structure of *Crenarchaeota* and *Euryarchaeota* in the highly compartmentalized gut of the soil-feeding termite *Cubitermes orthognathus*. Appl Environ Microbiol 67(10):4880–4890. https://doi.org/10.1128/AEM.67.10.4880-4890.2001

Gomati V, Ramasamy K, Kumar K, Sivaramaiah N, Mula R (2011) Greenhouse gas emissions from termite ecosystem. Afr J Environ Sci Technol 5:56–64

Gomez-Flores M, Nakhla G, Hafez H (2017) Hydrogen production and microbial kinetics of *Clostridium termitidis* in mono-culture and co-culture with *Clostridium beijerinckii* on cellulose. AMB Expr 7:84

Gonzalez-Martinez A et al (2016) Archaeal and bacterial community dynamics and bioprocess performance of a bench-scale two-stage anaerobic digester. Appl Microbiol Biotechnol 100:6013–6033. https://doi.org/10.1007/s00253-016-7393-z

Goria K, Singh HM, Singh A, Kothari R, Tyagi VV (2024) Insights into biohydrogen production from algal biomass: challenges, recent advancements and future directions. Int J Hydrog Energy 52:127–151. https://doi.org/10.1016/j.ijhydene.2023.03.174

Graber JR, Breznak JA (2004) Physiology and nutrition of *Treponema primitia*, an H_2/CO_2-acetogenic spirochete from termite hindguts. Appl Environ Microbiol 70:1307–1314

Hara K, Shinzato N, Oshima T, Yamagishi A (2004) Endosymbiotic *Methanobrevibacter* species living in symbiotic protists of the termite *Reticulitermes speratus* detected by fluorescent in situ hybridization. Microbes Environ 19:120–127. https://doi.org/10.1264/jsme2.19.120

Harun I, Jahim JM, Anuar N, Hassan O (2012) Hydrogen production performance by *Enterobacter cloacae* KBH3 isolated from termite guts. Int J Hydrog Energy 37:15052–15061

Hassa J et al (2018) Metagenome, metatranscriptome, and metaproteome approaches unravelled compositions and functional relationships of microbial communities residing in biogas plants. Appl Microbiol Biotechnol 102:5045–5063. https://doi.org/10.1007/s00253-018-8976-7

Hellemans S, Rocha MM, Wang M et al (2024) Genomic data provide insights into the classification of extant termites. Nat Commun 15:6724. https://doi.org/10.1038/s41467-024-51028-y

Inoue I, Saita K, Kudo T, Ui S, Ohkuma M (2007) Hydrogen production by termite gut protists: characterization of iron hydrogenases of parabasalian symbionts of the termite *Coptotermes formosanus*. Eukaryot Cell 6(10):1925–1932. https://doi.org/10.1128/EC.00251-07

Jamali H, Livesley SJ, Hutley LB, Fest B, Arndt SK (2013) The relationships between termite mound CH4:CO2 emissions and internal concentration ratios are species-specific. Biogeosciences 10:2229–2240. https://doi.org/10.5194/bg-10-2229-2013

Janusz G, Pawlik A, Swiderska-Burek U et al (2020) Laccase properties, physiological functions, and evolution. Int J Mol Sci 21:966.

Jones DT, Verkerk RHJ, Eggleton P (2005) Methods for Sampling Termites. In Insect Sampling in Forest Ecosystems, Leather SR (Ed.). pp. 221–253, Volume 1, Blackwell Science Ltd, Hoboken, New Jersey. https://doi.org/10.1002/9780470750513.ch10

Kane MD, Breznak JA (1991) *Acetonema longum* gen. nov. sp. nov., an H_2/CO_2 acetogenic bacterium from the termite *Pterotermes occidentis*. Arch Microbiol 156:91–98

Kazmi B, Sadiq T, Taqvi SAA, Nasir S, Khan MM, Naqvi SR, AlMohamadi H (2024) Towards a sustainable future: bio-hydrogen production from food waste for clean energy generation. Process Saf Environ Prot 183:555–567. https://doi.org/10.1016/j.psep.2024.01.045

Keskin T et al (2019) Effect of percolation frequency on biohydrogen production from fruit and vegetable wastes by dry fermentation. Int J Hydrog Energy 44:18767–18775. https://doi.org/10.1016/j.ijhydene.2018.12.099

Köhler T, Dietrich C, Scheffrahn RH, Brune A (2012) High-resolution analysis of gut environment and bacterial microbiota reveals functional compartmentation of the gut in wood-feeding higher termites (*Nasutitermes* spp.). Appl Environ Microbiol 78:4691–4701

Konishi T, Yamamoto D, Umezawa K, Itakura S (2020) Hydrogen production by microorganisms in the hindgut of the termite *Reticulitermes speratus* under anaerobic and aerobic conditions. J Environ Entomol Zool 31:51–55

Kumar G et al (2018) Insights into evolutionary trends in molecular biology tools in microbial screening for biohydrogen production through dark fermentation. Int J Hydrog Energy 43:19885–19901. https://doi.org/10.1016/j.ijhydene.2018.09.040

Kumar DJP, Mishra RK, Chinnam S, Binnal P, Dwivedi N (2024) A comprehensive study on anaerobic digestion of organic solid waste: a review on configurations, operating parameters, techno-economic analysis and current trends. Biotechnol Notes

Law SJ, Allison SD, Davies AB et al (2024) The challenge of estimating global termite methane emissions. Glob Chang Biol 30(6):e17390. https://doi.org/10.1111/gcb.17390

Lazuka A, Auer L, O'Donohue M, Hernandez-Raquet G (2018) Anaerobic lignocellulolytic microbial consortium derived from termite gut: enrichment, lignocellulose degradation and community dynamics. Biotechnol Biofuels 11:284

Li G, Yang Y, Yu Q et al (2024) Application of nanotechnology in hydrogen production from biomass: a critical review. Adv Compos Hybrid Mater 7:17. https://doi.org/10.1007/s42114-023-00815-0

Lim JW et al (2020) Chapter one—the microbiome driving anaerobic digestion and microbial analysis. In: Li Y, Khanal SK et al (eds) Advances in bioenergy. Elsevier, Amsterdam, pp 1–61

Mathew GM, Mathew DC, Lo SC, Alexios GM, Yang JC, Sashikumar JM, Shaikh TM, Huang CC (2013) Synergistic collaboration of gut symbionts in *Odontotermes formosanus* for lignocellulosic degradation and bio-hydrogen production. Bioresour Technol 145:337–344

Menzel T et al (2020) Role of microbial hydrolysis in anaerobic digestion. Energies 13:5555. https://doi.org/10.3390/en13215555

Mikaelyan A, Meuser K, Brune A (2017) Microenvironmental heterogeneity of gut compartments drives bacterial community structure in wood- and humus-feeding higher termites. FEMS Microbiol Ecol 93:1–11. https://doi.org/10.1093/femsec/fiw210

Mujtaba M, Fernandes FL, Fazeli M et al (2023) Lignocellulosic biomass from agricultural waste to the circular economy: a review with focus on biofuels, biocomposites and bioplastics. J Clean Prod 402:136815

Myer A, Forschler BT (2019) Evidence for the role of subterranean termites (*Reticulitermes* spp.) in temperate forest soil nutrient cycling. Ecosystems 22:602–618. https://doi.org/10.1007/s10021-018-0291-8

Odelson DA, Breznak JA (1985) Nutrition and growth characteristics of *Trichomitopsis termopsidis*, a cellulolytic protozoan from termites. Appl Environ Microbiol 49:614–621

Ohkuma M (2003) Termite symbiotic systems: efficient bio-recycling of lignocellulose. Appl Microbiol Biotechnol 61:1–9

Ohkuma M, Noda S, Kudo T (1999) Phylogenetic relationships of symbiotic methanogens in diverse termites. FEMS Microbiol Lett 171(2):147–153. https://doi.org/10.1111/j.1574-6968.1999.tb13425.x

Ozbayram EG, Kleinsteuber S, Nikolausz M (2020) Biotechnological utilization of animal gut microbiota for valorization of lignocellulosic biomass. Appl Microbiol Biotechnol 104(2):489–508. https://doi.org/10.1007/s00253-019-10239-w

Peng X, Wilken SE, Lankiewicz TS et al (2021) Genomic and functional analyses of fungal and bacterial consortia that enable lignocellulose breakdown in goat gut microbiomes. Nat Microbiol 6:499–511

Pester M, Brune A (2007) Hydrogen is the central free intermediate during lignocellulose degradation by termite gut symbionts. ISME J 1:551–565

Quevedo HD, Brandani CB, Bento CB et al (2021) Greenhouse gas emissions from termite mounds in a transition area between the *Cerrado Savanna* and the Atlantic Forest in Brazil. Acta Oecol 110:103690. https://doi.org/10.1016/j.actao.2020.103690

Rajeswari G, Jacob S, Chandel AK et al (2021) Unlocking the potential of insect and ruminant host symbionts for recycling of lignocellulosic carbon with a biorefinery approach: a review. Microb Cell Factories 20:107. https://doi.org/10.1186/s12934-021-01597-0

Ram C, Rani P, Kumar A (2024) Recent developments in biohydrogen production from wastewater: a review. Biocatal Biotransformation 42:1–18. https://doi.org/10.1080/10242422.2023.2181046

Reuß J, Rachel R, Kämpfer P et al (2015) Isolation of methanotrophic bacteria from termite gut. Microbiol Res 179:29–37. https://doi.org/10.1016/j.micres.2015.06.003

Rouland-Lefèvre C (2010) Termites as pests of agriculture. In: Bignell D, Roisin Y, Lo N (eds) Biology of termites: a modern synthesis. Springer, Dordrecht/Singapore, pp 499–517

Sadalage PS, Dar MA, Chavan AR, Pawar KD (2020) Formulation of synthetic bacterial consortia and their evaluation by principal component analysis for lignocellulose-rich biomass degradation. Renew Energy 148:467–477

Sahrin NT, Khoo KS, Lim JW et al (2022) Current perspectives, future challenges, and key technologies of biohydrogen production for building a carbon-neutral future: a review. Bioresour Technol 364:128088. https://doi.org/10.1016/j.biortech.2022.128088

Saravanan A et al (2021) Biohydrogen from organic wastes as a clean and environment-friendly energy source: production pathways, feedstock types, and future prospects. Bioresour Technol 342:126021. https://doi.org/10.1016/j.biortech.2021.126021

Saunois M, Stavert AR, Poulter B et al (2020) The global methane budget 2000–2017. Earth Syst Sci Data 12:1561–1623. https://doi.org/10.5194/essd-12-1561-2020

Scharf ME (2015) Termites as targets and models for biotechnology. Annu Rev Entomol 60:77–102

Scharf ME, Tartar A (2008) Termite digestive systems as sources for novel lignocellulases. Biofuels Bioprod Biorefin 2:540–552

Schmitt-Wagner D, Brune A (1999) Hydrogen profiles and localization of methanogenic activities in the highly compartmentalized hindgut of soil-feeding higher termites (*Cubitermes* spp.). Appl Environ Microbiol 65:4490–4496

Show BK, Banerjee S, Banerjee A et al (2022) Insect gut bacteria: a promising tool for enhanced biogas production. Rev Environ Sci Biotechnol 21:1–25. https://doi.org/10.1007/s11157-021-09607-8

Sieber JR et al (2012) Genomic insights into syntrophy: the paradigm for anaerobic metabolic cooperation. Ann Rev Microbiol 66:429–452. https://doi.org/10.1146/annurev-micro-090110-102844

Sikora A et al (2017) Anaerobic digestion: I. A common process ensuring energy flow and the circulation of matter in ecosystems. II. A tool for the production of gaseous biofuels. Ferment Processes 14:271. https://doi.org/10.5772/64645

Stephen JL, Periyasamy B (2018) Innovative developments in biofuels production from organic waste materials: a review. Fuel 214:623–633

Stolze Y et al (2016) Identification and genome reconstruction of abundant distinct taxa in microbiomes from one thermophilic and three mesophilic production-scale biogas plants. Biotechnol Biofuels 9:1–18. https://doi.org/10.1186/s13068-016-0565-3

Sun JZ, Chen CR (2010) Cellulolytic insects and their potentials for viable biofuels: a new frontier discipline in entomology and bioengineering. Chin Bull Entomol 47:1033–1042

Sun J, Zou XJ (2014) Lignocellulolytic systems of insects and their potential for viable biofuels. In: Sun J, Ding SY, Doran-Peterson J (eds) Biological conversion of biomass for fuels and chemicals: explorations from natural utilization systems, 1st edn. Royal Society of Chemistry, Cambridge, pp 195–222. https://doi.org/10.1039/9781849734738

Sun J, Ding SY, Doran-Peterson J (2014) Biological conversion of biomass for fuels and chemicals: exploration from natural utilization systems, 1st edn. Royal Society of Chemistry, Cambridge, pp 1–427

Tawfik A, Eraky M, Alhajeri NS, Osman AI, Rooney DW (2022) Cultivation of microalgae on liquid anaerobic digestate for depollution, biofuels, and cosmetics: a review. Environ Chem Lett 20:3631–3656. https://doi.org/10.1007/s10311-022-01481-2

Tholen A, Schink B, Brune A (1997) The gut microflora of *Reticulitermes flavipes*, its relation to oxygen, and evidence for oxygen-dependent acetogenesis by the most abundant *Enterococcus* sp. FEMS Microbiol Ecol 24:137–149

Tomar S, Agarwal S, Singh H et al (2023) Microalgae: a promising source for biofuel production. Biocatal Agric Biotechnol 53:102877. https://doi.org/10.1016/j.bcab.2023.102877

Tsegaye B, Balomajumder C, Roy P (2019) Isolation and characterization of novel lignolytic, cellulolytic, and hemicellulolytic bacteria from wood-feeding termite *Cryptotermes brevis*. Int Microbiol 22:29–39

Venkiteshwaran K et al (2015) Relating anaerobic digestion microbial community and process function: supplementary issue: water microbiology. Microbiol Insights 8:37–44. https://doi.org/10.4137/MBI.S33593

Wang J, Yin Y (2019) Progress in microbiology for fermentative hydrogen production from organic wastes. Crit Rev Environ Sci Technol 49:825–865. https://doi.org/10.1080/10643389.2018

Wang P et al (2018) Microbial characteristics in anaerobic digestion process of food waste for methane production—a review. Bioresour Technol 248:29–36. https://doi.org/10.1016/j.biortech.2017.06.152

Warnecke F, Luginbühl P, Ivanova N et al (2007) Metagenomic and functional analysis of hindgut microbiota of a wood-feeding higher termite. Nature 450:560–565

Xie R, Dong CC, Wang S et al (2023) Host-specific diversity of culturable bacteria in the gut systems of fungus-growing termites and their potential functions towards lignocellulose bioconversion. Insects 14:403

Xie R, Danso B, Sun J et al (2024) Biorefinery and bioremediation strategies for efficient management of recalcitrant pollutants using termites as an obscure yet promising source of bacterial gut symbionts: a review. Insects 15:908

Yacout DMM, Tysklind M, Upadhyayula VKK (2024) Socio-economic implications of forest-based biofuels for marine transportation in the Arctic: Sweden as a case study. Front Clim 6:1414813. https://doi.org/10.3389/fclim.2024.1414813

Yamin MA (1981) Cellulose metabolism by the flagellate *Trichonympha* from a termite is independent of endosymbiotic bacteria. Science 211:58–59

Zabed HM, Akhter S, Dar MA et al (2023) Enhanced fermentable sugar production in lignocellulosic biorefinery by exploring a novel corn stover and configuring high-solid pretreatment conditions. Bioresour Technol 386:129498

Zimmerman P, Greenberg J, Wandiga S, Crutzen P (1982) Termites: a potentially large source of atmospheric methane, carbon dioxide, and molecular hydrogen. Science 218:563–565

Chapter 8
Biomimics of the Insect Gut for Advanced Design of Lignocellulosic Biorefinery

Osikemekha Anthony Anani, Kenneth Kennedy Adama, Peter Mudiaga Etaware, Uduenevwo Francis Evuen, Arthur Efeoghene Essaghah, and Chukwudi Ogwu

Abstract Insects, one of the invertebrates, are well-known for their ability to pollinate plants; act as soil engineers, thereby aerating the soil; have symbiosis; or have a mutualistic effect with beneficial microbes and aid as pathogens for several diseases. Considering their ubiquitous nature in the ecosystem, their structural and functional ecological features, and their role with beneficial microbes, insects such as soil engineers and wood-feeding-like collaborators help break down rich lignocellulose biomass uniquely in their gut pathways. Studies have shown that the microbes in the gut system of insects can serve as a perfect model when utilized in the degradation of lignocellulose materials to produce biofuels using strategic biorefinery methods. This is a biomimicry or biomimetic way of helping humans solve the issues of climate change using a green biotechnological model. Because of the incidence of fossil fuel deterioration and depletion of the natural ecosystem, the

O. A. Anani (✉)
Laboratory for Hydrobiology and Ecotoxicology, Department of Animal and Environmental Biology, Delta State University, Abraka, Delta State, Nigeria
e-mail: osikemekhaa@delsu.edu.ng

K. K. Adama
Department of Chemical Engineering, Faculty of Engineering, Edo State University Uzairue, Edo State, Nigeria

P. M. Etaware
Departmment of Plant Science and Biotechnology, Faculty of Science, Delta State University of Science and Technology, Ozoro, Delta State, Nigeria

U. F. Evuen
Department of Biochemistry, Faculty of Science, Delta State University of Science and Technology, Ozoro, Delta State, Nigeria

A. E. Essaghah
Department of Urban and Regional Planning, Delta State University of Science and Technology (DSUST), Ozoro, Delta State, Nigeria

C. Ogwu
Department of Environmental Management and Toxicology, Delta State University of Science and Technology, Ozoro, Delta State, Nigeria

M. A. Dar et al. (eds.), *Recent Trends in Lignocellulosic Biofuels and Bioenergy*, Clean Energy Production Technologies,
https://doi.org/10.1007/978-981-96-4636-4_8

need to use alternative green technology is not negotiable. Hence, this chapter assesses the biomimics of the insect gut for the advanced design of lignocellulosic biorefinery. The ecology of the gut microbiome in insects and their role in the valorization of lignocellulose for biorefinery are discussed. The interplay between the insects and their gut microbes is also discussed. How gut microorganisms in insects break down lignocellulose and their mechanism of action serving as a potential biomimetic model for future biorefinery is highlighted. The application of microbes found in the gut of insects for biofuel production and the possible economic cost are highlighted. The environmental factors affecting the efficiencies of gut microbiomes in their utilization as tools for biorefinery are discussed.

Keywords Biofuel · Environmental efficiencies · Biomass · Economic cost · Biorefinery

8.1 Introduction

The negative incidences from fossil fuels, which have increased the levels of greenhouse gas emission into the environment, have caused key challenges to environmental scientists (Xie et al. 2024). In the renewable energy industry, the application of a green approach to improving the quality of biofuel through the bioconversion of biomass of lignocellulosic materials by insects has been heralded as a safe biotechnology for a green economy (Rajeswari et al. 2021). The main composition of insect lignin is lignin/noncarbohydrate polymers, hemicellulose, and cellulose/carbohydrate of percentages 10–25, 20–40, and 40–60%, respectively (Tye et al. 2016; Rajeswari et al. 2021). The components of lignocellulose have been reported by Dar et al. (2024a) as sustainable and green because of its non-noxious and renewable chattels. Plant lignocellulose materials are produced every year in large quantities by photosynthetic process and by valorization using enzymes that can mediate biological reactions by hydrolysis (Dashtban et al. 2009; Dar et al. 2024b). Some of the enzymes involved in this process include chitinases, mannanases, esterases, pectinases, amylases, xylanase, cellulases, manganese peroxidase, lignin-peroxidases, and laccase correspondingly.

Studies have shown that fluorescence excitation-emission matrix (EEM) can be used to characterize the lignocellulose materials of insects for their application in industry as materials in biofuel production (Engel and Moran 2013; Matthews et al. 2019; Rajeswari et al. 2021; Rossi et al. 2021). At times, the microorganisms found in the gut system of many mammals and insects, ruminants, and xylophagous kinds, which act as symbionts, aid in the digestion of lignocellulose materials ingested by the host organisms (Engel and Moran 2013; Matthews et al. 2019; Rajeswari et al. 2021). This distinct synergetic mode of digestion by the gut organism is an important strategy for biomimicry in the biofuel industry in the large biological degradation of lignocellulose materials for the production of biofuel (Rajeswari et al. 2021).

Insects gut systems (midgut, hindgut, and foregut) are a typical center for the activities of microorganism's colonization with several advantages that will benefit the entire metabolic pathway of the organism (Huang et al. 2010; Mondal et al. 2023). The microbes in the insect gut system aid in detoxifying noxious chemicals like pesticides and heavy metals, degradation of phytotoxins, and breakdown of food. The numbers of microbial consortia in the gut system of insects differ, and they all play different roles in the secretion of several enzymes effective for protection and offensive and defensive actions in adaptation purposes. The major enzymes found in insects used for this purpose are hemicelluloses and cellulose which contain large carbon chains (Huang et al. 2010). Huang et al. (2010) reported that the midgut system of insect larvae like the scarab beetles is highly alkaline and consists of viable enzymes like serine proteinase which when, secreted and sent to the hindgut, helps to stimulate the environmental activities of anaerobic microbiota. Meanwhile, other biological activities in the hindgut consist of fermentation and fatty acids presence which the anaerobic organisms use for sustenance in the absence of oxygen- methanogenesis (Huang et al. 2010). However, most of the microbes found in this region are obligated by nature.

Bioreactor simply means an engineering and biological design machine that sustains, houses, and utilizes in the culturing or growing of living entities like microbes (bacteria and viruses) and tissues (Delavar and Wang 2022). The environment of the bioreactor is not a media or biofluidic one but a mechanical one that uses a biological material made of musculoskeletal arrangement that encourages the proliferation of microbes to attain their biological function (Delavar and Wang 2022). The bioreactor consists of a substrate section that contains biological materials except cellular contents (blood cells) which aid in the proliferation and normal functioning of the microbes. The substrate here has a high tensile strength which allows the flexibility and porosity of both smooth and rough materials. A bioreactor plans to aid the removal of wastes and toxins via the detoxification of noxious chemicals and degradation of phytotoxins, which is a mimic of the normal physiological and biological removal of wastes and toxins in insects using the gut system (Mondal et al. 2023; Delavar and Wang 2022). Vikram et al. (2021) reported that termites have the potential to be used as a bioreactor tool in the conversion of large lignocellulosic materials from plants into useful resources like biomaterials and biofuels.

Nonetheless, in the background of a developing economy, the usual gut consortia of microorganisms are an important entity as an enrichment biological candidate for the functional and the enhancement of the cell factory of microorganisms in the recycling of lignin materials for the production of biofuels (Dar et al. 2024a, b). Biofuel production in different forms actually requires the metabolism of several microbial communities or consortia to mediate and fasten their breakdown for a smooth conversion process. Rajeswari et al. (2021) reported that gut microbes are recommended as an important candidate for the effective deployment in the biorefinery of lignocellulose materials to biofuel via biomimicry strategy.

Insect gut system is very complex, and the microflora found therein is of immense concern because they aid in deciphering carbohydrate active enzymes (CAZymes) in an unculturable state with a diverse plethora of hydrolytic enzymes found in the cells of the plant (Aves et al. 2019; Kala et al. 2019). Therefore, the mining of unexplored and unexploited hydrolytic/oxidative enzymes found in the gut system of microbial consortia—ecological niche—could be explored in the degradation of lignocellulose materials and could aid in the production of the biochemicals and biofuels, serving as a second-generation developmental production strategy for biological fuels in the biorefinery industry of the world (Dar et al. 2024b). So, in this scenario, the gut microbial consortia like the *Bacillus pumilus* SL8 possess the ability to valorize plant lignin by the application of anaerobic digestion, fermentation, saccharolytic enzymation, and pretreatment techniques for an effective biorefinery conversion from recalcitrant wastes to a bioeconomy of wealth (Rajeswari et al. 2021).

Tokuda et al. (1997) reported that some species of cockroaches and termites like the *Cryptocercus punctulatus* and *Reticulitermes flavipes*, respectively, are one of the basic wood-eating insects because each organism has a long phylogenetic relationship, and they both produce cellulolytic chemicals for the internal digestion of cellulose materials they degrade from the wood. In addition, they possess abundant protozoa in their paunch (hindgut) that assist in wood degradation through the secretion of enzymes like the β-glucosidase, endo-β-1,4-glucanase (Cx-cellulase) an endogenous form, and cellobiohydrolase (C_1-cellulase). Other examples of wood insects belong to the following families—Serritermitidae, Rhinotermitidae, Hodotermitidae, Termopsidae, Kalotermitidae, and Mastotermitidae (Vikram et al. 2021).

Examples of microbes found in the gut system of insects include yeasts, protozoans that are flagellates (eucarya), archaea, and bacteria (Vikram et al. 2021). These microorganisms are symbionts that play specific roles in humification, modification of polyaromatic hydrocarbons (PAHs), fixation of nitrogen in plants, hydrogen, and oxygen consumption, and the metabolism of nitrogen and carbohydrates (Brune and Ohkuma 2010; Vikram et al. 2021).

Zeng et al. (2023) reported that the major challenges of using insects as a model for reactor design are the issues of adaptation, tolerance, and natural selection. By using ants as an imperfect mimetic to enable *Siler collingwoodi* (spider) to adapt the reactor, Zeng et al. (2023) stated that some of the limitations of their study were that the venom, chemical repellents, and defensive traits of the ants were not put into cognizance before experimenting. More so, the antipredation examination was not carried out to determine the imperfection of the mimicry of the ant. The incidence of female polymorphism was also noticed.

The recent application of isolates from microorganisms from the rumen and gut of insects like the *Coptotermes formosanus* (wood-feeding termite) and the *Spodoptera litura* (oriental leafworm) (Dar et al. 2024a), has proven that they are efficient in the production of cellulosic biofuel: a second-generation of biofuel. Owing to this, their prospects in bioengineering, metabolic engineering, and imminent genetics are not in doubt. Thus, this chapter elucidates the potential application of the microbes in the insect gut as a biomimicry tool for the advanced design of lignocellulosic biorefinery.

8.2 Ecology of the Gut Microbiome in Insects

The gut microbiome of insects, such as termites, bees, aphids, roaches, and flies, comprise a diverse and complex community of microorganisms that play a crucial role in the well-being of the insect's (host) health, development, and ecological interactions (Engel and Moran 2013). A typical insect gut microbial community includes bacteria, fungi, protozoa, and viruses (Douglas 2015), which interact with each other and their insect host in a mutualistic symbiotic association that is beneficial to all. Some of the advantages associated with the complex ecology of the insect gut microbiome include the following:

8.2.1 Nutrient Acquisition and Digestion

The biochemical pathway for the digestion and extraction of nutrients from hemicellulose, cellulosic, and lignocellulosic substances varies greatly in different categories of insects. Therefore, special attention will be given to wood-boring beetles (longhorn and bark beetles), passalid beetles (bess beetles), locusts, silkworms, leafcutter ants, mites, bees, aphids, roaches, flies, and mosquitoes.

Termites Termites have a gut microbiome rich in bacteria and protozoa that help break down cellulose from wood; a typical example is *Treponema* sp., a bacterium found in the hindgut of termites (Douglas 2015). *Treponema* sp. produces enzymes such as cellulase and hemicellulase that can break down cellulose and hemicellulose into simple sugar, which the termites can then absorb and utilize for the production of energy and internal heat (Fig. 8.1).

Bees Honey bees have a specialized gut microbiome consisting of a few dominant bacterial species like *Gilliamella* sp. and *Snodgrassella* sp., which help in the digestion of pollen and nectar and contribute to the detoxification of harmful compounds found in their diet (Dillon and Dillon 2004).

Aphids This group of insects is renowned for engulfing the bacterium *Buchnera aphidicola* in specialized cells called bacteriocytes found in their guts. This mutualistic association is crucial for the aphids as *Buchnera* sp. provides essential amino acids that are uncommon in the aphid's phloem sap diet (Douglas 2015).

Cockroaches The gut microbiome of cockroaches includes bacteria such as *Blattabacterium*, which help in nitrogen recycling and provide essential nutrients, allowing cockroaches to survive on a varied and often poor diet (Douglas 2015).

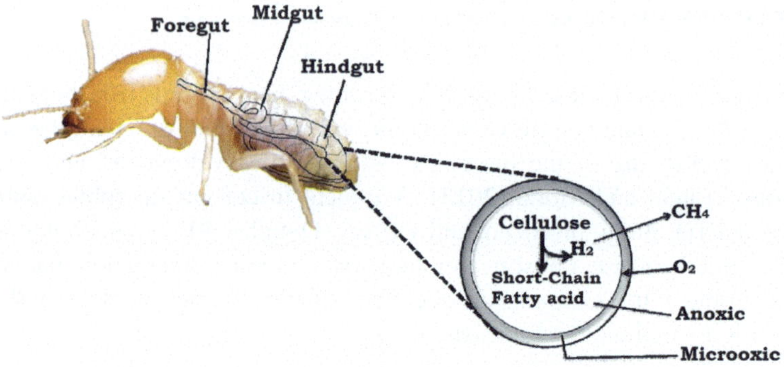

Fig. 8.1 The process of biodegradation of cellulose in the gut of termites

8.2.2 Environmental Adaptation of Microbes in Insect Gut System

The gut microbiome of insects is integral to their physiology and ecology, influencing their nutrition, immune function, development, and adaptation to various environments. Examples include the following:

Desert Locusts The gut microbiome of desert locusts adapts to their changing diet and environmental conditions. For example, when locusts switch from the solitary to gregarious phase, their gut microbiome adapts to help digest different types of plant materials consumed during mass migration.

Ants Different ant species harbor distinct gut microbiomes that reflect their diverse diets and ecological niches. Leaf-cutter ants, for example, have gut bacteria that help in breaking down plant material, complementing their fungus-farming lifestyle.

8.3 Examples of Gut Microbes in Insects and their Biological Role

Table 8.1 shows several gut microbes in insects, their biological role, and the enzymes secreted. It was shown that most of the gut microbes aid the insects of the order Coleoptera by using cellulase as their enzyme for breaking down biological materials, while the ones found in the gut of the Isoptera use ligninase as the enzymes for the biological degradation of food materials.

Table 8.1 Several gut microbes in insects, their biological role, and the enzymes secreted

S.no	Gut microorganism	Order	Enzymes secreted	References
1	*Burkholderia* sp.	Hemiptera	Manufactures aminase recycling and conversion of nitrogen in most larvae stage	Ohbayashi et al. (2019); Zhang et al. (2023)
2.	*Klebsiella oxytoca* and *Morganella morganii*	Diptera	Breaking down of nitrogen using HCL and a hydrolyzing agent	Ren et al. (2022); Zhang et al. (2023)
3.	*Bacillus licheniformis* and *Enterobacter hormaechei*	Isoptera	Manufacture ligninase	Suman et al. (2016); Zhang et al. (2023)
4.	*Klebsiella oxytoca* and *Bacillus pumilus*	Lepidoptera	Manufacture cellulase	Dantur et al. (2015); Zhang et al. (2023)
5.	Gut microbiota and *Stammera*	Coleoptera	Manufacture pectinase	Evangelista et al. (2015); Zhang et al. (2023)
6.	*Streptomyces* and *Pantoea*	Hymenoptera	Manufacture cellulase	Adams et al. (2011); Zhang et al., (2023)
7.	*Bacillus* sp. and *Providencia* sp.	Isoptera	Manufacture cellulase	Arfah et al. (2019); Zhang et al. (2023)
8.	*Bacillus safeness*	Coleoptera	Manufactures cellulase	Hatefi et al. (2017); Zhang et al. (2023)
9.	*Bacillus velezensis*	Coleoptera	Manufactures cellulase	Li et al. (2020); Zhang et al. (2023)

8.4 Biological Significance of Gut Microbes in Insects

The anatomy of the gut of insects showed that it's sectionalized into three main regions which are the hindgut, midgut, and the foregut (Jaffar et al. 2022). Studies by Chen and St Johnston (2022) showed that the anterior region of the hindgut and midgut are lined with epithelium cells of embryonic origin. These regions are protected from disease-carrying organisms because of the presence of integument glycol-proteins and chitin found in the exoskeleton section of the body of the insect. Meanwhile, the posterior midgut functions in the digestion and absorption of biological materials.

The biological significance of gut microbiome insects is that they offer a suitable environment that excites the diversification and proliferation of beneficial and symbiotic microbes which assist and facilitate the secondary metabolites that are important for the biological functioning of the cells (Bruno et al. 2019). The secondary metabolites also play important functions like the growth of the insects, food breakdown or digestion of lignin, immune defense, and detoxification of noxious chemicals like pesticides as illustrated in Fig. 8.2 (Dar et al. 2024a, b). Other roles or functions include the production of sterols, amino acids, and vitamins and the

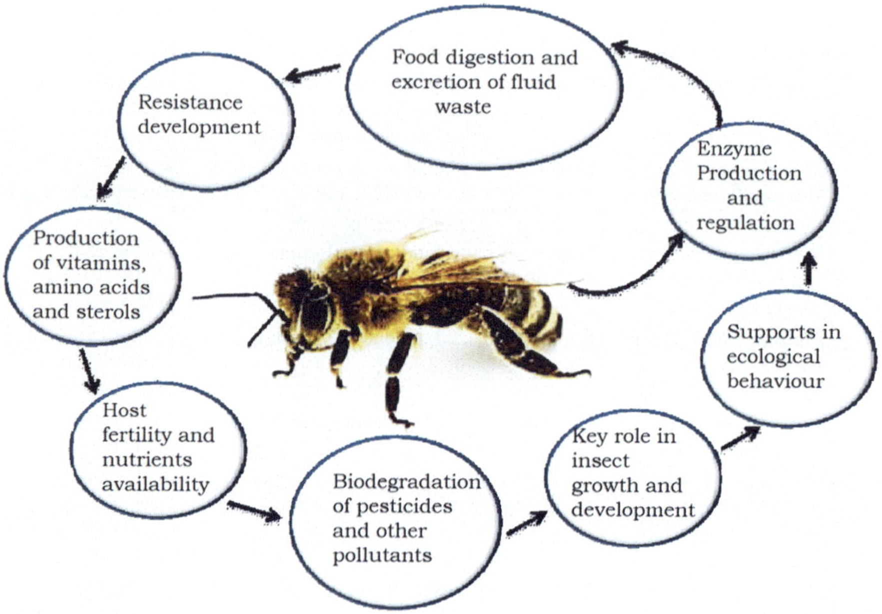

Fig. 8.2 Other basic role functions of the microbiome in insect

excretion of food wastes enzyme regulation and production, ecological behaviors, nutrient availability, and fertility (Jaffar et al. 2022).

8.5 The Mode of Interaction of Gut Microbes and Wood Digestion in Termites

Based on the level of benefits enjoyed by the microbes and the host, the mode of interaction of gut microbes can be broadly categorized into four types including mutualism, commensalism, competitive, and pathogenic. These interactions can significantly influence the host's health, development, behavior, and ecological roles. The mode of interaction of gut microbes includes the following:

8.5.1 Mutual Interactions/Mutualism

In mutualistic interactions, both the insect host and the gut microbes benefit. These relationships are often essential for the host's survival and well-being. Some essential aspects of the benefits of mutualistic interactions include:

1. ***Nutrient Provision and Digestion:*** On the aspect of nutrient provision, it is important to note that some essential amino acids, mineral elements, or other basic nutrients are not inherent in the raw food substances consumed by some of these insects; therefore, these gut microbes tend to convert or breakdown complex structures to form simpler units that are easily accessible and absorbable by the insect host. Based on digestion, these microbes help to breakdown complex recalcitrant food substances like hemicellulose, lignin, lignocellulose, and cellulose into simpler sugars (oligomers, dimers, and monomers) that are easily digested by the insects. Discussed below are some basic examples using specific illustrations:

Example 8.1 Termites rely on their gut microbes to digest cellulose. Protozoa and bacteria in the termite gut break down cellulose into simpler compounds, which the termite can absorb and use as a primary energy source (Engel and Moran 2013).

Example 8.2: Aphids and *Buchnera aphidicola* Aphids host *Buchnera* spp. (a bacterium) in specialized cells within their guts. *Buchnera aphidicola* provides essential amino acids, missing from the aphid's diet which consists of phloem sap, while the aphid provides a protected environment and nutrients for the bacteria (Engel and Moran 2013).

8.5.2 Digestion of Wood (Lignocellulose) in Termites

The digestion of wood in termites is a complex process that involves a symbiotic relationship between termites and a diverse community of microorganisms in their gut, including bacteria, protozoa, and archaea (Brune 2014). These microbes produce the necessary enzymes to break down lignocellulose (a combination of cellulose, hemicellulose, and lignin) in wood. The digestion process in termites occurs in several stages:

1. Mechanical breakdown of wood by termite mandibles
2. Enzymatic hydrolysis of cellulose and hemicellulose into sugars by termite-derived enzymes and gut symbionts
3. Fermentation of sugars into short-chain fatty acids by protozoa, bacteria, and archaea in the hindgut
4. Absorption of fatty acids (mainly acetate) for energy
5. Partial lignin degradation by gut microbes, allowing access to cellulose and hemicellulose
6. Nitrogen fixation by gut symbionts to provide necessary nutrients

8.5.3 Mechanical Breakdown

Mouthparts: Termites begin the digestion process by chewing wood with their strong mandibles, breaking it into smaller particles. This mechanical breakdown increases the surface area, making the material more accessible to enzymatic action (Slaytor 1992).

8.5.4 Enzymatic Hydrolysis of Cellulose and Hemicellulose

Salivary Glands Termites secrete endogenous enzymes (cellulases and hemicellulases) from their salivary glands. These enzymes initiate the breakdown of cellulose and hemicellulose into simpler sugars (mainly glucose and xylose) as the wood particles pass into the digestive tract (Brune 2014).

Gut Microbes Most of the actual breakdown, however, is carried out by the microorganisms in the termite gut, especially in the hindgut. Termites house a wide range of microbes, including protozoa, bacteria, and archaea, which play crucial roles in lignocellulose digestion. These microbes secrete additional cellulases, hemicellulases, and beta-glucosidases to hydrolyze the polysaccharides in wood (Ohkuma 2003).

8.5.5 Fermentation in the Hindgut

Protozoa and Bacteria: In lower termites (e.g., *Reticulitermes* species), protozoa dominate the gut microbiome. These protozoa ingest the wood particles and produce enzymes like cellulases and xylanases, which break down the cellulose and hemicellulose into sugars. These sugars are then fermented into short-chain fatty acids (acetate, propionate, and butyrate), which are the main energy sources for the termite (Inoue et al. 1997).

Acetogenic Bacteria: In the hindgut of both lower and higher termites, acetogenic bacteria ferment sugars and other byproducts into acetate. Acetate is absorbed by the termite and used as its primary energy source (Hongoh 2010).

Methanogenesis: Some termites, particularly lower termites, also have methanogenic archaea in their hindguts. These archaea consume hydrogen produced by fermentation and convert carbon dioxide into methane. The presence of these methanogens helps maintain efficient fermentation by removing excess hydrogen, which could inhibit microbial metabolism (Tokuda and Watanabe 2007).

8.5.6 Lignin Degradation

Incomplete Lignin Breakdown: Termites have limited ability to degrade lignin completely. While they can break it down partially through oxidative reactions, most lignin passes through their digestive system undigested. However, some bacteria in the termite gut produce enzymes like manganese peroxidases and laccases, which help break down lignin into smaller fragments, allowing better access to cellulose and hemicellulose (Brune 2014).

Higher Termites: Higher termites (e.g., *Nasutitermes*) lack gut protozoa and instead rely more on bacteria for lignocellulose breakdown. They exhibit a different lignin degradation pathway, with specialized bacteria breaking down lignin-derived products, though the process is still less efficient compared to cellulose and hemicellulose digestion (Tokuda and Watanabe 2007).

8.5.7 Absorption of Nutrients

After fermentation, the termite absorbs the short-chain fatty acids (mainly acetate) through its gut wall. These fatty acids are used as the primary energy source, providing the termite with the energy necessary to live. This highly specialized and symbiotic digestion mechanism allows termites to thrive on a diet of wood, a material that is indigestible to most animals. The efficiency of lignocellulose breakdown in termites is a subject of interest in biofuel research, where scientists aim to mimic this process to convert plant biomass into energy.

8.6 How Do Gut Microorganisms in Insects Break Down Lignocellulose?

Gut microbiome of insects specifically the herbivorous ones feed on plant parts thus digesting and breaking down large macromolecules (polymers) like lignocellulose materials using several enzymes such as ligninase, pectinase, and cellulase (Corinne et al. 2021; Zhang et al. 2023). They possess strong potentials for enzymatic production that when available to the insects, form amorphous or crystalline microfibers in the floral cell membrane/wall that when needed by them, are hydrolyzed into bits of sugar molecules via cellulase reactions: a process the microbiome found in the gut system of the insect mediate (Engel and Moran 2013). Studies by Adams et al. (2011), Dantur et al. (2015), Baharuddin (2016), Arfah et al. (2019), Ceja-Navarro et al. (2019), and Barbosa et al., (2020), showed that many gut insects like *Sirex noctilio*, *Macrotermes gilvus*, *Diatraea saccharalis*, *Cossus cossus*, *Odontotaenius disjunctus*, *Lepidiota mansueta*, and *Macrotermes gilvus* can breakdown larger cellulose and lignin materials in plants because of the mediated activities or functions of microorganisms like *Pantoea*, *Streptomyces*, *Klebsiella* spp., *Bacillus* spp., and *Providencia* sp., which also facilitate the acquisition of nutrients in their hosts. The breakdown of hemicellulose and cellulose fibers in the cell wall of plants is a function of the enzyme pectinase secreted by the microbiome in the gut system of insects like the *Cassida rubiginosa* (leaf beetle). Studies have shown that the beetle breaks down pectin because of the DNA code found in the gene of the organism (Salem et al. 2020, Kirsch et al. 2022; Zhang et al. 2023). Similarly, *Gilliamella apicola*, a bacterium that aids the honey bee, digests pectin in their gut system which was shown by a metagenomic research work by Engel et al. (2012). Lignin degradation is usually difficult to metabolize by most insects. Its degradation is mainly mediated by species of fungi and bacteria as demonstrated in the works of Suman et al. (2016), Zhou et al. (2017), and Tsegaye et al. (2019) found in the midgut of the insect-like the *Anoplophora glabripennis* (longhorn beetle) (Scully et al. 2014, 2018).

8.7 Biomimics of the Insect Gut Microbes Versus Biofuel Production

In the fight against the issues of environmental degradation that have resulted in global warming and climate change in our world today, innovators and scientists have been looking for a green way out to solve these issues (Adama and Anani 2023; Adama et al. 2023; Adentunji et al. 2022a, b). Studies have shown that the microbes in the gut system of insects can serve as a perfect model when utilized in the degradation of lignocellulose materials in the production of biofuels using strategic biorefinery methods. This is a biomimicry or biomimetic way of helping humans solve the issues of climate change using a green biotechnological model. Because of the incidence of fossil fuel deterioration and depletion of the natural ecosystem, the need to use alternative green technology is not negotiable. The practice of biomimicry is an emerging tool against combating the issues of climate change (Beciri 2011; Chetankumar 2023).

Enzymes from insects contain potentially degrading metabolites engineered by gut microbiomes. Studies by scientists from the University of Purdue have demonstrated that enzymes obtained from insects such as termites can be used to degrade the biomass of woody plants that can be utilized for biofuel production (Beciri 2011). However, the quantity of biofuels that could be produced from this biorefinery process was not yet estimated. Meanwhile, the scientist infused the gene in charge of facilitating the enzymes produced in the insect and introduced them as food pellets to the caterpillar of the same species which then generated a lot of enzymes. It was observed that the enzymes were effective in the release of sugar from the plant biomass as well as the effective degradation of the biomass for biofuel production. In line with this, it was found that three types of enzymes were active in different regions of the degraded biomass. Two of them released pentose and glucose, while the third aided in degrading lignin, the primary structural material found in the cell walls of plants (Beciri 2011). The scientists concluded that combining the three enzymes will be a potential tool to increase the biofuel amount from degraded biomass in biorefinery processes.

8.7.1 Insect Gut Functionalities: The Role of Microbial Enzymes

The deterioration and the depletion of the environment as a consequence of fossil fuels have caused serious issues in the sustainable and renewable of chemicals and resources from natural energies. Rajeswari et al. (2021), in a review, looked at the biorefinery technique in the utilization of microbes in the guts of ruminants and insects in the recycling of carbon locked in lignocellulosic materials as an alternative and sustainable management of ecosystem distortions. The first and second generations of biofuels have gained tremendous commercialization in the production of biofuels from food wastes and biochemicals. In today's view of green energy, different groups of enzymes and microorganisms have been explored for the treatment process and production of biofuels from lignocellulosic materials in consolidating the challenges faced in the various generations of biofuel production. The microbiota found in the gut system of ruminants and insects have been proving efficient because of their high potential as good degraders of lignocellulosic materials; in conjunction with the valorization of the lignocellulosic materials, the microbes intensify the biological diversities of microbiome thus enriching the molecular characteristics of the community in aiding the depolymerization of the materials by using enzymes such as the lignocellulose, lytic polysaccharide monooxygenase, and β-1,4 glucanase, a functionality in renewable and biofuels production of chemicals (Rajeswari et al. 2021). This process will yield the reprocessing framework of waste conversion from cradle to grave method.

In recent times, environmental sustainability has grown along sides with the emerging global population as regards the wastes generated which have been reported to have significant effects on the environment and humans. Using insects as an alternative to reduce the global carbon footprint has been the gainsay of many environmentalists. In line with this, Akgedik and Bayrakta (2023) evaluated the deployment of enzymes from microbes in the gut of insects in the biotransformation of wastes to fuels. Because of the microbes and enzymes found in the gut of insects, the biotransformation of wastes to useful products can be deployed for biofuels, pharma food, cosmetics, pharmaceuticals, and the industrial production sectors of the world. Some insect species have the ability the convert lignocellulosic materials to these valuable products because of their lifecycle which can serve as a green technology (Akgedik and Bayrakta 2023). Studies have shown that the biotransformation of materials can be obtained through the manipulation of the enzymes using metabolomics and the understanding of the metabolism of the insects so that it can be applied in the biorefinery of wastes. Fu et al. (2024) assessed the advanced method in the production of renewable energy by the exploration of microbes in the gut of insects. The gut system of insects consists of a plethora of consortia of microbes that play a mutualistic role in the digestion of food and solid materials. This region in the insect body can also play a biomimetic role in the metabolic and enzymic application in the renewable production of biomaterials like fuels. Insects like beetles, termites and, ants house plethoral communities of microbes that aid in the degradation of lignocellulosic compounds into simple 6-carbon sugars which could be fermented into methane and ethanol biofuels. The bioengineering of species of microbes could be employed to improve the production of effective enzymes that can be used in the biorefinery production of fuels.

The manufacture of renewable chemicals from plant biomass has been an emerging biorefinery process to aid address the issues of environmental pollution from fossil fuel overdependence. Cho et al. (2022) investigated the manufacture of chemicals from wastes using developed microbial cell factories (MCF). The cell factories of microorganisms that are engineered have been shown to produce chemicals that can be utilized in the breakdown of the carbon chains in lignocellulosic materials into useful materials like biofuels. This process consists of three streamlined pathways: created nonnative, nonnative-existing, and native-existing. These pathways aid in the performance, improvement, and construction of the MCF for an efficient metabolic system in the sustainable production of biofuels.

In recent years, the world has taken a shift in the global energy sector from fossil fuels to biofuels to cut down the effects of pollution caused by residue pollutants discharged into the environment. Iragavarapu et al. (2023) evaluated the biological processing of materials from wastes as a tool for the renewable of fuels and chemicals as a bioeconomy frontier. Biofuels have been reported to have low production costs because of their efficiency, green state, and less noxious form when compared to fossil fuels. Iragavarapu et al. (2023) stated that microalgae and bacteria play a vital role in the biorefinery production of fuels which can serve as a bioprocessing technique in improving the economy and attaining sustainable development goals.

8.7.2 Lignocellulose Breakdown and Potential Benefits of Different Processes

Ho et al. (2019) examined the potential utilization of alkaline H_2O_2 (hydrogen peroxide)-AHP in the valorization and pretreatment of lignocellulosic materials for the production of biofuels. AHP can attack ethylene and carbonyl groups selectively to improve the process of delignification of lignocellulosic materials. This potential by AHA is used to improve the process of valorization also. The authors reported that AHP is compatible with the hydrolysis of enzymes and biomass degradation. Its application in the pretreatment of lignocellulosic materials sequentially has been recounted to improve its biological products like lignin and silica, respectively. Ho et al. (2019) reported that this process is considered green because it is eco-friendly and produces less noxious chemicals. Similarly, Basera et al. (2024) investigated the enzymatic hydrolysis of lignocellulosic materials with key insights on economic sustainability. These lignocellulosic materials represent an abundant and sustainable feedstock with vast applications in bacterial fermentation processes for the manufacture of biochemical and biofuels. Plant biomass has been reported to be one of the alternative emerging tools to replace fossil fuels. Studies have shown that its degradation by the activities of nonenzymatic enzymes (hemicellulases and cellulases) from microbes has proven effective in the production of sustainable biofuels. In the aspect of the feasibility of the economy, the demand, availability of the substrates, and scalability are some of the significant obstacles in terms of costs which has also affected the demand in the downstream production of biofuels. To address these issues, Basera et al. (2024) reported that cost-effective optimization and the development of new microbial factories will aid in improving the viability of the conversion of lignocellulosic materials for a more resilient and sustainable green energy in the future.

The demand for bioenergy in the global market has taken a new switch. Al Zubi et al. (2023) in a study investigated the biorefinery of lignocellulose materials converted via nanobiotechnology. Lignocellulose materials are considered a source of renewable energy and an economical asset to any country that produces it sustainably. The nanotechnology of plant lignin materials involves the hydrolysis and enzymatic reaction of its components like hemicellulose, cellulose, and lignin itself. Al Zubi et al. (2023) reported that studies have indicated that nanomaterials have strong magnetic properties that can be harnessed to effectively immobilize, reuse, and recover enzymes that microbes from insects could secrete to degrade lignocellulose materials for biofuel production. Dutta et al. (2023) in a review examined the emerging developments of the use of nanotechnology alongside the lignocellulosic biomass for the production of biofuels. Biofuels play an immense role in helping to reduce climate change, global warming, and carbon influx in the environment. The demand for biofuel products like bioethanol has come to replace the petrochemical products used in the movements of vehicles because of the attendant environmental issues it causes. The application of nanomaterials to biofuels has been reported to effectively reduce the effects of climate change more when it is used alone. This is because of its efficiency in terms of energy, selectivity, and low economic costs. Lignocellulosic material interacts with nanoparticles at a very low concentration to give out carbohydrates to produce bioethanol.

8.7.3 Insect Gut as a Source of Biocatalysts for Biofuel Production

In the gut system of insects, like the *Reticulitermes flavipes*, *Reticulitermes speratus*, *Coptotermes formosanus*, and *Coptotermes gestroi*, enzymes such as ligninase, hemicellulase, and cellulase are used to degrade or breakdown lignocellulosic materials or plant biomass for biofuel production. About 70 species of insects have been identified for such importance in the biorefinery industry (Rajendran et al. 2018). Some termites lower in the classification hierarchy of the termes group have been identified to secrete β-glucanases, endoglucanase, and exo- and endoglucanases for the same purpose (Sethi and Scharf (2013). The valorization of wastes into useful ways of pollution reduction and the replacement of fossil fuels have resulted in the release of noxious chemicals into the environment. Wastes contain recalcitrant materials like lignin. Its breakdown into useful materials is very difficult. Karmee (2023) evaluated the utilization of biocatalysts in the breakdown of wastes from food in the refinery of biofuels. Many advanced techniques like valorization and catalytic using chemicals have been developed to convert wastes to biofuels. However, their application is not too efficient in the synthesis of biofuels, and the process and condition of application are quite harsh. An alternative method like the biocatalysts has been heralded recently because of its biodegradability, selectivity, recyclability, and mild settings of reaction (Karmee 2023). Lipase and monooxygenase are employed for the manufacture of biofuels like biomethanol and bioethanol through the oxidation of methane produced from degraded wastes obtained from food. The authors recommended that the economic and societal aspects should be looked into to prospect future limitations of biofuel usage.

In the biorefinery process, exploring the efficacy of green methods in the pretreatment of lignocellulose degraded by enzymic action to produce biofuel is pertinent in this era of environmental pollution from fossil fuels as an alternative. Zhang et al. (2022) investigated the breakdown of lignocellulose materials by enzymes secreted by microbes in the gut of an insect *Locusta migratoria manilensis*. Studies have shown that locust, an insect of the order Orthoptera, has the potential to break gramineous contents of lignocellulose more efficiently because of the enzymes (hemicellulase, ligninase, xylanase, and cellulase) it carries. The faces of locusts and the leaves of corn were characterized with FE-SEM, FTIR, XRD, and TG (scanning electron microscopy, Fourier transform infrared, X-ray diffraction pattern, and thermogravimetric). The results from this study showed that the highest activities of xylanase, filter paper cellulase (FPA), and carboxymethyl cellulase (CMCase) were observed in the insect foregut which is an indication that the foregut is the major point of the degradation of lignocellulose in the locust's segment. In addition, the main components of nutrition were absorbed in the mid-region of the gut system. The activities of MnPase (manganese peroxidase) and xylanase were significantly lower than CMCase. This might be linked to the locust's nutritional faculty being hemicellulose and cellulose except lignin in line with the findings from the results analysis of the TG, XRD, FTIR, and FE-SEM. In general, Zhang et al. (2022) stated that the findings of their study gave a basic insight into the mechanism of the degradation of lignocellulosic materials by active enzymes in the foregut of locust. This can be a mimicking application process in the large of biofuel production.

8.7.4 Examples of Insect for Biorefinery Designs for Biofuels Utilized

Plant biomass from agricultural products and wastes are considered as feedstock for biodiesel production to meet the growing need for global energy. Wang et al. (2017) assessed the biorefinery of crops using insects as a green method to produce biodiesel. The mealworm and soldier (*Tenebrio molitor* and *Hermetia illucens*) insects were used in this study. Rajendran et al. (2018) investigated the biorefinery of fuels using several insects as potentials for its design. Table 8.2 shows examples of potential insects used for biorefinery designs of biofuels:

Table 8.2 Example of some potential insects used as designs for utilizing biofuels

S.no	Types of insects	Biomass degraded	References
1	*Musca domestica*	Wood biomass	Cicˇkova´ et al. (2012); Rajendran et al. (2018)
2	*Chrysomya megacephala*	Wood biomass	Yang and Liu (2014); Rajendran et al. (2018)
3	*Hermetia illucens*	Wood biomass	Banks et al. (2014); Rajendran et al., (2018)
4.	*Boettcherisca peregrine*	Wood biomass	Yang et al. (2012)
5	*Tenebrio molitor L.*	Phytomass	Wang et al. (2017)
6	*Hermetia illucens L.*	Phytomass	Wang et al. (2017)
7	*Camponotus japonicus*	Not specified	
8	*Apis mellifera L.*	Not specified	Kim et al. (2024)
9	*Velarifictorus aspersus*	Not specified	
10	*Gesonula punctifrons*	Not specified	
11	*Ptecticus tenebrifer*	Not specified	
12	*Reticulitermes speratus*	Not specified	
13	*Cryptocercus kyebangensis*	Not specified	
14	*Galleria mellonella*	Not specified	
15	*Bombyx mori*	Not specified	
16	*Tenebrio molitor*	Not specified	
17	*Protaetia brevitarsis seutensis K*	Not specified	
18	*Allomyrina dichotoma*	Not specified	
19	*Dorcus titanus castanicolor*	Not specified	

8.8 Summary and Recommendations

This chapter discusses the potential application of the microbes in the insect gut as a biomimicry tool for the advanced design of lignocellulosic biorefinery. The negative incidences from fossil fuels, which have increased the levels of greenhouse gas emission into the environment, have caused key challenges to environmental scientists. In the renewable energy industry, the application of a green approach to improving the quality of biofuel through the bioconversion of biomass of lignocellulosic materials of insects has been heralded as a safe biotechnology for a green economy. At times, the microorganisms found in the gut system of many mammals and insects of the ruminant and xylophagous kinds, acting as symbionts, aid in the digestion of the lignocellulose materials ingested by the organisms. Insects, one of the invertebrates' arthropods, are well-known for their ability to pollinate plants; act as soil engineers, thereby aerating the soil; have symbiosis or have a mutualistic effect with beneficial microbes; and aid as pathogens for so many diseases. Considering their ubiquitous nature in the ecosystem, their structure, and their role with beneficial microbes, insects such as soil engineers and wood-feeding-like collaborators help break down rich lignocellulose biomass uniquely in their gut pathways. The ecology of gut microbiome found in insects, such as termites, bees, aphids, beetles, roaches, and flies, comprises a diverse and complex community of microorganisms that play a crucial role in the well-being of the insect's (host) health, development, and ecological interactions. A typical insect gut microbial community includes bacteria, fungi, protozoa, and viruses which interact with each other and their insect host in a mutualistic symbiotic association that is beneficial to all. Some examples of gut microbes in insects and their biological role are *Burkholderia* sp., which manufactures aminase, recycling, and conversion of nitrogen in most larvae stage; *Klebsiella oxytoca* and *Morganella morganii*, which break down of nitrogen using a hydrolyzing agent; *Bacillus licheniformis* and *Enterobacter hormaechei*, which manufacture ligninase; *Klebsiella oxytoca* and *Bacillus pumilus*, which manufacture cellulase; *gut microbiota* and *Stammera*, which manufacture pectinase; and *Streptomyces* and *Pantoea*, which manufacture cellulase.

The biological significance of gut microbes in insects is that they offer a suitable environment that excites the diversification and proliferation of beneficial and symbiotic microbes which assist and facilitate the secondary metabolites that are important for the biological functioning of the cells. The secondary metabolites also play important functions like the growth of the insects, food breakdown or digestion of lignin, immune defense, and detoxification of noxious chemicals like pesticides. Other roles or functions include the production of sterols, amino acids, and vitamins, excretion of food wastes enzyme regulation and production, ecological behaviors, nutrient availability, and fertility. The mode of interaction of gut microbes in insects can be broadly categorized into four types based on the level of benefits enjoyed by the microbes and the host. These interactions can significantly influence the host's health, development, behavior, and ecological roles. The modes of interactions of gut microbes include mutualistic, commensal, competitive, and pathogenic. In mutualistic interactions, both the insect host and the gut microbe benefit. These relationships are often essential for the host's survival and well-being. In commensal interactions, gut microbes benefit from the environment provided by the insect host without significantly affecting the host, positively or negatively. Competitive interactions occur when different microbial species vie for the same resources, often influencing the composition and stability of the gut microbiome. In pathogenic interactions, certain microbes can harm the insect host, leading to diseases or adverse health effects. Gut microorganisms in insects break down lignocellulose, and their mechanism of action includes several enzymes such as ligninase, pectinase, and cellulase. They possess strong potentials for enzymatic production that, when available to the insects, form amorphous or crystalline microfibers in the floral cell membrane/wall that, when needed by them, are hydrolyzed into bits of sugar molecules via cellulase reactions, a process the microbiome found in the gut system of the insect mediate. Lignin degradation is usually difficult to metabolize by most insects. Its degradation is mainly mediated by species of fungi and found in the midgut of the insect like the *Anoplophora glabripennis* (longhorn beetle).

Biomimics of the insect gut microbes versus biofuel production is essential in the fight against the issues of environmental degradation that have resulted in global warming and climate change in our world today. Enzymes from insects contain potentially degrading metabolites engineered by gut microbiomes. Meanwhile, it has been reported that specific genes are in charge of facilitating the enzymes produced in insects. It was observed that the enzymes were effective in the release of sugar from the plant biomass as well as the effective degradation of the biomass for biofuel production. In line with this, it was found that three types of enzymes were active in different regions of the degraded biomass. Two of them released pentose and glucose, while the third aided in degrading lignin, the primary structural material found in the cell walls of plants. The scientists concluded that combining the three enzymes will be a potential tool to increase the biofuel amount from degraded biomass in biorefinery processes.

The economic cost of using insect gut microbes for biofuel production can be broadly categorized into research and development, production, and implementation stages. Identifying and isolating effective gut microbes from insects involve significant investment in fieldwork and laboratory research. For example, isolating cellulolytic microbes from termite guts requires sophisticated equipment and skilled personnel. Engineering microbial strains to enhance their efficiency and stability for biofuel production can be expensive. This includes CRISPR/Cas9 editing, plasmid construction, and extensive trial and error. Gathering and preprocessing lignocellulosic biomass (e.g., agricultural residues, forestry waste) into a form suitable for microbial digestion can be costly. This includes transportation, grinding, and pretreatment processes to break down lignin and hemicellulose. Conducting assessments to ensure the biofuel production process is sustainable and has minimal environmental impact is required by law and adds to the overall cost. The economic cost of using insect gut microbes for biofuel production encompasses significant R&D investment, high production expenses, and substantial implementation and market penetration costs. Despite these challenges, the potential benefits, such as sustainable and renewable energy sources, can justify the investment. Advancements in biotechnology and favorable policy frameworks can help reduce these costs and make microbial biofuel production economically viable. However, the quantity of biofuels that could be produced from this biorefinery process is not yet estimated due to its economic cost and culturing of potential insects that can produce viable and long-lasting enzymes. It is recommended that stakeholders and research experts should invest in R&D from the global, regional, and local scales to make biofuel production economically viable and enormous for the present and future end users. This will end the fight against the issues of environmental degradation that have resulted in global warming and climate change in our world today.

References

Adama KK, Anani OA (2023) Experimental and theoretical assessment of phenomena linked with separation and purification of biodiesel from *Ricinus communis* seed oil. Heliyon 9:e16536. https://doi.org/10.1016/j.heliyon.2023.e16536

Adama KK, Ikhazuangbe PMO, Onyeachu IB, Aliu D, Anani OA (2023) Development of polyacrylamide from waste agricultural by-product for enhanced oil recovery process in the Nigerian oil industry: conceptual review. Nigerian Res J Eng Environ Sci 8(2):370–382

Adams AS et al (2011) Cellulose-degrading bacteria associated with the invasive woodwasp *Sirex noctilio*. ISME J 5:1323–1331

Adetunji CO, Olaniyan OT, Anani OA, Bodunrinde RE, Osemwegie OO, Ubi BE (2022a) Integrated processes for the production of pharmaceutical products from agro-waste. Biomass, biofuels, biochemicals. Elsevier. https://doi.org/10.1016/B978-0-323-89855-3.00021-2

Adetunji CO, Anani OA, Olaniyan OT, Bodunrinde RT, Osemwegie OO, Ubi BE (2022b) Sustainability of biofertilizers and other allied products from genetically modified microorganisms. Biomass, biofuels, biochemicals. Elsevier. https://doi.org/10.1016/B978-0-323-89855-3.00003-0

Akgedik AN, Bayrakta TBO (2023) Utilization of insect gut as a biosource for the development of future biotransformation processes converting wastes, organic materials, and residues into valuable products. Johnson Matthey Technol Rev 67(4):416–427. https://doi.org/10.1595/205651323X16732719244908

Al Zubi MA et al (2023) Conversion of lignocellulose biomass to bioenergy through nanobiotechnology. J Sustain Dev Energy Water Environ Syst 11(2):1110442. https://doi.org/10.13044/j.sdewes.d11.0442

Arfah RA et al (2019) Isolation and characterization of soil termites (*Macrotermes gilvus*) cellulolytic bacteria and activity determination of cellulase enzyme on newsprint substrates. J Phys Conf Ser 1341:032037

Aves SL et al (2019) Bioprospection of enzymes and microorganisms in insects to improve second-generation ethanol production. Ind Biotechnol 15:336–349

Baharuddin M (2016) Cellulase enzyme activity of *Bacillus circulans* from larvae *Cossus cossus* in lignocellulosic substratum. Am J Biomed Life Sci 4:21

Banks IJ, Gibson WT, Cameron MM (2014) Growth rates of black soldier fly larvae fed on fresh human feces and their implication for improving sanitation. Trop Med Int Health 19:14–22

Barbosa KL et al (2020) Bacterial cellulose from the intestinal tract of the sugarcane borer. Int J Biol Macromol 161:441–448

Basera P, Chakraborty S, Sharma N (2024) Lignocellulosic biomass: insights into enzymatic hydrolysis, influential factors, and economic viability. Discov Sustain 5:311. https://doi.org/10.1007/s43621-024-00543-5

Beciri D (2011) Termite digestive system biomimicry could lead to biofuel production. https://www.robaid.com/bionics/termite-digestive-system-biomimicry-could-lead-to-biofuel-production.htm

Brune A (2014) Symbiotic digestion of lignocellulose in termite guts. Nat Rev Microbiol 12(3):168–180

Brune A, Ohkuma M (2010) Role of the termite gut microbiota in symbiotic digestion. Biology of Termites: a Modern Synthesis:439–475. https://doi.org/10.1007/978-90-481-3977-4_16

Bruno D et al (2019) The digestive system of the adult *Hermetia illucens* (Diptera: Stratiomyidae): morphological features and functional properties. Cell Tissue Res 378:221–238. https://doi.org/10.1007/s00441-019-03025-7

Ceja-Navarro JA et al (2019) Gut anatomical properties and microbial functional assembly promote lignocellulose deconstruction and colony subsistence of a wood-feeding beetle. Nat Microbiol 4:864–875

Chen J, St Johnston D (2022) Epithelial Cell Polarity During Drosophila Midgut Development. Front Cell Dev Biol 10:886773. https://doi.org/10.3389/fcell.2022.886773

Chetankumar J (2023) Biomimicry of the biological pump: natural carbon sink. https://www.linkedin.com/pulse/biomimicry-biological-pump-natural-carbon-sink-jani-chetankumar/

Cho JS et al (2022) Designing microbial cell factories for the production of chemicals. ACS Au 2(8):1781–1799. https://doi.org/10.1021/jacsau.2c00344

Cicˇkova´ H et al (2012) Biodegradation of pig manure by the housefly, *Musca domestica*: a viable ecological strategy for pig manure management. PloS One 7:e32798

Corinne V et al (2021) Trees and insects have microbiomes: consequences for forest health and management. Currt Forest Rep 7:81–96

Dantur KI, Enrique R, Welin B, Castagnaro AP (2015) Isolation of cellulolytic bacteria from the intestine of *Diatraea saccharalis* larvae and evaluation of their capacity to degrade sugarcane biomass. AMB Express 5:15

Dar MA, Syed R, Pawar KD, Dhole NP, Xie R, Pandit RS, Sun J (2024a) Evaluation and characterization of the cellulolytic bacterium, *Bacillus pumilus* SL8 isolated from the gut of oriental leafworm, *Spodoptera litura*: an assessment of its potential value for lignocellulose bioconversion. Environ Technol Innov 27:102459

Dar MA, Xie R, Jing L, Qing X, Ali S, Pandit RS, Shaha CM, Sun J (2024b) Elucidating the structure and composition of bacterial symbionts in the gut regions of the wood-feeding termite, *Coptotermes formosanus*, and their functional profile towards lignocellulolytic systems. Front Microbiol 15:1395568. https://doi.org/10.3389/fmicb.2024.1395568

Dashtban M, Schraft H, Qin W (2009) Fungal bioconversion of lignocellulosic residues; opportunities and perspectives. Int J Biol 5:578–595

Delavar MA, Wang J (2022) Advanced methods and mathematical modeling of biofilms, chapter 7. In: Bioreactor concepts, types, and modeling. Academic Press, pp 195–245. https://doi.org/10.1016/B978-0-323-85690-4.00004-X

Dillon RJ, Dillon VM (2004) The gut bacteria of insects: nonpathogenic interactions. Annu Rev Entomol 49:71–92

Douglas AE (2015) Multiorganismal insects: diversity and function of resident microorganisms. Annu Rev Entomol 60:17–34

Dutta S et al (2023) Recent developments in lignocellulosic biofuel production with nanotechnological intervention: an emphasis on ethanol. Catalysts 13(11):1439. https://doi.org/10.3390/catal13111439

Engel P, Martinson VG, Moran NA (2012) Functional diversity within the simple gut microbiota of the honey bee. Proc Natl Acad Sci U S A 109:11002–11007

Engel P, Moran NA (2013) The gut microbiota of insects—diversity in structure and function. FEMS Microbiol Rev 37:699–735

Evangelista DE, de Paula FF, Rodrigues A, Henrique-Silva F (2015) Pectinases from *Sphenophorus levis* Vaurie, 1978 (Coleoptera: Curculionidae): putative accessory digestive enzymes. J Insect Sci 15:168

Fu C, Shah AA, Alissa M, Alsuwat MA, Ullah S, Khan MS, Zaman A (2024) Exploring the potential of insect gut microbes for advancing renewable energy production. J Appl Entomol 148:928–937. https://doi.org/10.1111/jen.13311

Hatefi A, Makhdoumi A, Asoodeh A, Mirshamsi O (2017) Characterization of a bifunctional cellulase produced by a gut bacterial resident of *Rosaceae* branch borer beetle, *Osphranteria coerulescens* (Coleoptera: Cerambycidae). Int J Biol Macromol 103:158–164

Ho C, Ong VZ, Wu TY (2019) Potential use of alkaline hydrogen peroxide in lignocellulosic biomass pretreatment and valorization—a review. Renew Sustain Energy Rev 112:75–86. https://doi.org/10.1016/j.rser.2019.04.082

Hongoh Y (2010) Diversity and genomes of uncultured microbial symbionts in the termite gut. Biosci Biotechnol Biochem 74:1145–1151

Huang SW, Zhang HY, Marshall S, Jackson TA (2010) The scarab gut: a potential bioreactor for biofuel production. Insect Sci 17:175–183. https://doi.org/10.1111/j.1744-7917.2010.01320.x

Inoue T, Murashima K, Azuma JI, Sugimoto A, Slaytor M (1997) Cellulose and lignin utilization by termites. Cellul 4:319–331

Iragavarapu GP, Imam SS, Sarkar O, Mohan SV, Chang YC, Reddy MV, Kim SH, Amradi NK (2023) Bioprocessing of waste for renewable chemicals and fuels to promote bioeconomy. Energies 16:3873. https://doi.org/10.3390/en16093873

Jaffar S, Ahmad S, Lu Y (2022) Contribution of insect gut microbiota and their associated enzymes in insect physiology and biodegradation of pesticides. Front Microbiol 13:979383. https://doi.org/10.3389/fmicb.2022.979383

Kala A, Kamra DN, Chaudhary LC, Agarwal N (2019) Metagenomics and CAZymes in rumen: a review. Indian J Anim Nutr 36:1

Karmee SK (2023) Moving towards the application of biocatalysis in food waste biorefinery. Fermentation 9:73. https://doi.org/10.3390/fermentation9010073

Kim JH, Kim HJ, Park SG, Lim SR (2024) Promising insects for sustainable food and biofuel. Int J Agric Nat Resour 51:2549. https://doi.org/10.7764/ijanr.v51i2.2549

Kirsch R et al (2022) Metabolic novelty originating from horizontal gene transfer is essential for leaf beetle survival. Proc Natl Acad Sci U S A 119:e2205857119

Li Y et al (2020) Genome sequencing of gut symbiotic *Bacillus velezensis* LC1 for bioethanol production from bamboo shoots. Biotechnol Biofuels 13:34

Matthews C, Crispie F, Lewis E, Reid M, O'Toole PW, Cotter PD (2019) The rumen microbiome: a crucial consideration when optimizing milk and meat production and nitrogen utilization efficiency. Gut Microbes 10:115–132. https://doi.org/10.1080/19490976.2018.1505176

Mondal S, Somani J, Roy S, Babu A, Pandey AK (2023) Insect microbial symbionts: ecology, interactions, and biological significance. Microorganisms 11:2665. https://doi.org/10.3390/microorganisms11112665

Ohbayashi T et al (2019) Comparative cytology, physiology, and transcriptomics of *Burkholderia insecticola* in symbiosis with the bean bug *Riptortus pedestris* and in culture. ISME J 13:1469–1483

Ohkuma M (2003) Termite symbiotic systems: efficient bio-recycling of lignocellulose. Appl Microbiol Biotechnol 61:1–9

Rajendran K, Surendra KC, Tomberlin JK, Khanal SK (2018) Insect-based biorefinery for bioenergy and bio-based products. Waste Biorefin:657–669. https://doi.org/10.1016/b978-0-444-63992-9.00022-7

Rajeswari G, Jacob S, Chandel AK, Kumar V (2021) Unlocking the potential of insect and ruminant host symbionts for recycling of lignocellulosic carbon with a biorefinery approach: a review. Microb Cell Fact 20:107. https://doi.org/10.1186/s12934-021-01597-0

Ren X et al (2022) Gut symbiotic bacteria are involved in nitrogen recycling in the tephritid fruit fly *Bactrocera dorsalis*. BMC Biol 20:201

Rossi G, Durek J, Ojha S, Schlüter OK (2021) Fluorescence-based characterisation of selected edible insect species: excitation emission matrix (EEM) and parallel factor (PARAFAC) analysis. Curr Res Food Sci 4:862–872. https://doi.org/10.1016/j.crfs.2021.11.004

Salem H et al (2020) Symbiont digestive range reflects host plant breadth in herbivorous beetles. Curr Biol 30:2875–2886.e4

Scully ED et al (2014) Functional genomics and microbiome profiling of the Asian longhorned beetle (*Anoplophora glabripennis*) reveal insights into the digestive physiology and nutritional ecology of wood-feeding beetles. BMC Genomics 15:1096

Scully ED et al (2018) Host-plant induced changes in microbial community structure and midgut gene expression in an invasive polyphage (*Anoplophora glabripennis*). Sci Rep 8:9620

Sethi A, Scharf ME (2013) Biofuels: fungal, bacterial, and insect degraders of lignocellulose. eLS. https://doi.org/10.1002/9780470015902.a0020374

Slaytor M (1992) Cellulose digestion in termites and cockroaches: what role do symbionts play? Comp Biochem Physiol B Comp Biochem 103:775–784

Suman SK et al (2016) Investigation of lignin biodegradation by *Trabulsiella* sp. isolated from termite gut. Int Biodeter Biodegr 112:12–17

Tokuda G, Watanabe H (2007) Hidden cellulases in termites: revision of an old hypothesis. Biol Lett 3:336–339

Tokuda G, Watanabe H, Matsumoto T, Noda H (1997) Cellulose digestion in the wood-eating higher termite, *Nasutitermes takasagoensis* (Shiraki): distribution of cellulases and properties of endo-β-1,4-glucanase. Zoolog Sci 14:83–93. https://doi.org/10.2108/zsj.14.83

Tsegaye B, Balomajumder C, Roy P (2019) Isolation and characterization of novel lignolytic, cellulolytic and hemicellulolytic bacteria from wood-feeding termite *Cryptotermes brevis*. Int J Microbiol 22:29–39

Tye YY, Lee KT, Wan Abdullah WN, Leh CP (2016) The world availability of nonwood lignocellulosic biomass for the production of cellulosic ethanol and potential pretreatments for the enhancement of enzymatic saccharification. Renew Sustain Energy Rev 60:155–172. https://doi.org/10.1016/j.rser.2016.01.072

Vikram S, Arneodo JD, Calcagno J, Ortiz M, Mon ML, Etcheverry C, Cowan DA, Talia P (2021) Diversity structure of the microbial communities in the guts of four neotropical termite species. PeerJ 9:e10959. https://doi.org/10.7717/peerj.10959

Wang H, Rehman KU, Liu X, Yang Q, Zheng L, Li W, Cai M, Li Q, Zhang J, Yu Z (2017) Insect biorefinery: a green approach for conversion of crop residues into biodiesel and protein. Biotechnol Biofuels 10:304. https://doi.org/10.1186/s13068-017-0986-7

Xie MA, Zabed R, Ali S, Zhu D, Sun J (2024) Termite microbial symbiosis as a model for innovative design of lignocellulosic future biorefinery: current paradigms and future perspectives. Biomass 4:180–201. https://doi.org/10.3390/biomass4010009

Yang S, Liu Z (2014) Pilot-scale biodegradation of swine manure via Chrysomya megacephala (Fabricius) for biodiesel production. Appl Energy 113:385–391

Yang S, Li Q, Zeng Q, Zhang J, Yu Z, Liu Z (2012) Conversion of solid organic wastes into oil via Boettcherisca peregrine (Diptera: Sarcophagidae) larvae and optimization of parameters for biodiesel production. PloS One 7:e45940

Zeng H, Zhao D, Zhang Z, Gao H, Zhang W (2023) Imperfect ant mimicry contributes to local adaptation in a jumping spider. iScience 26(6):106747. https://doi.org/10.1016/j.isci.2023.106747

Zhang H, Li Z, Zhang H, Li Y, Wang F, Xie H, Su L, Song A (2022) Biodegradation of gramineous lignocellulose by Locusta migratoria manilensis (Orthoptera: Acridoidea). Front Bioeng Biotechnol 10:943692. https://doi.org/10.3389/fbioe.2022.943692

Zhang Y, Zhang S, Xu L (2023) The pivotal roles of gut microbiota in insect-plant interactions for sustainable pest management. NPJ Biofilms Microbiomes 9:66. https://doi.org/10.1038/s41522-023-00435-y

Zhou H et al (2017) Screening and identification of lignin-degrading bacteria in termite gut and the construction of LiP-expressing recombinant Lactococcus lactis. Microorganism Pathog 112:63–69

Chapter 9
Technological Interventions to Bioengineer Extremophilic Microorganisms for Biofuel and Bioenergy Productions

Nida Maqsood, Wajid Ali, Wajeeha Ali, Fatima Atif, Trisanti Anindyawati, Apridah Cameliawati Djohan, Ika Rahmatul Layly, Marcelo Franco, and Muhammad Irfan

Abstract In this chapter, we explored advancements in biofuel and bioenergy productions using extremophilic microbes. The precise history of biofuels from different types of sources like crops and algae has been mentioned along with a precise description of extremophiles and their types. This chapter gives deep insight of technological interventions in bioengineering, and in this regard genome editing technologies like CRISPR have improved biofuel production by improving tolerance of microorganisms, optimizing utilization of substrate, and developing thermotolerant strains. The role of synthetic biology is also highlighted as it advanced the field by designing a robust microbial chassis. Biofuel production such as bioethanol, biobutanol, biodiesel, biogas, as well as biohydrogen has been explained well. So, this chapter gives a clear overlook of current and future potential of biofuel production by using extremophilic microbes by utilizing advanced bioengineering methods.

Keywords Extremophile · Dark fermentation · Lignocellulosic feedstocks · Acidogenesis · CRISPR technique · Biofuel

---

N. Maqsood · W. Ali · W. Ali · F. Atif
Department of Biotechnology, University of Sargodha, Sargodha, Pakistan

T. Anindyawati · A. C. Djohan · I. R. Layly
Research Center for Applied Microbiology, National Research and Innovation Agency (BRIN), Cibinong Bogor, Indonesia

M. Franco
Department of Exact Science, State University of Santa Cruz, Ilheus, Brazil

M. Irfan (✉)
Department of Biotechnology, University of Sargodha, Sargodha, Pakistan

Research Center for Applied Microbiology, National Research and Innovation Agency (BRIN), Cibinong Bogor, Indonesia
e-mail: irfan.ashraf@uos.edu.pk

M. A. Dar et al. (eds.), *Recent Trends in Lignocellulosic Biofuels and Bioenergy*, Clean Energy Production Technologies,
https://doi.org/10.1007/978-981-96-4636-4_9

9.1 Introduction

In the twentieth century, the research primarily focused on the development of refineries for natural gas petroleum as well as coal in order to meet global energy demand of world (Mukhtar and Aslam 2020). In recent decades, it's become clear that fossil fuel reserves are reducing, oil prices are rising, and supply is unstable. So this led to environmental concerns, which have pushed the need of new technologies in order to address our unsustainable energy consumption (Berger et al. 2014). The great challenge in this regard is to support a growing global population as well as decreasing oil reserves by finding renewable and sustainable energy sources (Mukhtar and Aslam 2020). All these understandings lead to the development of biofuels, which are renewable and environmentally friendly and considered as best alternative for reducing the effects caused by global warming (Berger et al. 2014). Hence this limits our reliance on fossil fuels as well as supports rural economies. Biofuels need to provide more energy than they consume and should be produced in large quantities without affecting the food supplies as a good alternative. They must provide benefit to the environment and should be economically competitive (Garg et al. 2023). However, most biofuels are currently expensive, and their production methods are also controversial (Sims et al. 2010). These renewable energy sources called biofuels are derived from biomass (organic material). It includes butanol, ethanol, and biogas which are used for cooking or heating. Hydrogen and biodiesel are used as substitutes or additives to regular fuels (Taylor et al. 2012). Biomass for biodiesel can be from a variety of sources like wheat, corn, sugarcane, and organic waste that can be derived from farms or cities, algae, and leftover plant material which is called lingo cellulosic waste (Taylor et al. 2012). Edible crops such as sugarcane in Brazil as well as corn in the United States have been used in the past to produce first-generation biofuels like bioethanol. Between 2005 and 2010, the production of bioethanol through these crops increased significantly high from approx 31 billion liters to over 85 billion liters in 2005 and 2010, respectively (Carriquiry et al. 2011). Hence concerns regarding the impact of these biofuel and sustainability have contributed to the production of second-generation biofuels which are mainly developed by biomass sources like agricultural, forestry residues, algae, as well as other nonedible parts of plants. So these biofuel productions are considered more promising as they offer higher energy efficiency and flexibility in contrast to first-generation biofuels (Havlík et al. 2011). They also minimize competition with food crops for land as well as with resources. Hence also offer greater flexibility in terms of feedstock (Havlík et al. 2011). Algae and genetically modified (GM) algae biomasses are called third generation and fourth generation, respectively (Garg et al. 2023). In the past, biofuels have been produced by using such organisms that can thrive in moderate conditions (mesophilic organisms) and their enzymes. In recent years, extremophiles, which can thrive in extreme environments, have also gained attention as promising alternative (Garg et al. 2023). Extremophiles are considered attractive for biofuel production as they are robust and adaptable to harsh conditions, and their versatility can be natural or enhanced through genetic engineering due to our increasing understanding of their metabolic pathways as well as availability of advanced genetic tools for manipulating their traits. An overview of different types of extremophiles and their unique features are discussed along with genetic engineering techniques like CRISPR. This chapter also covers diverse types of biofuels such as bioethanol, biogas, and biodiesel as well as how extremophilic microorganisms can be engineered for production of biofuel and bioenergy.

9.2 What Are Extremophiles?

Extremophiles are such microorganisms that grow in such environments which are considered too extreme for most life forms, including humans (Atif et al. 2024). They are found in places like hot springs in which temperature is near the boiling point (Mukhtar and Aslam 2020) and in the depth of sea where temperature is very low and under high water pressure (Atif et al. 2024). They can also survive under extreme conditions like extreme pH levels, high pressure, high salinity, radiation exposure, as well as severe dryness (Mukhtar and Aslam 2020). The term "extremophiles," which means "extreme lovers," (Mukhtar and Aslam 2020) was named by Mac Elroy (1974) (Atif et al. 2024). Acidophilic microorganisms flourish in highly acidic environments with pH levels between 0 and 5 (Mukhtar and Aslam 2020), while alkaliphiles are able to survive in alkaline environments having pH levels from 8.5 to 11.5 (Mukhtar and Aslam 2020). Each type of extremophiles has evolved specialized traits true to which they can thrive in intense habitats (Berger et al. 2014). Xerophiles have evolved to survive in such environments with very low water activity and radio-tolerant resistance to high levels of radiation (Mukhtar and Aslam 2020). Temperature is considered as the main factor influencing the major biological molecules and life processes (Mukhtar and Aslam 2020). Organisms are categorized based on their optimal temperature ranges such as psychrophiles that thrive below 15 °C (Mukhtar and Aslam 2020), mesophiles between 15 and 50 °C, and thermophiles above 50 °C (Berger et al. 2014). Thermophiles are further categorized into several types such as moderate thermophiles (50–70 °C), extreme thermophiles (70–85 °C), and hyperthermophiles (above 85 °C) (Atif et al. 2024). Extremophiles use various methods to protect their breakdown of biomolecules in harsh conditions as they store compatible solutes and salts in the cytoplasm of their cells while maintaining inherent stability of proteins and can rapidly replace damaged DNA, RNA, and protein (Berger et al. 2014). They have efficient DNA repair systems and elevated levels of reverse DNA gyrases (Garg et al. 2023). These microorganisms also stabilize their cell membranes due to special lipids like diether-linked lipids to endure extreme pH and temperatures (Mukhtar and Aslam 2020). These adaptations help extremophiles survive in extreme environments by maintaining cellular function in extreme environments on Earth (Berger et al. 2014). In the biopharmaceutical industry, enzymes have traditionally been engineered for better functionality and largely been succeeded (Demirjian et al. 2001). Traditional studies on extremophiles relied on culture-dependent methods and screening programs which cover almost less than 1% of viable organisms and require specialized costly (Cowan et al. 2004). Modern metagenomic technologies have overcome these limitations and revolutionized this field hence transforming the field by enabling the better expression of native and recombinant extremophilic enzyme more effectively (Burton et al. 2002). Early advancements such as improved fermentation processes, bioprocess design, bioreactor technology, and enzyme overproduction have improved biomass yields in extremophiles like *Pyrococcus furiosus* and *Sulfolobus* species (Garg et al. 2023), hence allowing the successful expression of genes that are previously inaccessible such as the thermostable maltodextrin phosphorylase from *Pyrococcus furiosus* for development of α-D-glucose-1-phosphate (Nahálka 2008). Modern Metagenomic techniques resulted the discovery of new biocatalysts by identification of an extended hydrocarbon chain degrading pathway in *Geobacillus thermodenitrificans* NG80-2 that can be beneficial for oil recovery as well as for bioremediation (Taylor et al. 2012).

9.2.1 Extremophilic Microbes Used in Biofuel Production

A thermophilic microorganism has a promising role in the production of biofuel but encounters challenges in commercial applications as well. Their unique property to withstand high temperatures and adaptability allows the effectively conversion of lignocellulosic biomass into bioethanol which lowers production costs as well. Furthermore, by modifying their genomes, we can enhance these traits, but it is considered a complex task and not yet widely implemented for commercial biofuel production (Garg et al. 2023). Different genetic tools like CRISPR/Cas9 can enhance thermotolerance in this microbe and can have our desired properties. There are certain extremophilic microbes that play an important role in biofuel production as precisely discussed in Table 9.1 (Garg et al. 2023).

9.2.1.1 Caldicellulosiruptor Bescii

Extremophilic microorganism *C. bescii* can thrive at temperature of about 78 °C and is able to consume variety of cellulosic materials and faces significant challenges while doing genome editing especially due to the activity of restriction enzymes. Cellulases, particularly hemicellulases, play a vital role of breaking down of cellulosic biomass of plants as present within microbe. Addition of exogenous cellulase enzymes does not improve the degradation of lignocellulose activity. Furthermore, overexpression of tapirin genes can boost *C. bescii's* capacity to breakdown the lignocelluloses more efficiently (Brunecky et al. 2020).

Table 9.1 Extremophilic microbes in biofuel production

Extremophiles	Growth conditions	Characteristics	Applications
Caldicellulosiruptor bescii	Survives at 78 °C	Utilizes variety of cellulosic biomass for its growth as well as for the production of various metabolites (Brunecky et al. 2020)	Having potential for increasing lignocellulose breakdown through tapirin genes
Thermococcus kodakarensis	Optimal growth at 85 °C.	Can produce hydrogen from starch and polysaccharides (Crosby et al. 2019)	Application in isoprenoid phytoene synthesis for industrial biofuels (Ramu et al. 2021)
Pyrococcus furiosus	Thrive optimally at temperature of about 100 °C	Specially engineered for production of 3-hydroxypropionate, ethanol, as well as n-butanol	Production of variety of biofuels like ethanol and n-butanol (Wang et al. 2021)
Thermotoga maritima	Anaerobic bacterium shows optimal growth at 80 °C	Can produce hydrogen with high efficiency (Singh et al. 2018)	Utilize both simple and complex sugars for biofuel production, genetic modification, can enhance hydrogen yield
Thermus thermophilus	Aerobic bacterium thrive at 70–80 °C (Wang et al. 2022)	Utilizes denitrification process for anaerobic growth	Engineering can enhance superoxide dismutase capacity and show potential in biotechnological applications
Sulfolobus sp.	Belongs to acidophilic archaea, can survive in sulfur-rich environments (Quehenberger et al. 2017)	Proficient in utilization of cellulose and overproduction of endoglucanase for efficient cellulose degradation	Limited industrial application due to lower yields and lengthy production times

Modified from Garg et al. (2023)

9.2.1.2 Thermococcus Kodakarensis

Thermococcus kodakarensis survives best at 85 °C and can produce hydrogen by using starch, polysaccharides, as well as pyruvate (Crosby et al. 2019). Genetically engineered strains can produce particularly more hydrogen as compared to non-engineered microbial strain which can be achieved via enhanced expression of chitinase genes and disruption of a glycolytic repressor (Crosby et al. 2019) present inside specific genomic regions of microbes. These microbes provide variety of industrially applications by producing hydrogen from renewable carbon sources and in biofuels through isoprenoid phytoene synthesis (Ramu et al. 2021). Ongoing efforts have main focus on further increasing hydrogen production by using CRISPR technology to protect various strains of microbes from viral contamination (Crosby et al. 2019).

9.2.1.3 Pyrococcus Furiosus

Pyrococcus furiosus survive best at optimum temperature 100 °C. Due to this property, *Pyrococcus furiosus* has been engineered to produce n-butanol, acetoin, 3-hydroxypropionate, as well as ethanol by using genetic modification techniques. But this microbe encounters challenges particularly due to its extreme heat as its enzymes can work well at lower temperatures around 70 °C (Wang et al. 2021). There are other hyper thermophilic bacteria named as *Thermotoga maritima* can grow optimally at 80 °C as described in Table 9.1. Due to this property, this microbe is also responsible for production of hydrogen through different genetic modifications that will specifically enhance their metabolic pathways while reducing growth and sugar consumption (Wang et al. 2021).

9.2.1.4 Thermotoga Maritima

Thermotoga maritima survives well in hot, oxygen-free environments about 80 °C and capable to produce hydrogen from sugars naturally (Singh et al. 2018). Researchers tried to increase its efficacy to produce hydrogen through different genetic engineering techniques by genetically modifying its genes like lactate dehydrogenase (Garg et al. 2023); as a result it will slow down its growth and will reduce sugar consumption. Furthermore, disruption of the ABC maltose transporter gene through genetic engineering can also pose growth challenges. Researchers also tried to engineer *Thermotoga* to produce cellulase enzymes for degradation of particularly cellulose (Singh et al. 2018).

9.2.1.5 Thermotoga Thermophilus

Thermus thermophilus is an aerobic bacteria which was identified in Japanese hot springs and able to survive best between 70 and 80 °C (Garg et al. 2023). This microbe has the ability to grow without oxygen by denitrification process and able to degrade sugars, lipids, peptides, as well as triglycerides (Wang et al. 2021). In particular environments with minimal glucose concentration, this microbe shows slower growth than *Escherichia coli* (Garg et al. 2023). Scientists enhanced its metabolism by altering specific genes like TTP0042 crucial for its genetic system in order to increase its resistance to variety of harmful molecules (Wang et al. 2021).

9.2.1.6 *Sulfolobus* Bacteria

Sulfolobus bacteria can grow well in acidic conditions while assimilating carbon for its growth, and some of these strains of microbes can even survive in sulfur-rich environments by producing sulfuric acid (Garg et al. 2023). Hence their industrial use is limited due to reduced yields and their slow production. Different fungi like *Pleurotus florida* are also used for genome editing in order to enhance biofuel efficiency by degrading lignin, in plant material (Quehenberger et al. 2017). *Spirogyra* algae can also be used as a promising biofuel feedstock with its genetic engineering that can further boost its productivity by optimizing sugar metabolism and fermentation pathways. Other fungi like *Aspergillus terreus* and *Trichoderma reesei* are also used for biofuel production as they have the capacity to degrade cellulose and detoxify plant material improving biofuel quality as well as its efficiency (Quehenberger et al. 2017).

9.3 Biofuel

First-generation biofuels are derived from easily accessible common crops which are rich in sugar, starch, and oil by using standard traditional methods (Garg et al. 2023). These biofuels are bioethanol, biodiesel, biogas, and biobutanol. Second-generation biofuels are derived from complex sources that are more difficult to decompose such as lignocellulosic materials. Nowadays most highly produced biofuels are biodiesel and bioethanol on an industrial scale that currently dominate the industrial biofuel market making up more than 90% of the market (Garg et al. 2023). Their market success is mainly based on certain chemical and physical properties including emerging technologies and their societal and environmental implications (Taylor et al. 2012). Different types of biofuels and their precise description have been explained in schematic diagram as shown in Fig. 9.1.

9.3.1 Bioethanol

Bioethanol is either used on its own or blended with gasoline as a replacement for petroleum mainly in transportation. As a result of programs in biotechnology, the methods for the biofuel production are particularly improved (Garg et al. 2023). The yeast *Saccharomyces cerevisiae* mainly ferments sugar cane molasses and starch from crops like wheat, corn, and barley in order to produce ethanol (Demain 2009). Primarily, this process involved separate hydrolysis and fermentation steps but now in order to increase yields up to 10% that can be performed in one step (Taylor et al. 2012). But the limitation is that *Saccharomyces cerevisiae* cannot ferment many sugars present in plant biomass which highly decrease its efficiency (Garg et al. 2023). So, researchers are paying attention genetic engineering and discovering

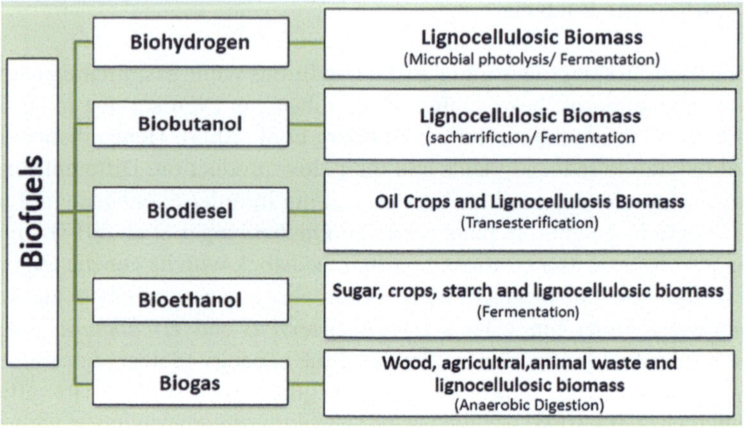

Fig. 9.1 Types of biofuel (Taylor et al. 2012)

different bacteria that are able to ferment such sugars, but such bacteria produce less ethanol as well as having poor ethanol tolerance (French 2009). Discovery of the most effective microorganism for high bioethanol production from a variety of biomass is considered a significant challenge. In this regard, discovery of extremophilic microbes and their unique characteristics plays vital role in development of biofuels (French 2009).

9.3.2 Biodiesel

Biodiesel is considered as safe, sulfur-free, biodegradable, and environmentally friendly renewable energy source that closely resembles to that of petroleum diesel. Biodiesel is highly suitable to use either in its own form referred to called as pure form or as a mixed form in unmodified diesel engines (Singh and Singh 2010). Biodiesel have certain properties like high flash point and low aromatic content that helps engines to last longer and decrease the cost of maintenance. The main characteristics of biodiesel is that it leads to the production of less harmful emissions as compared to that of conventional diesel like carbon monoxide and unburned hydrocarbons (Taylor et al. 2012). Biodiesel is produced through a process called as transesterification where chemical reactions occurs between vegetable oils (such as soybean, palm, or sunflower) with alcohols (like methanol or ethanol) using catalysts like sodium or potassium hydroxide (Al-Zuhair 2007). The main challenge is removing the catalyst and glycerol from the final product as well as treating alkaline wastewater. Moreover, the advancements in biotechnology lead the researchers to explore alternatives like using enzymes (like lipases) to catalyze their action, and this method is considered expensive (Al-Zuhair 2007).

9.3.3 Biobutanol

Biobutanol was initially developed since 1960s through industrial fermentation primarily used as a solvent. Butanol is a clear liquid with a strong odor and offers 25% more energy per volume of bioethanol. Biobutanol can be mixed either with gasoline or diesel without any need for vehicle modifications. Biobutanol production is considered expensive due to fermentation inhibition mainly when the concentrations of butanol reach 1–2% (Antoni et al. 2007). The biobutanol production is considered a two-step process using bacteria like *Clostridium* spp. called as ABE fermentation (Zaldivar et al. 2001). This process yields acetone, butanol, and ethanol (3:6:1) through different raw materials like corn or sugar beet. The key objective of recent study is to enhance fermentation efficiency and butanol yield by genetically modifying *Clostridium* species (Taylor et al. 2012).

9.3.4 Biogas

Biogas is generated from anaerobic breakdown of manure, organic household waste, garden wastes, and energy crops like poplar, maize, etc. This is called as organic waste. The main use of biogas is for electricity, heat, and engines (Demain 2009). The production of biogas is mainly a three-step microbial process; hydrolysis (degradation of polysaccharides into fermentable sugars); fermentation (producing acids and gases like carbon dioxide), which is considered as the slowest step due to the long generation times of the bacteria; and methanogenesis (production methane) (Demain 2009), which is mostly performed by archaea that can methane up to 70% and additionally also produce carbon dioxide, ammonia, and hydrogen sulfide (Taylor et al. 2012).

9.3.5 Biohydrogen

Biohydrogen is considered the most promising biofuel as it can be converted to electricity in fuel cells. Use of microbial methods in order to produce hydrogen is considered cost-effective as well as eco-friendly in contrast to current methods of productions, but not yet commercially in practice. Scientists are discovering different methods like using algae and cyanobacteria for biophotolysis, photofermentation organic substance by bacteria, as well as dark fermentation (Demain 2009) of organic substances by using anaerobic organisms (anaerobic fermentation) (Demain 2009). Various thermophilic microbes such as *Caldicellulosiruptor saccharolyticus* and *Thermotoga elfii* have shown particularly promising results in these processes (Taylor et al. 2012).

9.4 Lignocellulosic Feedstocks: Possible Sustainable Sources for the Production of Biohydrogen

The use of biofuels and ideas for their further use have sparked interest in the so-called "biom boom" sector. Historically, biomass—which comes from plants, animals, and microbes—has been regarded as one of the important preferable and promising energy sources for the creation of biofuels. Although these feedstocks based on carbon have the potential to replace the existing fossil fuel-based power lines, petroleum-based goods that are necessary for various applications across the sector will still need to be produced (Kucharska et al. 2018). In order to create items with more value, carbon-balanced biopolymers and green biofuels, lignocellulosic biomass has proven to be the most appropriate feedstock (Zhu et al. 2020). It is one of the world's biggest sources of plant biomass, renewable, and a woody feedstock. It belongs to the category of non-food crops or non-erroneous agricultural crops because it is derived from materials that are not usable as food crops (Table 9.2). More than 95% of the Lignocellulosic feed stock produced is generated from agricultural and forestry residues, while other potential sources include several other industrial and municipal wastes (Crosby et al. 2019). While LCFs of various kinds possess some characteristics that are similar to those of muscles, they are not identical in terms of their appearance and strength; LCFs are made up of three primary items which constitute 90% of their total weight. These consist of lignin (10–25 wt %), homopolymeric cellulose (36–56 wt%), and heteropolymeric hemicellulose (25–45 wt%) (Fig. 9.2). The development of linkages between the lignocellulosic polymers forms a mesh-like structure, and these bonds play an important role in determining the physical properties or configuration of the different LCFs (Chozhavendhan et al. 2020). Its principal element cellulose is a crystalline polymer-formulated D-glucose units joined by hydrogen bonds, van der Waals forces, and β-1,4 glycoside bonds. One of the most prevalent classes of organic polymers that are known to occur on Earth is this one, and together they provide the basic structure-based component of plant cell walls. Because of the microfibril structures that the long unbranched polymeric chains in cellulose's composition generated as a result, the resulting structure can be either crystalline or amorphous because the molecules pack closely together with intra- and intermolecular hydrogen bonding. Conversely,

Table 9.2 Components of few typical lignocellulosic feedstocks for the synthesis of biohydrogen

Lignocellulosic feedstocks	Lignin % of weight	Cellulose % of weight	Hemicellulose % of weight	Ash % of weight	Extractives% of weight
Cobs of corn	16	44	34	–	–
Stove for corn	19	42	23	–	–
Straw rice	17	33.1	25	–	–
Bagasse from sugarcane	21.8	41.4	36.5	2.6	–
Switchgrass	6–21	32–52	11–41	6.5	–
Wheatfield	17.4	38.1	25.1	2.3	18.2

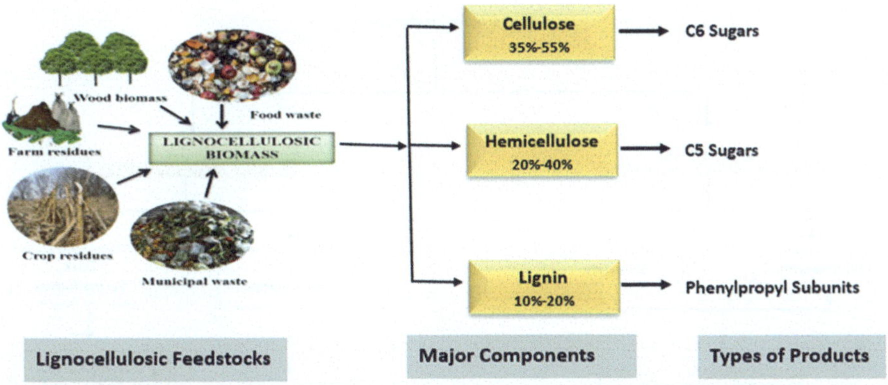

Fig. 9.2 Lignocellulosic composition and products generated upon hydrolysis

hemicellulose is a heteropolymer that shares similarities with cellulose but possesses a nonlinear structure consisting of a complex mixture of pentoses, Organic Acid sugar, and heterodisperse hexoses linked by much weaker α- and β-glycosidic links. This complex carbohydrate, found in both hardwood and softwood types, is essential for cross-linking with lignin and cellulose microfibril. The last component that makes up the framework of LCF is lignin, an insoluble heteropolymer with a large fraction of phenylpropanoid connections formed in various capacities by aryl-ether (C-O-C) and carbon-carbon (C-C) links. Tri-monolignol lignin is made up of sinapyl, coniferyl, and p-coumaryl alcohols. These alcohols serve as precursors for the lignin units that are synthesized to form the phenylpropane subgroups of guaiacyl, syringyl, and p-hydroxyphenyl units.

Due to its branched, polymeric, and cross-linked characteristics, lignin confers mechanical strength as well as hydrophobicity, in turn making it difficult to remove the cellulose/hemicellulose required for fermentable sugar production (Okolie et al. 2021; Boodhun et al. 2017). Lately, there is a presupposition to utilize LCFs in the synthesis of biohydrogen because of their high carbohydrate content in the form of hemicellulose and cellulose. According to some studies (Okolie et al. 2021; Boodhun et al. 2017), the most commonly used LCFs are corncob, maize stalk, corn stover, switchgrass, straw from rice, sugarcane bagasse, and wheat straw. Even while lignocellulosic materials can be found in LCFs, their utilization is restricted by complex structures, which in turn makes it more difficult to produce BioH_2. Because of the enzymatic action, control over the process, removal of the lignin polymer, and improved cellulose and hemicellulose digestibility during fermentation, a few physical, chemical, and biological pretreatment methods have been developed. Comparatively, chemical pretreatment methods include the use of hazardous chemicals and production of environmentally dangerous toxic reagents, whereas physical pretreatment methods used in businesses require significant energy consumption and high costs. However, the biological pretreatment process makes use of microorganisms and lignin's enzymatic hydrolysis, which is easier and uses less energy. It has also been demonstrated to have greater yields of the final

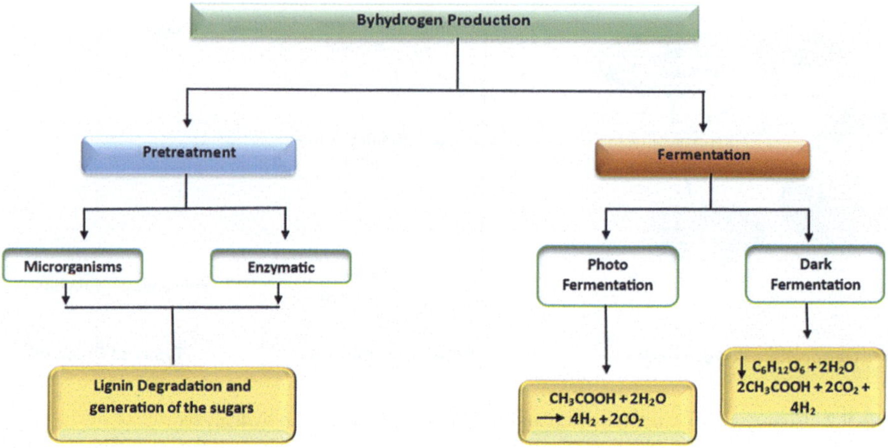

Fig. 9.3 Bioconversion process and the microbes used to produce biohydrogen

product and global applicability when compared to other cost-effective methods (Sharma et al. 2019). Moreover, to producing a number of valuable chemicals, the naturally occurring microbes that break down lignin polymer for extractions also help to synthesize BioH_2 following a fermentation process.

9.4.1 Lignocellulosic Biohydrogen Production Process

Low energy intake and ecofriendly characteristics that are associated with the properties of the biological process have ensured that the BioH2 production process is economically fruitful. Originally the microorganism and their own intrinsic biochemical complex have been established to be of significant significance in the process of the LCF bioconversion to BioH_2. Aspects of biology used in the creation of BioH2 are the biological pretreatment, Photo Fermentation, or Dark Fermentation processes (Fig. 9.3). Given Table highlights the summary of the main variations between the three biological mechanisms that produce BioH_2 from lignocellulosic biomass.

9.4.1.1 Biologic Pretreatment

Biological pretreatment involves the application of either ligninolytic bacteria or enzymes that have the ability to degrade LCFs. This first stage of the creation or manufacture of BioH_2 serves as an introduction to the fermentation and hydrolysis processes that come next. As of now, there are four distinct approaches used in biological pretreatment of biomass: (i) pretreatment using microorganisms such as bacteria and fungi, (ii) microbial consortia, (iii) ensiling, and (iv) enzyme pretreatment

(Zabed et al. 2018). The employed microorganisms depolymerize lignin by the secretion of extracellular ligninolytic enzymes, which are responsible for catalyzing several biochemical activities. Following this, the microbial inoculum proliferates to produce biomass, which ultimately causes the lignin to break down. However, not all microbe kinds will incubate at the same rate or for the same amount of time; in fact, it can take weeks or months (Sharma et al. 2019). Enzymes from fungus and bacteria are isolated and used directly on the pretreatment process in order to eradicate it. But in addition to purification, enzymatic synthesis and optimization could be expensive for the preparation, raising the process's overall expenses. The enhanced hydrolytic efficiency and high substrate adaptability of microbial consortiums make them advantageous over other forms of treatment, as they drastically shorten the pretreatment period. There is a synergistic activity to promote lignin degradation when enzymes from different species of bacteria and fungi are utilized sequentially or jointly (Yadav et al. 2020; Sağır and Hallenbeck 2019). Another cutting-edge method for biological pretreatment is ensilaging, which is created by using lactic acid bacteria in the process of anaerobic fermentation. This process breaks down macromolecules by producing various organic acids and prevents microbial acceptance. Nevertheless, throughout the ensilaging process, variables including dry matter, microbes, and biomass composition affect the silage yield content (Zabed et al. 2018). Of all the pretreatment techniques, biological treatment is one of the most systematic and cost-effective; yet, more study is required to shorten the process's total duration.

9.4.1.2 Photofermentation

Through the use of anaerobic photosynthetic bacteria, which are mainly made up of purple non-sulfur bacteria, PF is successful in the existence of sunlight (Yadav et al. 2020). Through the nitrogenase enzyme, which creates hydrogen and carbon disulfide from volatile organic acids, these bacteria take part in the PF process of ancient organic materials. Equation (Chozhavendhan et al. 2020) illustrates how the photosynthetic bacteria, which only possess photosystem I, use exogenous organic acids, such as lactate, acetate, butyrate, and malate, as acceptable electron donors to create hydrogen gas: $CH_3COOH + 2H2O \rightarrow 4H_2 + 2CO_2$. When protons and electrons are combined with ATP from the electron transport chain, the nitrogenase enzyme converts them to ammonia, hydrogen, adinosine diphosphate (ADP), and pi. The purple non-sulfur bacteria can now produce BioH2 from a diverse of waste biomass sources due to the organic substrates' employability (Sarangi and Nanda 2020). Under light conditions, these bacteria can use the nitrogenase enzyme to convert almost all of the electrons in the biomass to hydrogen and carbon dioxide. Moreover, the implementation of the PF process was made possible by advantages including high substrate conversion efficiency, adaptable metabolic capacity for light intensity, and equivalent $BioH_2$/FS generation (Zhang and Zhang 2018).

9.4.1.3 Dark Fermentation

DF is carried out indoors or during the night and employs obligate or facultative anaerobes that use organic substrates and waste biomass to generate hydrogen (Chozhavendhan et al. 2020). Hydrolysed biomass which accumulates reducing sugars as hydrolysate products and uses them as reactants for the anaerobic synthesis of hydrogen and other useful byproducts such as carbon dioxide, organic acids, and alcohol (Radhakrishnan et al. 2020). In terms of the electron oxidation process, it is known that the hydrogenase enzyme system is engaged in both the reduction of protons and the oxidation of electrons to synthesize hydrogen (Liu et al. 2020). Pyruvate is an important fermentation intermediary for the production of hydrogen and is effectively transformed enzymatically into acetyl CoA. Along with the hydrogen gas, other significant metabolites produced during fermentation include acetic and butyric acids (Bhatia et al. 2021) and anaerobic organisms $C_6H_{12}O_6 + 2H_2O$ $2CH_3COOH + 2CO_2 + 4H_2$. In contrast with PF, DF is significantly favored due to the following reasons: first, it does not cost a substantial amount of money; second, there is no need to maintain the light source; and third, the bioreactor to be used is relatively small in size (Sarangi and Nanda 2020). However, there has been an issue with substrate organic load where this is not completely transformed into the reagents during the reaction process hence resulting in lower H2 yields (Radhakrishnan et al. 2020).

9.4.2 Biogas Production

In order to produce biogas from lignocellulosic biomass, complex plant materials must be broken down anaerobically, releasing methane-rich gas through the breakdown of cellulose, hemicellulose, and lignin.

9.4.3 Bioreactions

AD is an intricate process that can be described by the sequential multistep process of degrading the organic matter in a given feedstock into methane under oxygen limitation. The four phases of the digestion process are the metabolic reactions finished by different taking part microbial populations. These include hydrolysis/liquefaction, acetogenesis, acidogenesis, and methanogenesis (Fig. 9.4).

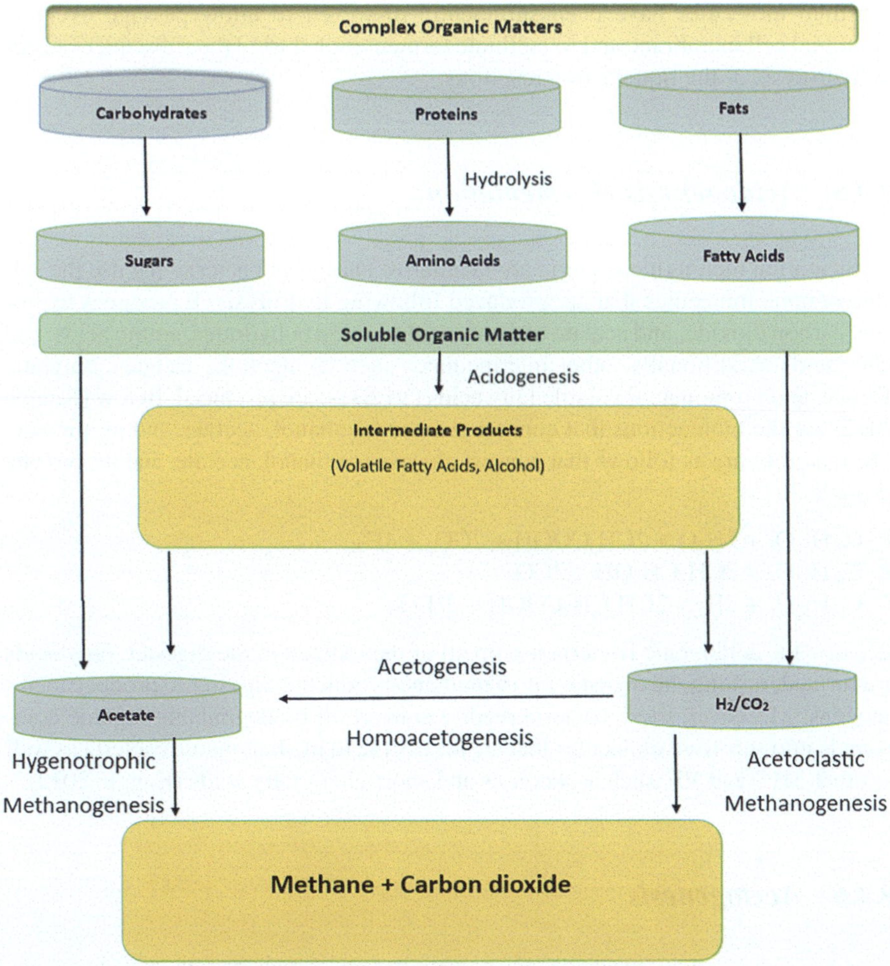

Fig. 9.4 Schematic diagram of anaerobic digestion

9.4.4 Hydrolysis

This step involves the hydrolysis of the organic biomass, which contains complex polymers, by facultative hydrolytic anaerobes that produce hydrolytic enzymes such cellulase, amylase, lipase, and protease. The hydrolysis process is referred to as the rate-limiting phase when the substrate particles are enlarging in size and hence provide an insufficient surface to volume ratio. Acetogenesis and methanogenesis themselves restrict the rate of reaction if the substrate particles are relatively degradable (Björnsson et al. 2001; Vavilin et al. 1996). Bacteriocides, *Clostridium*, *Enterobacteriaceae*, *Bifidobacteria*, and *Streptococci* are among the lactic and acetogenic bacteria that are active in this phase. By the time this phase concludes, the substrate molecules have been sufficiently dissolved to allow passage over the microbial cell membrane and to facilitate fermentation during the subsequent stages of hydrolysis or the liquefaction reactions.

9.4.5 *Acidogenesis (Fermentation)*

Fermentation meningitic bacteria are facultative anaerobic bacteria that use the soluble organic molecules that are produced following hydrolysis. It produces hydrogen, carbon dioxide, and acetate or acetic acid from carbohydrates, amino acids, and fatty acids. Additionally, other intermediates such as alcohols, lactates, butyrate, acetate, and propionate—volatile fatty acids (VFAs)—are produced. Below (Kangle 2012) are the bioreactions that convert glucose to ethanol, acetate, and propionate. The reactions are as follows that convert glucose to ethanol, acetate, and propionate (Kangle 2012):

1. $C_6\,H_{12}O_6 + 2H_2O > 2CH_3COOH + 2CO_2 + 4H_2$
2. $C_6\,H_{12}O_6 > 2CH_3CH_2OH + 2CO_2$
3. $C_6\,H_{12}O_6 + 2H_2 > 2CH_3CH_2COOH + 2H_2O$

Because the acidogenic bacteria use up all of the oxygen in the digester, only acids are formed, turning the digester into a methane-producing digester to produce methanogens. Moreover, these bacteria readily convert all biodegradable organic compounds into substrate molecules that are accessible to methanogenic bacteria as well as other SHC and VF, such as alcohols and short-chain fatty acids (Kangle 2012).

9.4.6 *Acetogenesis*

Acid formers are the collective term for all microorganisms involved in this stage. While homoacetogenic bacteria may create acetate from H_2 and CO_2, organotrophic bacteria, such as acetogenic bacteria, can synthesize acetyl-CoA from VFAs and alcohols and produce acetate, H_2, and CO_2 (Sterling et al. 2001). The bacteria implicated in these activities belong to the *Clostridium*, *Syntrophomonas*, *Syntrophobacter*, *Peptococcus*, *Actinomyces*, and *Lactobacillus* genera and are coupled to protons through reductive needs. Low H_2 partial pressures are associated with acetogenic processes, as proposed by Schink (1997), wherein syntrophic bacteria exhibit thermally beneficial activities. Butyric acid, ethanol, acetic acid, and propionic acid are the end products produced in this process.

9.4.7 *Methanogenesis*

Here, methane precursors maintain a low H_2 partial pressure while converting the byproducts of acetogenesis into biogas and carbon dioxide. Specialized methanogenic bacteria are obligatory facultative anaerobes that react to changes in the environment of the digester. These could be species that use acetate, like H_2/CO_2 group members or *Methanosarcina barkeri*, *Methanococcus mazei*, and *Methanothrix soehngenii*. According to the type of material that the methanogenic bacteria ingest, methanogenesis can be classified as follows (Kangle 2012). Therefore, methanogenesis can be classified as follows, depending on the kind of substrate that the methanogenic bacteria consume (Kangle 2012):

- Hydrogenotrophic methanogenesis:
- Methane is synthesized by using CO_2 and H_2 as mentioned below:

$$CO_2 + 4H > CH_4 + 2H_2O$$

- Acetotrophic process:
- Methane and CO_2 are synthesized by utilization of acetate as mentioned below:

$$CH_3COOH > CH_4 + CO_2$$

On an average 30% of the total methane produced in a stable biomethane production process comes from using H_2/CO_2, while 70% comes from the conversion of acetate. This is the slowest step in the process, and it is dependent upon various elements, including the type of feedstock, temperature, unloading rate, and pH level.

9.5 Biobutanol Production

Biobutanol was initially discovered in Clostridium species and is notable for its high density of 29.2 MJ/L, as compared to ethanol 19.6 MJ/L. This makes biobutanol a viable alternative fuel that can be mixed with gasoline and diesel without needing vehicle modifications. It has the ability to vaporize easily and does not increase the front-end volatility of gasoline, and it can be potentially upgraded to production of jet fuel, a feature that is not commonly related with biofuels (Knoshaug and Zhang 2009). The production process for biobutanol utilizes the same raw materials as bioethanol, and it can be distributed using the present facilities. However, a major challenge in the biobutanol industry is that most microorganisms have low tolerance to butanol. For example, ethanol-producing organisms like *Saccharomyces cerevisiae* can withstand ethanol concentrations up to 20% v/v, whereas butanol-producing organisms typically can only tolerate 2–3% v/v. This results in higher costs for downstream separation and the need for huge equipment production. The use of extremophiles and their genes in biobutanol production helps overcome this challenge. Unlike traditional extremophiles, certain organisms from the genera (Bibra et al. 2014) *Bacillus* and *Pseudomonas* can tolerate butanol concentrations in the range of 6–7%. This ability to withstand high solvent concentrations is an "extreme" behavior and could significantly enhance the efficiency of biobutanol production.

9.5.1 Role of Algae in the Production of Biofuel

Production of biodiesel from oil produced by algae is more stable than that from traditional crops used for oil production. Algae are capable of producing more oil and do not require fertile land for cultivation. Algae can be mass-produced through two main methods: open-air pond systems or closed systems like photobioreactors. To prevent any sort of contamination in open-air systems, like raceway ponds, extremophilic strains of algae are preferred as they can thrive in extreme conditions like high temperatures, low or high pH levels, and high salinity. Species like *Spirulina*, *Dunaliella*, and *Chlorella* have demonstrated the ability to grow in alkaline environment, halophilic environment, and environment containing nutrients, respectively, making them suitable for open-air systems. In contrast, closed photobioreactor systems are not constrained by contamination issues, allowing the use of any algal strain and achieving higher cell densities. Lipids and oil after extraction used for transesterification while the remaining algal biomass can be utilized as food source for animals, as an organic fertilizer or for human usage. Additionally, value-added products such as unsaturated fatty acids, polysaccharides, antioxidants, colors, food-coloring products, toxins, and stable isotopes can be simultaneously produced with the oil (Dragone et al. 2010).

Another source of extremophilic oil is the yeast able to live in cold environment *Rhodotorula glacialis* DBVPG 4785, which was isolated from an alpine glacial environment. This yeast is able to grow at zero and sub-zero temperatures, with optimal growth at 20 °C. Under nutrient limitation with excess carbon, the yeast produces a significant amount of lipids, up to 68% of its biomass. The fatty acids produced range from C_{14} to C_{18}, with traces of C_{12}, and include saturated, monounsaturated, and polyunsaturated forms. High glucose concentrations increase the making of monounsaturated fatty acids that form as a result of polyunsaturated ones. The lipid profile of this yeast is similar to that of soybean oil and rapeseed oil, indicating its high power as a source of biodiesel (Bibra et al. 2014).

9.5.2 Bioethanol

Bioethanol, primarily impacting the transportation sector, can be used in blends or as a complete substitute for fuels based on petroleum. Significant advancements in biotechnology have established reliable bioethanol production methods. The fermentation process, mainly utilizing *Saccharomyces cerevisiae*, typically involves molasses produced from sugarcane or the hydrolysis of starch by the use of various enzymes from sources like wheat, corn, cassava, rye, barley, and triticale. Traditionally, this production method follows a process that takes place in sequence where breakdown of cellulose and fermentation occur separately. Alternatively, a simultaneous use of saccharification and fermentation can be employed, where cellulases and fermentative microorganisms are added together. Optimizing *S. cerevisiae* and the fermentation process has increased ethanol produced by up to 10% v/v in huge amount by the production facilities derived from starch. However, despite these advancements, *S. cerevisiae*'s inability to ferment many pentose sugars, which are significant components of enzyme hydrolysates from lignocellulosic biomass, limits the conversion efficiency (Barnard et al. 2010). However, genetic engineering is currently addressing this issue, and alternative solutions are also being explored. Various microbes can utilize five carbon sugar as a source of carbon during the process of fermentation. These include *Mucor indicus*, *Chalara parvispora*, *Pachysolen tannophilus*, *Zymomonas mobilis*, *Streptococcus fragilis*, *Kluyveromyces fragilis*, *Clostridium thermocellum*, *Clostridium ljungdahlii*, *Escherichia coli*, *Pichia stipitis*, *Klebsiella oxytoca*, *Moorella* species, and *Carboxydocella* species. However, these bacteria generally produce lower ethanol yields compared to yeast fermentations due to competing reactions that generate other end products, and they typically exhibit very little ability to tolerate ethanol. Consequently, developing an effective fermentative microorganism remains a significant challenge in the making of bioethanol by utilizing biomass on commercial level (Wisselink et al. 2009).

9.6 Technological Interventions for Bioengineering

9.6.1 Genome Editing Technique CRISPR

Molecular biology techniques such as CRISPR have been essential in increasing the efficiency of extremophilic microbes for biofuel production. CRISPR-Cas technology helps in modifying genes of extremophiles to optimize their ability for biofuel production. Production of biofuel for commercial purpose from lignocellulosic waste is restricted due to different challenges. Genome editing via CRISPR-Cas technology is the promising approach to optimize the potential of extremophiles in biofuel production. Genes related with enzymatic activity and thermotolerance have been altered to advance the competence of intracellular enzymes, such as cellulase and hemicellulase in extremophilic fungi, bacteria, and microalgae. Extremophilic microbes such as *Thermococcus kodakarensis*, *Pyrococcus furiosus*, *Thermotoga maritima*, and *Sulfolobus* sp. are specifically used for production of biofuel (Garg et al. 2023). CRISPR/Cas9 system has evolved as strong tool for site-directed mutagenesis against phage attack in prokaryotic defense system. The predominant CRISPR system involves single guide RNA (sgRNA) and as well as a SpCas9 nuclease (also known as Cas9) (Garneau et al. 2010). Commonly, sgRNA binds with Cas9 and directs the complex to cut the target site (Jinek et al. 2012). Chimeric noncoding sgRNA comprises of CRISPR RNA array (crRNA) and trans-activating crRNA (tracrRNA). crRNA is a specific targeting sequence (Brouns et al. 2008), while tracrRNA guides recruitment of the Cas9 nuclease. Just like SpCas9, the 20 base pair target site sequence of sgRNA also requires specific protospacer adjacent motif (PAM) sequence usually 3 bp in length sequence on 3′ end (Sternberg et al. 2014). These nucleases identify diverse PAM sequences and operate through several mode of action (Sapranauskas et al. 2011).

Cas9 works as RNA-guided DNA endonuclease, cleaves the desire DNA, and creates DNA double-strand break by generating nick in each strand of DNA via RuvC1 and HNH nuclease domains (Jinek et al. 2012; Sapranauskas et al. 2011). These breaks can be repaired using various mechanisms (Fig. 9.5). Non-homologousend joining (NHEJ) is a DNA repair pathway that directly rejoins broken ends through insertion/deletion, while the homology-directed repair (HDR) approach requires additional components such as homology-directed repair (HDR) enzyme, DNA template, and sgRNA. tracrRNA and crRNA combine to make a structure similar to hairpin loop, which serve as guide for Cas9 enzyme that cleave DNA by identifying the crRNA (Schiermeyer et al. 2019). A plasmid or virus-based

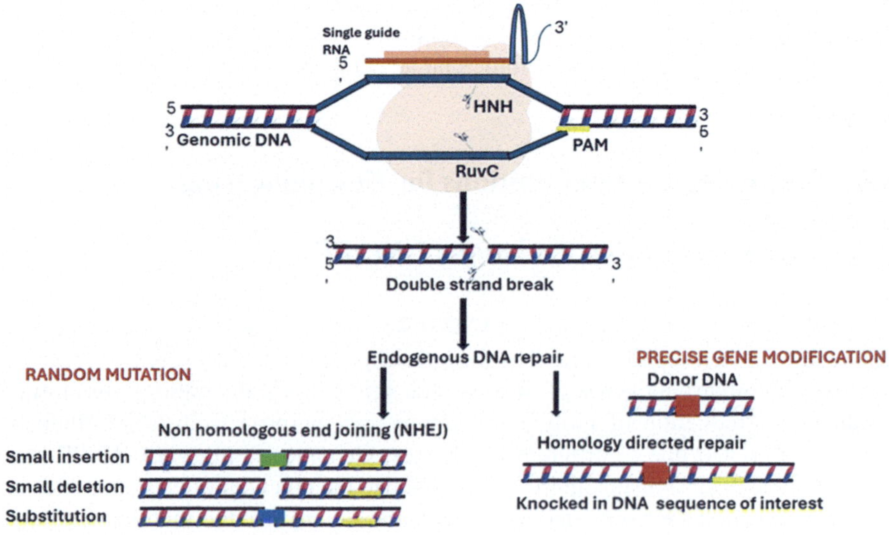

Fig. 9.5 Schematic representation Of CRISPR-mediated genome editing. (Adapted from Garg et al. (2023))

Table 9.3 Example of few extremophilic microbe's genome modification

Organism	Microbial type	Optimal temperature	Metabolic ability	Genomic modification	Enzymatic expression	Reference
Thermotoga maritima	Anaerobic, hyperthermophilic	80 degrees Celsius	Utilizes both simple and complex sugars for producing hydrogen	Lactate dehydrogenase is disrupted for increased hydrogen generation but lessen growth and sugar intake; disturbance of ABC maltose transporter affected growth	Recombinant expression of cellulases through native signal peptides; cellulolytic activity detected in supernatant	(Singh et al. 2018)
Thermus thermophilus	Aerobic, thermophilic	70–80 degrees Celsius	Denitrification utilization for anaerobic growth on sugars, lipids, peptides, and triglycerides	Disrupted for heat-stable β-galactosidase (TTP0042) and improved superoxide dismutase action	Enhanced superoxide dismutase capacity	(Wang et al. 2022)

vector is commonly used to insert the CRISPR/Cas9 system into target cells. The single guide RNA (sgRNA) is designed to direct the Cas9 enzyme for site-specific cleavage in genome. Different successful studies are conducted that involve CRISPR/Cas9 technology to engineer microbes for improved characteristics. For example, increased production of b-cyclodextrin glycosyl transferase in *Bacillus subtilis* and improved resistance to spore formation. Likewise, *Saccharomyces cerevisiae*'s genome has been changed to enhance production of mevalonate and (R, R)-2,3-butanediol and improve consumption of xylose (Vervoort et al. 2017). Genetic modification of some extremophilic microbes is shown in Table 9.3.

9.6.2 Enhance Biofuel Production Via CRISPR-Cas Genome Editing

SgRNA guides the Cas9 and the targeted sequence of genome inside the microbial cells via biolistic gene gun, electroporation, or heat shock transformation. The T7 promoter controls the cassette having Cas9, while the U6 promoter regulates gRNA scaffold. DNA strand separation is stimulated by Cas9 and allows binding of sgRNA for targeted DNA cleavage with a specific sequence in the target gene. The CRISPR/Cas9 system puts the desire DNA in the Cas9 active site with specific orientation to a PAM binding sequence and directs each nuclease domains of Cas9 to accurately cleave each strand of the desired DNA sequence 3 base pair upstream of the PAM site. The double-stranded DNA distorts and results in either error-prone

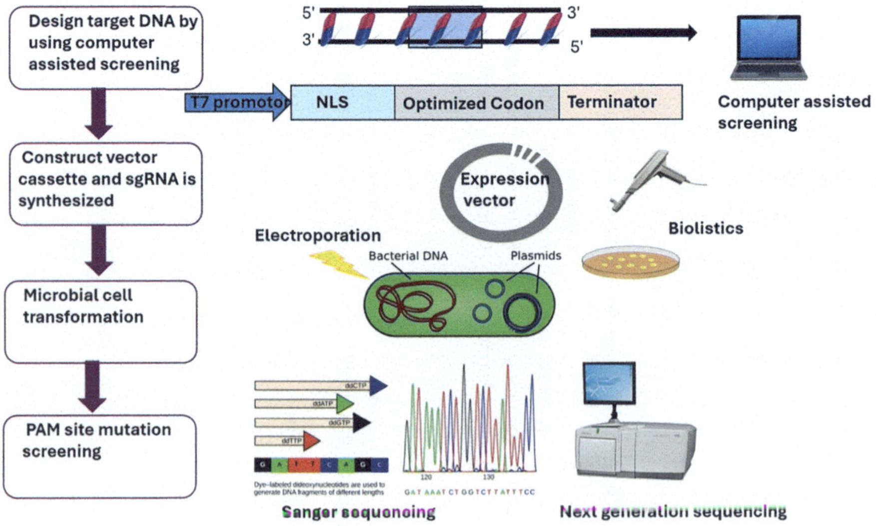

Fig. 9.6 Schematic representation of the enhanced biofuel production via CRISPR-Cas genome editing in microbial cells (Garg et al. 2023; Javed et al. 2019)

nonhomologous end-joining (NHEJ) or homology-directed repair (HDR) DNA repair. Almost one-third indels that repair normal gene function can be restored by using the correct reading frame in the coding region. Real-time PCR, Sanger sequencing, and next-generation sequencing are employed to find out the expression profile and mutations nature in the target gene (Nymark et al. 2016; Ulaganathan et al. 2017). Schematic representation of the improved biofuel production via CRISPR-Cas genome editing in microbial cells is shown in Fig. 9.6.

9.6.3 Improving Biofuel Tolerance by Genome Engineering

The biofuel products due to their antimicrobial activities can influence metabolic pathways of many microbes. Consequently, microorganisms demand genome editing for enhancing tolerance to these finish products (Dunlop 2011). Both targeted and random mutageneses have identified many genes where the mutation lead to either improved or bring down ethanol tolerance such as NADH dehydrogenase enzyme is present on mitochondrial membrane of many microbes, and it plays role of catalyzing electrons transfer from NADH to coenzyme Q (CoQ). Frame-shift mutation of this gene, especially generating a stop codon at codon 41, has improved ethanol tolerance and its manufacture capability (Ulaganathan et al. 2017). The transfer of the NADH flux from respiration toward ethanol production was found to be in association with mutation-stimulated aerobic growth. In addition, numerous techniques have been introduced to lessen the antimicrobial toxicity of biofuels. This involves changing membrane proteins, activating stress-responsive genes, and heterologously expressing efflux pumps or heat shock proteins (Mukhopadhyay 2015). Heat-shock proteins are primarily engaged in protein synthesis, transport, and proper folding. Heat shock protein upregulation plays a crucial role in restricting the protein agglutination under stress condition. Microbial cell wall composition was found to be altered on exposing butanol, preventing resistance to its entrance by using cis-trans isomerase. This enzyme allows the conversion of cis-fatty acids to trans fatty acids, thereby membrane fluidity and solvent tolerance are altered and influenced. To implement abovementioned genome modifications efficiently and flexibly, the CRISPR/Cas9 technique is used (Bhatia et al. 2017).

9.6.4 Enhancing Inhibitor Tolerance

The key objective of gene editing is to enhance the leniency of microbes to extremely toxic compounds such as fermentation process inhibitors and specific growth inhibitor. The pretreatment of lignocellulosic biomass using physiochemical approaches produces a large number of cellulolytic enzymes and microbial growth inhibitors during fermentation, which ultimately leads to low ethanol yields. Modification of single amino in multiple RpoD protein locations enhanced furfural resistance in

Z. mobilis. Insertion of transposon altered the SSK2 gene increased furfural tolerance in *S. cerevisiae*'s low ethanol yields (Sriariyanun et al. 2016).

9.6.5 Enhancing Biofuel Host Specificity and Capacity to Use Substrates

Expression of recombinant protein Cas9 is highly toxic to chromosomes of prokaryotes, leading to lethal DNA damage, and results in reduced transformation efficiency and genetic manipulation loss. The potential application of CRISPR-Cas technology for *Clostridium* helps to mitigate Cas9 toxicity and transformation abilities (Arroyo-Olarte et al. 2021). Carbon catabolite inhibition can be reduced by modifying genes related to sugar absorption, which enabled *Clostridium* species to use a broad range of carbohydrates. Fast carbon uptake for bacterial fermentation prior to lignin integration has been found a feasible strategy. Acetogenic anaerobic bacteria may alter carbon obtained from gaseous sources like syngas—H_2, CO, and CO_2 into liquid fuels through the Wood-Ljungdahl pathway (Benevenuti et al. 2021).

9.6.6 Developing Thermotolerance by Genome Engineering

Various constraints reduced the employ of thermophilic microbes in industrial lignocellulosic-based bioethanol manufacture even with their abundance. The modifications of thermostable microbes genome to enhance enzyme activity and stability are very difficult compared to editing microbial strains that are previously employed in bioethanol production to advance thermostability (Ulaganathan et al. 2017). Thermotolerance is a complex aspect that comprises many genes. Bioethanol production generally involves high temperature which results in the reduced growth of ethanol-producing microorganisms (Ulaganathan et al. 2017). Accumulations of trehalose and heat shock proteins are associated with enhance thermotolerance. Advancements in thermotolerance and ethanol production in *Z. mobilis* and *S. cerevisiae* have been obtained through single amino acid alteration in the pyruvate kinase and NADH dehydrogenase (Caspeta et al. 2014). By use of possible microbes from various sources, CRISPR/Cas9 may specifically use to alter proteins concerned in thermal resistance.

9.6.7 Role of Synthetic Biology in Bioengineering Extremophilic Microbes

Over the last 20 years, synthetic biology has grown as a significant multidisciplinary field combining genetic and metabolic engineering to modify genetic circuits for the manufacturing of desired molecules by using *E. coli* and *S. cerevisiae*. Synthetic biology research has expanded when scientists started using different extremophilic organisms like *Deinococcus*, *Geobacillus*, *Halomonas*, *Pyrococcus*, *Thermococcus*, and *Thermus* species. These extremophiles have adaptations to survive in extreme environments that can be altered through synthetic biology techniques. Synthetic genetic circuits enabled these organisms to flourish in altered or desired environment, enabling the manufacture of various bioactive compounds with major industrial and pharmaceutical applications (Li et al. 2017). Research has been conducted to use the thermoacidophilic *Sulfolobus* species as one of potential sources for metabolic engineering and artificial biology due to its versatility and stability. It is promising to alter regulatory systems and redirect metabolic processes to enhance development in *Sulfolobus acidocaldarius* and *Sulfolobus islandicus* by whole genome sequencing, marker-free in-frame deletion mutants, and homologous expression of protein during ectopic incorporation of foreign genes (Schreck & Grunden 2014). Synthetic genetic circuit having superoxide dismutase and heat shock protein HB8 and MB4 of *Thermoanaerobacter tengcongensis* and *Thermus thermophiles*, respectively, enabled the *S. cerevisiae* INVSc1 strain to thrive at 42 °C. This strain has produced more ethanol as compared to the wild type. Extremophile alcohol dehydrogenases are noteworthy for their definite efficiency as catalysts for butanol generation in noncellular systems. Additionally, research has found that alkali lignin can depolymerized into vanillin and syringate by using metabolically altered *E. coli* strains. These compounds can afterward bioconverted into cis, cis-muconic acid and pyrogallol, respectively. By using hydrothermally depolymerized lignin aromatics as a feedstock, cis and cis-muconic acid are created (Konings et al. 2009). Metabolically modified *Pseudomonas putida* KT2440 MA-9 is polymerized into nylon after hydrogenation to form adipic acid. It has been revealed that deleting vanillin dehydrogenase from commercially important strains enhanced the vanillin quantity produced from lignocellulose wastes (Liu et al. 2019). Thermoregulated genetic structure is employed as an alternative, which comprises two essential enzymes expressed heterologously in *E. coli*, like as enoyl-CoA hydratase/aldolase (Ech) and feruloyl-CoA synthetase (Fcs) of the thermophilic actinomycete *Amycolatopsis thermoflava* N1165. With the help of source ferulic acid, this system enables *E. coli* to produce vanillyl alcohol at 30 °C. Subsequently vanillin is synthesized at 50 °C by the enzymatic actions of Fcs and Ech from vanillyl alcohol (Ohta et al. 2004). Employment of renewable sources for the making of biofuels and bioenergy is a great demand industrial zone to deal with global warming and demanding of fossil fuels. Current improvement in multi-omics technologies; the innovation of very proficient lignin-degrading enzymes; the incorporation of physical, chemical, and biological methods in lignin depolymerization; and the need for cost-effective, feasible, and environmentally friendly production of biofuel and bioenergy necessitates researchers to engineer industrially significant microorganisms through this approach.

9.6.8 Synthetic Biology Chassis

In synthetic biology (SB), the word "chassis" refers to a microbial species competent of serving as adaptable and efficient carrier for processing genetic information from outside sources that codes for the production of particular metabolites. An organism to be used for industrial production intended to undergo iterative improvement processes. A chassis organism should have these characteristics:

- Ease of genome manipulation.
- Fully sequenced genome.
- Standardized genetic parts accessibility, with conventional and reported traits (Danchin 2012).

Thermophilic bacteria are generally more restricted for genetic manipulations. However, little number of genera from them can be widely manipulated to incorporate heterologous sequences through either plasmids or incorporation into their genomic DNA. *T. thermophilus* is an efficient source of thermophilic enzymes (including thermostable polymerases) and has importance in the thermic degradation of biomass. Although the synthetic biology toolset of *T. thermophilus* is restricted, investigation groups have expanded the toolset with fundamentals for indels, a modular vector toolbox for the production of genes, comprising thermosensors, replication origins, promoters, and thermostable CRISPR Cas9 systems that function at 65 °C. *Geobacillus* spp. recognized for their capability to use a wide variety of carbohydrates as feedstock compared to other thermophiles are promising candidates for bioproduction in biorefinery applications. Genetic modification of *Geobacillus* usually involves the conjugative transfer of self-replicating vectors, which relics a known technique regardless of its drawbacks. These drawbacks consist of reduced transformation efficiency and the failure to accurately target the integration of heterologous DNA into the genome, reducing the ability to do genetic studies such as gene knockouts. Reeve et al. made strides in 2016 by characterizing a collection of bacterial promoters and ribosome binding sites particularly for *Geobacillus thermoglucosidasius*. They achieved elevated transformation efficiency by means of electroporation with numerous self-replicating vectors, reporting transformation efficiencies in the range of 10^3–10^4 colony-forming units per microgram of DNA, compared to the 10^6 reported for *E. coli* controls (Reeve et al. 2016). There has been partial improvement in increasing the biotechnological capabilities of thermophiles. New advancements comprise the improvement of a set of expression vectors and riboswitches enabling inducible transcription. On the other hand, the deficiency of tools and standardized laboratory practices continues to deter the

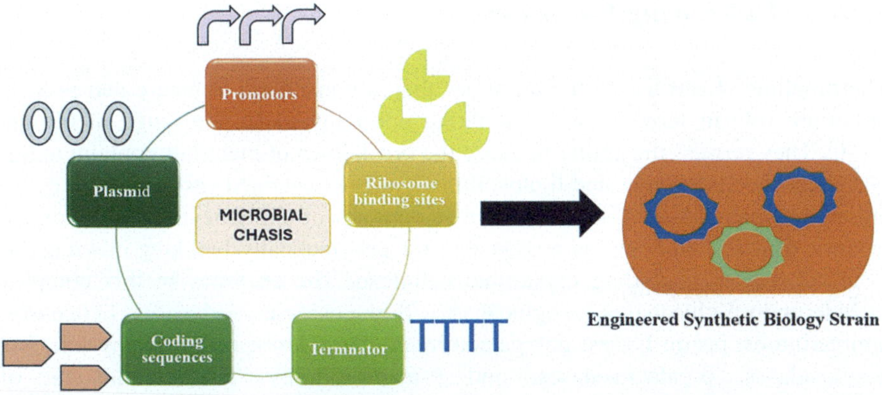

Fig. 9.7 Characteristics of a synthetic biology chassis is represented. (Modified from Vavitsas et al. (2022))

research area from fully harnessing the potential of thermophiles. Tools of a synthetic biology chassis is represented in the (Fig. 9.7) (Vavitsas et al. 2022).

9.7 Metabolic Engineering of Extremophiles for Biofuel Production

Biofuels are fuels made from biodegradable industrial and municipal trash, as well as organic resources such as sugar cane, corn, beets, wheat, sorghum, rapeseed, sunflower, soybean, palm, coconut, and *Jatropha*. With the proalcol program, Brazil launched the first significant attempt to develop biofuels on a commercial scale in 1975. It is anticipated that transportation will account for 80% of the rise in the world's liquid fuel consumption between 2006 and 2030. The high cost of crude oil, the decreasing supply of fossil fuels, and global warming from greenhouse gas emissions have driven the interest in biofuels as an alternative energy source. Although chemical and thermochemical methods are currently used to produce biofuels, using microorganisms to convert biomass into biofuel is more cost-effective and has become more popular recently. Researchers are also exploring the use of extremophiles (organisms that live in extreme conditions) and their hardy enzymes in this process (Barnard et al. 2010). A number of microbes have been engineered for the production of large number of biofuels are discussed below.

9.7.1 Heat-Loving Clostridia

Thermophilic clostridia are the microbes that are anaerobic in nature and play an important role in fermentation that thrive optimally under the range of 60 and 65 °C. They possess the ability to cause the breakdown of materials containing lignin, such as cellulosomes and lignocellulosic waste contained vast variety of cellulase and hemicellulase. The cellulosome, a complex of multienzyme located on the exterior of cell membrane, is responsible for enzymatically breaking down cellulosic substances, including crystalline cellulose. The enzymes in this complex include endo-β-glucanases, exoglucanases, β-glucosidases, xylanases, lichenases, laminarinases, pectin lyases, polygalacturonate hydrolases, pectin methylesterase, β-xylosidases, β-galactosidases, and β-mannosidases. The cellulosome of *Clostridium thermocellum* enables the degradation of cellulose into cellobiose and cellodextrins, and hemicellulose into xylose, xylobiose, and other pentose sugars (Barnard et al. 2010).

Another problem with biofuel production is the low yield of ethanol, caused by the creation of end products like lactate and acetate. This issue might be solved by genetically modifying strains to remove the genes responsible for acetate kinase and lactate dehydrogenase. The techniques for modifying thermophilic clostridia (a type of bacteria) are becoming more advanced. Scientists have already created a genetically engineered strain of *C. thermocellum* that can tolerate up to 60 g/L of ethanol and produce up to 26 g/L of ethanol. In different areas of studies, researchers created a mutant strain of *C. thermocellum* without the lactate dehydrogenase (ldh) gene, which resulted in higher ethanol production and improved tolerance to the product (Tailliez et al. 1989).

The cost of production of bioethanol from lignocellulose could be halved by using thermophilic steps. The process of converting plant material into ethanol using different organisms involves several steps. First, the bacterium *Clostridium thermocellum* produces enzymes called cellulase and hemicellulase. These enzymes break down cellulose into smaller sugars like cello-oligomers and cellobiose and hemicellulose into sugars like xylans, xylobiose, and other pentose sugars. Next, *C. thermocellum* absorbs these sugars and ferments hexose sugars (six-carbon sugars) to produce ethanol. For fermenting pentose sugars (five-carbon sugars), a different mixed culture of thermophilic anaerobic organisms is used. Complete breakdown and utilization of plant material (lignocellulose) to produce ethanol require five different organisms. *C. thermocellum* is often used in mixed fermentations with various thermophilic organisms, including *Thermoanaerobacterium, Thermosaccharolyticum, thermohydrosulfuricum, Thermoanaerobacterium saccharolyticum, Thermoanaerobacter brockii, Thermoanaerobacter ethanolicus*, and *Geobacillus stearothermophilus*. This combination of organisms helps in efficiently converting plant material into ethanol (Lynd 2005).

9.7.2 Geobacillus

Geobacillus are heat-loving bacilli known for their high activity to catabolize and potential for metabolic engineering. Some forms have the ability to perform the fermentation of sugar like D-xylose, D-glucose, and L-arabinose around 55–70 °C, producing a mixture of lactate, formate, acetate, and ethanol from glucose. Additionally, certain *Geobacillus* strains can degrade complex carbohydrates like xylan due to the presence of xylanases. At 70 °C, *Geobacillus stearothermophilus* can make ethanol at a level similar to that of *S. cerevisiae*. It has been determined that a strain of *Geobacillus thermoglucosidasius* is capable of withstanding ethanol concentrations of up to 10% (v/v) without experiencing growth. Consequently, these organisms are of significant interest for industrial bioethanol production (Wu et al. 2006).

Saccharomyces cerevisiae and *Zymomonas mobilis* used in the production of ethanol using the pyruvate decarboxylase (pdc) and alcohol dehydrogenase (adh) genes. These enzymes cause the conversion of pyruvate and NADH to ethanol, along with pyruvate decarboxylase turning pyruvate into acetaldehyde, which alcohol dehydrogenase then converts to ethanol. Bacterial pdc genes are rare, but the ones that have been found have unique characteristics in terms of their reaction rates and heat stability. These features could be useful for creating biocatalysts for the ethanol industry (Witt and Heilmeyer 1966). Scientists have tried to introduce this pathway in various organisms, including *Geobacillus* species. Although the pdc activity in *G. thermoglucosidasius* wasn't stable above 54 °C, this research shows promise for future development. TMO Renewables Ltd. has engineered a *Geobacillus* strain by removing the lactate dehydrogenase (ldh) and pyruvate formate lyase genes and increasing the pyruvate dehydrogenase gene activity. This modified strain can cause the fermentation of both pentose and hexose sugars to produce ethanol at nearly maximum theoretical yields at above 60 °C (Hübner et al. 1988).

9.7.3 Thermoanaerobacterium Saccharolyticum

Thermoanaerobacterium is a heat-loving anaerobic bacterium that can break down hemicellulose and utilize pentose carbohydrates, like as xylose, to generate acidic compounds and ethanol. The pyruvate enzymes ferredoxin oxidoreductase (POR), phosphate acetyltransferase (Pta), and acetate kinase (Ack) are used to make acetic acid, while L-lactate dehydrogenase (Ldh) is used to make lactic acid. To boost ethanol production, scientists created a mutant strain of *Thermoanaerobacterium saccharolyticum* by knocking out the lactate dehydrogenase gene. This mutant produced less lactate and four times more ethanol. The new strain, TD1, efficiently uses xylose. Recently, in another strain, ALK1, engineering can be done only to generate its organic product, ethanol (Shaw et al. 2009). In this activity, the genes that produce acetate (ack/pta) and lactate (ldh) were eliminated. The POR process, which is distinct from the mechanisms in other microorganisms with homo-ethanol fermentation, uses pyruvate to produce ethanol. *T. saccharolyticum* strain ALK2 was created by expanding the xylose concentration in a continuous culture of strain ALK1. With continuous culture, this novel strain could consume xylose, glucose, mannose, and galactose to produce 33 g/L of ethanol and 37 g/L in fed-batch cultures. According to Shaw et al. (2008), these are the highest ethanol yields yet recorded by anaerobic heat-loving mircrobes.

9.8 Conclusion

The pursuit for renewable energy sources has catalyzed the development of biofuel and bioenergy production and regarding this extremophile play its particular key role because of their key property to adapt themselves in severe extreme conditions. This property makes them invaluable for bioengineering applications. Technologies like CRISPR and synthetic biology play its great role for transforming bioengineering, and these technologies aided to improve biofuel production by enhancing tolerance to inhibitors and developing thermotolerance in these extremophilic microbes. In order to design robust biofuel-producing organisms, synthetic biology plays its great role. In this chapter we studied various biofuels like ethanol, butanol, and biodiesel along with the sources of bioenergy such as biogas and bioelectricity. Various successful examples showed that engineered extremophiles can be very effective for future research and development. Bioenergy production methods like biogas and biohydrogen have shown high contribution in global energy landscape, but it is very important to consider the environmental and social impacts. Continuous research in bioengineering and synthetic biology plays its vital role in order to overcome present challenges. Extremophiles offer bright future for renewable energy and will lead us to more sustainable solutions.

References

Al-Zuhair S (2007) Production of biodiesel: possibilities and challenges. Biofuels, Bioproducts and Biorefining: Innovation for a sustainable economy 1(1):57–66

Antoni D, Zverlov VV, Schwarz WH (2007) Biofuels from microbes. Appl Microbiol Biotechnol 77:23–35

Arroyo-Olarte RD, Bravo Rodriguez R, Morales-Ríos E (2021) Genome editing in bacteria: CRISPR-Cas and beyond. Microorganisms 9(4):844

Atif F, Maqsood N, Ali W, Ali W, Irfan M (2024) Extremophiles and their enzymatic diversity and biotechnological potential. Syst Microbiol Biomanufacturing 4:1–17

Barnard D, Casanueva A, Tuffin M, Cowan D (2010) Extremophiles in biofuel synthesis. Environ Technol 31(8–9):871–888

Benevenuti C, Amaral P, Ferreira T, Seidl P (2021) Impacts of syngas composition on anaerobic fermentation. Reactions 2(4):391–407

Berger E, Ferreras E, Taylor MP, Cowan D (2014) Extremophiles and their use in biofuel synthesis. In: Industrial biocatalysis. Pan Stanford Publishing Pte Ltd., Singapore, pp 239–282

Bhatia SK, Jagtap SS, Bedekar AA, Bhatia RK, Rajendran K, Pugazhendhi A, Rao CV, Atabani A, Kumar G, Yang YH (2021) Renewable biohydrogen production from lignocellulosic biomass using fermentation and integration of systems with other energy generation technologies. Sci Total Environ 765:144429

Bhatia SK, Kim SH, Yoon JJ, Yang YH (2017) Current status and strategies for second-generation biofuel production using microbial systems. Energy Convers Manag 148:1142–1156

Bibra M, Wang J, Squillace P, Pinkelman R, Papendick S, Schneiderman S, Wood V, Amar V, Kumar S, Salem D (2014) Biofuels and value-added products from extremophiles. In: advances in biotechnology. Nawani NN, Khetmalas M, Razdan PN, Pandey a (eds), pp 268

Björnsson B, Steinarsson A, Oddgeirsson M (2001) Optimal temperature for growth and feed conversion of immature cod (*Gadus morhua* L.). ICES J Mar Sci 58(1):29–38

Boodhun BSF, Mudhoo A, Kumar G, Kim SH, Lin CY (2017) Research perspectives on constraints, prospects and opportunities in biohydrogen production. Int J Hydrog Energy 42(45):27471–27481

Brouns SJ, Jore MM, Lundgren M, Westra ER, Slijkhuis RJ, Snijders AP, Dickman MJ, Makarova KS, Koonin EV, Van Der Oost J (2008) Small CRISPR RNAs guide antiviral defense in prokaryotes. Science 321(5891):960–964

Brunecky R, Subramanian V, Yarbrough JM, Donohoe BS, Vinzant TB, Vanderwall TA, Knott BC, Chaudhari YB, Bomble YJ, Himmel ME (2020) Synthetic fungal multifunctional cellulases for enhanced biomass conversion. Green Chem 22(2):478–489

Burton SG, Cowan DA, Woodley JM (2002) The search for the ideal biocatalyst. Nat Biotechnol 20(1):37–45

Carriquiry MA, Du X, Timilsina GR (2011) Second-generation biofuels: economics and policies. Energy Policy 39(7):4222–4234

Caspeta L, Chen Y, Ghiaci P, Feizi A, Buskov S, Hallström BM, Petranovic D, Nielsen J (2014) Altered sterol composition renders yeast thermotolerant. Science 346(6205):75–78

Chozhavendhan S, Rajamehala M, Karthigadevi G, Praveenkumar R, Bharathiraja B (2020) A review on feedstock, pretreatment methods, influencing factors, production and purification processes of bio-hydrogen production. Case Stud Chem Environ Eng 2:100038

Cowan DA, Arslanoglu A, Burton SG, Baker G, Cameron R, Smith JJ, Meyer Q (2004) Metagenomics, gene discovery and the ideal biocatalyst. Biochem Soc Trans 32(2):298–302

Crosby JR, Laemthong T, Lewis AM, Straub CT, Adams MW, Kelly RM (2019) Extreme thermophiles as emerging metabolic engineering platforms. Curr Opin Biotechnol 59:55–64

Danchin A (2012) Scaling up synthetic biology: do not forget the chassis. FEBS Lett 586(15):2129–2137

Demain AL (2009) Biosolutions to the energy problem. J Ind Microbiol Biotechnol 36(3):319–332

Demirjian DC, Morís-Varas F, Cassidy CS (2001) Enzymes from extremophiles. Curr Opin Chem Biol 5(2):144–151

Dragone G, Fernandes BD, Vicente AA, Teixeira JA (2010) Third-generation biofuels from microalgae

Dunlop MJ (2011) Engineering microbes for tolerance to next-generation biofuels. Biotechnol Biofuels 4:1–9

French CE (2009) Synthetic biology and biomass conversion: a match made in heaven? J R Soc Interface 6(Suppl_4):S547–S558

Garg D, Samota MK, Contis N, Patel N, Bala S, Rosado AS (2023) Revolutionizing biofuel generation: unleashing the power of CRISPR-Cas mediated gene editing of extremophiles. Microbiol Res 127443:127443

Garneau JE, Dupuis MÈ, Villion M, Romero DA, Barrangou R, Boyaval P, Fremaux C, Horvath P, Magadán AH, Moineau S (2010) The CRISPR/Cas bacterial immune system cleaves bacteriophage and plasmid DNA. Nature 468(7320):67–71

Havlík P, Schneider UA, Schmid E, Böttcher H, Fritz S, Skalský R, Aoki K, De Cara S, Kindermann G, Kraxner F (2011) Global land-use implications of first and second-generation biofuel targets. Energy Policy 39(10):5690–5702

Hübner G, König S, Schellenberger A (1988) The functional role of thiol groups of pyruvate decarboxylase from brewer's yeast. Biomed Biochim Acta 47(1):9–18

Javed MR, Noman M, Shahid M, Ahmed T, Khurshid M, Rashid MH, Ismail M, Sadaf M, Khan F (2019) Current situation of biofuel production and its enhancement by CRISPR/Cas9-mediated genome engineering of microbial cells. Microbiol Res 219:1–11. https://doi.org/10.1016/j.micres.2018.10.010

Jinek M, Chylinski K, Fonfara I, Hauer M, Doudna JA, Charpentier E (2012) A programmable dual-RNA–guided DNA endonuclease in adaptive bacterial immunity. Science 337(6096):816–821

Kangle L. (2012) All-solid-state front-illuminated titania nanotube-based dye-sensitized solar-cells. PhD, Materials Science and Engineering, National University of Singapore, Singapore

Knoshaug EP, Zhang M (2009) Butanol tolerance in a selection of microorganisms. Appl Biochem Biotechnol 153:13–20

Konings WN, Albers SV, Koning SM, Driessen AJ (2009) Survival strategies and membrane properties of extremophiles. Extremophiles-Vol I 3:98

Kucharska K, Rybarczyk P, Hołowacz I, Łukajtis R, Glinka M, Kamiński M (2018) Pretreatment of lignocellulosic materials as substrates for fermentation processes. Molecules 23(11):2937

Li X, Kappler U, Jiang G, Bond PL (2017) The ecology of acidophilic microorganisms in the corroding concrete sewer environment. Front Microbiol 8:248017

Liu C, Baffoe DK, Zhan Y, Zhang M, Li Y, Zhang G (2019) Halophile, an essential platform for bioproduction. J Microbiol Methods 166:105704

Liu Y, Min J, Feng X, He Y, Liu J, Wang Y, He J, Do H, Sage V, Yang G (2020) A review of biohydrogen productions from lignocellulosic precursor via dark fermentation: perspective on hydrolysate composition and electron-equivalent balance. Energies 13(10):2451

Lynd LR (2005) Production of ethanol from lignocellulosic materials using thermophilic bacteria: critical evaluation of potential and review. In: Lignocellulosic materials. Springer, pp 1–52

Mukhopadhyay A (2015) Tolerance engineering in bacteria for the production of advanced biofuels and chemicals. Trends Microbiol 23(8):498–508

Mukhtar S, Aslam M (2020) Biofuel synthesis by extremophilic microorganisms. In: Biofuels Production—Sustainability and Advances in Microbial Bioresources. Springer, Cham, pp 115–138

Nahálka J (2008) Physiological aggregation of maltodextrin phosphorylase from *Pyrococcus furiosus* and its application in a process of batch starch degradation to α-d-glucose-1-phosphate. J Ind Microbiol Biotechnol 35(4):219–223

Nymark M, Sharma AK, Sparstad T, Bones AM, Winge P (2016) A CRISPR/Cas9 system adapted for gene editing in marine algae. Sci Rep 6(1):24951

Ohta Y, Hatada Y, Nogi Y, Miyazaki M, Li Z, Akita M, Hidaka Y, Goda S, Ito S, Horikoshi K (2004) Enzymatic properties and nucleotide and amino acid sequences of a thermostable β-agarase from a novel species of deep-sea *Microbulbifer*. Appl Microbiol Biotechnol 64:505–514

Okolie JA, Nanda S, Dalai AK, Kozinski JA (2021) Chemistry and specialty industrial applications of lignocellulosic biomass. Waste Biomass Valor 12:2145–2169

Quehenberger J, Shen L, Albers SV, Siebers B, Spadiut O (2017) Sulfolobus–a potential key organism in future biotechnology. Front Microbiol 8:313898

Radhakrishnan S, Prasad S, Kumar S, Subramanian D (2020) Production of biohydrogen from lignocellulosic feedstocks. In: Lignocellulosic biorefining technologies. Wiley, pp 47–67

Ramu SM, Dinesh GH, Thulasinathan B, Thondi Rajan AS, Ponnuchamy K, Pugazhendhi A, Alagarsamy A (2021) Dark fermentative biohydrogen production from rice mill wastewater. Int J Energy Res 45(12):17233–17243

Reeve B, Martinez-Klimova E, de Jonghe J, Leak DJ, Ellis T (2016) The Geobacillus plasmid set: a modular toolkit for thermophile engineering. ACS Synth Biol 5(12):1342–1347

Sağır E, Hallenbeck PC (2019) Photofermentative hydrogen production. In: Biohydrogen. Elsevier, pp 141–157

Sapranauskas R, Gasiunas G, Fremaux C, Barrangou R, Horvath P, Siksnys V (2011) The *Streptococcus thermophilus* CRISPR/Cas system provides immunity in Escherichia coli. Nucleic Acids Res 39(21):9275–9282

Sarangi PK, Nanda S (2020) Biohydrogen production through dark fermentation. Chem Eng Technol 43(4):601–612

Schiermeyer A, Schneider K, Kirchhoff J, Schmelter T, Koch N, Jiang K, Herwartz D, Blue R, Marri P, Samuel P (2019) Targeted insertion of large DNA sequences by homology-directed repair or non-homologous end joining in engineered tobacco BY-2 cells using designed zinc finger nucleases. Plant Direct 3(7):e00153

Schink B (1997) Energetics of syntrophic cooperation in methanogenic degradation. Microbiology and molecular biology reviews, 61(2), 262-280.

Schreck SD, Grunden AM (2014) Biotechnological applications of halophilic lipases and thioesterases. Appl Microbiol Biotechnol 98:1011–1021

Sharma HK, Xu C, Qin W (2019) Biological pretreatment of lignocellulosic biomass for biofuels and bioproducts: an overview. Waste Biomass Valor 10:235–251

Shaw AJ, Jenney FE Jr, Adams MW, Lynd LR (2008) End-product pathways in the xylose fermenting bacterium, *Thermoanaerobacterium saccharolyticum.* Enzym Microb Technol 42(6):453–458

Shaw AJ, Podkaminer KK, Desai SG, Bardsley JS, Rogers SR, Thorne PG, Hogsett DA, Lynd LR (2009) Metabolic engineering of a thermophilic bacterium to produce ethanol at high yield. Int Sugar J 111(1323):164–171

Sims RE, Mabee W, Saddler JN, Taylor M (2010) An overview of second-generation biofuel technologies. Bioresour Technol 101(6):1570–1580

Singh R, White D, Demirel Y, Kelly R, Noll K, Blum P (2018) Uncoupling fermentative synthesis of molecular hydrogen from biomass formation in *Thermotoga maritima.* Appl Environ Microbiol 84(17):e00998–e00918

Singh S, Singh D (2010) Biodiesel production through the use of different sources and characterization of oils and their esters as the substitute of diesel: a review. Renew Sust Energ Rev 14(1):200–216

Sriariyanun M, Tantayotai P, Yasurin P, Pornwongthong P, Cheenkachorn K (2016) Production, purification and characterization of an ionic liquid tolerant cellulase from *Bacillus* sp. isolated from rice paddy field soil. Electron J Biotechnol 19:23–28

Sterling JD, Quigley HF, Orellana A, Mohnen D (2001) The catalytic site of the pectin biosynthetic enzyme α-1,4-galacturonosyltransferase is located in the lumen of the Golgi. Plant Physiol 127(1):360–371

Sternberg S, Redding S, Jinek M, Greene EJA, Doudna J (2014) DNA interrogation by the CRISPR RNA-guided endonuclease Cas9. Nature 507:62–67. https://www.ncbi.nlm.nih.gov/pubmed/24476820. https://doi.org/10.1038/nature13011

Tailliez P, Girard H, Millet J, Beguin P (1989) Enhanced cellulose fermentation by an asporogenous and ethanol-tolerant mutant of *Clostridium thermocellum.* Appl Environ Microbiol 55(1):207–211

Taylor M, Bauer R, Mackay S, Tuffin M, Cowan D (2012) Extremophiles metabolic engineering and process integration. Appl Microbiol Biotechnol 56:17–23

Ulaganathan K, Goud S, Reddy M, Kayalvili U (2017) Genome engineering for breaking barriers in lignocellulosic bioethanol production. Renew Sust Energ Rev 74:1080–1107

Vavilin V, Rytov S, Lokshina LY (1996) A description of hydrolysis kinetics in anaerobic degradation of particulate organic matter. Bioresour Technol 56(2–3):229–237

Vavitsas K, Glekas PD, Hatzinikolaou DG (2022) Synthetic biology of thermophiles: taking bioengineering to the extremes? Appl Microbiol 2(1):165–174

Vervoort Y, Linares AG, Roncoroni M, Liu C, Steensels J, Verstrepen KJ (2017) High-throughput system-wide engineering and screening for microbial biotechnology. Curr Opin Biotechnol 46:120–125

Wang J, Wei J, Li H, Li Y (2022) High-efficiency genome editing of an extreme thermophile *Thermus thermophilus* using endogenous type I and type III CRISPR-Cas systems. mLife 1(4):412–427

Wang Y, Song Y, Ma C, Kang Z, Zhu Z (2021) A heterologously-expressed thermostable *Pyrococcus furiosus* cytoplasmic [NiFe]-hydrogenase I used as the catalyst of H2/air biofuel cells. Int J Hydrog Energy 46(4):3035–3044

Wisselink HW, Toirkens MJ, Wu Q, Pronk JT, van Maris AJ (2009) Novel evolutionary engineering approach for accelerated utilization of glucose, xylose, and arabinose mixtures by engineered *Saccharomyces cerevisiae* strains. Appl Environ Microbiol 75(4):907–914

Witt I, Heilmeyer L (1966) Regulation of pyruvate decarboxylase (EC 4.1.1.1) synthesis by coenzyme induction in *Saccharomyces cerevisiae*. Biochem Biophys Res Commun 25(3):340–345

Yadav M, Paritosh K, Vivekanand V (2020) Lignocellulose to bio-hydrogen: an overview on recent developments. Int J Hydrog Energy 45(36):18195–18210

Zabed H, Sultana S, Sahu JN, Qi X (2018) An overview on the application of ligninolytic microorganisms and enzymes for pretreatment of lignocellulosic biomass. In: Recent advancements in biofuels and bioenergy utilization. Springer, Singapore, pp 53–72

Zaldivar J, Nielsen J, Olsson L (2001) Fuel ethanol production from lignocellulose: a challenge for metabolic engineering and process integration. Appl Microbiol Biotechnol 56:17–34

Zhang Q, Zhang Z (2018) Biological hydrogen production from renewable resources by photofermentation. In: Advances in bioenergy, vol 3. Elsevier, pp 137–160

Zhu P, Abdelaziz OY, Hulteberg CP, Riisager A (2020) New synthetic approaches to biofuels from lignocellulosic biomass. Curr Opin Green Sustainable Chem 21:16–21

Chapter 10
Global Policies and Perspectives of Biofuels and Bioenergy: Efforts to Combat Climate Change

Kossivi Fabrice Dossa and Yann Emmanuel Miassi

Abstract Supported by ambitious energy policies, biofuels and bioenergy play a central role in the transition to sustainable energy systems. This article aims to identify the types of biofuels and bioenergy as well as the energy policies that underpin their implementation, while highlighting their advantages and limitations, including the issues surrounding these energy sources. A systematic review guided by the Reporting Standards for Systematic Evidence Syntheses (ROSES) approach was conducted, allowing the selection of 93 relevant documents after excluding several others. The results highlight a steady increase in publications on these energies, reflecting a growing scientific interest in the global energy transition. Five forms of biofuels and bioenergy (solid biofuels, biogas, biodiesels, biogasoline, and others) were identified, each reflecting climate and economic benefits and negative impacts such as biodiversity loss. International (Kyoto Protocol, RED I and II, and others) and regional energy policies have catalyzed progress by promoting strategic frameworks to reduce greenhouse gas (GHG) emissions and promote innovation. However, barriers remain, including high production costs, technological and logistical challenges, and the lack of incentive policies. As a perspective, the study highlights the need for investments in infrastructure, as well as the implementation of

K. F. Dossa (✉)
Faculty of Forestry, Geography and Geomatics, Laval University, Quebec, QC, Canada

Action-Research for Sustainable Development NGO, Department of Research Project, Cotonou, Benin

Department of Agricultural Economics, Faculty of Agriculture, University of Nigeria, Nsukka, Nigeria

Y. E. Miassi
Faculty of Forestry, Geography and Geomatics, Laval University, Quebec, QC, Canada

Action-Research for Sustainable Development NGO, Department of Research Project, Cotonou, Benin

Department of Agricultural Economics, Faculty of Agriculture, Çukurova University, Adana, Turkey

M. A. Dar et al. (eds.), *Recent Trends in Lignocellulosic Biofuels and Bioenergy*, Clean Energy Production Technologies,
https://doi.org/10.1007/978-981-96-4636-4_10

strengthened incentive policies to maximize the potential of these energies in the global energy transition.

Keywords Biofuels · Bioenergy · Energy policies · Climate change · Perspectives

10.1 Introduction

Taking urgent action to mitigate climate change and its impacts has become a global goal (Hriţcan 2023). This phenomenon is first reflected in the gradual evolution of global temperatures, extreme changes in precipitation, and significant impacts on biodiversity (Rial 2024). These changes in agroecological conditions accentuate the severity of weather events, such as heat, floods, and droughts, thus weakening human lives, food production, economic growth, and livelihoods including income distribution (Rial 2024; Semde et al. 2021). At the same time, we are increasingly witnessing a significant increase in greenhouse gas (GHG) emissions characterized mainly by a significant amount of carbon dioxide (CO_2) (Ikevuje et al. 2024). For most scientists, the consumption of fossil fuels remains the major cause of GHG (Ikevuje et al. 2024; Raihan et al. 2023). At the same time, these emissions are also identified as the main factor in global warming (Akhmieva et al. 2024; Raihan 2023). Indeed, the interaction between GHG and global warming is no longer in doubt (Bajoria et al. 2024; Rial 2024; Wolniak and Skotnicka-Zasadzień 2023). The increase in GHGs, resulting in a significant amount of CO_2, methane (CH_4), and nitrogen (N) in the atmosphere, is increasingly observed around the world, accentuating changes in climatic conditions (Rial 2024). Faced with these environmental concerns, all social classes of the world are converging towards a particular action: the transition to a low-emission economy characterized mainly by the use of various renewable energy sources (Suman 2018; Wolniak and Skotnicka-Zasadzień 2023). Thus, they gradually recognize that relying heavily on fossil fuels is not sustainable and presents significant risks to their well-being and prosperity (Wolniak and Skotnicka-Zasadzień 2023).

To address these challenges, biofuels and bioenergy are identified as alternative energies capable of ensuring a transition to more sustainable systems (Dudley 2018; Ebadian et al. 2020; Rial 2024). Indeed, nowadays, the question of using these energy sources as an alternative in the mechanism for reducing the effects of climate change is increasingly the subject of scientific debate (Calvin et al. 2021; Demirbas 2009; Scarlat et al. 2015; Ussiri and Lal 2017). These are renewable energy sources that offer considerable advantages compared to fossil fuels (Wolniak and Skotnicka-Zasadzień 2023). Specifically, renewable energies are abundant and sustainable and produce fewer or no GHG emissions during use (Scarlat et al. 2015; Wolniak and Skotnicka-Zasadzień 2023). To promote the use of these energy sources and accelerate the energy transition, policies are being developed at both international and regional levels (Ebadian et al. 2020). These are policies aimed at preserving biofuels from direct competition with fossil fuels, stimulating investment, and facilitating technological development and organization around the adoption of biofuels worldwide (Costantini et al. 2014; Nemet 2009). However, despite the development of these policy frameworks, the adoption of biofuels and bioenergy still encounters technical, socioeconomic, and environmental constraints, which are important to elucidate for better optimization of their effects (Rial 2024). Understanding these constraints requires an in-depth analysis of these policies as well as possible solutions. Given this context, the present study was initiated, aiming firstly to synthesize knowledge on energy policies and their impacts. Secondly, this study aims to highlight innovative avenues in a context of growing but still scattered research on these renewable energies. By highlighting the successes and shortcomings of current policies, the study provides a solid basis for future programs as well as for the possible development of new, more appropriate policies. After a clear development of the methodological approach used, the main results (characteristics of the studies, state of play of biofuels, global policies, and their climatic contributions, perspectives) obtained are provided in five subsections.

10.2 Materials and Methods

10.2.1 Data Collection

The study was conducted following the approach proposed by Berrang-Ford et al. (2015) based on the Reporting Standards for Systematic Evidence Syntheses (ROSES) protocol. This process has also been used by several other authors (Kpadé et al. 2024; Drepper et al. 2021; Sánchez et al. 2021) to ensure transparency and ease in reproducing the study. The steps followed in implementing this study are as follows: (a) identification of adequate literature sources on these energy sources (biofuel and bioenergy), (b) structuring of research concepts and in-depth description of the implementation technique, (c) development of eligibility criteria, and (d) documentation of the identified studies and elimination of irrelevant works.

10.2.2 Research Strategy

Several databases that are highly recognized as the most used in the scientific world were used, including Scopus and Web of Science (AlRyalat et al. 2019). However, other databases such as CrossRef and Google Scholar were explored to ensure that most relevant works were included. In terms of periods, the study took into consideration the works conducted in the interval from 1999 to 2024 and particularly addressed the political themes around biofuels and bioenergy. In addition, the thematic search strategy was conducted using keywords combined with Boolean operators to capture all relevant studies. Our search equation was defined as follows: ("biofuels" OR "bioenergy") AND ("climate change" OR "emission reduction")

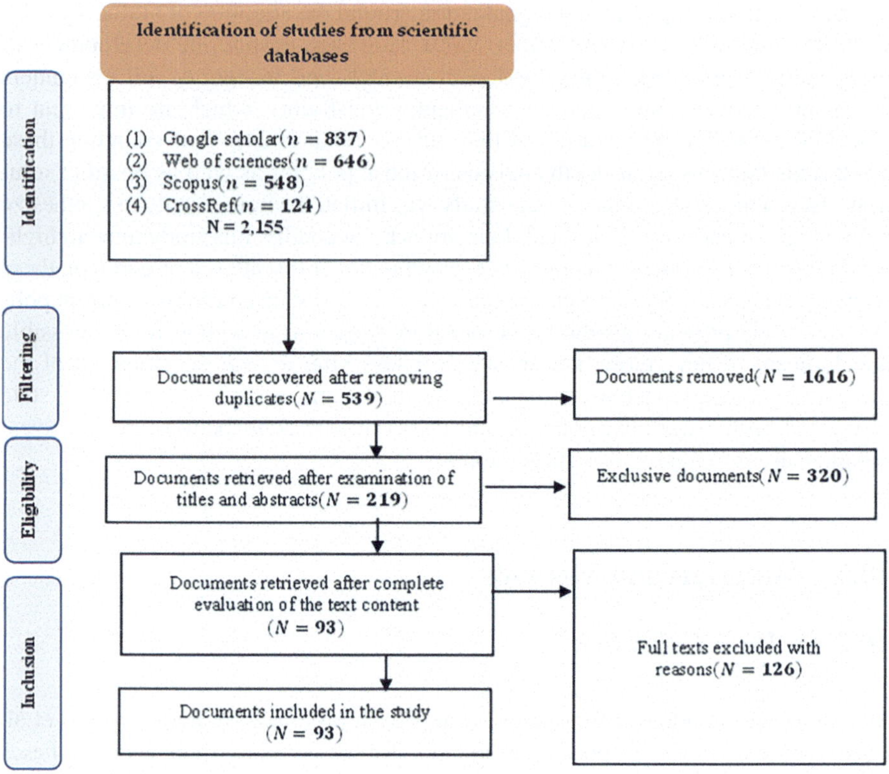

Fig. 10.1 Diagram describing the document selection process

AND ("energy policies" OR "global perspectives" OR "international guidelines"). This search strategy was applied to the titles, abstracts, and keywords of the publications. A complementary search was performed to include institutional documents and technical reports to fill possible gaps in traditional scientific databases. All this methodological approach allowed us to retrieve 2155 references, with a total of 539 after removing duplicates between the four databases. Figure 10.1 details how the selection of titles/abstracts and full texts allowed us to identify 93 references, consisting only of articles published in scientific journals.

10.2.3 Description and Eligibility Criteria (Inclusion and Exclusion)

After conducting a preliminary search, we exported the identified papers for initial screening. After removing duplicates and irrelevant publication types, we screened titles, keywords, and abstracts against predefined inclusion and exclusion criteria. Subsequently, relevant publications underwent full-text review to confirm their

Table 10.1 Inclusion and exclusion criteria considered in the study

Sr. no	Inclusion criteria	Exclusion criteria
1	Text written in English or French	Text written in a language other than English and French
2	Journal article (empirical data)	Article review, book, book chapters, book series, conference proceedings
3	Studies dealing with biofuels, bioenergy, and their links with climate change	Studies not addressing the topics of biofuels, bioenergy, and their links to climate change
4	The text includes sufficient detail to perform data analysis	The text does not include enough detail to perform data analysis
5	Studies carried out between 2000 and 2024	Studies carried out before 2000

eligibility for inclusion. Any inaccessible or excluded material was noted, and the final curated list included publications that met the relevance and accessibility criteria.

All publications on biofuel and bioenergy policies developed in the context of climate change mitigation identified during the preliminary search were compiled into a comprehensive database. This database contains essential information such as author contact details, year of publication, title, country, source, policies, and alternative energy sources. This systematic organization of the data facilitates a thorough understanding of the literature on this topic, allowing us to explore specific aspects of the relationships between these indicators (policies, biofuels, bioenergy, climate change). In addition, the inclusion and exclusion criteria guiding the selection process are described in Table 10.1, which brings transparency and clarity to our methodology. Subsequently, information was collected in detail from relevant publications, including (a) authors (year), (b) country, (c) policies, (d) energy types (biofuels and bioenergy), (e) challenges (opportunities, obstacles), (f) relationship with climate change (advantages, limitations), and (g) prospects and innovations around these energy sources. In terms of energy policy, in addition to international policies, some regional policies were identified as case studies. These are mainly policies developed in countries such as the United States, Canada, Germany, and Brazil. The choice of these countries is justified by their position among the world's top ten energy consumers over the past 20 years, reflecting their key role in global energy dynamics (Chen et al. 2022; Wang et al. 2024).

10.2.4 Analysis Methods

In this study, descriptive statistics were used to present quantitative trends, such as the frequency of publications over time and their geographical distribution. To examine in depth the main areas of interest of the works considered, the study used thematic content analysis based on a review of titles, abstracts, keywords and main contents.

The status of the main biofuels and bioenergy as well as the most used raw materials was assessed and visualized using a Sankey diagram. This is a methodological approach commonly used in the literature that has also addressed the energy issues of biofuels (Daniel et al. 2019; Wu and Pfenninger 2023) or not (Harpet and Gully 2013). The R software version 4.3.0 and Microsoft Excel were used for the various statistical analyses.

10.3 Results

10.3.1 Characterization and Description of Selected Works

The first analyses carried out aim to assess the dynamics of publications produced on issues related to renewable energy sources such as biofuels and bioenergy. This preliminary analysis thus makes it possible to assess the interest of the scientific world in the issues of energy transition, while highlighting the years of high scientific intensity. Figure 10.2 shows the evolution of scientific work in this field during the period 1990 and 2024.

The results show a notable increase in the number of scientific publications on issues related to energy sources over the years. Between 2000 and 2005, interest was still limited, with annual frequencies ranging from 1 to 2 articles (i.e., 1.08% and 2.15%). However, from 2010 onwards, a gradual increase is observed, marking a growing interest in this field. For example, in 2010 and 2012, 4 articles were published (i.e., 4.30% of the articles collected), compared to only two or three articles per year in previous years. This trend accelerated further from 2015, when six

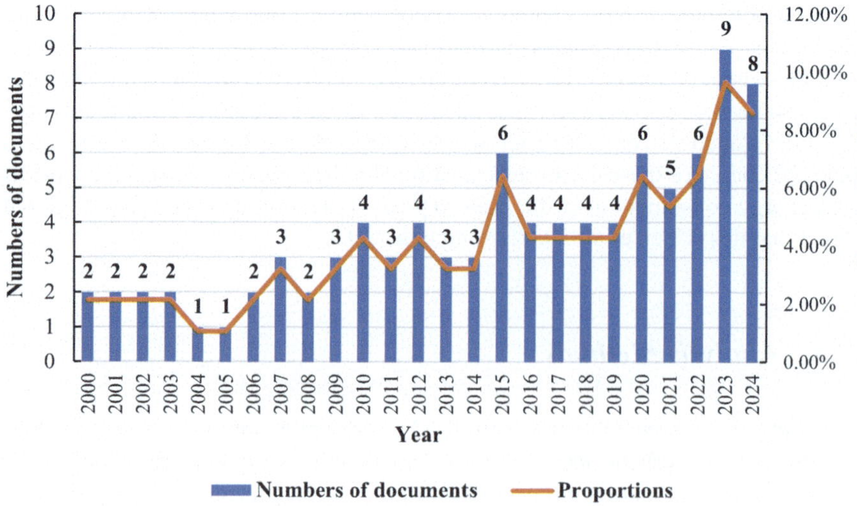

Fig. 10.2 Evolution of scientific publications on biofuel and bioenergy policies (2000–2024)

articles were identified (6.45%), before reaching a peak of nine publications in 2023 (9.68%) and eight in 2024 (8.60%). These data indicate a continued increase in power, reflecting the growing importance given to this topic in the context of global energy and climate challenges.

10.3.2 *Geographic Distribution of Selected Studies on Biofuels and Bioenergy*

Figure 10.3 illustrates the geographical distribution of scientific studies conducted on biofuels and bioenergy issues, highlighting the most studied countries and regions as well as the overall interest in these topics. Analysis of the results reveals a marked geographical distribution of scientific publications on biofuels and bioenergy issues. A clear predominance emerges from studies conducted at the global level, with 21 publications, followed by those covering all European Union countries (19 publications).

These results show that the interest in these issues is global, going beyond the framework of a regional scale. At the same time, it also raises the urgency around

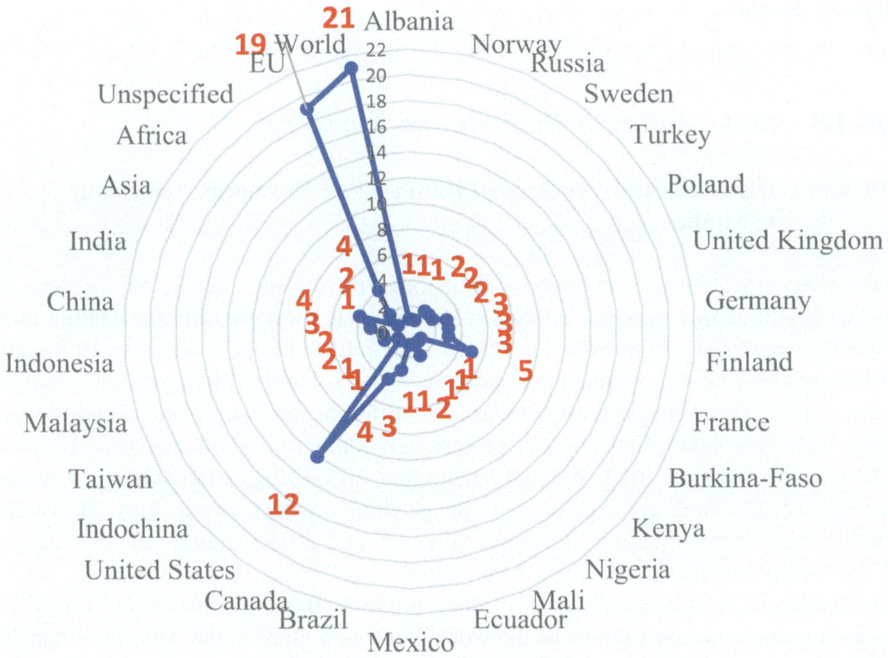

Fig. 10.3 Radar chart illustrating the geographical distribution of empirical studies on biofuels and bioenergy issues

energy challenges, integrating other dimensions such as climate change and sustainable development.

Regarding the work carried out on specific regions, we find in the foreground, the studies conducted in the United States on biofuels and bioenergies. This observation is linked mainly to the potential interest of this nation in the minimization of its dependence on fossil fuels. Then come Canada and India with each four important publications. This also underlines the commitment of these two nations in the path of energy transition through approaches of solutions appropriate to their needs.

On the other hand, at the European level, France stands out with five publications, followed by other countries such as Germany, Finland, and the United Kingdom, each with three scientific works. The work was also carried out specifically on the Swedish context (two publications) and Poland (two publications), thus reflecting a deep commitment in European countries on energy transition issues.

Finally, several African countries, such as Mali (two publications), Nigeria, and Burkina Faso have onlyone publication each. Although modest, these contributions also highlight a growing interest in these topics in Africa, particularly in regions where access to energy remains a major issue. These results illustrate the global dynamics around the issues of biofuels and bioenergy. The joint efforts of researchers from around the world highlight not only the growing interest in these themes but also their cross-cutting nature, integrating economic, social, and environmental considerations. This diverse geographical distribution shows that, even if some countries or regions are at the forefront, research on biofuels and bioenergy is a universal issue.

10.3.3 State of Play of Biofuels and Bioenergy

10.3.3.1 Definition and Typology of Biofuels and Bioenergy According to Studies

Bioenergy, i.e., the use of biomass feedstocks to provide energy, has become a growing source of renewable energy in Europe. It is used not only for heating and cooling (increasing from 66% to 90% of total renewable heat between 1990 and 2018) but also for bio-blended transport fuels and subsidized bioelectricity (Banja et al. 2019). There are currently at least five different forms of biofuels and bioenergy, namely, solid biofuels, biogas, biodiesel, bioethanol, and others (Amigun et al. 2011; Cavelius et al. 2023; Wu and Pfenninger 2023) (Fig. 10.4). The most well-known solid biofuels are organic materials such as sawdust, wood chips, firewood, bark, and animal matter (Hurmekoski et al. 2018; Jabari et al. 2022; Kiehbadroudinezhad et al. 2023; Manandhar et al. 2022). These resources, often derived from forest or agricultural residues, are more present in the energy transformation sector in many regions of the world but much more in the Americas than in Europe (Sahoo et al. 2022). Indeed, in Europe, although the share of solid biofuels in overall bioenergy use has decreased considerably, from 81% in 2000 to 60% in

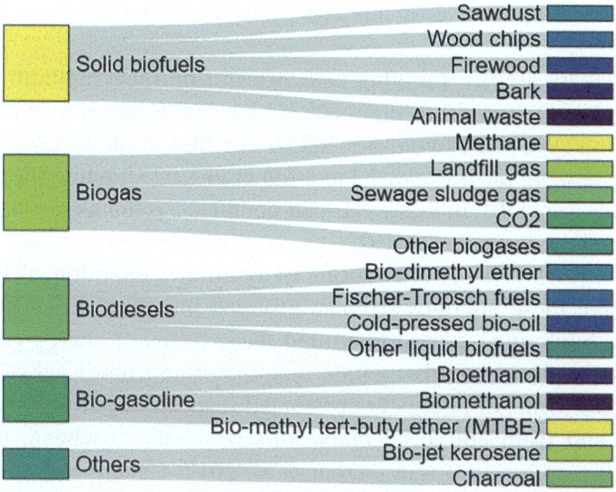

Fig. 10.4 Classification of the main types of biofuels and bioenergy and their products and raw materials

2017, their absolute volume continues to grow (Wu and Pfenninger 2023). They are mainly used for domestic and industrial heating, their availability being favored by well-established infrastructures for their collection and treatment.

Biogas, mainly from anaerobic fermentation, includes products such as methane, landfill gas, sewage sludge gas, carbon dioxide, and other biogases (Liebetrau et al. 2019; Rasi et al. 2007; Wu and Pfenninger 2023). However, the valuable component of biogas is methane, with a proportion of 55–70%, compared to about 35–40% for carbon dioxide, and a very small percentage of nitrogen and other chemical compounds (hydrogen sulfide) (Jameel et al. 2024). The quality of biogas is highly dependent on the methane content. The higher the methane content, the better the quality of biogas, and the more economical its production (Jameel et al. 2024). It is a category of bioenergy that is greatly appreciated for its ability to be used both to produce electricity and to power cars (Wu and Pfenninger 2023). In Europe, the use of biogas has grown significantly, from 3% in 2000 to 10% in 2017, while it remains underexploited in Africa due to technical challenges (Amigun et al. 2011; Wu and Pfenninger 2023).

Regarding biodiesel, the raw materials used mainly concern Fischer-Tropsch fuel and cold-pressed bio-oils (Himabindu and Ravikrishna 2010; Kokkinos et al. 2024; Wu and Pfenninger 2023). This form of energy is exploited particularly as an alternative to fossil diesel, especially for its compatibility with existing transport infrastructures (Huang et al. 2012; Murugesan et al. 2009). In terms of spatial distribution, biodiesel is one of the two most exploited forms of biofuels in Africa, particularly to power vehicle engines (Amigun et al. 2011). A lthough its production is evolving considerably in the European area (11% of total bioenergy in 2017 compared to 1% in 2000), this energy is still significantly dependent on imports (Wu and Pfenninger 2023). Precisely, more than 33% of this energy comes from outside the

EU, reflecting the importance of this energy in the global biofuel trade (Wu and Pfenninger 2023).

In addition, there are several other categories of biofuels including biokerosene (useful for airplanes), as well as charcoal, which is fundamental for domestic and industrial heating (Wu and Pfenninger 2023). The first (biokerosene) has the specificity of decarbonizing air transport (Jacobsen et al. 2023), while the second (charcoal), dominant especially in many African countries, remains a pillar of the local economy (Nabukalu and Gieré 2019).

However, biofuels follow other forms of typology, including the generational typology. Indeed, these energies are also grouped into four different generations, marking the technological evolution around these forms of energy (Cavelius et al. 2023). First-generation biofuels include bioethanol and biodiesel (Cavelius et al. 2023). These two forms of energy are produced respectively from food crops (corn or sugarcane) and from edible vegetable oils (rapeseed, soybean, and palm oil) depending on the region of origin (Azhar et al. 2017; Cavelius et al. 2023; Favaro et al. 2019; Tse et al. 2021). However, the production of biofuels of this generation faces significant concerns especially regarding competition with food crops and excessive exploitation of agricultural land. This has led to the birth of the following generations (Cavelius et al. 2023).

It is in this context that second-generation biofuels have been developed using non-food lignocellulosic resources and agricultural or industrial waste, such as crop residues, animal fats, or used cooking oils (Cavelius et al. 2023; Kumar and Sharma 2008). They are produced from materials such as *Jatropha curcas* or other drought-resistant plants, capable of surviving on uncultivated land. This choice thus aims to minimize the impact of the adoption of renewable energies on food production (Cavelius et al. 2023; Kumar and Sharma 2008). Unfortunately, despite their benefits, the production of these biofuels remains limited by inefficient processes, including low butanol production and product inhibition, leading to research on even more advanced solutions (Jiang et al. 2019), hence the birth of third- generation biofuels.

The latter focus on the use of microalgae and cyanobacteria, which naturally produce lipids and alcohols that can be transformed into biodiesel or other high-energy fuels (Slade and Bauen 2013). These biofuels do not require the use of agricultural land and thus allow the complete decoupling of biofuel production from food (Cavelius et al. 2023). However, their production is energy-intensive, making their downstream transformation more expensive compared to other generations (Cavelius et al. 2023). Faced with these challenges, fourth-generation biofuels integrate genetic engineering to improve production, for example, by optimizing photosynthesis or increasing lipid synthesis in the organisms used (Dexter and Fu 2009). Although this approach promises significant advances, it is still limited by poorly developed genetic tools for certain biofuel-producing organisms, thus hampering their large-scale adoption (Cavelius et al. 2023).

10.3.3.2 Main Areas of Interest of Selected Studies by Country

This subsection explores the main research areas addressed in the selected studies, highlighting the predominant themes developed according to the world regions and their specific contexts. Indeed, the analysis of the areas of interest reveals a high concentration of studies on biofuels, bioenergy, and energy policies (BBPE), which obtain a very high score (20 articles). This theme is particularly predominant in Europe (20 articles), followed by America (13) and Asia (9), reflecting the importance of regulatory frameworks and energy strategies for the transition to sustainable energy sources.

The state of play and climate impacts of biofuels and bioenergy (ELECBB) comes next with a high score (16 articles), mainly addressed in Europe (16) and America (15), as well as at the global level (11). These results reflect a growing concern to assess the climate impact of biofuels, a major issue to align energy policies with global emission reduction targets. On the other hand, the theme around economic, environmental, and social challenges, as well as opportunities related to the energy transition (DEESOTE), records an average score (five articles) at the regional level, with notable contributions in Europe (five) and Asia (four). This work explores the multidimensional implications of energy transitions in various economic and social contexts (Fig. 10.5).

The socioeconomic impacts of bioenergy (BISE) score average are mainly explored in Europe (10) and America (8). This attention reflects a desire to better understand the socioeconomic effects of bioenergy, particularly in terms of job creation and impact on local communities. Also, the themes around biofuels and their environmental impacts (BIE) as well as those showing the relationship between bioenergy and water (RBE) record average scores, with significant contributions in Europe (6) and America (6). These themes reflect a growing concern for the sustainable use of natural resources and the environmental impacts associated with bioenergy production. In addition to these areas of interest, there are also themes highlighting the relationships between climate, land, energy and water (RCTEE), bioenergy, and land use (BUT), as well as issues surrounding biofuels and policy impacts (BIP). These areas of interest are poorly addressed (maximum two articles per region), mainly in Africa and America. These themes, although less explored, highlight critical issues related to environmental trade-offs and energy policies.

Similarly, studies addressing climate policies and bioenergy sources (PCSB) and the adoption or development of biofuels and bioenergy (ADBB) remain marginal, with four and five articles, respectively, distributed mainly in America (three) and Africa (two). These topics require further attention to fill the gaps in their implementation and acceptance by communities. The least addressed area of interest is the assessment of climate change impacts on bioenergy supply systems (ICCSAB). Finally, other themes, such as energy perspectives or public attitudes, are developed particularly in Asia (eight articles) and Europe (six articles), reflecting an interest in energy transitions in response to high energy demand and pollution concerns.

These results highlight a prioritization of strategic and environmental themes in Europe and America, while socioeconomic themes remain behind, particularly in

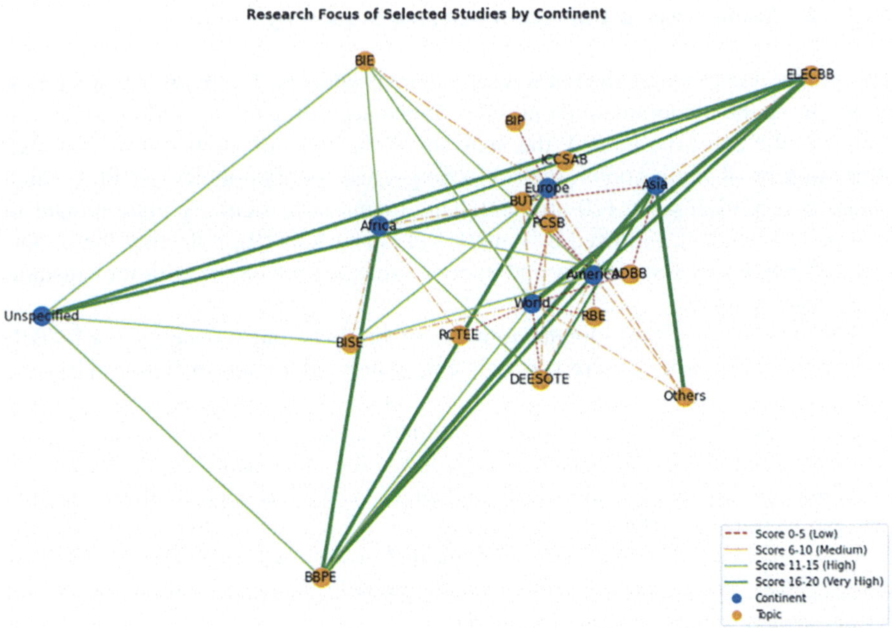

Fig. 10.5 Focus of selected studies. *ELECBB* status and climate effects of biofuels and bioenergy, *DEESOTE* challenges (economic, environmental, and social) and opportunities related to the energy transition (bioenergy), *BBPE* biofuels, bioenergy, and energy policies, *PCSB* climate policies and bioenergy source, *BISE* bioenergy and socioeconomic impacts, *ICCSAB* impacts of CC on bioenergy supply systems, *BUT* bioenergy and land use, *BIP* biofuels and policy impacts, *BIE* biofuels and environmental impacts, *RCTEE* relationship between climate, land, energy, and water, *RBE* relationship between bioenergy and water, *ADBB* adoption or development of biofuels and bioenergy, *others* other interests (energy perspectives, public attitudes or visions towards biofuels, biofuel marketing, energy production forecasting)

Africa and Asia. The overall balance of studies could be improved by paying greater attention to underrepresented regions and emerging themes for an inclusive and sustainable energy transition.

10.4 Global Policies Around Biofuels and Bioenergy in Combating Climate Change

10.4.1 Analysis of International Guidelines (Global Policies)

10.4.1.1 Biofuel Production Incentive Policy

In 2003, a directive (Directive 2003/96/EC70) was introduced authorizing Member States to use fiscal tools for the development of biofuels (Akbi 2013). This policy aimed to stimulate biofuel production in Europe through different financial and fiscal mechanisms (Akbi 2013). Thus, to reduce production costs, agricultural subsidies were introduced under the CAP, such as the aid for energy crops (ACE) set at €45/hectare until 2008 (Persillet Shonkwiler 2012). Farmers also benefited from tax incentives, authorized by Directive 2003/96/EC, allowing total or partial exemptions under certain conditions, thus promoting the competitiveness of biofuels (Akbi 2013). At the same time, industrial investment relief mechanisms (capital grants, guaranteed loans, tax exemptions) have been put in place to support infrastructure creation, with notable examples such as in Austria, where subsidies covering up to 55% of costs have been granted. However, subsequent reforms (draft Directive 2011) have reoriented aid according to environmental performance, setting minimum taxes based on CO_2 emissions and energy content, marking a shift towards more conditional and sustainable support for biofuels (Table 10.2).

10.4.1.2 Border Protection Policy

This policy aims to protect the European biofuels market from international competition through differentiated customs duties. Ethanol, classified as an agricultural product, is subject to high customs duties (25% to 45% ad valorem), while biodiesel, considered a chemical product, benefits from lower tariffs (0% for some oils such as palm oil and up to 3.2% for others) (Akbi 2013; Baumber 2017). However, practices such as importing ethanol blended with gasoline to reduce customs duties have helped circumvent these measures, highlighting the flaws in this system. Indeed, limited import controls have long allowed subsidized products, particularly from the United States, to enter the European market at competitive prices (Akbi 2013). It was only in 2009 that anti-dumping measures were taken to limit these imbalances, particularly regarding American biodiesel producers (Lorne 2011; Morency 2011).

This policy also includes preferential regimes for developing countries, facilitating their access to the European market (Stevens 2008). These agreements, such as the Generalized System of Preferences (GSP) and the ACP agreements in the framework of economic partnerships, make it possible to exempt or reduce customs duties for major biofuel exporters such as Brazil, Argentina, Indonesia, and Malaysia (Akbi 2013). Since 2008, this openness has strengthened trade with these countries, while promoting their economic development. However, this dual approach (protection and openness) reflects a certain tension between the protection of European producers and the promotion of fair trade with international partners.

10.4.1.3 Directive 2009/28/EC or RED I

The Renewable Energy Directive 2009/28/EC, known as RED I, marked an important step in the promotion of renewable energy within the European Union. Adopted in 2009, this policy aimed to diversify the energy mix, reduce greenhouse gas (GHG) emissions, and strengthen energy security (Mai-Moulin et al. 2021). It

Table 10.2 Key policies around biofuels and bioenergy: objectives and results

Policies	Main objectives	Results obtained	References
Policy on incentives for the production of biofuels (Directive 2003/96/EC)	**Boosting biofuel production in Europe**: Encourage the development of the sector via financial and fiscal mechanisms to reduce GHG emissions in the energy and agricultural sectors	**Reduction of biofuel production costs**: Thanks to tax exemptions and subsidies, the competitiveness of biofuels has been strengthened **Expansion of production infrastructure**: Establishment of infrastructure supported by financial mechanisms promoting industrial investments **Transition to sustainable support**: The 2011 reforms marked a shift towards incentives more aligned with environmental objectives, strengthening the sustainability of the sector	Akbi (2013) and Persillet Shonkwiler (2012)
Border protection policy	Protect the European biofuels market against international competition while promoting trade with developing countries	**Strengthening international trade**: Since 2008, opening of the European market to developing countries, stimulating their economy **Limited protection for European producers**: Imports of blended ethanol and delayed control of imports have allowed subsidized products, particularly American ones, to compete with European products **Impact of anti-dumping measures:** In 2009, they reduced the imbalances caused by US subsidies, improving the competitiveness of European producers	Akbi (2013), Baumber (2017), Lorne (2011), Morency (2011) and Stevens (2008)

(continued)

Table 10.2 (continued)

Policies	Main objectives	Results obtained	References
European Union Renewable Energy Directives RED I ((Directive 2009/28/EC)	**Promoting renewable energy**: Reach 20% of their share in the EU's gross final energy consumption by 2020 **Reduce greenhouse gas (GHG) emissions**: Promote a transition to a low-carbon economy **Strengthening energy security**: Diversify energy sources and reduce dependence on fossil fuels **Boosting the integration of renewable energies in transport**: Ensure that at least 10% of the energy consumed in transport comes from renewable sources	**Advances**: Strong growth in bioenergy and partial reduction of EU energy dependence **Challenges:** Lack of sustainability criteria for solid and gaseous biomass; heterogeneous implementation due to the development of national criteria adapted to local specificities	Akbi (2013), Mai-Moulin et al. (2021), Masjuki et al. (2013) and Montini (2024)
European Union Renewable Energy Directives RED II ((Directive 2018/2001)	**Setting new goals for 2030**: Achieve at least 32% of renewable energy in gross final energy consumption, with a sub-target of 14% in transport **Strengthening environmental sustainability**: Introduce binding criteria for large solid and gaseous biomass installations **Improving resource efficiency**: Promote optimal use of biomass and limit negative environmental impacts **Reducing risks associated with indirect land-use change**: Encourage the use of residues and waste as raw materials **Supporting the climate commitments of the Paris Agreement**: Contribute to limiting the increase in global temperature to well below 2 °C	**Catalyst for energy communities**: The RED II directive has boosted the development of renewable energy communities (RECs) in Italy, growing from 35 operational initiatives to 41 planned projects and 24 in the initial phase in 2022, thanks to adapted national and regional policy frameworks **Establishment of legal and institutional frameworks**: Adoption of several legislative decrees (notably Decree No. 199 of 2021) to transpose RED II provisions, including criteria such as voluntary participation and autonomy, while relaxing certain capacity restrictions to encourage the development of RECs	Hennenberg et al. (2018), Krug et al. (2022, 2023); Montini (2024), Searchinger et al. (2018) and Webster (2020)

(continued)

Table 10.2 (continued)

Policies	Main objectives	Results obtained	References
Kyoto Protocol and Paris Agreements	Reduce global greenhouse gas (GHG) emissions to limit global warming to 2 °C, with an ambition of 1.5 °C for the Paris Agreement	**Limited progress in reducing emissions**: Disparities in the implementation of national commitments and technical challenges related to the sustainability of biofuels **Impact of agricultural and forestry practices**: Intensive practices, forest fires, and social impacts have reduced the environmental benefits of biofuels, increasing upstream GHG emissions **Global Perspectives**: An acceleration of global decarbonization efforts between 2022 and 2030 remains imperative to achieve the objectives of the Paris Agreement	Kumar (2018), Sarika et al. (2023), Kumar et al. (2020), Santos et al. (2022), Costa et al. (2020) and Sen and Dabi (2021)
Global Bioenergy Partnership (GBEP) Policy	**Promoting sustainable bioenergy through international partnerships and providing tools**, including 24 sustainability indicators, to help countries integrate bioenergy into their energy policies	**Limitations in implementation**: Weak institutional and financial capacities in some countries to assess impacts **Progress in coordination**: Difficulties in coordinating partners and integrating bioenergies into global policies **Platform for harmonization**: GBEP remains an important framework for harmonizing international efforts despite persistent challenges	Ramirez-Contreras and Faaij (2018), Van de Graaf and Westphal (2011), Naiki (2016), Pirelli et al. (2021) and Coelho and Goldemberg (2013)

(continued)

Table 10.2 (continued)

Policies	Main objectives	Results obtained	References
Sustainable Energy for All (SE4All) Initiative	Significantly increase the contribution of renewable energy to global energy consumption by 2030	**Progress**: Increasing interest in bioenergy at the international level **Challenges**: Spatial disparity in progress; low investment, low adoption of sustainable technologies, and governance issues	Gallagher and Wykes (2014), Rogelj et al. (2013), Chirambo (2018), Anindya (2015) and Bellanca and Wilson (2012)
Sustainable Development Goals (SDG 7)	Ensure universal access to affordable, reliable, sustainable, and modern energy by 2030, integrating biofuels as a lever for the energy transition	**Advances:** Increasing the level of access to rural electricity and increasing renewable energy in some countries **Challenges:** Challenges related to the sustainability of biofuels; land- use conflicts and environmental impacts associated with biomass cultivation	Nam-Chol and Kim (2019), Acheampong et al. (2017), Gaffney and Marley (2009), Scovronick and Wilkinson (2014), Kothari et al. (2023), Stoeglehner (2020), Trinh and Chung (2023), Li et al. (2022), Olabi et al. (2023), Swain and Karimu (2020) and Vlek et al. (2017)

imposed binding targets on Member States: to reach 20% renewable energy in gross final energy consumption and to integrate at least 10% renewable energy in the transport sector by 2020 (Akbi 2013; Mai-Moulin et al. 2021; Masjuki et al. 2013). This is a directive that responds to the commitments of EU countries regarding the issues surrounding climate change and the energy transition in Europe.

However, this directive, despite having promoted the adoption of bioenergy, has considerable shortcomings. On the one hand, in terms of limitations, it is a policy that does not clearly highlight the sustainability criteria for solid and gaseous biomass used for electricity, heating, and cooling (Mai-Moulin et al. 2021). It is in this context that the European Commission (EC) has suggested the establishment of national sustainability criteria adapted to local specificities, thus leading to a certain heterogeneity in implementation (Mai-Moulin et al. 2021).

10.4.1.4 RED II

To address the shortcomings of the first directive (RED I), the EU developed the RED II directive in 2018, revising and strengthening the RED I measures for the period 2021–2030 (Krug et al. 2023; Webster 2020). As a specific feature, this new directive set as its main goal to reach 32% renewable energy in gross final energy consumption by 2030 (Mai-Moulin et al. 2021). This directive introduced binding environmental criteria for large bioenergy plants, while encouraging sustainable forest management and limiting negative environmental impacts. RED II also aimed to improve resource efficiency by setting thresholds for bioheating installations and power plants, but it has been criticized for some definitions deemed incomplete, notably on sustainable forest management and the use of agricultural or forestry residues (Hennenberg et al. 2018; Searchinger et al. 2018). Although the directive has enabled significant progress, concerns remain about the unsustainable mobilization of raw materials and the risk of carbon stock depletion, particularly for biofuels certified as having a low indirect impact on land use (Mai-Moulin et al. 2021). Results obtained up to 2020 have demonstrated significant progress towards energy targets, but limitations remain in terms of coherence and uniformity in the application of sustainability criteria at the European level (Mai-Moulin et al. 2021).

10.4.1.5 Kyoto Protocol and Paris Agreement

The Kyoto Protocol (1997) and the Paris Agreement (2015) are major initiatives in the fight against climate change, integrating the promotion of biofuels as a solution to reduce greenhouse gas (GHG) emissions (Kumar 2018; Sarika et al. 2023). The main objective is to limit global warming to 2 °C (with an ambition of 1.5 °C for Paris) by promoting incentive mechanisms such as the Clean Development Mechanism (CDM) (Kumar 2018; Kumar et al. 2020). However, the Kyoto Protocol policy only sets binding emission reduction targets for industrialized countries and economies in transition and the European Union (Santos et al. 2022). Thus, this policy allows developed countries to invest in projects reducing GHGs, notably through the production and use of biofuels in developing countries, while obtaining carbon credits.

Despite their ambition, limited progress is observed due to the disparity in the implementation of national commitments and technical challenges related to the sustainability of biofuels (Costa et al. 2020). Some countries (France, Germany, etc.) have invested in local programs to integrate biofuels into their energy mix. However, the gains in terms of emission reductions are often compromised by intensive agricultural practices, wildfires, and social challenges for indigenous and local communities in developing countries increasing upstream GHG emissions (Costa et al. 2020; Sen and Dabi 2021). These agreements therefore show their limits, and rapid global decarbonization between 2022 and 2030 is still mandatory to achieve the Paris targets (Santos et al. 2022).

10.4.1.6 Global Bioenergy Partnership (GBEP)

Launched in 2006 by the G8, the Global Bioenergy Partnership (GBEP) policy aims to promote sustainable bioenergy through international partnerships (Ramirez-Contreras and Faaij 2018; Van de Graaf and Westphal 2011). The main objective is to provide tools to guide countries in the development of energy policies integrating bioenergy. Among the flagship actions, GBEP has defined 24 sustainability indicators divided into three categories: environmental, social, and economic, facilitating the assessment of the impacts of bioenergy projects (Naiki 2016; Pirelli et al. 2021; Ramirez-Contreras and Faaij 2018). These are sustainability indicators for bioenergy and biomass, which can contribute to the reduction of GHG emissions and facilitate access to bioenergy (Pirelli et al. 2021).

Indeed, one of the issues that concern the GBEP is food security because food production has a complex and multifaceted relationship with bioenergy (Ramirez-Contreras and Faaij 2018). Thus, this policy also aims to demonstrate that the production and sustainable use of bioenergy can contribute to both energy and food security. For this reason, the main indicators of the GBEP are related to food security: (1) price and supply of a national food basket, (2) land use and LUC, (3) land allocation and tenure, (4) income change, (5) bioenergy used to expand access to modern energy services, and (6) infrastructure and logistics for the distribution of bioenergy (Ramirez-Contreras and Faaij 2018). This set of indicators is complemented by other indicators that affect food security, such as soil quality, landscape biodiversity, water use and efficiency, and jobs in the bioenergy sector (Ramirez-Contreras and Faaij 2018).

However, progress in implementing these indicators remains limited, particularly in countries where institutional and financial capacities are lacking to properly assess impacts (Pirelli et al. 2021). Coordination among partners remains a challenge, as does the integration of bioenergy into global energy policies (Coelho and Goldemberg 2013). However, this initiative remains an important platform to harmonize international efforts.

10.4.1.7 Sustainable Energy for All (SE4All)

Launched by the UN in 2012, the Sustainable Energy for All (SE4All) initiative aims to double the share of renewable energy, including bioenergy, in the global energy mix by 2030 (Gallagher and Wykes 2014; Rogelj et al. 2013). It focuses on access to sustainable energy, through three energy-related ambitions, namely, (1) ensuring access to modern energy services for all, (2) doubling the contribution of renewable energy to global energy consumption, and (3) doubling the global rate of improvement in energy efficiency (Chirambo 2018). To achieve this, several strategies have been developed. This mainly involves support for various bioenergy projects and programs aimed at improving energy efficiency, through the promotion of public-private partnerships (Anindya 2015; Bellanca and Wilson 2012; Gallagher and Wykes 2014).

However, despite the importance of this policy, there is a spatial disparity in progress towards achieving the objectives related to the energy transition (Chirambo 2018). As major constraints encountered, we still find insufficient investment, the low level of use of renewable energies, and governance problems in some countries (Chirambo 2018). As a result, SE4All remains a flagship initiative but requires increased efforts to accelerate the achievement of its objectives.

10.4.1.8 Sustainable Development Goals (SDGs)

Another international policy implemented in this context is SDG 7. Established in 2015, this policy also seeks to guarantee access for all to reliable, modern, and sustainable energy (Nam-Chol and Kim 2019). Indeed, biofuels and bioenergies have been placed at the heart of this policy as the main key levers that can promote the energy transition to the detriment of fossil fuels (Acheampong et al. 2017; Gaffney and Marley 2009; Scovronick and Wilkinson 2014). The strategies developed include the introduction of these renewable energies into national approaches, support for bioenergy infrastructures, and the establishment of financial mechanisms aimed at attracting private investments (Kothari et al. 2023; Stoeglehner 2020; Trinh and Chung 2023).

Although this policy has positively influenced access to electricity in rural areas and the adoption of renewable energies in several countries, progress is still hampered by constraints associated with the question of sustainability of these energy sources (Li et al. 2022; Olabi et al. 2023; Swain and Karimu 2020). At the same time, objections continue to arise, especially regarding land- use conflicts and environmental concerns related to biomass cultivation (Vlek et al. 2017). Thus, SDG 7 remains a global benchmark, but requires more integrated approaches to balance development and sustainability.

10.4.2 Analysis of Regional Directives or Policies: Some Case Studies of Regional Policies (Countries)

10.4.2.1 Case of the United States

The United States has issued several impact policies for the promotion of renewable energy and biofuels, in order to reduce greenhouse gas (GHG) emissions and support the transition to energy independence (Table 10.3). To support the transition, we note significant contributions such as the Energy Independence and Security Act (EISA) of 2007, aiming to absorb 36 billion gallons of biofuels by 2022, and the 2008 Farm Bill, revised in 2014, which through the BCAP program participates in the production of biomass and the development of biorefineries (Bacovsky et al. 2016). The Portfolio Standards (RPS) support states in the diversification of their energy source through renewable energy quotas unlike the State Energy Program (SEP) which acts for the development of an economy based on clean energy.

Financial resources, including loan guarantees, biodiesel tax credits, and biomass research grants, reinforce this contribution, while supporting private investment and technological innovation (Bacovsky et al. 2016). However, the development of these policies faces enormous challenges, including mismanagement between the federal and state levels and the lack of transparency on the sustainability of energy resources (Rossi 2013). However, although there have been some challenges, these contributions have marked the trigger for the development of bioenergy production

Table 10.3 Case studies of some energy policies: the United States, Canada, Germany, and Brazil

Country	Main objectives targeted	Policies	Results (Success)	References
The United States	– Reducing GHG emissions through the adoption of renewable energy – Boosting the production and use of biofuels, with a goal of 36 billion gallons of biofuels consumed by 2022, including 21 billion gallons of advanced biofuels – Promoting energy independence and security while stimulating innovation and the clean energy economy	**Key policies include:** – The Energy Independence and Security Act (EISA 2007) – The agricultural law of 2008 and 2014 – The State Energy Program (SEP) – Renewable Energy Portfolio Standards (RPS) – Financial incentives such as loan guarantees, biodiesel tax credits, and the BCAP program are also being implemented	– Advances in biofuel production and development of biomass projects via BCAP – Diversity in the use of renewable energies through RPS – Limited progress in coordinating energy, environmental, and security policies, hampering a coherent approach at the federal and state levels	Bacovsky et al. (2016), Sciortino (2010) and Rossi (2013)
Canada	Promote the sustainable energy transition by reducing greenhouse gas emissions, while promoting the development of renewable energies and technological innovation	– **Federal Renewable Fuels Regulations (2010):** Impose a minimum renewable content in fossil fuels – **Energy Research and Development program (ERDP)**: Develop knowledge and tools on biomass for energy uses – **ecoENERGY Programs**: Promote biofuels, energy efficiency, and renewable energy – **Green Energy Act, 2009 (Ontario)**: Supporting the energy transition and the green economy – **Climate Change Strategy (Alberta, 2008)**: Reduce GHG emissions from industries – **Regional initiatives (Quebec, British Columbia):** Promoting bioenergy considering local needs	– Reduction of GHG emissions – Support for the biofuel market – Increased availability of biomass resources and improved conversion technologies – Increased use of agricultural biogas and biomass; support for farmers in their energy transition – Strengthening local capacities and supporting bioenergy projects	Bacovsky et al. (2016), Laan et al. (2009), Smith et al. (2013), Bradley (2012) and Madiraju (2014)

(continued)

Table 10.3 (continued)

Country	Main objectives targeted	Policies	Results (Success)	References
Germany	Increase renewable energy to 18% of gross final energy consumption by 2020 Reduce GHG emissions: 40% (2020), 55% (2030), 70% (2040), 80–95% (2050), base 1990 Achieve 80% renewable energy in the electricity sector by 2050	– EEG (Renewable Energy) Act with: Guaranteed feed-in tariffs to promote renewable energy; gradual transition to auctions (from 2017); 2014 amendment (reduction of tariffs for biomass, ceiling of 100 MWel/year for new installations) – Cogeneration Act (KWKG): Ordinances on biomass (BiomasseV) and its sustainability (BioSt-NachV)	– In 2019, renewable energy accounted for 17.4% of gross energy consumption – Reduction of GHG emissions by approximately 35% in 2019 compared to 1990 – In 2020, approximately 46% of electricity came from renewable sources – Constant progress in improving the competitiveness of renewable energies and reducing dependence on fossil fuels	Bacovsky et al. (2016), Fell (2019), Ohlhorst (2020), Bauermann (2016), Cansino et al. (2010) and Schomerus (2012)

Brazil	**Promoting a sustainable energy transition**	– **Promotion of sugarcane-based ethanol** (implementation of the ethanol and gasoline blending program "gasohol" with a mandatory 27.5% ethanol blend; strategic use of vast agricultural resources for sugarcane cultivation; reducing dependence on imported fossil fuels through ethanol as a transportation fuel) – **Exploitation of biomass for energy production** (development of biomass power plants using agricultural, forestry, and organic waste; cogeneration and production of electricity from biomass; contribution to rural development and job creation in agricultural districts; management of organic waste to limit environmental impacts) – **National Biofuels Policy (RenovaBio)** (setting targets to reduce greenhouse gas emissions in the transport industry; increasing the consumption of biofuels in the national energy matrix; creating an open market for carbon emission reduction credits) – **Financial support and incentives** (tax breaks for bioenergy projects; preferential interest rate credit lines to encourage investment in the sector; blending requirements to promote the use of biofuels in transport) – **Sustainable agricultural practices** (promotion of sustainable sugarcane cultivation, with environmentally friendly practices; protection against deforestation and compliance with strict environmental regulations; soil conservation, carbon sequestration and protection of biodiversity through sustainable agricultural methods)	– **Reducing dependence on fossil fuels**: Supportive policies have enabled massive adoption of sugarcane-based ethanol and biomass-based electricity production, thereby diversified the energy mix and strengthened energy security – **Global Leadership and Energy Sustainability**: Brazil has become a major player in the bioenergy industry thanks to its natural resources, sustainable agricultural practices, and an energy transition focused on reducing greenhouse gas emissions	Falcone (2023), Gravina et al. (2021), Hernandez et al. (2021), Tenelli et al. (2021), Borges et al. (2021), Grangeia et al. (2022) and Khartukov and Starostina (2022)

and have reinforced the diversification of the national energy mix (Bacovsky et al. 2016).

10.4.2.2 Case of Canada

Canada has also issued a number of policies in promoting renewable energy and reducing greenhouse gas emissions. Minimum limits for renewable content in fossil fuels (5% bioethanol and 2% biodiesel) have been established by the federal Renewable Fuels Regulations (Bacovsky et al. 2016; Laan et al. 2009; Smith et al. 2013). However, programs such as ecoENERGY and IRAP encourage biomass research and technological innovation to contribute to more efficient energy and ensure its progression towards clean energy (Bradley 2012; Madiraju 2014). Also, provinces such as Ontario and Alberta, issue specific policies in promoting biomass and agricultural biogas, involving financial support for the support in the adoption of renewable energy and the reduction of some tax payments (Bacovsky et al. 2016). Provincial actions, such as the Agricultural Adaptation Program or the actions of the RECs in Quebec, incorporate local and regional needs into the development of bioenergy (Bacovsky et al. 2016). For example, British Columbia has made available 16.1 million Canadian dollars in bioenergy projects, and the province of Ontario provides support in the management of agricultural biogas production. These issued actions facilitate the diversification of energy sources, activate local economies, and help reduce GHG emissions, although some programs require more support to have a significant impact.

10.4.2.3 Case of Germany

Germany has developed an upstart strategic energy concept, which consists of incorporating renewable energies, and at the same time involving the reduction of greenhouse gas (GHG) emissions. This concept, based on the 2009 energy theory, defines as objectives the addition of the proportion of renewable energies in gross energy consumption to 18% by 2020 and the reduction of 40% of GHGs by the same year compared to 1990 (Bacovsky et al. 2016), a commitment that is reinforced with a continuous objective in the reduction of 80 to 95% of GHGs by 2050 (Bacovsky et al. 2016). The Renewable Energy Act (EEG), formerly referred to as the buy-back and currently attributed to the auction system, is a significant means to migrate towards the significant proportion of 80% renewable energy in the electricity sector by the 2050s (Fell 2019; Ohlhorst 2020). However, there has been a slowdown in the development of this field since the 2014s following the decrease in biomass tariffs and the setting of the annual cap on new capacities (Bacovsky et al. 2016).

Programs such as the Market Incentive Program (MAP) provide financial means for the promotion of renewable energies in the areas of heating and cooling (Bauermann 2016; Cansino et al. 2010). In addition, some laws such as the

Cogeneration Act (KWKG) and the Biomass Ordinances set rules for the energy transition (Schomerus 2012). In general, the German concept aims to promote efficient, competitive, and environmentally friendly energy, with a view to facilitating the competitive spirit of the industry, and also considering these policies on the climate commitments of the coalition agreement.

10.4.2.4 Case of Brazil

Brazil, for its part, has implemented upstart actions thanks to its numerous agricultural and forestry resources for the promotion towards a sustainable energy transition (Falcone 2023). This strategy mainly depends on ethanol (chemical compound made from sugarcane); the "gasohol" program sets a mixture of 27.5% ethanol in gasoline (Gravina et al. 2021; Hernandes et al. 2021). This theory significantly reduces the use of imported fossil fuels in favor of local resources (Tenelli et al. 2021). In addition, the use of biomass from agricultural and forestry waste has facilitated the progress of cogeneration and electricity production for the benefit of sustainable management of organic waste and rural employment (Borges et al. 2021). To strengthen this dynamic, the National Biofuels Policy (RenovaBio) sets a number of ambitious objectives in helping to reduce greenhouse gas emissions and in creating a carbon credit market accompanied by an increase in the proportion of biofuels in the energy mix (Falcone 2023; Grangeia et al. 2022).

Due to these contributions, Brazil has positioned itself as a global leader in bioenergy (Tenelli et al. 2021). Financial incentives, such as tax breaks and preferential loans, support the sector's numerous investments (Falcone 2023). Also, the emphasis on sustainable agricultural practices, such as soil conservation, protection against deforestation, and carbon sequestration, have accentuated the positive environmental impact of energy policies (Khartukov and Starostina 2022; Tenelli et al. 2021). The results highlight a successful diversification of the energy mix and a significant restoration of energy security, confirming Brazil as an example for an economically efficient, as well as environmentally friendly, energy transition (Falcone 2023).

10.5 Contributions of Biofuels to the Fight Against Climate Change

10.5.1 Environmental and Economic Benefits of Biofuels and Bioenergy

10.5.1.1 Environmental Impacts

Biofuels and bioenergy occupy a strategic position in the fight against climate change thanks to their unique roles and contributions to the decarbonization of energy sectors (Wu and Pfenninger 2023) (Fig. 10.5). Unlike other renewable energy sources such as wind or solar, bioenergy offers the capacity to stabilize energy intermittency, thus strengthening the resilience of energy systems (Thrän 2015; Szarka et al. 2013). In Europe, where wind energy and solar energy dominate, bioenergy can balance fluctuations through technical solutions such as intermediate biomass storage or the conversion of biogas to biomethane (Thrän 2015). Furthermore, bioenergy is essential in sectors that are difficult to decarbonize, such as industry or heavy transport (Wu and Pfenninger 2023). It also enables negative emissions through technologies such as BECCS (bioenergy with carbon capture and storage), which is considered essential to achieve the emission reduction targets set by the 1.5 °C and 2 °C scenarios (Muri 2018). However, these environmental benefits depend on sustainable resource management, to avoid net losses of terrestrial carbon or overuse of freshwater resources (Wu and Pfenninger 2023).

At the same time, Birdsey et al. (2018) demonstrate that the use of biomass from forest residues, such as tops, stumps, or branches, leads to net reductions in greenhouse gas emissions in the short term, particularly when these residues would otherwise have been burned or left to decompose (Fig. 10.5). For example, in Canada, the incorporation of these residues into bioenergy not only reduces emissions but also reduces the risk of wildfires (Birdsey et al. 2018). However, the choices made in the use of these biomasses strongly influence their climate impact. Using biomass for long-life products, such as composite panels, could offer greater climate benefits than its direct use for bioenergy. This impact is also confirmed by Akbi (2013), when he emphasizes that maximizing the proportion of harvested wood intended for these products while reserving the residues for bioenergy significantly increases the mitigation benefits (Akbi 2013).

A final aspect is that the production of biofuels and bioenergy helps prevent the loss of valuable biodiversity while limiting undesirable changes in land use. In this context, Demirbas (2009) claims that they contribute to more efficient management of land and water resources. These balanced approaches highlight the importance of bioenergy in reducing greenhouse gas emissions while optimizing the use of available resources.

10.5.1.2 Socioeconomic Impacts

The increase in global demand for bioenergy, particularly wood pellets, has significant economic implications for forest product markets (Fig. 10.5). According to Johnston and Van Kooten (2014), this increase is stimulating demand for logs, leading to higher global industrial roundwood prices (+1.1%) and increased production. In Canada, for example, demand for roundwood is expected to grow by 1.4%, while in Europe it is expected to increase by 1.7% due to strong demand for wood pellets (Johnston and Van Kooten 2014). At the same time, this dynamic has positive effects on by-products such as sawmill residues and plywood, which are becoming more valuable (Johnston and Van Kooten 2014). Although the exact impact on producers is difficult to predict, consumers benefit from increased supply and more competitive prices (Birdsey et al. 2018). These developments indicate a beneficial economic dynamic for wood-related industries, despite some uncertainties about the final impact on producers (Birdsey et al. 2018). On the other hand, bioenergy also contributes to economic growth and sustainable development through its integration into rural areas and agricultural value chains. Many studies (Kalogiannidis et al. 2023; Bhuiyan et al. 2022) affirm the power of bioenergy in reducing poverty and restoring livelihoods in rural areas through the establishment of jobs and the provision of accessible energy (Fig. 10.6). The improvement of energy security for the use of local economies is observed through the diversification of energy resources, mainly renewable energies. These significant contributions within the framework of the Sustainable Development Goals (SDGs), participate in the development of more efficient and resilient energy systems (Cirstea et al. 2019). In sub-Saharan Africa,

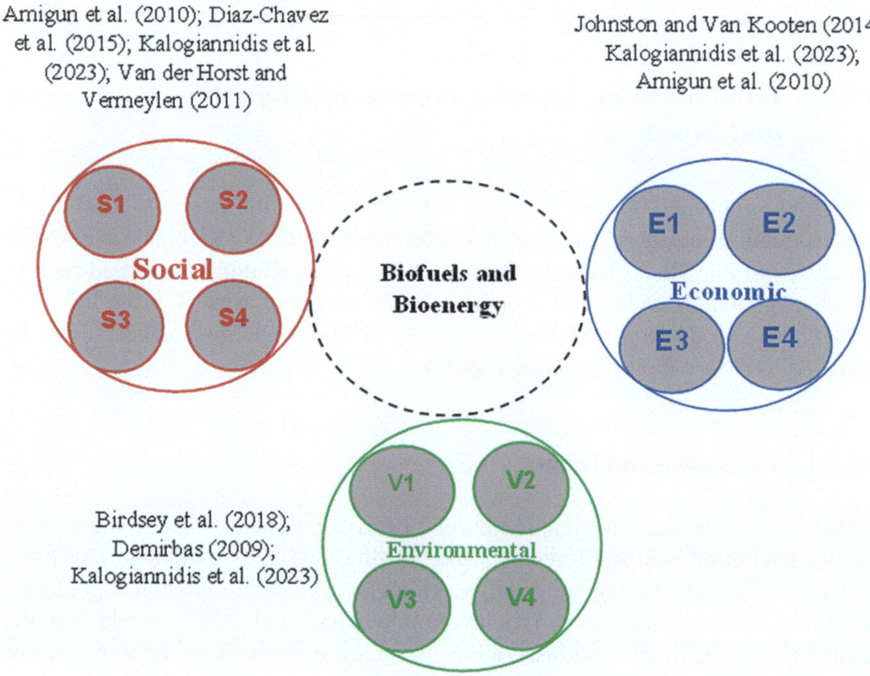

Fig. 10.6 Summary of the environmental and socioeconomic impacts of biofuels and bioenergy. *Social impacts: S1* providing employment opportunities in rural areas, *S2* minimizing poverty levels by improving incomes of rural communities from economic opportunities, *S3* strengthening living conditions through access to sustainable and affordable energy, *S4* improving skills levels and access to infrastructure in rural areas. *Economic impacts*: *E1* increase in global prices of industrial roundwood and stimulation of wood pellet production, *E2* valorization of timber and plywood residues, promoting more efficient use of raw materials, *E3* improvement of local economic conditions, particularly in sub-Saharan Africa, by reducing dependence on oil, *E4* support for economic growth and sustainable development through more efficient and resilient energy systems. *Environmental impacts*: *V1* reduction of greenhouse gas (GHG) emissions through the use of low-carbon bioenergy, *V2* contribution to carbon sequestration through sustainable forest management, *V3* limitation of the use of fossil fuels and support for the energy transition, *V4* better use of land and water

for example, biofuels represent a major opportunity to limit the use of oil and activate rural economies, although their potential is still poorly exploited (Kalogiannidis et al. 2023; Maltsoglou and Khwaja 2010). Finally, the progress of bioenergy has considerable social and economic impact, particularly in reducing poverty. In countries such as Mozambique and South Africa, biofuels provide support for national rural development programs. Bioenergy restores the quality of life of poor populations through the provision of renewable energy and by increasing farmers' survival funds (Amigun et al. 2011). However, there are some difficulties, particularly in the sharing of gains and the introduction of local communities into development strategies. These positive economic impacts demonstrate the potential of bioenergy to improve socioeconomic structures in the vision towards a sustainable energy transition.

10.5.2 Limitations or Negative Impacts of Biofuels and Bioenergy

Although biofuels and bioenergy appear to be promising alternatives to fossil fuels, their production and use face several important limitations (Kiehbadroudinezhad et al. 2023) (Table 10.4). These limitations can be grouped into three broad categories: environmental, social, and economic. These negative aspects, often linked to the intensive production and increasing use of these energy sources, must be addressed to ensure their long-term viability.

10.5.2.1 Environmental Limits

Biofuel production can have major environmental consequences due to the conversion of agricultural land and forests for the exploitation of raw materials. This practice leads to the reduction of soil fertility, which leads to environmental degradation and also to a loss of biodiversity (Kiehbadroudinezhad et al. 2023). In addition, the massive use of fertilizers and insecticides to increase biomass productivity causes groundwater pollution and soil degradation, impacting human health and ecosystems (Beig et al. 2021; Li et al. 2019; Nair and Paulose 2014). Finally, the increase in carbon dioxide emissions may be due to the intensity of activity carried out in connection with biofuel production, generating trade-offs in the expected climate benefits (Alalwan et al. 2019).

10.5.2.2 Social and Societal Limits

The concern for biofuels also refers to social issues. The rise of crops attributed to biofuels can lead to direct competition with food crops, thus widening the rate of food insecurity in certain regions (Kiehbadroudinezhad et al. 2023). The reduction

Table 10.4 Limitations and negative impacts of the use of biofuels and bioenergy

Categories	Boundaries	References
Environmental limitations	1. Decreased soil fertility and loss of biodiversity due to conversion of agricultural land and forests 2. Groundwater pollution and soil degradation caused by the massive use of fertilizers and insecticides 3. Indirect addition of carbon dioxide emissions despite expected climate benefits	Kiehbadroudinezhad et al. (2023), Beig et al. (2021), Li et al. (2019), Nair and Paulose (2014) and Alalwan et al. (2019)
Social limits	1. Competition with food crops, increasing food insecurity 2. Soil degradation impacting traditional agricultural production and the livelihoods of local populations 3. Environmental repercussions affecting the living conditions of communities close to production areas	Kiehbadroudinezhad et al. (2023), Beig et al. (2021) and Li et al. (2019)
Economic limitations	1. High costs of production, transportation, and storage of biofuels 2. Lack of competitiveness compared to fossil fuels due to the price gap and the absence of subsidies 3. Significant investments are needed to develop production and distribution infrastructure	Kiehbadroudinezhad et al. (2023), Alalwan et al. (2019) and Correa et al. (2019)

in soil quality associated with intensive agricultural practices causes a drop in traditional agricultural production, compromising the means of survival of local communities (Beig et al. 2021). In addition, the environmental consequences listed above have impacts on the lifestyles of populations, particularly those living near production sites (Li et al. 2019).

10.5.2.3 Economic Limits

Economically, biofuel production faces high payments, especially in terms of production, transportation, and storage. For example, biogas and biomethane, although they have a certain amount of potential, are associated with technical and economic constraints, in particular the lack of essential components such as CO_2, H_2, and CH_4 to achieve their energy efficiency (Kiehbadroudinezhad et al. 2023). Also, when faced with a lack of funds, the competitiveness of biofuels is limited compared to that of fossil fuels. Both energy sources constitute a barrier to their large-scale adoption due to the diversity of prices (Alalwan et al. 2019). In addition, the development of infrastructure essential for the production and distribution of biofuels constitutes a significant investment, which complicates their incorporation into existing systems (Correa et al. 2019). Therefore, biofuels and bioenergy, despite representing ambitious energy solutions, present a significant number of limitations that remain considerable for equitable and sustainable use. The proper management of environmental, social, and economic consequences, as well as the innovative technological approach, will be essential to overcome these challenges.

10.6 Challenges and Prospects in the Field of Biofuels and Bioenergy

10.6.1 Main Challenges Related to Biofuels and Bioenergy (Obstacles)

10.6.1.1 Raw Materials and Sustainability Challenges

Biofuel production relies on the supply of biological feedstocks, which must be available, accessible, and environmentally sustainable. These feedstocks must also guarantee a high yield while requiring minimal fertilizer and water. However, access to these resources is limited, and their cost is often high, which can pose profitability challenges for producers (Ambaye et al. 2021) (Table 10.5). At the same time, the intensive cultivation of these feedstocks, aimed at meeting growing demand, contributes to the pressure on agricultural land (Kiehbadroudinezhad et al. 2023). This situation creates direct competition with food crops and may call into question the long-term sustainability of these practices. In addition, the thermal stability of biofuels remains a significant technical obstacle, making their storage complex and exposing operators to risks related to their degradation (Long et al. 2021).

10.6.1.2 Technological and Logistical Obstacles

Technological development remains a major challenge for the biofuel and bioenergy sector (Kiehbadroudinezhad et al. 2023) (Table 10.5). Indeed, the processes for transforming biomass into high calorific biofuels, suitable for engines, are complex and costly. These technologies, still in development, require significant investments, which hinders their large-scale adoption (Kiehbadroudinezhad et al. 2023). In addition, existing infrastructures, often designed for fossil fuels, are not adapted to manage the production, transport, and storage of biofuels (Suparmaniam et al. 2019). The lack of infrastructures makes their incorporation into current energy networks difficult and causes high additional funds for their implementation. To this end,

Table 10.5 Main challenges or obstacles related to the use of biofuels and bioenergy

Categories	Challenges	References
Raw materials and sustainability challenges	1. Limited budget and expensive organic raw materials 2. Competition with food crops and pressure on agricultural land 3. Difficulty relating to thermal stability and conservation of biofuels	Ambaye et al. (2021), Kiehbadroudinezhad et al. (2023) and Long et al. (2021)
Technological and logistical obstacles	1. High cost and complexity of transformation processes into high calorific biofuels 2. Inadequate production, transport and storage infrastructure 3. Limited flexibility of raw materials and inefficiencies of distribution systems	Kiehbadroudinezhad et al. (2023), Suparmaniam et al. (2019), Correa et al. (2019) and Sharma et al. (2020)
Lack of economic and social incentives	1. Lack of powerful financial mechanisms and activating policies 2. Resources for fossil fuels 3. Negative social views and insufficient awareness of the benefits of biofuels	Mai-Moulin et al. (2021) and Ganesan et al. (2020)
Questions of sustainability and political uncertainty	1. Unresolved environmental impacts (deforestation, soil depletion) 2. Political uncertainty and lack of coherent regulatory frameworks	Mai-Moulin et al. (2021) and Junginger (2007)

logistical constraints such as the limited flexibility of raw materials and the inefficiency of distribution systems further increase the problems faced by this field.

10.6.1.3 Lack of Economic and Social Incentives

The lack of strong political strategies and financial support means represents a major challenge to the progress of biofuels and bioenergy (Mai-Moulin et al. 2021) (Table 10.5). Resources and investments oriented towards fossil fuels continue to hamper the competitiveness of biofuels (Ganesan et al. 2020). In this sense, the high production budget, associated with a low level of short-term productivity, does not convince private and public actors to make investments in this sector. In addition, the absence of general social agreement constitutes a major constraint (Mai-Moulin et al. 2021). This particularly characterizes a certain number of negative apprehensions linked to undesirable environmental consequences or a limited contribution to economic growth, fueling these limits.

10.6.1.4 Lack of Economic and Social Incentives

The constraints associated with the sustainability of biofuels involve the issues related to unresolved environmental consequences, such as soil depletion or deforestation. These difficulties generate a bias in political and commercial actions, causing an unstable environment for investment and strategic decision-making (Mai-Moulin et al. 2021). Moreover, current political actions usually considered insufficient based on scientific data also cause the agreement and progress of bioenergy (Junginger 2007; Mai-Moulin et al. 2021).

10.6.2 Prospects for the Use of Biofuels and Bioenergy

In terms of prospects for increasing the use of biofuels and bioenergy, the literature review highlights valuable opportunities both on a global and regional scale. For example, at the EU country level, scientists converge on the possibility of short- and long-term growth (Proskurina and Vakkilainen 2024). They also note that changes to existing policies are essential to consolidate its place as a world leader in terms of energy (Proskurina and Vakkilainen 2024). Among the proposed changes, the emphasis is mainly on increasing the number of raw materials exploited, strengthening strategies to promote biofuels and bioenergy, and completely eradicating financing around fossil fuels (Proskurina and Vakkilainen 2024).

Furthermore, countries in Asia (China, India, Japan) and Latin America such as Brazil are identified as potential areas to produce biofuels and bioenergy, especially because of their strong capacity to ensure agricultural biomass production and the intensification of investments around bioenergy infrastructures (Junginger et al. 2020). Meanwhile, in Africa, the energy transition from the use of traditional energies to modern renewable energy (biofuels and bioenergy) is also considered the main option to ensure the development of the energy sector (Amigun et al. 2011; Dabat et al. 2014; Dahunsi et al. 2020).

At the international level, ideas converge towards a need to continue developing strategies and policies to support biofuels and bioenergy to further promote the energy transition. On the one hand, a scientific work such as that conducted by the World Bioenergy Association (WBA) has specified that direct subsidies, carbon pricing against a complete eradication of taxes and barriers on the use of biofuels and bioenergy, remain the main perspectives that can better stimulate investments in these forms of energy (Proskurina and Vakkilainen 2024). On the other hand, some calls for a readjustment in the coordination strategies between policies affecting the energy, agricultural, and environmental sectors, to guarantee a harmonious and long-term increase (Proskurina and Vakkilainen 2024).

However, before considering each of these perspectives, Gerard and Jayet (2023) emphasize the need for a basic clarification around the concepts of "sustainable bioenergy" to limit the level of divergence in understandings and ensure a consistent implementation of policies in all countries. Finally, as perspectives, other works have also added that automation, innovation, and increased financing in these renewable energies can increase the economy while limiting environmental impacts (Fotourehchi 2017; Kalogiannidis et al. 2023). Each of these proposals is essential to meet not only the challenges of carbon removal but also of optimizing climate benefits while maintaining competitiveness in global markets.

10.7 Discussion

The results of this study show a gradual evolution of scientific work around biofuels and bioenergy. This was also highlighted by Welfle et al. (2020), thus testifying to the interest of the scientific world in questions relating to the energy transition towards more sustainable energies. This remark also reflects a global recognition of the challenges of the energy sector, in particular the need for more resilient and more environmentally beneficial systems. This global recognition may also be the factor that can explain the higher intensity of studies carried out in a global context than those conducted at the regional or local scale, thus suggesting the global nature of the challenges around the energy sector (Nazir 2023). This global priority is explained by the interdependence of energy systems, where the transition to renewable sources requires international collaboration to share innovations, harmonize policies, and strengthen technological capacities (Kawabata 2024).

Regarding the biofuel and bioenergy categories, the results highlight the constant evolution of these energy sources in response to global environmental, energy, and economic concerns. Indeed, the transition from first-generation biofuels, controversial for their competition with food crops, to more advanced generations shows a desire to reconcile technological innovation and sustainability (Cavelius et al. 2023). Second- and third-generation biofuels, particularly those derived from lignocellulosic residues or microalgae, demonstrate efforts to reduce pressure on agricultural resources while diversifying energy feedstocks (Kumar and Sharma 2008; Slade and Bauen 2013). However, the technological challenges associated with these second and third-generation biofuels, such as process efficiency and high production costs, highlight the need for further investment in research and development (Sindhu et al. 2019). At the same time, fourth-generation energies, well- known for their innovations, are also facing ethical and technical challenges, considered as limiting factors of their use (Nozzi et al. 2013). All these situations testify to the complexity between technological advances and economic and environmental viability while highlighting the determining role of a political framework to facilitate the integration of these energies at the global level.

Regarding policies developed at the global level, the study highlights in particular the United Nations SDGs, the Paris Agreement (2015), and many other policy frameworks such as RED I, RED II, the Kyoto Protocol, and the GBEP. In addition to the strategies put in place for the mitigation of GHG emissions, these policies ensure better energy efficiency and stimulate investments in renewable energies (Akbi 2013; Mai-Moulin et al. 2021). These policy frameworks are in phase with each other in their interest to ensure a change in energy systems towards more modern and resilient energies (Elkhatat and Al-Muhtaseb 2024). They mainly seek to stimulate states and energy production units to undertake decarbonization approaches, while generating opportunities from modern markets that are essential for a fair and inclusive energy transition (Otto and Oberthür 2024).

The cases of the United States, Canada, Germany, and Brazil illustrate well the variability of regional strategies implemented to support these forms of energy and minimize GHG emissions (Bacovsky et al. 2016). As a particularity, the United States and Canada favor policies oriented towards the promotion of biofuels and biomass by relying on tax incentives and loan guarantee mechanisms (Fell 2019; Sciortino 2010). Meanwhile, other countries such as Germany use an inclusive strategy, emphasizing the reduction of GHGs, starting with the implementation of guaranteed feed-in tariffs and subsidies for renewable energies (Fell 2019; Sciortino 2010). On the other hand, Brazil relies on its agricultural resources to build a competitive biofuels industry, while promoting environmental sustainability (Tenelli et al. 2021). However, the implementation of these policies within these nations still faces common challenges. These include weak coordination among different stakeholders, weak assurance of the sustainability of the resources used, and unsecured long-term financial support (Klass and Wilson 2012).

In terms of benefits, the study highlights the contribution of biofuels and bioenergy in the fight against climate change. This is by providing approaches to effective energy decarbonization solutions, particularly in the transport and industry sectors, where electrification remains complex (Wu and Pfenninger 2023). Their contribution in stabilizing energy systems (wind, solar) increases their utility while such as BECCS offer negative emissions essential to minimize global warming to 1.5 °C or 2 °C (Hanssen et al. 2020). At the same time, the use of forest residues (branches and stumps) reduces GHG emissions, reduces the risk of fire, and optimizes climate benefits (Birdsey et al. 2018; Smyth et al. 2014). Economically, the continued adoption of bioenergy, especially wood pellets, increases forest market opportunities, thereby strengthening value chains in rural areas (Johnston and Van Kooten 2014). In summary, the use of these forms of energy not only ensures energy security and minimizes poverty levels but also fits with the SDGs, although a balanced management of benefits is useful for a better optimization of their potential (Amigun et al. 2011; Kalogiannidis et al. 2023).

Furthermore, the use of these forms of energy not only has advantages but also negative impacts on the environmental, social, and economic levels. In terms of environmental impacts, the production of these energies leads to a transformation of agricultural or forest land, leading to a deterioration of biodiversity and soils (Kiehbadroudinezhad et al. 2023; Beig et al. 2021). Added to this are the pollution of groundwater and the increase in CO_2 in nature, especially due to the excessive use of chemical inputs (Alalwan et al. 2019; Beig et al. 2021; Kiehbadroudinezhad et al. 2023). Socially, their uses lead to competition with food crops, thus accentuating the issue of food insecurity (Li et al. 2019). From an economic point of view, the progressive use of these forms of energy is hampered by high production, transport, and storage costs, as well as their reduced competitiveness compared to fossil fuels in the absence of subsidies (Correa et al. 2019).

In terms of challenges, the study highlights several aspects, especially the challenges of sustainability, logistics, policies, and social perception. Thus, questions about the environmental and economic viability of biofuels and bioenergy are increasingly raised due to the restricted and expensive access to raw materials and pressures around agricultural land (Abdullah et al. 2019; Kiehbadroudinezhad et al. 2023). In addition, the study highlights the technological barriers and the lack of appropriate infrastructure that hinder their integration into current energy networks (Suparmaniam et al. 2019; Correa et al. 2019). At the same time, others raise concerns about the lack of solid financial mechanisms and incentive policies, accentuated by an insufficient awareness-raising strategy among decision-makers and the public (Mai-Moulin et al. 2019).

However, the outlook in the literature remains encouraging, with strong growth potential, especially in Asia, Latin America, and Africa where biomass resources are more accessible and investments in bioenergy infrastructure are becoming more important (Junginger et al. 2020; Proskurina and Vakkilainen 2024). Also, the lack of financing programs and incentive policies, accentuated by low awareness among stakeholders, are obstacles to be taken into consideration for optimizing the economic and environmental benefits of these energies (Fotourehchi 2017; Kalogiannidis et al. 2023).

10.8 Conclusion

This study shows the considerable progress made in the field of biofuels and bioenergy while highlighting the major challenges to be overcome to optimize their benefits. On the one hand, second- and third-generation biofuels, derived from lignocellulosic residues and microalgae, represent promising solutions to reduce pressure on agricultural land and diversify raw materials. Indeed, these forms of energy (biofuels and bioenergy) offer environmental benefits, including the reduction of greenhouse gas emissions, but their production also raises social and environmental concerns, including competition with food crops and impacts on biodiversity.

On the other hand, although international policies have facilitated an energy transition to renewables, challenges remain, such as multi-level policy coordination, resource sustainability, and continued financial support. At the same time, previous studies have identified technological barriers, including high production costs and process efficiency, requiring sustained investments in research and development.

In this context, to ensure a sustainable future for these energy sources, it is essential to strengthen supportive public policies, including carbon pricing and direct subsidies. Furthermore, greater international and regional collaboration in sharing technologies and good practices could accelerate the integration of biofuels into existing energy systems. Finally, it is crucial to continue to encourage technological innovation and automation in the bioenergy sector to overcome technical and logistical barriers, while ensuring the sustainability of biofuel production and better management of agricultural and forest resources. These perspectives offer promising avenues to address the challenges of the energy transition while meeting sustainability and carbon neutrality objectives.

References

Abdullah B, Muhammad SAFAS, Shokravi Z, Ismail S, Kassim KA, Mahmood AN, Aziz MMA (2019) Fourth generation biofuel: a review on risks and mitigation strategies. Renew Sustain Energy Rev 107:37–50

Acheampong M, Ertem FC, Kappler B, Neubauer P (2017) In pursuit of sustainable development goal (SDG) number 7: will biofuels be reliable? Renew Sustain Energy Rev 75:927–937

Akbi A (2013) The implications of biofuel development: what impact on developing countries? Doctoral dissertation, Université Nice Sophia Antipolis

Akhmieva R, Vagapova A, Gumaev I (2024) Assessing the impact: escalating greenhouse gas emissions and their effects. E3S Web Conf 537:03014

Alalwan HA, Alminshid AH, Aljaafari HA (2019) Promising evolution of biofuel generations. Subject review. Renew Energy Focus 28:127–139

AlRyalat SAS, Malkawi LW, Momani SM (2019) Comparing bibliometric analysis using PubMed, Scopus, and web of science databases. J Vis Exp 152:e58494

Ambaye TG, Vaccari M, Bonilla-Petriciolet A, Prasad S, van Hullebusch ED, Rtimi S (2021) Emerging technologies for biofuel production: a critical review on recent progress, challenges and perspectives. J Environ Manage 290:112627

Amigun B, Musango JK, Stafford W (2011) Biofuels and sustainability in Africa. Renew Sustain Energy Rev 15(2):1360–1372

Anindya B (2015) An integrated analytical framework of sustainable energy for all: developing Asia perspective. Kyoto University

Azhar SHM, Abdulla R, Jambo SA, Marbawi H, Gansau JA, Faik AAM, Rodrigues KF (2017) Yeasts in sustainable bioethanol production: a review. Biochem Biophys Rep 10:52–61

Bacovsky D, Ludwiczek N, Pointner C, Verma VK (2016) IEA bioenergy Countries' report: bioenergy policies and status of implementation (no. IEA-Bioenergy-796-TR-N41029016-01). Bioenergy 2020+ GmbH, Graz, Austria

Bajoria A, Kanpariya J, Bera A (2024) Greenhouse gases and global warming. In: Advances and technology development in greenhouse gases: emission, capture and conversion. Elsevier, Amsterdam, pp 121–135

Banja M, Sikkema R, Jgard M, Motola V, Dallemand JF (2019) Biomass for energy in the EU–the support framework. Energy Policy 131:215–228

Bauermann K (2016) German Energiewende and the heating market–impact and limits of policy. Energy Policy 94:235–246

Baumber A (2017) Enhancing ecosystem services through targeted bioenergy support policies. Ecosyst Serv 26:98–110

Beig B, Riaz M, Naqvi SR, Hassan M, Zheng Z, Karimi K, Pugazhendhi A, Atabani AE, Chi NTL (2021) Current challenges and innovative developments in pretreatment of lignocellulosic residues for biofuel production: a review. Fuel 287:119670

Bellanca R, Wilson E (2012) Sustainable energy for all and the private sector Available at:https://www.iied.org/g03383
Berrang-Ford L, Pearce T, Ford JD (2015) Systematic review approaches for climate change adaptation research. Reg Environ Chang 15:755–769
Bhuiyan MA, Dinçer H, Yüksel S, Mikhaylov A, Danish MSS, Pinter G, Uyeh DD, Stepanova D (2022) Economic indicators and bioenergy supply in developed economies: QROF -DEMATEL and random forest models. Energy Rep 8:561–570
Birdsey R, Duffy P, Smyth C, Kurz WA, Dugan AJ, Houghton R (2018) Climate, economic, and environmental impacts of producing wood for bioenergy. Environ Res Lett 13(5):050201
Borges CP, Sobczak JC, Silberg TR, Uriona-Maldonado M, Vaz CR (2021) A systems modeling approach to estimate biogas potential from biomass sources in Brazil. Renew Sustain Energy Rev 138:110518
Bradley D (2012) IEA bioenergy task 40 country report Canada 2011. Canadian Bioenergy Association, Ottawa, ON
Calvin K, Cowie A, Berndes G, Arneth A, Cherubini F, Portugal-Pereira J, Grassi G, Johnson F, Popp A, Rounsevell M, Slade R, Smith P (2021) Bioenergy for climate change mitigation: scale and sustainability. GCB Bioenergy 13(9):1346–1371
Cansino JM, Pablo-Romero MD, Román R, Yñiguez R (2010) Tax incentives to promote green electricity: an overview of EU-27 countries. Energy Policy 38(10):6000–6008
Cavelius P, Engelhart-Straub S, Mehlmer N, Lercher J, Awad D, Brück T (2023) The potential of biofuels from first to fourth generation. PLoS Biol 21(3):e3002063
Chen C, Lu Y, He G (2022) Driving mechanisms for decoupling CO2 emissions from economic development in the ten largest emission countries. Ecosyst Health Sustain 8(1):2059016
Chirambo D (2018) Towards the achievement of SDG 7 in sub-Saharan Africa: creating synergies between power Africa, sustainable energy for all and climate finance in-order to achieve universal energy access before 2030. Renew Sustain Energy Rev 94:600–608
Cristea M, Noja GG, Marcu N, Siminică M, Țîrcă DM (2020) Modelling EU bioeconomy credentials in the economic development framework: The role of intellectual capital. Technol Econ Dev Econ 26(6):1139–1164.https://doi.org/10.3846/tede.2020.13159
Coelho ST, Goldemberg J (2013) Energy access: lessons learned in Brazil and perspectives for replication in other developing countries. Energy Policy 61:1088–1096
Costantini V, Crespi F, Curci Y (2014) A keyword selection method for mapping technological knowledge in specific sectors through patent data: the case of the biofuels sector. Econ Innov New Technol. https://doi.org/10.1080/10438599.2014.942583
Correa DF, Beyer HL, Fargione JE, Hill JD, Possingham HP, Thomas-Hall SR, Schenk PM (2019) Towards the implementation of sustainable biofuel production systems. Renew Sustain Energy Rev 107:250–263
Costa H, De Rigo D, Libertà G, Houston Durrant T, San-Miguel-Ayanz J (2020) European wildfire danger and vulnerability in a changing climate: towards integrating risk dimensions. Publications Office of the European Union, Luxembourg, p 59
Dabat MH, Gautier D, Gazull L, Pinta F (2014) Energy challenges: threats or opportunities? In: Family farming and the world to come, pp 183–200
Dahunsi SO, Fagbiele OO, Yusuf EO (2020) Bioenergy technologies adoption in Africa: a review of past and current status. J Clean Prod 264:121683
Daniel M, Rivière G, Couture N (2019) A shape-shifting cylindrical histogram to display energy availability forecasts. J Hum Syst Interact 8(1):1–13
Demirbas A (2009) Political, economic and environmental impacts of biofuels: a review. Appl Energy 86:S108–S117
Dexter J, Fu P (2009) Metabolic engineering of cyanobacteria for ethanol production. Energ Environ Sci 2(8):857–864
Diaz-Chavez R, Johnson FX, Richard TL, Chanakya H (2015) Biomass resources, energy access and poverty reduction. In: Souza GM, Victoria R, Joly C, Verdade L (eds) Bioenergy

and sustainability: bridging the gaps. Scientific Committee on Problems of the Environment (SCOPE), Paris

Drepper B, Bamps B, Gobin A, Van Orshoven J (2021) Strategies for managing spring frost risks in orchards: effectiveness and conditionality—a systematic review protocol. Environ Evid 10:1–13

Dudley B (2018) BP statistical review of world energy 2018. Energy econ, center for energy economics research and policy. British Petroleum. https://www.bp.com/en/global/corporate/energy-economics/statistical-review-of-world-energy/electricity.html

Dusser P (2019) The European energy policy for 2020–2030 RED II: what future for vegetable oil as a source of bioenergy? OCL 26:51

Ebadian M, van Dyk S, McMillan JD, Saddler J (2020) Biofuels policies that have encouraged their production and use: an international perspective. Energy Policy 147:111906

Elkhatat A, Al-Muhtaseb S (2024) Climate change and energy security: a comparative analysis of the role of energy policies in advancing environmental sustainability. Energies 17(13):3179

Falcone PM (2023) Sustainable energy policies in developing countries: a review of challenges and opportunities. Energies 16(18):6682

Favaro L, Jansen T, van Zyl WH (2019) Exploring industrial and natural *Saccharomyces cerevisiae* strains for the bio-based economy from biomass: the case of bioethanol. Crit Rev Biotechnol 39(6):800–816

Fell HJ (2019) The shift from feed-in-tariffs is hindering the transformation of the global energy supply to renewable energies. In: EnergyWatch Group Policy Paper 1, pp 1–24

Fotourehchi Z (2017) Renewable energy consumption and economic growth: a case study for developing countries. Int J Energy Econ Policy 7(2):61–64

Gaffney JS, Marley NA (2009) The impacts of combustion emissions on air quality and climate—from coal to biofuels and beyond. Atmos Environ 43(1):23–36

Gallagher M, Wykes S (2014) Civil society participation in the sustainable energy for all initiativeAvailable at: https://www.iied.org/g03878

Gérard M, Jayet PA (2023) European farmers' response to crop residue prices and implications for bioenergy policies. Energy Policy 177:C

Ganesan R, Manigandan S, Samuel MS, Shanmuganathan R, Brindhadevi K, Chi NTL, Duc A, Pugazhendhi A (2020) A review on prospective production of biofuel from microalgae. Biotechnol Rep 27:e00509

Grangeia C, Santos L, Lazaro LLB (2022) The Brazilian biofuel policy (RenovaBio) and its uncertainties: an assessment of technical, socioeconomic and institutional aspects. Energy Convers Manag X 13:100156

Gravina OS, Santos GG, Correchel V, Silva GCD, Medrado LDC, Flores RA, Severiano EDC (2021) Physical attributes of ferralsol in fertigated sugarcane production environments for bioethanol in the Midwest of Brazil. Agronomy 11(8):1641

Hanssen SV, Daioglou V, Steinmann ZJN, Doelman JC, Van Vuuren DP, Huijbregts MAJ (2020) The climate change mitigation potential of bioenergy with carbon capture and storage. Nat Clim Chang 10(11):1023–1029

Harper AB, Powell T, Cox PM, House J, Huntingford C, Lenton TM, Shu S (2018) Land-use emissions play a critical role in land-based mitigation for Paris climate targets. Nat Commun 9(1):2938

Harpet C, Gully E (2013) Industrial and territorial ecology: what decision-making tools? From flow analysis to an integrated approach. Waste Sci Tech 63:44–50

Hennenberg KJ, Böttcher H, Bradshaw CJ (2018) Revised European Union renewable-energy policies erode nature protection. Nat Ecol Evol 2(10):1519–1520

Hernandes TAD, Duft DG, dos Santos Luciano AC, Leal MRLV, Cavalett O (2021) Identifying suitable areas for expanding sugarcane ethanol production in Brazil under conservation of environmentally relevant habitats. J Clean Prod 292:125318

Himabindu M, Ravikrishna RV (2010) Potential of bio-DME as a transportation fuel for India. J Renew Sustain Energy 2(5):052701

Hriţcan CC (2023) Uncovering the real challenges in the fight against climate change. Atmos Clim Sci 13(3):401–414

Huang D, Zhou H, Lin L (2012) Biodiesel: an alternative to conventional fuel. Energy Process 16:1874–1885

Hurmekoski E, Jonsson R, Korhonen J, Jänis J, Mäkinen M, Leskinen P, Hetemäki L (2018) Diversification of the forest industries: role of new wood-based products. Can J For Res 48(12):1417–1432

Ikevuje AH, Kwakye JM, Ekechukwu DE, Benjamin O (2024) Optimizing the energy mix: strategies for reducing energy dependence. Open Access Res J Multidiscip Stud 8(01):094–104

Jabari F, Arasteh H, Sheikhi-Fini A, Ghaebi H, Bannae-Sharifian MB, Mohammadi-Ivatloo B, Mohammadpourfard M (2022) A biogas-steam combined cycle for sustainable development of industrial-scale water-power hybrid microgrids: design and optimal scheduling. Biofuels Bioprod Biorefin 16(1):172–192

Jacobsen E, Skov SM, Singlitico A, Frandsen HL (2023) Techno-economic analysis of green aviation fuel production using an integrated electrolyzer and a "biomass-battery" storage system. Int J Hydrogen Energy 48:37314

Jameel MK, Mustafa MA, Ahmed HS, Jassim Mohammed A, Ghazy H, Shakir MN, Lawas AM, Mohammed SK, Idan AH, Mahmoud ZH, Sayadi H, Kianfar E (2024) Biogas: production, properties, applications, economics and challenges: a review. Results Chem 7:101549

Jiang Y, Lv Y, Wu R, Sui Y, Chen C, Xin F, Zhou J, Dong W, Jiang M (2019) Current status and perspectives on biobutanol production using lignocellulosic feedstocks. Bioresour Technol Rep 7:100245

Johnston CM, Van Kooten GC (2014) Economic consequences of increased bioenergy demand. For Chron 90(5):636–642

Junginger HM (2007) Lessons from (European) Bioenergy Policies—results of a literature review for IEA Bioenergy Task 40. Utrecht University, Utrecht

Junginger M, Koppejan J, Goh CS (2020) Sustainable bioenergy deployment in east and South East Asia: notes on recent trends. Sustain Sci 15(5):1455–1459

Kalogiannidis S, Chatzitheodoridis F, Kontsas S, Syndoukas D (2023) Impact of bioenergy on economic growth and development: an European perspective. Int J Energy Econ Policy 13(3):494–506

Kawabata T (2024) Network analysis on energy transition cooperation between countries. Energy Sustain Dev 81:101503

Khartukov EM, Starostina EE (2022) Brazil's green energy: today and tomorrow. In: Industry 4.0: fighting climate change in the economy of the future. Springer International, Cham, pp 149–162

Kiehbadroudinezhad M, Merabet A, Ghenai C, Abo-Khalil AG, Salameh T (2023) The role of biofuels for sustainable microgrids: a path towards carbon neutrality and the green economy. Heliyon 9(2):e13407

Klass AB, Wilson EJ (2012) Interstate transmission challenges for renewable energy: a federalism mismatch. Vand L Rev 65:1801

Kokkinos NC, Emmanouilidou E, Sharma SK (2024) Waste-to-biofuel production for the transportation sector. In: Intelligent transportation system and advanced technology. Springer Nature, Singapore, pp 99–122

Kothari R, Goria K, Bharti A, Singh HM, Pathak VV, Pathak A, Tyagi VV (2023) Sustainable development goals (SDGs-7) for bioeconomy with bioenergy sector. In: Sustainable butanol biofuels. CRC, Boca Raton, FL, pp 29–56

Kpadé CP, Tamini LD, Pepin S, Khasa DP, Abbas Y, Lamhamedi MS (2024) Evaluating multicriteria decision-making methods for sustainable management of forest ecosystems: a systematic review. Forests 15(10):1728

Krug M, Di Nucci MR, Caldera M, De Luca E (2022) Mainstreaming community energy: is the renewable energy directive a driver for renewable energy communities in Germany and Italy? Sustain For 14(12):7181

Krug M, Di Nucci MR, Schwarz L, Alonso I, Azevedo I, Bastiani M, Dylag A, Laes E, Hinsch A, Klavs G, Kudrenickis I, Maleki P, Massa G, Meynaerts E, Pappa S, Standal K (2023) Implementing European Union provisions and enabling frameworks for renewable energy communities in nine countries: progress, delays, and gaps. Sustain For 15(11):8861

Kumar A (2018) Global warming, climate change and greenhouse gas mitigation. In: Biofuels: greenhouse gas mitigation and global warming: next generation biofuels and role of biotechnology, pp 1–16

Kumar A (2020) Climate change: challenges to reduce global warming and role of biofuels. In: Climate change, photosynthesis and advanced biofuels: the role of biotechnology in the production of value-added plant bio-products, pp 13–54

Kumar A, Sharma S (2008) An evaluation of multipurpose oil seed crop for industrial uses (*Jatropha curcas* L.): a review. Ind Crop Prod 28(1):1–10

Kumar A, Yau YY, Ogita S, Scheibe R (2020) Climate change, photosynthesis and advanced biofuels, vol 490. Springer, Singapore

Laan T, Litman TA, Steenblik R (2009) Biofuels at what price? Government support for ethanol and biodiesel in Canada. International Institute for Sustainable Development, Winnipeg, MB

Li P, Sakuragi K, Makino H (2019) Extraction techniques in sustainable biofuel production: a concise review. Fuel Process Technol 193:295–303

Li D, Bae JH, Rishi M (2022) Sustainable development and SDG-7 in sub-Saharan Africa: balancing energy access, economic growth, and carbon emissions. Eur J Dev Res 35(1):112

Liebetrau J, Sträuber H, Kretzschmar J, Denysenko V, Nelles M (2019) Anaerobic digestion. In: Biorefineries, pp 281–299

Long F, Liu W, Jiang X, Zhai Q, Cao X, Jiang J, Xu J (2021) State-of-the-art technologies for biofuel production from triglycerides: a review. Renew Sustain Energy Rev 148:111269

Lorne D (2011) Panorama 2012-biofuels update: growth in national and international markets

Madiraju KS (2014) The Canadian context: energy. In: Global sustainable communities handbook, pp 255–291

Mai-Moulin T, Hoefnagels R, Grundmann P, Junginger M (2021) Effective sustainability criteria for bioenergy: towards the implementation of the European renewable directive II. Renew Sustain Energy Rev 138:110645

Maltsoglou I, Khwaja Y (2010) Environment and natural resources management working paper the BEFS analysis for Tanzania bioenergy. https://www.fao.org

Manandhar A, Mousavi-Avval SH, Tatum J, Shrestha E, Nazemi P, Shah A (2022) Solid biofuels. In: Biomass, biofuels, biochemicals. Elsevier, Amsterdam, pp 343–370

Martinez Romera B (2016) The Paris agreement and the regulation of international bunker fuels. J Eur Comp Int Environ Law 25(2):215–227

Masjuki HH, Kalam MA, Mofijur M, Shahabuddin M (2013) Biofuel: policy, standardization and recommendation for sustainable future energy supply. Energy Procedia 42:577–586

Montini M (2024) Addressing the conflicts between climate-related renewable energy goals and environmental protection interests under the RED directive. Eur Law Open 3(1):209–219

Morency PO (2011) The potential for valorizing forest biomass for energy purposes in Quebec: state of the situation. Doctoral dissertation, Université Laval

Muri H (2018) The role of large-scale BECCS in the pursuit of the 1.5 C target: an earth system model perspective. Environ Res Lett 13(4):044010

Murphy F, McDonnell K (2017) Investigation of the potential impact of the Paris agreement on national mitigation policies and the risk of carbon leakage: an analysis of the Irish bioenergy industry. Energy Policy 104:80–88

Murugesan A, Umarani C, Subramanian R, Nedunchezhian N (2009) Bio-diesel as an alternative fuel for diesel engines—a review. Renew Sustain Energy Rev 13(3):653–662

Nabukalu C, Gieré R (2019) Charcoal as an energy resource: global trade, production and socioeconomic practices observed in Uganda. Resources 8(4):183

Naiki Y (2016) Trade and bioenergy: explaining and assessing the regime complexity for sustainable bioenergy. Eur J Int Law 27(1):129–159

Nair S, Paulose H (2014) Emergence of green business models: the case of algae biofuel for aviation. Energy Policy 65:175–184

Nam-Chol O, Kim H (2019) Towards the 2 C goal: achieving sustainable development goal (SDG) 7 in DPR Korea. Resour Conserv Recycl 150:104412

Nazir U (2023) The winds of change are blowing: globalization's impact on renewable energy and environmental challenges. Arch Soc Sci 2(1):78–93

Nemet G (2009) Demand-pull, technology-push, and government-led incentives for non-incremental technical change. Res Policy 38:700–709

Nozzi NE, Oliver JW, Atsumi S (2013) Cyanobacteria as a platform for biofuel production. Front Bioeng Biotechnol 1:7

Ohlhorst D (2020) Germany: from feed-in-tariffs to auctions and the question of various actors. In: Decentralised energy: a global game changer, pp 82–100

Olabi AG, Elsaid K, Obaideen K, Abdelkareem MA, Rezk H, Wilberforce T, Maghrabie H, Sayed ET (2023) Renewable energy systems: comparisons, challenges and barriers, sustainability indicators, and the contribution to UN sustainable development goals. Int J Thermofluids 20:100498

Otto S, Oberthür S (2024) International cooperation for the decarbonization of energy-intensive industries: unlocking the full potential. Clim Policy:1–17

Persillet Shonkwiler V (2012) First generation biofuels: a mixed assessment. In: INRA sciences Sociales (1/2012), p 8

Pirelli T, Chiumenti A, Morese MM, Bonati G, Fabiani S, Pulighe G (2021) Environmental sustainability of the biogas pathway in Italy through the methodology of the global bioenergy partnership. J Clean Prod 318:128483

Proskurina S, Vakkilainen E (2024) Perspectives on challenges to bioenergy use in the EU. Biofuels Bioprod Biorefin 18(4):938–951

Raihan A (2023) A review of the global climate change impacts, adaptation strategies, and mitigation options in the socio-economic and environmental sectors. J Environ Sci Econ 2(3):36–58

Raihan A, Rashid M, Voumik LC, Akter S, Esquivias MA (2023) The dynamic impacts of economic growth, financial globalization, fossil fuel, renewable energy, and urbanization on load capacity factors in Mexico. Sustain For 15(18):13462

Ramirez-Contreras NE, Faaij AP (2018) A review of key international biomass and bioenergy sustainability frameworks and certification systems and their application and implications in Colombia. Renew Sustain Energy Rev 96:460–478

Rasi S, Veijanen A, Rintala J (2007) Trace compounds of biogas from different biogas production plants. Energy 32(8):1375–1380

Rial RC (2024) Biofuels versus climate change: exploring potentials and challenges in the energy transition. Renew Sustain Energy Rev 196:114369

Rogelj J, McCollum DL, Riahi K (2013) The UN's 'Sustainable Energy for All' initiative is compatible with a warming limit of 2°C. Nat Clim Chang 3(6):545–551

Rossi J (2013) Maladaptive federalism: the structural barriers to coordination of states sustainability initiatives. Case W Rsrv L Rev 64:1759

Sahoo G, Sharma A, Dash AC (2022) Biomass from trees for bioenergy and biofuels—a briefing paper. Mater Today Proc 65:461–467

Sánchez AC, Estrada-Carmona N, Juventia SD, Jones SK (2021) The impact of diversified farming practices on terrestrial biodiversity outcomes and agricultural yield worldwide: a systematic review protocol. Methods Protocols 4(1):8

Santos FD, Ferreira PL, Pedersen JST (2022) The climate change challenge: a review of the barriers and solutions to deliver a Paris solution. Climate 10(5):75

Sarika AA, Meena R, Mina U, Shukla A, Sharma A (2023) Adoption of green energy technology for greenhouse gas emission mitigation: embracing the goals of the Paris Agreement. In: Nanomaterials and nanoliquids: applications in energy and environment. Springer Nature, Singapore, pp 47–72

Scarlat N, Dallemand JF, Monforti-Ferrario F, Nita V (2015) The role of biomass and bioenergy in a future bioeconomy: policies and facts. Environ Dev 15:3–34
Schomerus T (2012) German climate and energy legislation: an ambitious but fragmented framework. In: Climate law in EU member states. Edward Elgar, Cheltenham, pp 178–202
Sciortino M (2010) States stepping forward: best practices for state-led energy efficiency programs. American Council for an Energy-Efficient Economy, Washington, DC
Scovronick N, Wilkinson P (2014) Health impacts of liquid biofuel production and use: a review. Glob Environ Chang 24:155–164
Searchinger TD, Beringer T, Holtsmark B, Kammen DM, Lambin EF, Lucht W, Raven P, van Ypersele JP (2018) Europe's renewable energy directive weighed to harm global forests. Nat Commun 9(1):3741
Semde I, Neya T, Yonkeu S, Pare S (2021) Impacts of climate risks on food security in Burkina Faso: identification of resilience paths. Int J Innov Sci Res 55:168
Sen A, Dabi N (2021) Tightening the net: net zero climate targets—implications for land and food equity. Oxfam, Oxford
Sharma S, Kundu A, Basu S, Shetti NP, Aminabhavi TM (2020) Sustainable environmental management and related biofuel technologies. J Environ Manage 273:111096
Sindhu R, Binod P, Pandey A, Ankaram S, Duan Y, Awasthi MK (2019) Biofuel production from biomass: toward sustainable development. In: Current developments in biotechnology and bioengineering. Elsevier, Amsterdam, pp 79–92
Slade R, Bauen A (2013) Micro-algae cultivation for biofuels: cost, energy balance, environmental impacts and future prospects. Biomass Bioenergy 53:29–38
Smith AL, Klenk N, Wood S, Hewitt N, Henriques I, Yan N, Bazely DR (2013) Second generation biofuels and bioinvasions: an evaluation of invasive risks and policy responses in the United States and Canada. Renew Sustain Energy Rev 27:30–42
Smyth CE, Stinson G, Neilson E, Lemprière TC, Hafer M, Rampley GJ, Kurz WA (2014) Quantifying the biophysical climate change mitigation potential of Canada's forest sector. Biogeosciences 11(13):3515–3529
Stevens C (2008) Biofuels and development: will the EU help or hinder? Overseas Development Institute, London
Stoeglehner G (2020) Integrated spatial and energy planning: a means to achieve sustainable development goals. Evol Inst Econ Rev 17(2):473–486
Suman S (2018) Hybrid nuclear-renewable energy systems: a review. J Clean Prod 181:166–177
Suparmaniam U, Lam MK, Uemura Y, Lim JW, Lee KT, Shuit SH (2019) Insights into the microalgae cultivation technology and harvesting process for biofuel production: a review. Renew Sustain Energy Rev 115:109361
Swain RB, Karimu A (2020) Renewable electricity and sustainable development goals in the EU. World Dev 125:104693
Szarka N, Scholwin F, Trommler M, Jacobi HF, Eichhorn M, Ortwein A, Thrän D (2013) A new role for bioenergy: a flexible, demand-oriented power supply. Energy 61:18–26
Tenelli S, Bordonal RO, Cherubin MR, Cerri CEP, Carvalho JLN (2021) Multilocation changes in soil carbon stocks from sugarcane straw removal for bioenergy production in Brazil. GCB Bioenergy 13(7):1099–1111
Thrän D (2015) Smart bioenergy: technologies and concepts for a more flexible bioenergy provision in future energy systems. Springer International, Cham
Trinh VL, Chung CK (2023) Renewable energy for SDG-7 and sustainable electrical production, integration, industrial application, and globalization. Clean Eng Technol 15:100657
Tse TJ, Wiens DJ, Reaney MJ (2021) Production of bioethanol—a review of factors affecting ethanol yield. Fermentation 7(4):268
Ussiri DA, Lal R (2017) The role of bioenergy in mitigating climate change. In: Carbon sequestration for climate change mitigation and adaptation, pp 433–495
Van de Graaf T, Westphal K (2011) The G8 and G20 as global steering committees for energy: opportunities and constraints. Global Pol 2:19–30

Van der Horst D, Vermeylen S (2011) Spatial scale and social impacts of biofuel production. Biomass Bioenergy 35(6):2435–2443
Vlek PL, Khamzina A, Tamene LD (2017) Land degradation and the sustainable development goals: threats and potential remedies. International Center for Tropical Agriculture (CIAT), Nairobi
Wang X, Lu Y, Chen C, Yi X, Cui H (2024) Total-factor energy efficiency of ten major global energy-consuming countries. J Environ Sci 137:41–52
Webster E (2020) Transnational legal processes, the EU and RED II: strengthening the global governance of bioenergy. Rev Eur Comp Int Environ Law 29(1):86–94
Welfle A, Thornley P, Röder M (2020) A review of the role of bioenergy modeling in renewable energy research & policy development. Biomass Bioenergy 136:105542
Wolniak R, Skotnicka-Zasadzień B (2023) Development of wind energy in EU countries as an alternative resource to fossil fuels in the years 2016–2022. Resources 12(8):96
Wu F, Pfenninger S (2023) Challenges and opportunities for bioenergy in Europe: national deployment, policy support, and possible future roles. Bioresour Technol Rep 22:101430

Chapter 11
Economics and Policies of the Developed Nations for the Promotion of Lignocellulosic Biofuels and Bioenergy

Babbiker Mohammed Taher Gorish, Mudasir A. Dar, Waha Ismail Yahia Abdelmula, Rongrong Xie, Farhan Ahmad, Jianzhong Sun, and Daochen Zhu

Abstract The advancement of lignocellulosic biofuels in developed countries is crucial for achieving a sustainable energy transition, particularly in light of increasing energy demands and environmental issues. Lignocellulosic biomass, including agricultural residues and forestry byproducts, presents a renewable substitute for fossil fuels, significantly lowering greenhouse gas emissions and decreasing dependence on unstable energy markets. Nevertheless, the progress of biofuels derived from this biomass faces numerous economic, regulatory, and technological obstacles, especially concerning elevated production costs and the intricate nature of adhering to diverse sustainability regulations. Innovations in conversion technologies, such as Simultaneous Saccharification and Co-Fermentation (SSCF) and Consolidated Bioprocessing (CBP), are vital for improving the efficiency of processing lignocellulosic materials. This chapter aims to delineate the current economic policies that support biofuel development, analyze the specific challenges faced, and provide recommendations to enhance the feasibility of lignocellulosic biofuels within the renewable energy sector.

Keywords Lignocellulosic biomass · Biofuels · Sustainability · Conversion technologies · Renewable energy policies

---

B. M. T. Gorish · M. A. Dar · W. I. Y. Abdelmula · R. Xie · F. Ahmad · J. Sun · D. Zhu (✉)
International Joint Laboratory on Synthetic Biology and Biomass Biorefinery, School of Emergency Management, School of the Environment and Safety Engineering, Biofuels Institute, Jiangsu University, Zhenjiang, People's Republic of China

Jiangsu Collaborative Innovation Center of Technology and Material of Water Treatment, Suzhou University of Science and Technology, Suzhou, People's Republic of China
e-mail: dczhucn@ujs.edu.cn

M. A. Dar et al. (eds.), *Recent Trends in Lignocellulosic Biofuels and Bioenergy*, Clean Energy Production Technologies,
https://doi.org/10.1007/978-981-96-4636-4_11

11.1 Introduction

Global energy consumption has experienced a substantial rise due to rapid industrialization, unplanned urbanization, and economic growth over recent decades. Approximately 86% of the global energy supply is derived from fossil fuels, including coal, crude oil, and natural gas, which are extensively utilized to meet the increasing energy demands (Shahbeik et al. 2022). Historically, these traditional energy sources have underpinned global energy systems; however, ongoing dependence on them presents considerable challenges, such as resource depletion, escalating costs, and adverse environmental impacts, including greenhouse gas emissions and global warming (Shafizadeh et al. 2023). The International Energy Agency (IEA) reports that global energy consumption doubled from 1970 to 2000 and rose by 26% from 2000 to 2010. Projections indicate a 28% rise in global energy demand by 2040, emphasizing the urgency for alternative and sustainable energy solutions (Elgarahy et al. 2021). Economic, environmental, geopolitical, and strategic factors drive renewable energy targets worldwide. These targets aim to enhance energy security, foster innovation, combat climate change, and establish leadership in the green economy (Hassan et al. 2024). Countries like China, the EU, the United States, and Costa Rica have made significant strides, motivated by concerns like air pollution, energy independence, and economic benefits. For example, by 2020, China exceeded its renewable capacity goals, the EU reached 19.7% renewable energy, and Costa Rica achieved over 98% renewable electricity. In contrast, countries like Russia have been slower, reflecting reliance on fossil fuels. These varying approaches highlight the role of renewables in shaping sustainable and resilient energy futures tailored to national priorities (Hassan et al. 2024).

Lignocellulosic biofuels derived from plant-based biomass are becoming crucial in our transition to a sustainable energy future. These biofuels utilize the renewable and abundant nature of lignocellulosic biomass (LCB), which comprises cellulose, hemicellulose, and lignin. Globally, LCB is produced in vast quantities—around 1.3×10^{10} metric tons annually—and includes agricultural byproducts like wheat straw, sugarcane bagasse, and cotton stalks (Dar et al. 2024). We can achieve several key objectives by harnessing these materials for biofuel production: reducing our reliance on fossil fuels, lowering CO_2 emissions, and promoting a circular bioeconomy (Mishra et al. 2022). Biofuels from LCB, such as bioethanol and biodiesel, have a distinct advantage as the only renewable energy sources that can produce liquid fuels compatible with our current energy infrastructure. Compared to petroleum-based fuels, lignocellulosic biofuels demonstrate better thermodynamic and oxidative stability, making them suitable for various applications (El-Araby 2024).

Lignocellulosic biomass can also be sourced from agricultural and forestry byproducts, which mitigates competition with food crops and minimizes waste. By 2050, energy derived from biomass could lead the renewable energy sector, potentially representing approximately two-thirds of global renewable energy consumption (Hassan et al. 2024). Converting lignocellulosic biomass into biofuels entails intricate thermochemical and biochemical processes, including pretreatment, saccharification, fermentation, and purification of the final products. These sophisticated technologies offer high efficiency, zero waste generation, and significant economic benefits, facilitating their growing implementation (Shahbeik et al. 2022). Global initiatives, such as those advocated by the IEA, underscore the critical role of biofuels in achieving net-zero emissions by 2050, highlighting their vital contribution to climate change mitigation (Shahzad et al. 2024). Developing biofuels from lignocellulosic sources represents a sustainable strategy for enhancing energy security while safeguarding the environment.

11.2 Economic Policies for Promoting Sustainable Bioenergy

Table 11.1 provides a comprehensive overview of the various economic policies that can be implemented to promote the development and deployment of sustainable bioenergy. These policies are categorized into five key areas: financial incentives, carbon pricing mechanisms, feedstock development support, market development and mandates, and research and development funding (Falcone 2023). Financial support is essential for reducing the economic barriers associated with lignocellulosic biofuel production. Subsidies and grants can offset the costs of feedstock cultivation, processing technology, and infrastructure development. Tax incentives, such as production and investment tax credits, encourage companies to adopt biofuel technologies by lowering operational expenses. Loan guarantees further assist biofuel producers in accessing capital for scaling up their operations, particularly in the early stages of commercial production (Nuhu et al. 2023).

Carbon pricing, through mechanisms such as carbon taxes and cap-and-trade systems, internalizes the environmental cost of fossil fuels. By making fossil fuels less economically attractive, these policies enhance the competitiveness of lignocellulosic biofuels. For instance, carbon taxes directly increase the cost of fossil fuel consumption, incentivizing industries to switch to cleaner alternatives. Cap-and-trade systems set emission caps, allowing companies with lower emissions to sell credits, indirectly benefiting biofuel producers. From the perspective of policy shocks, carbon trading has increased non-hydro renewable electricity generation by 73.32%, while carbon tax has increased it by 31.79%. This indicates that the overall impact of carbon trading on renewable energy is more significant than that of carbon tax. However, the elasticity coefficients of renewable energy to carbon trading prices and carbon tax rates are 0.1801 and 0.1845, respectively, suggesting a slightly more significant marginal effect of carbon tax on renewable energy than carbon trading. Both carbon tax and carbon trading have mitigated the growth of fossil electricity and encouraged public investment in renewable energy, thereby fostering its development. The influence of carbon pricing on renewable energy varies by income level; notably, implementing these policies in high-income countries has diminished their promotional effect on renewable energy. The contribution of technological innovation to renewable energy development is smaller than that of

Table 11.1 Overview of economic policies for sustainable bioenergy

Policy area	Description	Examples and impact	References
Financial incentives	Support to mitigate economic barriers and enhance adoption	Subsidies and grants: Offset costs for feedstock cultivation and infrastructure development Tax incentives: Production and investment tax credits reduce operational expenses for biofuel producers Loan guarantees: Facilitate access to capital, especially for early-stage commercialization	Nuhu et al. (2023)
Carbon pricing mechanisms	Internalize environmental costs associated with fossil fuels	Carbon taxes: Directly increase fossil fuel prices, encouraging a shift to biofuels Cap-and-trade systems: Allow companies to trade emissions credits, promoting cleaner technologies Significant influence on renewable energy generation	Xu and Yang (2024)
Feedstock development support	Policies to cultivate and stabilize biomass supply chains	Subsidies for energy crop cultivation: Financial support for farmers growing as bioenergy crops Government procurement programs: Assure stable demand for crops such as switchgrass and agricultural residues	van Dijk et al. (2024)
Market development and mandates	Establish frameworks for sustainable bioenergy markets	Renewable fuel standards (RFS): Mandate specific biofuel shares in energy mix Renewable portfolio standards (RPS): Ensure demand stability for biofuels	Skeer (2017)
Research and Development funding	Financial investment to enhance bioenergy technologies	Government funding: US DOE's BETO supports research into biomass conversion to biofuels EU funding: Supports innovations related to bio-based processes Continuous R&D investment is critical for overcoming competition for limited resources and development timelines	International Energy Agency (2024)

policies including carbon trading and carbon tax, indicating that renewable energy development during the sample period was predominantly driven by policy measures (Xu and Yang 2024).

Policies supporting the cultivation and supply chain of lignocellulosic feedstocks, such as switchgrass, agricultural residues, and forestry byproducts, are critical. This includes subsidizing farmers for growing energy crops and ensuring stable market demand through government procurement programs. There are various pathways by which biomass can be converted to bioenergy, depending on the type and source of biomass feedstock, the conversion technology, the particular form of energy, and how energy is used. Feedstock for energy use can come from different organic materials, each with methods of energy production to generate a product with a particular (energy) use (van Dijk et al. 2024). Market development for promoting sustainable bioenergy involves creating policies and frameworks that support the establishment of biomass supply chains, particularly for energy use in heating plants. Recent strategies emphasize aligning bioenergy efforts with the sustainable development goals (SDGs) and require systematic monitoring of soil contributions. Collaborations, such as those with development banks like KfW, are essential for fostering sustainable practices and technologies in bioenergy (Skeer 2017). Renewable energy mandates and blending targets create assured markets for lignocellulosic biofuels. Policies like renewable fuel standards (RFS) or renewable portfolio standards (RPS) mandate a specific share of biofuels in the energy mix, ensuring demand stability. Additionally, trade policies that protect domestic biofuel industries from international competition further strengthen market development (Skeer 2017).

Research and Development (R&D) funding for promoting sustainable bioenergy is available through various sources. Continuous R&D investment is necessary to improve the cost-effectiveness and efficiency of lignocellulosic biofuels. Governments can fund academic and industrial research initiatives, focusing on optimizing conversion technologies, enhancing feedstock yield, and minimizing production costs (IEA 2024). For example, the US Department of Energy's Bioenergy Technologies Office (BETO) supports applied research to convert biomass into biofuels and bioproducts. Additionally, the European Union provides funding for research and innovation related to bio-based products and processes (Bioenergy Technologies Office | Department of Energy 2025). These initiatives focus on developing renewable hydrocarbon fuels and maximizing non-food biomass and waste for sustainable energy production. However, challenges such as competition for limited resources, lengthy development timelines, and market acceptance persist (IEA 2024). Continued investment and collaboration among stakeholders are essential to overcome these hurdles and promote innovations that significantly contribute to energy sustainability.

11.3 Economic Rationale for Lignocellulosic Biofuels

11.3.1 Reducing Dependence on Fossil Fuels and Enhancing Domestic Energy Production Capabilities

Lignocellulose-based biofuels play a pivotal role in reducing dependence on fossil fuels while bolstering domestic energy production capabilities. By utilizing agricultural byproducts such as sugarcane bagasse, with an annual production of 279 million metric tons, and rice straw, which generates 370–520 million tons globally, countries can tap into abundant, locally available resources to produce renewable energy (Areepak et al. 2022; Ajala et al. 2021). This shift decreases reliance on volatile and often imported fossil fuels, improving energy security and reducing susceptibility to global oil price fluctuations. For instance, black liquor, a byproduct of the pulp and paper industry, provides a valuable domestic feedstock for lignocellulosic biorefineries, further enhancing self-sufficiency in energy production (Haile et al. 2021). It is estimated that lignocellulosic biofuels could replace more than 30% of global fossil fuel consumption in the coming decades, significantly cutting greenhouse gas emissions while driving localized economic growth and reducing trade deficits related to energy imports (Okolie et al. 2021).

Furthermore, developing domestic biofuel industries promotes energy resilience by creating a decentralized and sustainable energy infrastructure. Lignocellulosic biorefineries can utilize regionally specific biomass, such as corn stover in the Americas or wheat straw in Europe, to tailor energy production to local needs and resources. By reducing reliance on fossil fuel imports, countries can mitigate geopolitical risks associated with energy supply chains, fostering long-term energy independence. This transition to lignocellulose-based biofuels not only aligns with global decarbonization goals but also catalyzes innovation in renewable energy technologies, creating jobs and enhancing energy access in rural and industrial sectors. Ultimately, the widespread adoption of lignocellulosic biofuels offers a pathway to achieving a cleaner, more secure, and self-reliant energy future.

11.3.2 Environmental Benefits to Combat Climate Change

The transition from fossil fuels to renewable energy sources like biomass offers a powerful solution for GHG emissions, a primary driver of climate change. Biomass-based biofuels, derived from agricultural residues, forest waste, and other organic matter, provide a sustainable alternative that significantly reduces carbon emissions. Unlike fossil fuels, whose combustion releases ancient carbon stored for millions of years, biomass fuels operate within a closed carbon cycle. Through photosynthesis, plants absorb CO2 from the atmosphere, which is then released back during biomass fuel combustion, maintaining a neutral impact on atmospheric carbon levels (www.berkeleyearth.org). This process contrasts sharply with fossil fuel emissions, where CO2 accumulates in the atmosphere, exacerbating global warming. For instance, the production of sustainable aviation fuel (SAF) from agricultural residues reduces lifecycle GHG emissions by up to 80%, making it a transformative solution for sectors like aviation that are hard to decarbonize (Safarian et al. 2020). Additionally, the global push for renewable energy adoption contributed to 27% of electricity generation in 2019, supported by policies and innovations like those seen in India's solar and wind energy sectors (www.yearbook.enerdata.net n.d.; www.mnre.gov.in).

The adoption of biomass fuels not only mitigates GHG emissions but also addresses the broader challenges of climate change by decreasing reliance on carbon-intensive fossil fuels. Currently, sectors like electricity, transportation, and industry contribute over 70% of global GHG emissions, primarily from fossil fuel combustion (www.weforum.org). By integrating technologies such as biomass gasification and the production of hydro-treated vegetable oil (HVO) and cellulosic ethanol, biomass fuels offer viable pathways to decarbonize these energy-intensive industries. Additionally, reducing in situ burning of agricultural residues prevents the release of particulate matter, enhancing air quality during harvest seasons. Companies like EcoCeres are spearheading these efforts, converting low-value biomass waste into high-value sustainable fuels with over 90% reductions in lifecycle emissions (www.weforum.org 2023). Scaling such innovations, supported by robust policies and infrastructure, holds the potential to replace fossil fuels with cleaner, renewable alternatives, paving the way for a carbon-neutral energy future and a significant reduction in climate change impacts (Bock 2025).

11.3.3 Rural Development and Job Creation

Recent advancements in biofuel production demonstrate significant social and environmental benefits, with economic viability contingent on factors such as feedstock availability, technology options, and project management (Pasha et al. 2021; Sheth et al. 2021; Lal et al. 2022). The economic rationale for lignocellulosic biofuels is particularly compelling in terms of rural development and job creation. By fostering the cultivation of energy crops on underutilized agricultural lands, biofuel production creates new markets for farmers, enhancing their financial stability and diversifying income streams (Rathmann et al. 2010; Ambaye et al. 2021). Additionally, the biofuel value chain drives rural economic growth by generating employment opportunities across production, transportation, and processing sectors (Hartley et al. 2019). For instance, in 2021, the US biofuel and renewable diesel sector supported 65,000 jobs and contributed $17 billion in economic impacts (Ambaye et al. 2021). These activities not only boost household incomes in rural areas but also promote labor mobility and socioeconomic well-being by expanding agricultural markets (Datta 2022; Yimam 2022). Thus, lignocellulosic biofuels present a robust pathway to achieving sustainable development goals, particularly poverty alleviation, while reducing dependency on fossil fuels and supporting a circular bioeconomy.

11.4 Policy Frameworks Supporting Lignocellulosic Biofuels

11.4.1 Regulatory Policies

11.4.1.1 Overview of Mandates and Standards (USA)

The renewable fuel standard (RFS) program, established by the Energy Policy Act of 2005 and subsequently enhanced by the Energy Independence and Security Act of 2007, serves as a crucial regulatory framework promoting the use of lignocellulosic biofuels (Center 2007). Its primary objectives are to diminish reliance on fossil fuels and to lower GHG emissions. The RFS requires the incorporation of renewable fuels into various sectors, including transportation fuel, home heating oil, and jet fuel. It identifies four principal categories of renewable fuels: biomass-based diesel, cellulosic biofuel, advanced biofuel, and total renewable fuel (Ofori-Boateng 2022). Each category is mandated to meet specific lifecycle GHG reduction criteria, which range from 20% for conventional biofuels to 60% for cellulosic biofuels, using a 2005 petroleum baseline as a reference. The program encourages innovation by prioritizing advanced biofuels, such as cellulosic ethanol produced from lignocellulosic biomass, to meet annual renewable fuel volume requirements and stimulate investments in sustainable biofuel technologies (Bock 2025).

Central to the RFS program is a compliance system utilizing Renewable Identification Numbers (RINs) to track the production and blending of renewable fuels. Obligated entities, including fuel refiners and importers, fulfill their Renewable Volume Obligations (RVOs) by retiring an adequate number of RINs each year. RINs are classified by D-codes, with each code representing a specific type of biofuel, thereby providing flexibility and transparency in achieving GHG reduction goals (IEA 2024). To tackle issues such as fluctuating cellulosic biofuel production, the program incorporates waiver provisions and mechanisms like Cellulosic Waiver Credits (CWCs). Following 2022, the Environmental Protection Agency (EPA) has the authority to establish volume targets while taking into account economic, environmental, and energy security considerations (Bock 2025). By creating a robust market for lignocellulosic biofuels and enforcing stringent compliance measures, the RFS program exemplifies a progressive policy aimed at promoting renewable energy and addressing climate change (USEPA 2024b).

The final rule issued by the EPA on June 21, 2023, strengthens the RFS framework by establishing biofuel volume mandates for the years 2023 to 2025 (Dorman 2023). These mandates encompass specific targets for various categories, including cellulosic biofuel (ranging from 0.84 to 1.38 billion RINs), biomass-based diesel (between 2.82 and 3.35 billion gallons), advanced biofuel from 5.94 to 7.33 billion RINs, and total renewable fuel set at 20.94 to 22.33 billion RIN (USEPA 2024b). Additionally, there is a supplemental requirement of 250 million gallons of renewable fuel for the year 2023. This regulation is designed to promote consistent growth in the biofuel sector while considering essential factors such as costs, air quality, climate change, energy security, and infrastructure development (Bock 2025; Dorman 2023). By emphasizing sustainable energy solutions, the EPA seeks to expedite the integration of renewable fuels, mitigate greenhouse gas emissions, and enhance energy independence in the United States. This strategy ensures alignment with long-term environmental and economic objectives while remaining responsive to the evolving needs of the industry.

11.4.1.2 Sustainability Criteria Established by the European Union (EU) and Other Regions

Regulatory policies are essential in the advancement and implementation of lignocellulosic biofuels, particularly in setting sustainability standards that encourage environmentally responsible production methods. In the EU, the Renewable Energy Directive (RED) specifies particular sustainability benchmarks that biofuels must fulfill to receive support under EU legislation (Jeswani et al. 2020). These benchmarks include a requirement for a 50% reduction in GHG emissions relative to fossil fuel alternatives for facilities established before October 2015, with more stringent standards for newer operations. Additionally, the criteria evaluate the effects of land-use changes, ensuring that biofuels do not contribute to deforestation or the conversion of high-biodiversity areas for feedstock production (Austin et al. 2022). Such measures demonstrate the EU's dedication to embedding sustainability within its energy policies, thereby promoting the production of second-generation biofuels sourced from lignocellulosic materials, which do not compete with food crops and can utilize agricultural residues or dedicated energy crops (Xu and Yang 2024).

In other parts of the world, analogous policy frameworks have been established, creating consistency in sustainability standards for lignocellulosic biofuels. For instance, Brazil supports lignocellulosic biofuels through its National Ethanol Program, which encourages the development and incorporation of sustainable biofuels into national energy policies. The establishment of these diverse regulatory frameworks underscores the increasing acknowledgment of lignocellulosic biofuels as a sustainable energy solution that can lessen reliance on fossil fuels while reducing GHG emissions (Bock 2025). Through comprehensive sustainability standards and supportive policies, regions worldwide are collaboratively striving towards a more sustainable biofuel economy, addressing climate change challenges while enhancing energy security.

11.4.2 Financial Incentives

11.4.2.1 Subsidies, Grants, and Tax Incentives for Biofuel Production

Subsidies, grants, and tax incentives are essential in facilitating biofuel production by enhancing its financial feasibility and attractiveness to both producers and investors. Numerous nations have adopted these financial strategies to encourage the use of renewable energy sources, aiming to decrease GHG emissions and improve energy security. For example, the United States has implemented the RFS, which offers financial benefits for the production and consumption of biofuels, including grants and tax credits for manufacturers in the biofuel sector (Suali and Suali 2022). Additionally, initiatives such as the Biomass Crop Assistance Program (BCAP) provide direct payments to farmers for the cultivation of biomass feedstocks, thereby further invigorating the biofuel industry (USEPA 2024a). These subsidies not only promote the expansion of the biofuel sector but also bolster rural economies by generating employment opportunities and encouraging agricultural diversification.

Moreover, tax incentives have played a pivotal role in advancing biofuel utilization. The federal Volumetric Ethanol Excise Tax Credit (VEETC), for instance, offers a per-gallon tax credit for ethanol production, significantly enhancing domestic biofuel output. Numerous states also provide income and property tax incentives specifically designed for biofuel production facilities, which assist in alleviating the initial capital expenditures required for establishing biorefineries (IRENA 2024a). These financial incentives are crucial for promoting innovation in advanced biofuels, such as cellulosic ethanol, which necessitate considerable investment in research and development. By creating a stable economic framework through these subsidies and tax incentives, governments can enable biofuel producers to expand their operations and ultimately contribute to a more sustainable energy future while decreasing dependence on fossil fuels.

11.4.2.2 Investment in Research and Development for Advanced Biofuels

Investment in R&D for advanced biofuels is essential for meeting sustainable energy objectives and addressing climate change. The global market for advanced biofuels was valued at USD 64.3 billion in 2022 and is anticipated to grow to USD 1499.8 billion by 2031, indicating an impressive compound annual growth rate (CAGR) of 41.9% from 2023 to 2031 (USEPA 2024a; Bock 2025). Such investments are vital for overcoming the cost competitiveness challenges that advanced biofuels encounter when compared to traditional fossil fuels, with biodiesel production being 70–130% more expensive in the wholesale market (IRENA 2024b). Furthermore, it is imperative for governments worldwide to implement stringent regulations and supportive policies, as the effective adoption of advanced biofuels is closely linked to frameworks like the RFS in the United States, which establishes annual production targets for advanced biofuels (Straits Research 2024). Funding for research is critical for enhancing production technologies, with US agencies such as the Department of Energy (DOE) making substantial investments to promote innovation and lower production costs (DOE and EPA Announce $9.4 Million to Spur Development of 2024).

The strategic enhancement of R&D investments is crucial not only for advancing technology but also for fostering economic growth. Research suggests that timely investments in R&D can provide substantial economic advantages by alleviating obstacles to the growth of advanced biofuel technologies and creating job opportunities in associated fields. To meet the climate targets outlined in the Paris Agreement, a projected 64% increase in cumulative low-carbon R&D investments is necessary by the middle of the century, underscoring the need for a solid funding base (Aleluia Reis et al. 2023). This increase is vital, as it has been noted that investments in clean energy technologies, including advanced biofuels, often lead to a beneficial cycle of learning-by-doing, which reduces technology costs and accelerates the transition to net-zero emissions (Austin et al. 2022; Aleluia Reis et al. 2023). Consequently, targeted and sustained investment in advanced biofuels, supported by coordinated national and international initiatives, will empower the industry to create scalable and economically feasible biofuel solutions that can satisfy future energy requirements while contributing to global climate goals.

11.4.3 Market Development Initiatives

11.4.3.1 Strategies to Enhance Market Access for Lignocellulosic Biofuels

Enhancing market access for lignocellulosic biofuels necessitates a comprehensive strategy that tackles economic, regulatory, and infrastructural obstacles. A pivotal approach is to invest in infrastructure development, which is essential for the efficient production and distribution of these biofuels. Recent analyses suggest that substantial improvements in supply chain infrastructure will be required to satisfy the surging demand for lignocellulosic bioenergy; projections indicate that investments could reach up to USD 33.6 billion from 2024 to 2028 (Srivastava et al. 2023). Furthermore, the creation of biorefinery hubs that combine the processing and distribution of lignocellulosic feedstocks can enhance operational efficiency and lower costs, thereby increasing the competitiveness of biofuels relative to fossil fuels. It is imperative for governments and private sector stakeholders to work together to foster an environment conducive to such investments, which may include grants and loan guarantees to mitigate the financial risks associated with the establishment and operation of biorefineries.

Another essential strategy for improving market access for lignocellulosic biofuels is regulatory streamlining. The simplification and modernization of current regulations can significantly diminish barriers to entry in this sector. For example, the recent initiatives by the US EPA to consolidate and simplify fuel quality regulations aim to alleviate the compliance burden on biofuel producers and can serve as a benchmark for other regions (USEPA 2024a). Additionally, the introduction of tax incentives and subsidy programs can stimulate growth by making lignocellulosic biofuels more appealing to both consumers and producers. It is estimated that tax credits and grants could enhance the economic viability of lignocellulosic biofuel production by as much as 25%, thereby significantly promoting market adoption (Suali and Suali 2022). By addressing both infrastructural and regulatory challenges, stakeholders can effectively advance the market for lignocellulosic biofuels.

11.4.3.2 Promotion of Biofuel Blending Mandates

The advancement of biofuel blending mandates is crucial for promoting the use of renewable energy sources in the transportation industry. Recent studies indicate that nations adopting mandatory biofuel blending targets can markedly boost their biofuel usage. For instance, the European Union's RED has set a goal of achieving 14% renewable energy in transport fuels by 2030, a significant rise from the 5.5% blending requirement established in 2010 (IEA 2023). Additionally, Brazil's initiative to incorporate 27% biodiesel into its diesel supply by 2024 serves as a model of effective policy execution, anticipated to lower greenhouse gas emissions by around ten million tons each year, which is comparable to the carbon absorption of 16 million acres of forest (IEA 2023). Moreover, the United States has established ambitious objectives under its RFS, aiming for 22.33 billion gallons of renewable fuel blending by 2025, a notable increase from the 20.94 billion gallons projected for 2023 (Barbose 2024). These mandates not only bolster energy security but also stimulate rural economic growth and encourage technological advancements in the biofuels sector, thereby facilitating the transition to a more sustainable energy landscape.

11.5 Case Studies on the Effective Use of Lignocellulosic Biofuels

Table 11.2 summarizes the case studies highlighting the successful deployment of lignocellulosic biofuels in developed nations, effectively showcasing their dedication to renewable energy, the reduction of greenhouse gas emissions, and the enhancement of energy security. The distinctive initiatives undertaken by each nation illustrate their strategic approaches to utilizing local resources, establishing benchmarks for sustainable biofuel production. This collective effort not only emphasizes the potential of lignocellulosic biofuels in addressing environmental challenges but also underscores the importance of innovation and collaboration in the pursuit of a more sustainable energy future.

Table 11.2 Case studies of developed nations in lignocellulosic biofuels

Country	Initiative	Key features	Key outcomes
United Kingdom	Biofuel production from lignocellulosic biomass	Focuses on using agricultural byproducts (straw, grass) Supported by UK research and innovation	Estimated production: 800 million liters of ethanol/year Carbon reduction: ~1.5 million tons CO_2 annually Goal of 9.75% renewable fuel usage by 2020
Sweden	Sustainable biofuels and biorefinery	Uses forestry resources for biofuel production Strong bioenergy market (13% of total consumption)	Produces ~3.5 million cubic meters of biofuels/year Reduces greenhouse gas emissions by ~4 million tons
Brazil	National alcohol program (ProAlcool)	Transforms sugarcane biomass into renewable energy Launched in 1975 to decrease oil reliance	Became largest producer of bioethanol globally (30 billion liters by 2020) ~18% of national energy from biofuels 80% of vehicles are flexible-fuel
United States	Project liberty	Transforms corn stover into cellulosic ethanol Managed by POET-DSM in Iowa	Generates 25 million gallons of ethanol/year Carbon footprint is ~60% lower than gasoline Attracts over $200 million in investments

11.5.1 European Union

The United Kingdom has achieved notable progress in the effective deployment of lignocellulosic biofuels, primarily through various governmental initiatives and funding for projects that encourage the use of renewable energy sources. A prominent example is the "Biofuel Production from Lignocellulosic Biomass" initiative, which aims to convert agricultural byproducts, particularly straw and grass, into advanced biofuels like ethanol (Sadaqat et al. 2025). This initiative, bolstered by significant financial support from UK Research and Innovation, seeks to improve energy security and mitigate greenhouse gas emissions within the transportation sector (Gómez-Marín and Bridgwater 2021). From a quantitative perspective, the potential benefits of lignocellulosic biofuels in the United Kingdom are considerable. Recent projections suggest that the utilization of straw from wheat and other cereal crops could yield around 1.6 million tons of biomass each year, equating to an estimated production of 800 million liters of ethanol annually. This capacity could replace nearly 650 million liters of fossil fuels, leading to a reduction in carbon dioxide emissions by approximately 1.5 million tons each year. Furthermore, the UK government's Renewable Transport Fuel Obligation (RTFO) has established a goal for renewable fuel usage in transportation to reach a minimum of 9.75% by 2020, with biofuels being essential to meet this objective (Holland and Shapira 2024). By advancing lignocellulosic biofuels, the United Kingdom not only reinforces its commitment to renewable energy but also aids in achieving broader climate objectives while fostering rural economies through enhanced agricultural productivity and job creation in the bioenergy sector.

Sweden has emerged as a prominent European nation alongside the United Kingdom in the effective adoption of lignocellulosic biofuels. The "Sustainable Biofuels and Biorefinery" initiative illustrates how Sweden has leveraged its forestry resources to generate advanced biofuels. The country's ability to produce lignocellulosic biofuels is bolstered by a robust bioenergy market, which constituted approximately 13% of its total energy consumption in 2020. This capacity has enabled Sweden to generate around 3.5 million cubic meters of biofuels each year, predominantly from woody biomass, resulting in a reduction of greenhouse gas emissions by about four million tons (IEA 2024). The successful initiatives undertaken by both the United Kingdom and Sweden highlight the feasibility of lignocellulosic biofuels in meeting energy transition goals, showcasing both environmental advantages and economic opportunities. By prioritizing research and enhancing production technologies, these nations are well-positioned to pioneer sustainable biofuel solutions that could serve as a benchmark for other developed countries seeking to improve their renewable energy strategies (Pan et al. 2024).

11.5.2 United States

The United States has taken a leading role in the development of lignocellulosic biofuels, driven by technological innovations and government initiatives that support renewable energy. A prominent example of this effort is the "Project Liberty" initiative, managed by the biorefinery firm POET-DSM in Emmetsburg, Iowa. This facility is dedicated to transforming lignocellulosic materials, particularly corn stover, into cellulosic ethanol, thereby establishing a successful framework for lignocellulosic biofuel production in the country (Austin et al. 2022). In quantitative terms, Project Liberty aims to generate around 25 million gallons of cellulosic ethanol each year from approximately 770,000 tons of corn stover. This project not only enhances the diversity of the US biofuel sector but also plays a significant role in reducing greenhouse gas emissions. It is estimated that the lifecycle carbon footprint of cellulosic ethanol derived from corn stover is roughly 60% lower than that of conventional gasoline, leading to a reduction of about 400,000 tons of carbon dioxide emissions annually (Usmani et al. 2023). Additionally, the successful operation of Project Liberty has attracted investments surpassing $200 million, including both federal and state funding, highlighting the economic potential of lignocellulosic biofuels (US Department of Energy Awards $118 Million to Accelerate 2023).

11.5.3 Brazil

Brazil has established itself as a prominent global player in the production and application of lignocellulosic biofuels, particularly through the innovative utilization of sugarcane biomass. The Brazilian biofuels industry has effectively capitalized on its agricultural capabilities, transforming lignocellulosic byproducts from sugarcane processing into sustainable energy sources, especially ethanol (Vandenberghe et al. 2022). The initiatives of the Brazilian government, along with historical developments and technological progress, have facilitated the country's leading position in this sector. A significant illustration of successful lignocellulosic biofuel application in Brazil is the launch of the National Alcohol Program, known as "ProAlcool," which commenced in 1975. This initiative aimed to enhance ethanol production from sugarcane to decrease reliance on oil and bolster the domestic sugar industry (Vandenberghe et al. 2022). By 2020, Brazil had become the largest producer of bioethanol globally, with an output of approximately 30 billion liters derived solely from sugarcane (Viva Energy Group Limited 2025). Given that sugarcane biomass primarily consists of lignocellulosic material, including bagasse—the fibrous residue left after juice extraction—this sector exemplifies the successful conversion of agricultural waste into valuable biofuels. The lignocellulosic byproducts not only facilitate the generation of bioelectricity to power the mills but also play a crucial role in Brazil's overall energy matrix, which accounted for around 18% of the national energy supply from biofuels by 2021 (Viva Energy Group Limited 2025).

Brazil has also experienced a significant increase in the use of flexible-fuel vehicles, capable of operating on various blends of ethanol and gasoline. By 2020, approximately 80% of the vehicle fleet in Brazil was comprised of flex-fuel vehicles, demonstrating the effective incorporation of lignocellulosic biofuels into the nation's transportation system (Martinez et al. 2021). This shift not only reduces the nation's dependence on fossil fuels but also establishes Brazil as a leading example of sustainable biofuel production and use on a global scale.

11.6 Challenges and Future Perspectives

As shown below, Fig. 11.1. outlines the critical challenges encountered in the production of lignocellulosic biofuels while also emphasizing potential pathways for future development. To effectively advance the commercialization and sustainability of lignocellulosic biofuels, it is imperative to tackle these challenges with innovative solutions and strategic investments.

11.6.1 Economic Challenges

11.6.1.1 High Production Costs Associated with Lignocellulosic Biofuels

The economic viability of lignocellulosic biofuels is heavily influenced by the substantial production costs associated with their creation, which continue to pose a significant obstacle to their widespread adoption. As of 2023, the expense of converting lignocellulosic biomass into biofuels can reach over USD 3.00 per gallon in certain instances, rendering them much more costly than conventional fossil fuels, where gasoline and diesel prices generally range from USD 2.00 to USD 3.00 per

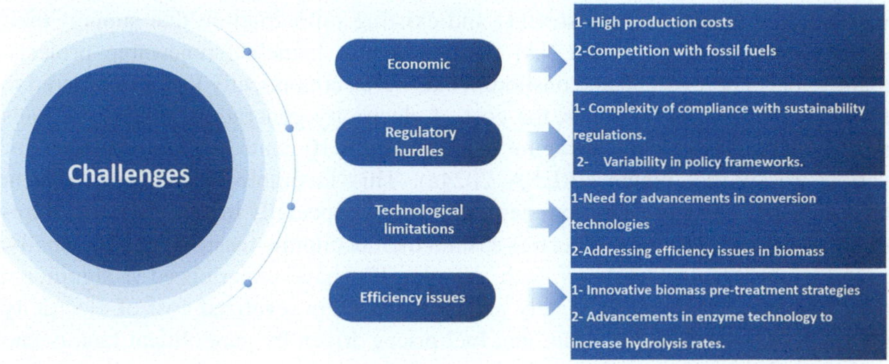

Fig. 11.1 Challenges faced in the production of lignocellulosic biofuels

gallon (Paul et al. 2024). These elevated costs are primarily due to the intricate nature of the production process, which encompasses several demanding stages, including pretreatment, enzymatic hydrolysis, and fermentation. Additionally, the fluctuating costs of feedstock procurement, influenced by agricultural output and market demand, significantly affect overall production costs (Singh et al. 2022). Furthermore, a report from the US Department of Energy highlighted that capital expenditures for biorefineries that process lignocellulosic biomass must be reduced by at least 50% to attain commercial feasibility and compete effectively with fossil fuels (Broda et al. 2022).

Looking forward, tackling these economic challenges will require innovative strategies and targeted policy measures to reduce production costs and improve the economic viability of lignocellulosic biofuels. Recent investments in research and development aimed at advanced bioprocessing technologies hold significant promise for substantially lowering costs (Broda et al. 2022). For example, the application of synthetic biology to create more efficient enzymes can optimize chemical processes and decrease dependence on costly catalysts. Furthermore, supportive measures such as the recently suggested federal tax incentives for biorefineries could stimulate investment and mitigate the financial risks linked to large-scale production. Adopting a circular economy model to effectively utilize agricultural residues and minimize waste can also contribute to lowering feedstock expenses (Singh et al. 2022). These improvements, along with ongoing collaboration between the public and private sectors, will be essential in cultivating a competitive market for lignocellulosic biofuels and progressing towards a more sustainable energy landscape in the future.

11.6.1.2 Competition with Fossil Fuels and Other Renewable Energy Sources

The economic environment for renewable energy is increasingly influenced by intense competition from fossil fuels and other renewable technologies. As of 2023, fossil fuels continue to play a crucial role due to their established market presence, which leads to ongoing investments and existing infrastructure that support their utilization. According to IRENA, 81% of newly launched renewable projects, mainly solar photovoltaic and onshore wind, were economically viable compared to fossil fuels in 2023. However, the cost of electricity generation from fossil fuels remains attractively low, averaging approximately 10 cents per kilowatt-hour for various fossil fuel plants (IRENA 2024a). This favorable pricing environment obstructs the swift transition to renewable energy, especially in areas where government subsidies for fossil fuel sectors sustain their economic feasibility. Additionally, while renewable technologies such as solar PV have experienced a significant reduction in costs—approximately a 60% decrease in levelized cost of electricity since 2010—variations in traditional fuel prices driven by geopolitical factors and supply chain challenges present further obstacles to the competitiveness of renewables (IRENA 2024b).

In a market for renewable energy that is becoming increasingly diverse, competition now extends beyond traditional fossil fuels to include various forms of renewable energy. Recent statistics indicate that different renewable sources have distinct costs and efficiencies, which in turn influence their market share and appeal to investors. For example, while solar photovoltaic systems are achieving low level sized costs of energy (LCOEs), offshore wind technologies face higher associated expenses, averaging $230 per megawatt-hour for fixed installations, rendering them less competitive despite a rising interest (Austin et al. 2022). As governments enhance their goals for renewable energy integration, calling for significant increases in capacity by 2030, including a tripling of solar and wind installations, renewable technologies must consistently innovate and lower costs to sustain a competitive advantage. Therefore, the economic challenges posed by this shifting competitive environment require a strategic emphasis on technological advancements and a more customized policy framework that can effectively address both market dynamics and investment hurdles within the renewable sector.

11.6.2 Regulatory Hurdles

11.6.2.1 Complexity of Compliance with Sustainability Regulations

The intricate nature of adhering to sustainability regulations poses a considerable challenge for the implementation of renewable energy initiatives. In the United States, a multitude of regulatory frameworks oversee the development, financing, and operation of renewable energy projects, resulting in a complex environment that project developers must navigate. These regulations frequently differ across federal, state, and local levels, creating a fragmented set of requirements that can significantly prolong timelines and escalate costs for renewable energy endeavors (Olabi et al. 2023). Research indicates that more than 79% of the wind energy pipeline, along with a comparable percentage of solar projects, are currently stalled in the permitting stage, underscoring the substantial obstacles created by overlapping regulatory demands. As developers endeavor to meet various safety, environmental, and operational standards, the ensuing administrative challenges can dissuade investment and hinder project implementation, ultimately impeding progress toward sustainability objectives (Olabi et al. 2023).

The enforcement of regulations such as the National Environmental Policy Act (NEPA) introduces additional complexities that can impede progress. The environmental review processes mandated by NEPA may extend over several years, with transmission projects alone averaging timelines of approximately 4.3 years. These protracted review periods not only create significant delays in project schedules but also generate uncertainties concerning financial feasibility and market competitiveness (NEPA 2023; Suali and Suali 2022). In recent times, there has been an increasing demand for the simplification of regulatory frameworks, highlighting the necessity for clearer and more straightforward guidelines that facilitate compliance while still ensuring essential environmental protections and community involvement. As sustainability regulations continue to develop, it is essential that they are crafted with the foresight needed to support, rather than obstruct, the advancement of renewable energy initiatives (Susskind et al. 2022). This approach would help reshape the regulatory environment to encourage investment and the expansion of clean energy solutions, rather than constraining them with excessive complexity.

11.6.2.2 Variability in Policy Frameworks Across Regions

The diversity of policy frameworks across various regions poses a significant obstacle to the progress of renewable energy initiatives. This inconsistency stems from the varying local, state, and national regulations that oversee the implementation of renewable energy, resulting in a fragmented environment where project developers must navigate a multitude of compliance obligations (Usmani et al. 2023). For example, in the United States, individual states possess substantial authority in determining their renewable energy policies, leading to a mosaic of regulations that can be either highly supportive or quite restrictive. A report from the National Renewable Energy Laboratory indicates that states with ambitious renewable portfolio standards have effectively enhanced their renewable energy capacity, whereas those with less robust policies face challenges in attracting investments (Morone et al. 2023). This variation generates uncertainty, potentially discouraging investors or developers who may hesitate to participate in markets characterized by unpredictable regulatory conditions and fluctuating compliance costs.

The variability in renewable energy strategies is not limited to the United States; it also encompasses international scenarios where countries implement significantly different approaches to renewable energy integration. For instance, while Europe largely focuses on regulatory harmonization, notable differences persist among member states, particularly concerning feed-in tariffs and grid access regulations. Germany's progressive renewable energy legislation has spurred considerable sector growth, whereas nations like Poland face challenges with slower advancement due to unresolved regulatory issues that deter investment. These disparities can create competitive disadvantages for regions that do not optimize their policy frameworks, ultimately impacting global renewable energy supply chains (Feindt et al. 2020). To address these challenges, international cooperation and the standardization of regulations are essential for fostering a more unified strategy for renewable energy deployment. This alignment allows regions to coordinate efforts and streamline processes for cross-border projects. By harmonizing policies and cultivating collaborative regulatory environments, regions can improve their appeal to investors and support the sustainable development of renewable energy technologies.

11.6.3 Technological Limitations

11.6.3.1 Need for Advancements in Conversion Technologies

The pursuit of sustainable biofuels from lignocellulosic biomass encounters considerable technological challenges, particularly regarding the efficiency of conversion technologies. Two key approaches to biofuel production—Simultaneous Saccharification and Co-Fermentation (SSCF) and Consolidated Bioprocessing (CBP)—underscore the urgent need for progress in this area. SSCF facilitates the concurrent hydrolysis of cellulose and the fermentation of both hexose and pentose sugars, thereby enhancing biomass utilization; however, its overall efficiency is hindered by various technical obstacles. Existing microbial strains frequently exhibit difficulties in effectively fermenting pentose sugars, which constrains the conversion efficiency of lignocellulosic resources (Sharma et al. 2023). To address these challenges, improvements in enzyme functionality and the creation of genetically engineered microorganisms capable of metabolizing a wider array of sugars are crucial, as they would lead to increased yields and a more efficient process.

Additionally, CBP, which seeks to merge saccharification and fermentation into a unified step, introduces its own array of technological hurdles. This approach holds the promise of lowering overall production costs and simplifying processes by incorporating enzyme production directly within the fermentation stage. Nevertheless, the effective realization of CBP necessitates the creation of highly specialized microorganisms that can proficiently execute all required functions, which continues to pose a considerable challenge in the sector (Rodionova et al. 2022). Recent progress in synthetic biology and metabolic engineering presents encouraging avenues for overcoming these obstacles, as it allows for the design of customized microbial strains capable of efficiently handling both hydrolysis and fermentation, thus optimizing production workflows. Prioritizing research aimed at enhancing current technologies while also investigating novel solutions will be essential for fully harnessing the potential of lignocellulosic biomass and achieving improved commercial feasibility for biofuels.

11.6.3.2 Addressing Efficiency Issues in Biomass Processing

The advancement of biofuels derived from lignocellulosic biomass encounters considerable technological challenges, particularly regarding the efficiency of biomass processing. A significant obstacle is the recalcitrance of lignocellulosic materials, which complicates the ability of enzymes and microorganisms to break down these intricate carbohydrates into fermentable sugars. Typically, the commercial production of biofuels from lignocellulosic biomass consists of several stages, including pre-treatment, hydrolysis, and fermentation, each of which can introduce inefficiencies and increase production costs. Recent research indicates that enhancing the efficiency of these processes can substantially improve biofuel yields, thereby making lignocellulosic feedstocks more competitive with fossil fuels (Khan et al. 2022; Jain and Kumar 2024).

To tackle these efficiency challenges, innovative strategies in biomass pretreatment are crucial. Effective pretreatment techniques, such as steam explosion or dilute acid hydrolysis, are designed to dismantle the lignin and hemicellulose components, thereby facilitating easier access to cellulose for enzymatic hydrolysis (Hoang et al. 2023). Nevertheless, existing pretreatment methods frequently result in the generation of inhibitory byproducts that can impede subsequent fermentation. Studies have shown that optimizing pretreatment conditions can reduce these inhibitors and enhance sugar yields, which is vital for the large-scale viability of lignocellulosic biofuels (Shukla et al. 2023). Furthermore, advancements in enzyme technology, particularly the creation of more resilient and efficient cellulases, are essential for improving hydrolysis rates. Recent progress in high-throughput screening methods enables the swift identification of engineered enzymes that exhibit superior performance under diverse conditions, potentially increasing the overall efficiency of the biomass-to-biofuel conversion process (Abolore et al. 2024).

11.7 Recommendations and Future Directions

The shift towards lignocellulosic biofuels as a viable sustainable energy alternative necessitates the resolution of economic, technological, and policy-related challenges (Fig. 11.2). The advantages of biofuels are evident, including a decrease in greenhouse gas emissions and improved energy security; however, the realization of widespread adoption requires economically viable production methods, strong regulatory frameworks, and ongoing technological advancements. Present economic hurdles, such as elevated production costs and competitive market dynamics, necessitate focused interventions. Investments in bioprocessing technologies, synthetic biology, and circular economy initiatives can substantially lower costs. Additionally, the formation of public-private partnerships and the provision of tax incentives may enhance investment and stimulate market expansion.

Equally important are regulatory frameworks that support the advancement of lignocellulosic biofuels. Streamlining compliance procedures and aligning policies across different regions can expedite the development and approval processes for biofuel initiatives. Engaging stakeholders transparently, including partnerships with environmental organizations and local communities, is vital for fostering trust and minimizing delays. Furthermore, the international standardization of policies can promote cross-border renewable energy projects, providing consistent and reliable guidelines for developers and investors.

Technological progress is crucial for realizing the full potential of lignocellulosic biofuels. Improved biomass pretreatment techniques, such as steam explosion and acid hydrolysis, can enhance sugar yields while reducing inhibitors. Concurrently, advancements in enzyme engineering and microbial strain optimization can boost conversion efficiencies. The integration of automation and digital technologies, including high-throughput screening and data analytics, will further refine production processes. Establishing pilot projects to evaluate and enhance these technologies can facilitate the pathway to large-scale commercialization.

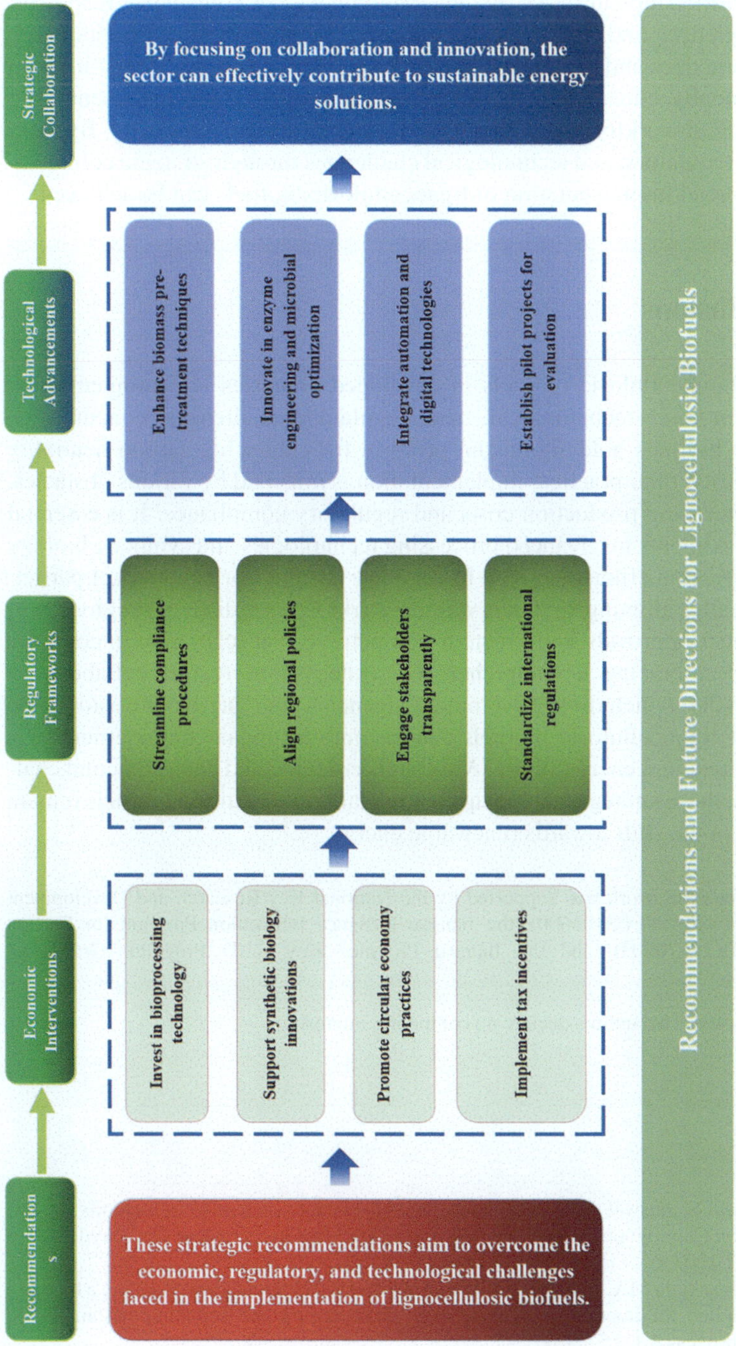

Fig. 11.2 Recommendations and proposed future directions pave the way for the effective implementation of lignocellulosic biofuels

Strategic collaboration plays a crucial role in realizing these objectives. International partnerships in research and development can consolidate resources, exchange knowledge, and expedite advancements. Platforms that provide open access to best practices and data would enhance knowledge-sharing across different regions. Additionally, educational initiatives and training programs are essential for cultivating a skilled workforce capable of advancing the biofuels industry. By tackling economic, regulatory, and technological challenges through strategic collaboration, the widespread implementation of lignocellulosic biofuels can be achieved.

11.8 Conclusions

The future of lignocellulosic biofuels in developed countries is contingent upon addressing economic, technological, and regulatory challenges concurrently. Although these biofuels hold significant promise for enhancing carbon neutrality and energy security, their practical implementation is impeded by various obstacles, particularly concerning production costs and regulatory compliance. It is essential to prioritize investments in advanced processing technologies and synthetic biology to improve conversion efficiencies and lower entry barriers for new market participants. Additionally, aligning regulatory frameworks across different regions will streamline project approvals and create a competitive landscape that encourages innovation. By embracing a comprehensive strategy that fosters collaboration among stakeholders, including government agencies and private investors, the development of lignocellulosic biofuels can become a fundamental element of a sustainable and resilient energy future. As global economic and environmental challenges progress, these strategies will help ensure that lignocellulosic biofuels remain a viable option in the shift towards renewable energy.

Acknowledgments This work was supported by the National Key Research and Development Program of China (2023YFC3403600), the Jiangsu Province Innovation Program for Carbon Neutralization (BK20220003), and the Jiangsu Province Key R&D Program (Grant No. BE2021691).

Competing Interests The authors declare no competing interests.

References

Abolore RS, Jaiswal S, Jaiswal AK (2024) Green and sustainable pretreatment methods for cellulose extraction from lignocellulosic biomass and its applications: a review. Carbohydr Polym Technol Appl 7:100396

Ajala EO, Ighalo JO, Ajala MA, Adeniyi AG, Ayanshola AM (2021) Sugarcane bagasse: a biomass sufficiently applied for improving global energy, environment and economic sustainability. Bioresour Bioprocess 8:1–25

Aleluia Reis L, Vrontisi Z, Verdolini E, Fragkiadakis K, Tavoni M (2023) A research and development investment strategy to achieve the Paris climate agreement. Nat Commun 14:3581. https://doi.org/10.1038/s41467-023-38620-4

Ambaye TG, Vaccari M, Bonilla-Petriciolet A, Prasad S, van Hullebusch ED, Rtimi S (2021) Emerging technologies for biofuel production: a critical review on recent progress, challenges and perspectives. J Environ Manage 290:112627

Areepak C, Jiradechakorn T, Chuetor S, Phalakornkule C, Sriariyanun M, Raita M, Laosiripojana N (2022) Improvement of lignocellulosic pretreatment efficiency by combined chemo-mechanical pretreatment for energy consumption reduction and biofuel production. Renew Energy 182:1094–1102

Austin K, Jones J, Clark C (2022) A review of domestic land use change attributable to U.S. biofuel policy. Renew Sustain Energy Rev 159:112181. https://doi.org/10.1016/j.rser.2022.112181

Barbose G (2024) US State Renewables Portfolio & Clean Electricity Standards: 2024 Status Update. Lawrence Berkeley National Laboratory (LBNL), Berkeley, CA (United States)

Bioenergy Technologies Office | Department of Energy. (2025). Bioenergy Technologies Office. Retrieved from https://www.energy.gov/eere/bioenergy/bioenergy-technologies-office

Bock T (2025) Sustainability aspects of powerfuels and their certification. In: Powerfuels. Springer, Cham, pp 977–1023

Broda M, Yelle DJ, Serwańska K (2022) Bioethanol production from lignocellulosic biomass—challenges and solutions. Molecules 27(24):8717. https://doi.org/10.3390/molecules27248717

Center AFD (2007) Energy independence and security act of 2007. US Department of Energy, Energy Efficiency and Renewable Energy Office. https://afdc.energy.gov/laws/eisa.html. Accessed 29 Mar 2021

Dar MA, Xie R, Zabed HM, Ali S, Zhu D, Sun J (2024) Termite microbial symbiosis as a model for innovative design of lignocellulosic future biorefinery: current paradigms and future perspectives. Biomass 4:180–201

Datta B (2022) An economic analysis of biofuels: policies, trade, and employment opportunities. In: Handbook of biofuels. Academic Press, New York, pp 3–29

DOE and EPA Announce $9.4 Million to Spur Development of Advanced Biofuels as Part of President Biden's Investing in America Agenda (2024). Energy.gov. https://www.energy.gov/eere/bioenergy/articles/doe-and-epa-announce-94-million-spur-development-advanced-biofuels-part

Dorman S (2023) The top percent: revising the renewable fuel standard program to include percentage-based volumes. LSU J Energ Law Resour 11(1):12

El-Araby R (2024) Biofuel production: exploring renewable energy solutions for a greener future. Biotechnol Biofuels Bioprod 17(1):129

Elgarahy AM, Hammad A, El-Sherif DM, Abouzid M, Gaballah MS, Elwakeel KZ (2021) Thermochemical conversion strategies of biomass to biofuels, technoeconomic and bibliometric analysis: a conceptual review. J Environ Chem Eng 9:106503. https://doi.org/10.1016/j.jece.2021.106503

Enerdata (n.d.) Renewables in Electricity Production | Statistics Map by Region | Enerdata. https://yearbook.enerdata.net/renewables/renewable-in-electricity-production-share.html

Falcone PM (2023) Sustainable energy policies in developing countries: a review of challenges and opportunities. Energies 16(18):6682

Feindt PH, Proestou M, Daedlow K (2020) Resilience and policy design in the emerging bioeconomy—the RPD framework and the changing role of energy crop systems in Germany. J Environ Policy Plann 22(5):636–652

Gómez-Marín N, Bridgwater AV (2021) Mapping bioenergy stakeholders: a systematic and scientometric review of capabilities and expertise in bioenergy research in the United Kingdom. Renew Sustain Energy Rev 137:110496

Haile A, Gelebo GG, Tesfaye T, Mengie W, Mebrate MA, Abuhay A, Limeneh DY (2021) Pulp and paper mill wastes: utilizations and prospects for high value-added biomaterials. Bioresour Bioprocess 8:1–22

Hartley F, van Seventer D, Tostão E, Arndt C (2019) Economic impacts of developing a biofuel industry in Mozambique. Dev South Afr 36(2):233–249

Hassan Q, Viktor P, Al-Musawi TJ, Ali BM, Algburi S, Alzoubi HM, Jaszczur M (2024) The renewable energy role in the global energy transformations. Renew Energy Focus 48:100545

Hoang AT, Nguyen XP, Duong XQ, Ağbulut Ü, Len C, Nguyen PQP, Chen WH (2023) Steam explosion as sustainable biomass pretreatment technique for biofuel production: characteristics and challenges. Bioresour Technol 385:129398

Holland C, Shapira P (2024) Building the bioeconomy: a targeted assessment approach to identifying biobased technologies, challenges and opportunities. Eng Biol 8(1):1–15

IEA (2023) Renewable energy market update—June 2023. IEA, Paris

IEA (2024) IEA Bioenergy. https://www.ieabioenergy.com/. Accessed 12 Dec 2024

IRENA (2024a) Record growth drives cost advantage of renewable power. https://www.irena.org/News/pressreleases/2024/Sep/Record-Growth-Drives-Cost-Advantage-of-Renewable-Power

IRENA (2024b) Record renewables growth fuels cost competitiveness—IRENA report. https://www.reuters.com/business/energy/record-renewables-growth-fuels-cost-competitiveness-irena-report-shows-2024-09-24/

Jain S, Kumar S (2024) A comprehensive review of bioethanol production from diverse feedstocks: current advancements and economic perspectives. Energy 296:131130

Jeswani HK, Chilvers A, Azapagic A (2020) Environmental sustainability of biofuels: a review. Proc R Soc A 476(2243):20200351

Khan MU, Usman M, Ashraf MA, Dutta N, Luo G, Zhang S (2022) A review of recent advancements in pretreatment techniques of lignocellulosic materials for biogas production: opportunities and limitations. Chem Eng J Adv 10:100263

Lal R, Kumar S, Naidu S (2022) A cost-benefit analysis of small biofuel projects in Fiji: lessons and implications. J Clean Prod 340:130812

Martinez CLM, Saari J, Melo Y, Cardoso M, de Almeida GM, Vakkilainen E (2021) Evaluation of thermochemical routes for the valorization of solid coffee residues to produce biofuels: a Brazilian case. Renew Sustain Energy Rev 137:110585

Mishra K, Rathore M, Tickoo J, Kumar Singh A (2022) Production of bioethanol from fruit waste. Mater Today Proc 68:1167–1171. https://doi.org/10.1016/j.matpr.2022.09.324

Morone P, Cottoni L, Giudice F (2023) Biofuels: technology, economics, and policy issues. In: Handbook of biofuels production. Woodhead Publishing, Sawston, pp 55–92

NEPA Permitting Process Crucial to Renewable Infrastructure... (2023). https://www.americanprogress.org/article/nepa-permitting-process-crucial-to-renewable-infrastructure-project-success/

Nuhu SK, Emmanuel EO, Manan ZA (2023) Global challenges to sustainable renewable energy expansion: the funding and incentives structures. Chem Eng Trans 106:553–558

Ofori-Boateng C (2022) Global profile and market potentials of the third-generation biofuels. In: 3rd generation biofuels. Woodhead Publishing, Sawston, pp 745–756

Okolie JA, Nanda S, Dalai AK, Kozinski JA (2021) Chemistry and specialty industrial applications of lignocellulosic biomass. Waste Biomass Valor 12:2145–2169

Olabi AG, Elsaid K, Obaideen K, Abdelkareem MA, Rezk H, Wilberforce T et al (2023) Renewable energy systems: comparisons, challenges and barriers, sustainability indicators, and the contribution to UN sustainable development goals. Int J Thermofluids 20:100498

Pasha MK, Dai L, Liu D, Guo M, Du W (2021) An overview to process design, simulation and sustainability evaluation of biodiesel production. Biotechnol Biofuels 14:1–23

Paul S, Mazumder C, Mukherjee S (2024) Challenges faced in commercialization of biofuel from biomass energy resources. Biocatal Agric Biotechnol 60:103312

Pan T, Ocłoń P, He L, Van Fan Y, Zhang S, Wang B, Nowak-Ocłoń M, Tóth Á, Varbanov PS (2024) Strategic integration of residential electricity: An optimisation model for solar energy utilisation and carbon reduction. Energy 310:133227

Rathmann R, Szklo A, Schaeffer R (2010) Land use competition for production of food and liquid biofuels: an analysis of the arguments in the current debate. Renew Energy 35(1):14–22

Rodionova MV, Bozieva AM, Zharmukhamedov SK, Leong YK, Lan JCW, Veziroglu A et al (2022) A comprehensive review on lignocellulosic biomass biorefinery for sustainable biofuel production. Int J Hydrogen Energy 47(3):1481–1498

Sadaqat B, Dar MA, Xie R, Sun J (2025) Drawbacks of first-generation biofuels: challenges and paradigm shifts in technology for second- and third-generation biofuels. In: Zhu D, Dar MA, Shahnawaz M (eds) Biofuels and sustainability: life-cycle assessments, systems biology, policies and emerging technologies. Elsevier Science, Amsterdam, pp 33–47. https://doi.org/10.1016/B978-0-443-21433-2.00003-7

Safarian S, Unnthorsson R, Richter C (2020) Simulation and performance analysis of integrated gasification–syngas fermentation plant for lignocellulosic ethanol production. Fermentation 6(3):68

Shafizadeh A, Rastegari H, Shahbeik H, Mobli H, Pan J, Peng W, Li G, Tabatabaei M, Aghbashlo M (2023) A critical review of the use of nanomaterials in the biomass pyrolysis process. J Clean Prod 400:136705. https://doi.org/10.1016/j.jclepro.2023.136705

Shahbeik H, Peng W, Kazemi Shariat Panahi HK, Dehhaghi M, Guillemin GJ, Fallahi A, Amiri H (2022) Synthesis of liquid biofuels from biomass by hydrothermal gasification: a critical review. Renew Sustain Energy Rev 167:112833

Shahzad S, Faheem M, Muqeet HA, Waseem M (2024) Charting the UK's path to net zero emissions by 2050: challenges, strategies, and future directions, vol 7. IET Smart Grid, p 716

Sharma P, Sangwan S, Kaur H, Patra A, Mehta S (2023) Microbial remediation of agricultural residues. In Sustainable Agriculture Reviews 60: Microbial Processes in Agriculture (pp. 325–358). Cham: Springer Nature Switzerland

Sheth A, Sarkar D, Mukhopadhyay I (2021) Social benefit cost and life cycle cost analysis of sustainable biodiesel bus transport in India. Int J Sustain Eng 14(2):123–136

Shukla A, Kumar D, Girdhar M, Kumar A, Goyal A, Malik T, Mohan A (2023) Strategies of pretreatment of feedstocks for optimized bioethanol production: distinct and integrated approaches. Biotechnol Biofuels Bioprod 16(1):44. https://doi.org/10.1186/s13068-023-02295-2

Singh N, Singhania RR, Nigam PS, Dong CD, Patel AK, Puri M (2022) Global status of lignocellulosic biorefinery: challenges and perspectives. Bioresour Technol 344:126415

Skeer J (2017) Bioenergy for sustainable development. Int Renew Energy Agency (IRENA). https://www.irena.org/-/media/Files/IRENA/Agency/Presentations/Energy-sources/2017/Nov/COP23-Bioenergy-Event-20171024-Bioenergy-for-Sustainable-Development-for-COP23-BONN-1107.pdf

Srivastava A, Rai PK, Choure K (2023) Sustainable biofuels and carbon footprints. Asian Biotechnol Dev Rev 25(3)

Straits Research (2024) Advanced biofuel market size, trends and growth analysis report 2031. straitsresearch.com. https://straitsresearch.com/report/advanced-biofuel-market

Suali E, Suali L (2022) Impact assessment of global biofuel regulations and policies on biodiversity. In: Environmental sustainability of biofuels, pp 137–161. https://doi.org/10.1016/B978-0-323-91159-7.00012-6

Susskind L, Chun J, Gant A, Hodgkins C, Cohen J, Lohmar S (2022) Sources of opposition to renewable energy projects in the United States. Energy Policy 165:112922

US Department of Energy Awards $118 Million to Accelerate... (2023). https://www.energy.gov/eere/bioenergy/articles/us-department-energy-awards-118-million-accelerate-domestic-biofuel

USEPA (2024a) Fuels Regulatory Streamlining | USEPA. https://www.epa.gov/diesel-fuel-standards/fuels-regulatory-streamlining

USEPA (2024b) Overview of the Renewable Fuel Standard Program | USEPA. https://www.epa.gov/renewable-fuel-standard-program/overview-renewable-fuel-standard-program

Usmani RA, Mohammad AS, Ansari SS (2023) Comprehensive biofuel policy analysis framework: a novel approach evaluating the policy influences. Renew Sustain Energy Rev 183:113403

Van Dijk M, Goedegebure R, Nap JP (2024) Public acceptance of biomass for bioenergy: the need for feedstock differentiation and communicating a waste utilization frame. Renew Sustain Energy Rev 202:114670

Vandenberghe LPS, Valladares-Diestra KK, Bittencourt GA, Torres LZ, Vieira S, Karp SG et al (2022) Beyond sugar and ethanol: the future of sugarcane biorefineries in Brazil. Renew Sustain Energy Rev 167:112721

Viva Energy Group Limited. (2025). Appendix 4E and 2024 Annual Report. Retrieved from https://www.listcorp.com/asx/vea/viva-energy-group-limited/news/appendix-4e-and-2024-annual-report-3155864.html

World Economic Forum (2023) Biomass waste: sustainable fuels for carbon and climate change. https://www.weforum.org/stories/2023/12/biomass-waste-sustainable-fuels-carbon-climate-change/

Xu L, Yang J (2024) Carbon pricing policies and renewable energy development: analysis based on cross-country panel data. J Environ Manage 366:121784

Yimam A (2022) Contextual analysis of the biofuel sector in Ethiopia: a comprehensive review focusing on sustainability. Biofuels Bioprod Biorefin 16(1):290–302

Chapter 12
Assessing the Market and Policies for Biofuels/Bioenergy Across Asian Countries

Kenneth Kennedy Adama, Osikemekha Anthony Anani, Uduenevwo Francis Evuen, Peter Mudiaga Etaware, Arthur Efeoghene Essaghah, Prosper Monday Ohien Ikhazuangbe, Festus Oamen Isaac, and Chukwudi Ogwu

Abstract The use of biofuels across Asian countries has provided the means of meeting the emission reduction commitment under the Kyoto Protocol agreement and a means of earning foreign exchange and reducing energy imports. The interplay of market forces and institutional policies in the biofuel and bioenergy sectors of Asian countries has made it imperative to advance knowledge which would galvanize the different players. Across Asian countries, there have been increased bio-

K. K. Adama · P. M. O. Ikhazuangbe
Department of Chemical Engineering, Faculty of Engineering, Edo State University Uzairue, Uzairue, Edo State, Nigeria

O. A. Anani (✉)
Laboratory for Ecotoxicology and Hydrobiology, Department of Animal and Environmental Biology, Delta State University Abraka, Abraka, Nigeria
e-mail: osikemekhaa@delsu.edu.ng

U. F. Evuen
Department of Biochemistry, Faculty of Science, Delta State University of Science and Technology, Ozoro, Delta State, Nigeria

P. M. Etaware
Departmment of Plant Science and Biotechnology, Faculty of Science, Delta State University of Science and Technology, Ozoro, Delta State, Nigeria

A. E. Essaghah
Department of Urban and Regional Planning, Delta State University of Science and Technology, Ozoro, Delta State, Nigeria

F. O. Isaac
Department of Mechanical Engineering, Faculty of Engineering, Edo State University, Uzairue, Edo State, Nigeria

C. Ogwu
Department of Environmental Management and Toxicology, Delta State University of Science and Technology, Ozoro, Delta, Nigeria

M. A. Dar et al. (eds.), *Recent Trends in Lignocellulosic Biofuels and Bioenergy*, Clean Energy Production Technologies,
https://doi.org/10.1007/978-981-96-4636-4_12

fuel production activities which has further created greater bioenergy value chain in the downstream sector. This chapter elucidates the ways and means through which biofuels and bioenergy markets and policies have impacted Asian countries in terms of technological processes, applications, and bioenergy audits. An overview of the different feedstocks, production technologies applicable, and consideration of utilization as regards sustainability, market, and policies, and investment atmosphere are highlighted. Furthermore, the individual regimes and directions of the different Asian countries are presented as regards the source and production of the different biofuels and related bioenergy activities. Also, present available policies domiciled within individual countries are discussed. Demand and supply activities within the context of market forces are enunciated and evaluated, and the market projections as regards biofuel/bioenergy production are presented. Assessment of the impacts of the policies on local and institutional players is further evaluated, and improvements on individual country's biofuel/bioenergy policies are also provided. In conclusion, the present challenges and future perspectives on the possibilities of having a unified market and policies across Asian countries on biofuels/bioenergy are presented with appropriate recommendations.

Keywords Biofuel production · Technological capacities · Value chain auditing · Biofuel strategies · Market forces

12.1 Introduction

In the last few years, there have been intense debates, interest, and discussions around biofuels globally, as a result of the biofuel boom which started around 2004/2005 (USAID/ASIA 2009). Different policies and incentives were announced by several governments the world over especially the United States and European Union to support and encourage greater production of biofuels from various feedstocks while increasing trade among countries in biofuels. As a result of the global trends towards the use of biofuels, several Asian countries went on to announce plans to promote biofuel production for domestic use as well as for export purposes (APERC 2006; ADB 2006; ADB 2009). Thus, between 2004/2005 and the present, biofuel production has become a globally traded commodity with estimated investments approaching the billion-dollar market value. In Asia, the total volume of biofuel production has grown in leaps and bound from 2004 when it was typically 2 billion liters to 2008 when it had attained 89 billion liters for bioethanol and 12 billion liters for biodiesel and still increasing progressively (OECD-FAO 2008; OECD 2008a, b; Bennett and Anex 2009). Notwithstanding the astronomical growth in production volumes of biofuels, the transportation and energy mix of biofuels across Asian countries is still relatively low (Koizumi and Ohga 2007; OECD 2008a, b). Across Asia, the average transport mix involving the use of biofuel is about 3%. Though the volume is quite small, biofuels still provide different prospects and, thus, would require considerable balances and major financial and ecological threats (IEA 2023). Asian countries experience massive problems associated with ecological conservation and the supply of energy. The enlarged request and consumption of nonrenewable fuels have caused deterioration in air quality, astronomical rise in emission of Greenhouse gases (GHG), and made the countries to be import-dependent nations for their energy needs (Canabarro and Silva-Ortiz 2023).

Similarly, over the last few years, there have been intense concerns about the market profitability of biofuels across Asian countries due majorly to the extreme fluctuation in production of goods and services and crude oil demand and supply globally and in-country production-capacity decline due to facilities operating at less than optimal levels (Collins 2008; Fabian 2008). Across Asian countries, there have been significant challenges related to the environment and energy production, utilization, and sustainability. Most rural communities across Asia are deficient in terms of choice in energy varieties though prospects for biofuel developments in a cluster-wise arrangement within each community exist (Fabian 2008; IEA 2023; Canabarro and Silva-Ortiz 2023). Different studies revealed that large-scale manufacturing of biofuels is likely to contribute significantly to Asia's future transport energy demand but is dependent on each country's ability to achieve her national policy mandates on the different biofuels applicable in each country (Koizumi and Ohga 2007; Canabarro and Silva-Ortiz 2023; IEA 2023). These studies further revealed that many of the biofuels produced in Asia had limited GHG emissions or net energy benefits and that most of the large-scale production systems were not viable economically except with the provision of subsidies and incentives by the different governments (IEA 2023). Thus, it was inferred that the decentralization of biofuel production facilities would provide much-needed benefits to biofuel in Asia (Yosiyana 2015; Chang et al. 2016; Tun 2019). Additionally, it was also reported that sustainable biofuel policies were needed to safeguard food security and mitigate any negative feelings about the biofuel and bioenergy sectors being harbored by certain segments of Asian society. Furthermore, the study inferred that the impacts of production and utilization of biofuel on the environment are largely depended on the types of feedstocks, production systems, and practices employed in land cultivation across Asia (ADB 2015; IRENA and ACE 2016; ASEAN-MARD 2019). Table 12.1 shows the various subsidies provided for the development of biofuels across selected Asian countries as part of the policy framework.

Asian countries are among some of the emerging economic giants of the five continents (Datta et al. 2019). These countries believe in the concept of inclusivity, shared national development, vision, technological advancement, economic growth, capacity building, equity, environmental posterity, and healthy well-being of their citizens (Blanco-Canqui 2016). The standard of living of all citizens across Asia is a critical factor, and energy is important for achieving this in all ramifications. The energy strategy incorporates such activities as electrification, capacity for renewable energy, energy discharges decline, and reduction in use of fossil energy volume in electricity mix by 2030 (Adama et al. 2021; Mahapatra et al. 2021). However, there is still visible and possible growth in energy additions of gas, crude oil, nuclear, hydro, and green energy sources though it is still evident that nonrenewable energy

Table 12.1 Biofuel subsidies provided in carefully chosen countries in Asia (USAID/ASIA 2009)

Nation	Total biofuels subsidies
China	In 2007, total subsidies were about USD 220 million; 1373 Chinese Yan (CNY) was provided as blending credit in addition to low-interest loans provided for investment in capital infrastructure and tax holidays. However, subsidies for biodiesel were not provided
Malaysia	Low-interest loans of about USD 16 million were provided in addition to federal grants amounting to about USD 3.3 million
Thailand	Blending credit amounting to 24.9 THB/L (Thai baht/ liter) was provided. Also, E20 blend of ethanol was retailed at 6 baht/ liter, far cheaper than that sold for premium motor spirit (gasoline). E85 vehicles were provided tax incentives while B2 manufacturers were given 30 THB per liter as blending credit
Indonesia	Interest rate subsidy to the tune of 1 trillion Indonesia rupiah (IDR) was provided for farmers growing sugar cane, cassava, jatropha, and oil palm. Biofuels distribution clusters received about IDR 563–572 per liter, and there were drastic reductions in excise duty as part of the subsidies provided
India	Subsidies amounting to about 90% were provided to enhance the irrigation of jatropha cultivation while production facilities for oilseeds were enhanced to about 30% in addition to providing subsidies for fertilizers
Philippines	Favorable loan policies and value-added tax (VAT) exceptions were some of the subsidies provided

sources continue to contribute enormous source or share in the energy market of most Asian countries (Hassan and Kalam 2013; Demirbas et al. 2016). Notwithstanding the pivotal role of nonrenewable fuel in combination with different forms of energy, several reports stated that the supply of resources is limited, not renewable, environmentally unfriendly, and polluting and requires efficient utilization to reduce environmental hazards. In contrast, green sources of energy are local, benign, and essentially infinite (Adama et al. 2024).

Increased request for fuel energy and continual rise in price of fossil fuels occasioned by inflation across the globe has continued to put pressure financially on most countries of the world including Asia. There have been environmental concerns due to the activities of exploration and exploitation of these energy sources (Raju 2006; Chang et al. 2016). The interrelationship between energy consumption, food production, improved livelihood, and environmental quality has assumed substantial recognition due to the need for energy security across Asian countries. More so, Asian countries with their population explosion are more prone to energy crises than other countries of the world (Raju 2006; IES 2019) because, across Asian countries, the energy needs of the population for electricity and transportation has resulted in increased importation of petroleum products. Consequently, biofuels across Asian countries can play vital roles as indigenous and renewable energy sources in reducing the dependency on imported petroleum products thereby catalyzing the economic development of rural Asian communities (Raju 2006; DAEDEME 2012).

World market crude oil prices have kept on fluctuating globally in the last few decades. World economies including that of Asia have been severely affected by these vagaries in the international oil markets (Mizik and Gyarmati 2021). In most countries across Asia, the land transportation sector alone accounts for about 6.7% of fossil-diesel demand which is a substantial component of gross domestic product (GDP). Demand for energy as regards the need for automotive gas oil (AGO), compressed natural gas (CNG), liquefied petroleum gas (LPG), and premium motor spirit (gasoline) continues to rise steadily as a result of the demand constraint viz-a-viz the supply (Saladini et al. 2016; Jeswani et al. 2020). Thus, energy security in most Asian countries would remain vulnerable except with the provision or availability of alternative sources of fuel to substitute or supplement petrol-based fuels developed from renewable resources or feedstock (Arshad et al. 2018; Oumer et al. 2018).

At the international level, the development of biofuels has grown astronomically over the last few decades (Ruan et al. 2019). Across Asian countries, biofuel policies are aimed at bringing renewed attention which takes account of global outlooks and domestic situation in exploitation of home-grown feedstocks for biofuels manufacture and marketing. These policies also consider current and future trends in the expansion of subsequent biofuel generation know-how grounded on novel input materials while promoting the domestic availability of feedstocks and exploring the biodiversity of the individual country's potential. The technological framework, institutional interventions, and financial involvement as well as enabling mechanisms are all encapsulated in the policy documents for biofuels and bioenergy across individual Asian countries (Sajid et al. 2017; 2018). Recently in many Asian countries, development of bioenergy and biofuel production activities has seen astronomical rise. Most Asian countries consider biofuels and bioenergy activities as avenue of enhancing energy security, development of economies through trade balance enhancement, and agricultural sector expansion, as well as poverty mitigation. However, across these countries, their biofuel and bioenergy policies were concentrated majorly on products obtained from agricultural activities and innovative commercial prospects while less attention was placed on equally important issues such as eco-friendly and communal effects of bioenergy and biofuel production actions. Essentially, the astronomical growth of the economy across Asian countries has enhanced the energy demand, and this has heightened cultivation of arable lands for biofuel crops. However, this growth is not without its attendant environmental strain on the land and climatic conditions (Li and Chan-Halbrendt 2009; Matsumoto et al. 2009; Silalertruksa et al. 2009; Tan et al. 2009; Zhou and Thomson 2009). Thus, there is a necessity to carefully confront the societal and ecological effects associated with biofuel manufacturing due to their intricate interrelationship. Figure 12.1 shows the driving force and challenges of biofuel development as it relates to the ecological and communal influence on society. The environmental and societal issues raised include the utilization of energy, biodiversity, greenhouse gas emission, land-use modifications, water consumption, air quality, food security, use of nutrients and insecticide, feedstocks, creation of job, and reduction in poverty (Hassan and Kalam 2013). These issues could either negatively or positively benefit the environment and society, but there would be a need to make trade-offs among the different impacts.

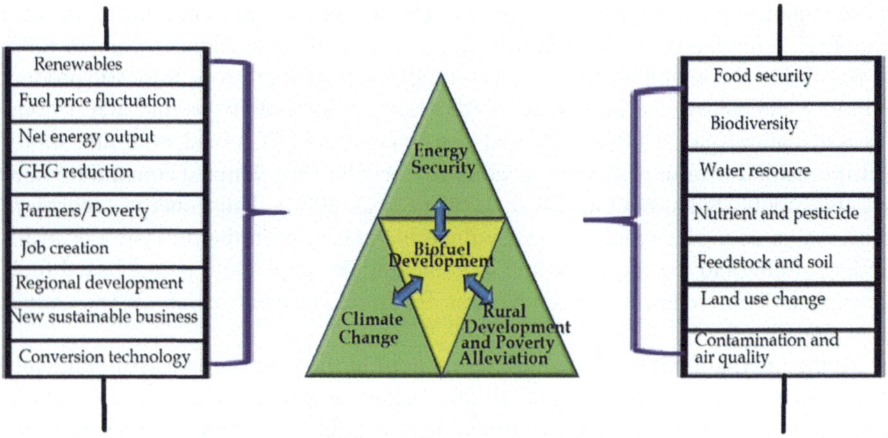

Fig. 12.1 Biofuel expansion challenges and prevailing factors

Consequently, there have been concerted efforts across Asian countries to decrease the importation of crude oil and natural gas as energy avenues that are not renewable through the introduction of different policies which include growing production locally, improvement in refining capacity, enhanced energy productivity protocols (standards), adaptation of biofuels and renewables, and demand substitution processes. Thus, the policy predicts an intended direction for Asian energy activity as it concerns biofuels (Abdullah et al. 2019; Sharma et al. 2019).

12.2 Overview of Market and Policy Issues in Biofuel/Bioenergy across Asia

The influence of biofuels and bioenergy in the energy space has improved tremendously within the last decades, especially as regards power generation and vehicular transportation industries. It is generally believed that the biofuels/bioenergy sector will be greatly involved in the transportation industry as much as 25% of transportation fuel by 2050 (Dafnomilis et al. 2017; Arshad et al. 2018). The alarming worldwide climatic warming issue (IPCC 2014) and its attendant release of CO_2 (carbon dioxide) and many hazardous gases from nonrenewable fuel combustion have greatly generated interest in biofuel uses and its production in meeting energy needs of mankind as vital avenues for bioenergy (Kumar et al. 2013; Orsi et al. 2016; Walker et al. 2016). To stem the increasing climatic change associated with these processes, it has become expedient to develop sustainable production of biofuels as a mitigating tool which would have a multiplier effect on different economies in terms of boosting local economies of developing countries such as countries across Asia through enhanced energy security (Van Eijck et al. 2014; Creutzig et al. 2015). A report by Smith (2013) revealed that fossil energy usage reduction and drastic drop in fossil fuel reserves in addition to eco-system pollution related to fossil fuel exploration, exploitation, and utilization can be mitigated by deliberate advancement into exploitation of different sources of renewable biofuels. Also, Dale et al. (2014) opined that the exploration of environmentally friendly renewable energy sources would help provide economically feasible biofuels and bioenergy options in mitigating energy crises in the future. The environmental concerns raised in various studies (Dale et al. 2014; Ho et al. 2014; Karwat et al. 2014) could be reduced through deliberate and concerted attention in expansion or exploration of bio-based substitutes of biofuels that includes BD (biodiesel), BG (biogas), and BE (bioethanol) whose application and domestication forms different policies of application in the assessment of biofuel and bioenergy activities across Asian countries. Table 12.2 presents the biofuel production approach and different raw materials adopted across Asian countries as reported by Koizumi and Ohga (2007). To achieve the much-needed power requirements of population of Asians living in cities and villages, renewable biomass and waste materials including plastics, solid waste from municipal activities, and gaseous waste must be converted to biofuels. National energy security is the foremost interest in the conversion process of renewable biomass and waste regeneration for environmental sustainability. The process thereby supplements conventional energy needs enhances decreasing dependency on fuel importations and mitigates the continent's energy needs optimally (Gonzalez-salazar et al. 2016; Ruan et al. 2019).

The biofuel market and policies in Asia have assumed a life of their own primarily due to environmental concerns and energy security for the individual countries within the region. There have been different mechanisms, incentives, and subsidies applied by the different countries in meeting their domestic energy needs, as tools for effective rural development and as an agency for employment generation

Table 12.2 Current raw materials and production dynaics of biofuels in Asia. Modified from Koizumi and Ohga (2007)

Country	Bioethanol		Biodiesel	
	Annual production (1000 KL)	Current raw materials	Annual production (1000 KL)	Current raw materials
People's republic of China (PRC)	1292	From wheat and corn	76	Vegetable oils
Malaysia	–	–	200	Palm oil
India	1100	From sugar cane molasses	38	Oils from vegetable origins
Japan	0.03	From cane sugar molasses, and wheat that are unsuitable for food consumption	3	Used vegetable oil
Thailand	3.09	From cassava tubers and cane sugar molasses	76	Palm oil
Indonesia	–	–	3.5	Palm oil

(Demirbas 2007; Gonzalez-Salazar et al. 2016). However, most of the biofuel policies are tailored towards the promotion of indigenous feedstocks (Dahman et al. 2019).

Biofuel is a liquid, gas, or solid fuel obtained from biological materials otherwise called biomass. In most cases, biofuels are also discussed as fuels obtained from gaseous and liquid origins. The biomass could be animal waste, algae material, or plant materials depending on feedstock composition (Adama and Anani 2023; Adama et al. 2024). Biofuel feedstock materials are readily replenished since they are from renewable energy sources (Adama 2021; Adama and Anani 2023). Biofuels are economical and ecologically pleasant sources of renewable fuel compared to petroleum, natural gas, and coal, because of issues related to global warming associated with fossil-based fuels (Adama 2021). Some of the major advantages of biofuels include sustainability, greenhouse gas emission mitigation, and supply security occasioned by the volatile global oil and gas market interplay. The major benefits associated with biofuels for sustainability across Asia are presented in Table 12.3 while Table 12.4 presents biofuel production and feedstock availability across selected Asian countries.

The BD indicated in Table 12.4 implies and includes renewable diesel; regarded as hydrogenated vegetable oil (HVO) which was used to compute the outlook although both products are quite different. Similarly, the potential biofuel impacts on mitigation of weather alteration analysis across Asian countries about the transportation sector provides a significantly improved position for biofuel activities in de-carbonization. This analysis requires different sets of strategy motivations with cost-effective outlooks taking into consideration the effects on sustainability of using biomass assets and consequence on safety of sustenance and livelihood.

Table 12.3 Summary of the benefits of advanced biofuels produced from the lignocellulosic biomass

Evaluation	Beneficial applications
Indigenous impacts on society	Domestic fuel target achievement and sustainable availability. Consistency and reliability in supply Reduction in the use of fossil-based oils Indigenous distribution system
Effective climate	Greenhouse gas emission reduction Minimization of air pollution and pollutant Biodegradability Efficient combustion ability Enhanced carbon sequestration
Market value	Minimal level of competition from international players Imported petroleum independency Development of farming and agricultural sectors of the economy Increases in investment into industrial plants and equipment Institutional improvements in terms of taxes and royalties to the government Job and employment openings within rural communities Sustainability in production Improved availability in the fuel mix and type thereby reducing in monopoly of the market

Table 12.4 Biofuel production and feedstock availability across selected Asian countries

Country	Production status		Available feedstock	
	Ethanol	Biodiesel	Ethanol	Biodiesel
PRChina	3 (8%)	8 (3%)	Maize	Waste oils
India	5 (2%)	15 (0.5%)	Molasses	Palm oil
Indonesia	23 (0.2%)	4 (10%)	Molasses	Palm oil
Thailand	7 (1.5%)	6 (4%)	Molasses/cassava	Palm oil

Source: OECD/FAO (2019); *NB* Numbers in percentage refers to production share

Biofuels and bioenergy relevance across Asian countries cannot be overemphasized. The implications are far-reaching when considered from the viewpoint of environmental trepidations, energy haven, savings from foreign investments, and issues related to the socioeconomic perspective of rural communities within the Asian States (Arshad et al. 2018). Biomass resources are widespread within these countries, and thus, bio-based biofuel technologies are a potential source of employment for a large majority of the rural dwellers than technology adopted to fossil fuel. Secondly, in view of gradual exhaustion of traditional nonrenewable energy opportunities like coal, natural gas, and petroleum, it has become imperative to engage the larger population in biomass fuel and bioenergy development as renewable and evolving environmentally friendly alternatives. Thirdly, in most Asian countries, the environmental benefits of producing biofuels from biological material would create a niche for biofuel in the automotive fuel market which is developing at a very rapid pace within Asian countries. Finally, the inherent benefits of biofuels across Asian countries and worldwide further include their biodegradability and contributions to the sustainability of life (Datta et al. 2019). In this regard, different scenarios have led to conclusions that biofuels would find application widely across Asian countries shortly energy systems because its application would facilitate the switch from economy that utilizes hydrocarbon to one that is dependent on biological materials to manufacture different grades of renewable energy systems that serves as replacement, substitute, or alternatives for old-fashioned fossil fuels and feedstocks; biofuels would also produce equivalent rates of growth in gross domestic products (GDP) and per capita income thereby reducing intensities of fossil energy and GDP and, thus, reducing importation of crude oil in different Asian countries (Anto et al. 2020). Essentially, the biofuel sector across these Asian countries has benefits and relevance in terms of growth rate in GDP, CO_2 emission mitigation, production process energy ratio, direct job creation, and employment opportunities, requirement for biomass plantation growth production system feasibility, and biofuel production from available biomass.

According to Kumar et al. (2013), across most Asian countries, the bioenergy sector has been considered a renewable and sustainable energy source whose utilization would greatly mitigate and serve as a source of price control for the rising conventional energy prices. This sector is also envisioned to serve as a means of providing additional income to rural communities that are predominantly subsistent farmers in various Asian countries. The use of bioenergy in these countries has seen deliberate government and institutional support to help reduce rising petrol/gas prices, meet growing energy demand, mitigate against growing global concerns on greenhouse gas emissions, and increase encouragement and incentives on use of sources of energy that is renewable for domestic and locally developed security energy thereby creating enabling environment for extension to newer areas for bio-based produces to meet-up and challenge existing market imbalances globally (Creutzig et al. 2015; Scarlat et al. 2015; Foster et al. 2017).

Various researchers (Jeswani et al. 2020; Mahapatra et al. 2021; Mizik and Gyarmati 2021) noted that despite keen interest in the bioenergy sector with its attendant benefits and relevance, there has been very limited international and domestic players associated with the bioenergy sector across Asian countries even with increasing prices of fossil fuel, energy demand, and global warming issues connected with traditional energy bases. As recent as 5 years ago, biofuel was considered a cheap alternative to fossil fuel worldwide. Their enormous reduction in carboneous material discharges in comparison to traditional nonrenewable sources and impacts on rural expansion in addition to current high cost of crude oil in the transnational marketplace are some key elements encouraging the market development of biofuel and bioenergy across most Asian countries. Different researchers have reported on the development of feedstock varieties with high oil-yielding potential from non-edible plants and waste materials to generate lucrative, large production, and environmentally benign biofuel with minimal emission potentials (Arshad et al. 2018; Datta et al. 2019; Anto et al. 2020).

In terms of specifics, Hassan and Kalam (2013) and Anto et al. (2020) reported that India imports large quantities of energy in form of ethanol, biodiesel, and petrol diesel. Due to its large volume of diesel demand, there could be adverse ecological and well-being effects of its uninterrupted burning on population, and its utilization has greatly increased in the last few years. However, the biofuel supply has continued to increase by many folds and is largely ascribed to blending policies among several others that had encouraged application and production across Asia. Other policy directions included the deliberate encouragement of the production of biofuels using animal and plant materials with the continuous fuel versus food and feed competition still influencing the market and affecting policy guidelines and operations.

The policy consideration for biofuels and bioenergy across most Asian countries plays pivotal roles in the determination of the growth trajectories of the sector and is still likely to continue shortly. According to Datta et al. (2019), the biofuel policies incorporated in individual Asian countries could encourage or impede sustainable approaches to the development of the sector through conscious barrier reduction and highlighting funding needs and information sharing among individual countries. The key policy approaches for biofuel-producing countries in Asia emphasize the blending requirements for individual biofuels, rules for individual biofuel mix within a stepped time frame, subsidy for biofuel production for cost balancing on production across the different countries, tax incentives for biofuel-converted vehicles viz-a-viz gasoline-only fueled vehicles, tax exemptions and incentives to encourage and promote intensive biofuel production, and social inclusion among many others (Alam and Hasan 2017). Arshad et al. (2018) reported that in the PRC, bioenergy is a strategic energy policy whose implementation has existed since the early 2000s and was based on direct subsidies for biofuels obtained from conventional grains. In her policy, China has a 15% biofuel target by 2030 with a 10% incremental mandate yearly for the next 10 years. Blending at least 10% is part of the policy directives in the mandate policy of China in her attempt to encourage greater participation in the biofuel value chain by 2030 (Anto et al. 2020). China similarly plays an active role in curbing CO_2 emissions and has continuously encouraged the development of friendly State policies for renewable energy including biofuels. Similarly, Jonker et al. (2015), PTI (2016), Alam and Hasan (2017), Araújo et al. (2017), and Anto et al. (2020) reported the deployment of biofuel national policy of Indian government in 2009. The program encouraged deployment of biofuels in the transportation sector such that it was estimated that about 20% of fossil fuel-based fuel would be replaced with biofuels in a 5-year policy plan. In achieving the policy directive, 10% biofuel (ethanol) in gasoline requirement for blending was introduced in the appropriate pricing regime. Other policy changes were envisaged to incorporate the blending of biodiesel with gasoline.

12.3 Production Techniques for Biofuels/Bioenergy Across Asian Countries

Orsi et al. (2016) reported that biofuels of different types are available across different Asian countries. They reported that these biofuels included bio-alcohols (ethanol, methanol, and butanol), BD (biodiesel), GD (green diesel), and solid biofuels and are produced using different techniques that are similarly applied across Asian countries to different degrees. In biofuel production, the conversion process of feedstock to final biofuel is a salient step production value chain for biofuel (Yan and Lin 2009; Adama et al. 2021; Adama and Anani 2023). In considering the competitiveness of biofuels viz-a-viz to fossil fuels in Asian markets, much emphasis is placed on factors like yields and energy consumption in the choice of production technique. Figure 12.2 presents different processes for converting biomass to different biofuels. Using various production techniques such as biological, chemical, and thermal processes, and dependent on types of biological material utilized and biofuel to be produced, several approaches are employed to produce biofuels across Asian countries (Zhou and Thomson 2009; Arshad et al. 2018; Anto et al. 2020). Different feedstocks for production of biofuels can include fat and oil, starches, sugars, cellulose, hemicellulose, lignin, and algae. The process of transformation could be by chemical methods which are esterification/transesterification and hydro-treating; biological techniques such as orthodox alcohol fermentation, hydrolysis by enzyme reaction, and aerobic fermentation; and digestion using anaerobic means and innovative approaches that include thermochemical technology such as catalysis to liquid fuels, gasification, and pyrolysis. All the products obtained are

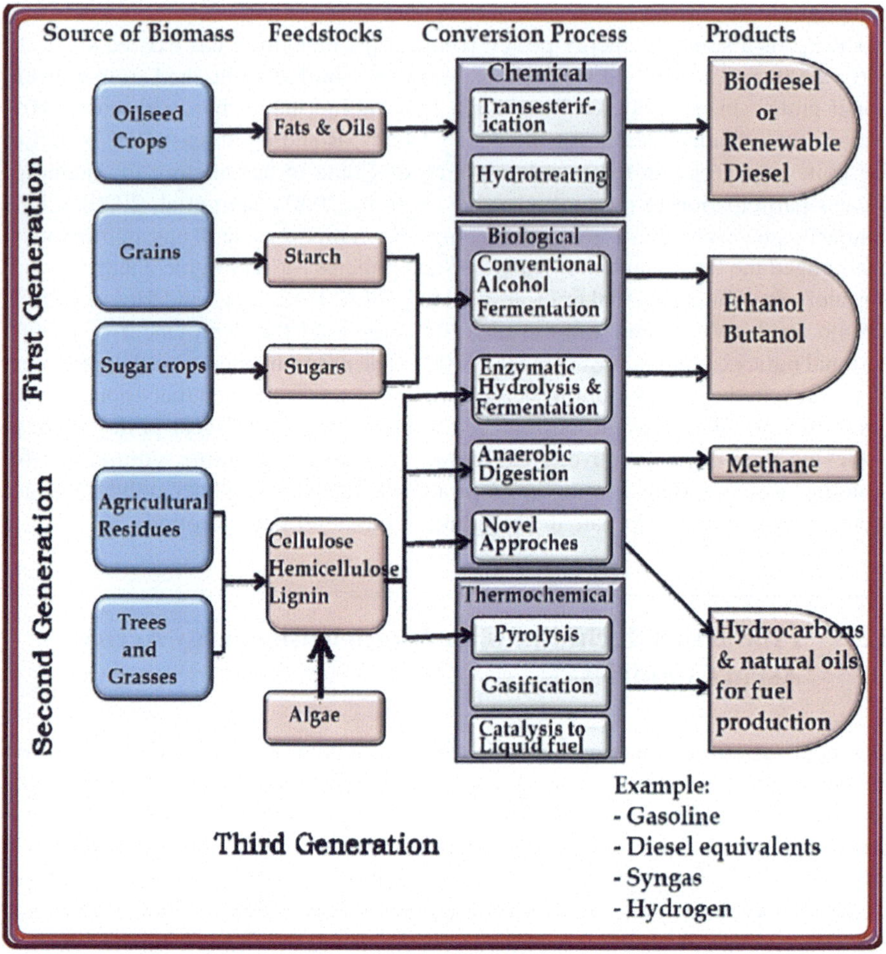

Fig. 12.2 Biomass to biofuel conversion process diagram

dependent on the conversion routes taken ranging from renewable diesel, ethanol, methane, hydrocarbons, gasoline or diesel equivalent, syngas, and hydrogen (Arshad et al. 2018; Adama 2021; Adama and Anani 2023).

In selecting different production techniques, process optimization of biofuel production is dependent on feedstock types, the efficiency of conversion, cost implications, and greenhouse gas (GHG) emission mitigation (Adama et al. 2021; Adama 2021; Adama and Anani 2023). As technologies continue to evolve across Asian countries and globally, there will be greater focus on using lignocellulosic conversion technology due to the large volume of available feedstocks, inherent climatic change benefits, and issues of food security (Scarlat et al. 2015; Dafnomilis et al. 2017; Foster et al. 2017; Arshad et al. 2018; Datta et al. 2019; Adama and Anani 2023; Adama et al. 2024).

12.4 Effects of Environmental Conditions on Production and Policies

Production of biofuels from different feedstocks represents a complex activity whose implications on the environment could be adverse (Mofijur et al. 2012a, b). Life cycle reductions in hazardous or toxic gases which include CO_2 (carbon dioxide) and CO (carbon monoxide) emissions within the ecosystem are dependent firstly on source of feedstock, secondly on production techniques utilized, and thirdly on expectations presented for different applications of the acreage for biological raw materials production (Sarin et al. 2006). Greenhouse gas emission balancing in most cases is practically impossible to evaluate where the feedstock history is uncertain, and this uncertainty is rather associated with the growing history and potential of the parent plant with most oil-bearing plants containing or producing nitrous oxide emissions due to the quantity of nitrogenous fertilizers applied during growth. For favorable balance of energy in a system, output energy must be greater than the input energy. The production of most biofuels from biomass and/or oil-bearing plants has positive energy in terms of drought sustainability, waste utilization, marginal rainfall needs, and soil fertility requirement (Divakara et al. 2010; Kumar and Sharma 2011). Owing to increasing biofuels demand across Asian countries, the required feedstocks need to be readily available in the right quantity and consistency, and as such, there is increased demand for irrigation water for biofuel crops, and this has progressively affected water demand and use for energy purposes and water application in agricultural practices. The ability to grow most biofuel crops under harsh environmental conditions and provide adequate vegetation covers for degraded lands has provided the much-needed channel for the utilization of water for other positive purposes (Axelsson and Franzen 2010; Achten et al. 2007). Several benefits in the production of biofuels from crops residues have shown greater adaptation leading to environmental protection which in turn leads to lesser emission of GHG when compared to fossil fuels. Prueksakorn and Gheewala (2006) stated total emissions from GHG systems based on life cycle assessment were about 90% which was observed to affect the final application of such fuel. The reports further estimated that the potential of biofuels (especially biodiesel) to cause global warming was about 23% of that of fossil fuel (Prueksakorn and Gheewala 2006). However, it has been reported that certain biofuel feedstocks have toxic constituents that may induce environmental and public health concerns (Buddenhagen et al. 2009; German Technical Cooperation 2010). In assessing the biofuel market and policies on effects of environmental conditions on production, a holistic policy developmental approach and implementation must be cognizance of forest reserve conservations and environment in addition to focusing on higher production rate of biofuel feedstocks so that greenhouse gas (GHG) savings by biofuel activities would not be an illusion but a practical activity through undertaking comprehensive total life cycle analysis of biofuels from farmstead to actual utilization of biofuel. In this regard, Hassan and Kalam (2013) reported that it was essential to carry out an all-inclusive parameters' evaluation study of renewable biofuel and nonrenewable fossil fuel life cycle analysis to ascertain apparent advantages and/or disadvantages inherent in/GHG savings. In appraising biofuel feedstocks and production techniques, an important tool to utilize is life cycle analysis. However, few life cycle analysis (LCA) studies are available across Asian countries for feedstocks and locations. LCAs would greatly assist in policy-making, investment options, project development, and community organizations to make informed decisions that would facilitate sustainable production of biofuel via scaling-up options at the right time (Tiziana et al. 2018; Tun et al. 2019). Figure 12.3 presents a detailed life cycle analysis of GHG emissions for fossil fuels and biofuels.

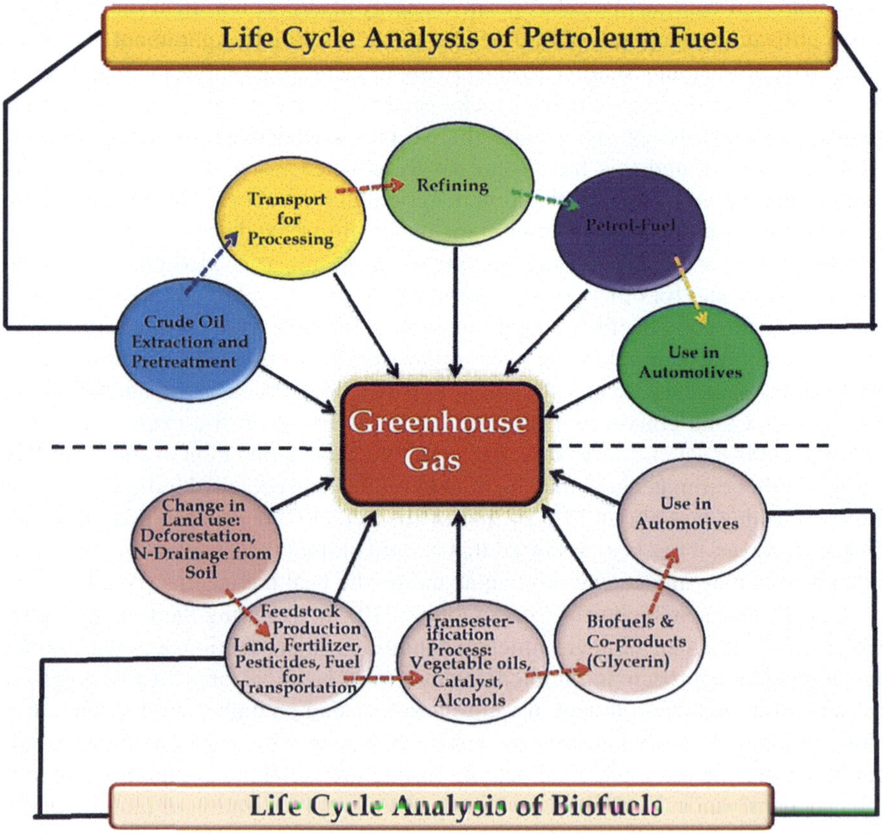

Fig. 12.3 Life cycle analysis of petroleum fuels and biofuels

12.5 Policy Direction on Biofuels and Bioenergy across Asia

According to several studies (Hoff 2011; Foster et al. 2017; Kibler et al. 2018; Anto et al. 2020; Benites-Lazaro et al. 2020), biofuels and bioenergy have assumed importance across Asian countries due to growing security needs for energy and associated concerns on the environment from fossil fuel activities in recent times. Based on the domestic requirements of individual Asian countries, there have been concerted efforts geared towards the establishment of several mechanisms, aids, and inducements to promote the growth and expansion of biofuels and bioenergy and bioenergy sectors of their different economies. This policy direction entailed using biofuels and bioenergy activities to create employment and develop rural communities through the promotion of indigenous feedstock development. Furthermore, Elbehri et al. (2013) reported that most Asian governments had increased the levels of interventions in biofuel promotion through the establishment of different layers of structured programs such as blending of ethanol with petrol program, national vision on biodiesel, and blending biodiesel programs. Based on past experiences and to meet the demand-supply nexus, most governments across Asia have tried to strengthen these programs to be able to effectively cope with pricing, incentives, biodiesel sales to bulk and retail customers, opening new routes for ethanol production, and engaging more in developmental research and innovations (RD&I) activities regarding biofuel and bioenergy. Essentially, there has been a positive impact on the biofuel sector due to the implementation of these programs across different Asian countries (Demirbas 2009a, b; Carriquiry et al. 2011; Ho et al. 2014; Önal and Núñez 2014; Jonker et al. 2015; Alam and Hasan 2017). These policies have shown benefits in terms of offering great opportunities for integration and also assisting in raising the income of rural agriculturalists, reducing fossil fuel importation, generating employment, and creating wealth from waste biomass. Thus, the different country's existing biodiversity is utilized effectively for wealth creation for the rural population and also contributes to sustainable development across all facets of the economy (Demirbas 2009a, b; Araújo et al. 2017; Khan et al. 2021; Saini et al. 2021; Adama and Anani 2023).

PTI (2016) and Moioli et al. (2018) reported that owing to the attention biofuels and bioenergy activities have received in the last few decades globally, it has become imperative to keep in touch with developmental trends in the field of biofuel across Asian countries. Hence, the policy further seeks to increase the level of technological knowledge worldwide in biofuel activities across Asia to meet up with the international perspectives on biofuel development using local indigenous raw materials (Datta et al. 2019; Mahapatra et al. 2021; Adama and Anani 2023). The policy further considers different developmental aspects regarding future-generation biofuel transformation techniques that are founded on newer input sources and promotion of available domestic feedstocks through exploration and utilization of the biodiversity available in individual Asian countries (Hassan and Kalam 2013; Demirbas et al. 2016). Table 12.5 shows the biofuel policies of selected Asian countries. These policy directions are enunciated and captured in their different technical

Table 12.5 Policies in the development of biofuels across selected Asian countries

Nation state	Target for first-generation biofuels and plans for second-generation biofuels	Blending mandates	Economic measures
People's republic of China (PRC)	Utilization of non-grain path to biofuel development	Ethanol: Trial periods requiring 10% blending mandates in some regions	Ethanol: Incentives, subsidies, and tax exemptions for production purposes. Biodiesel: Tax exemptions for biodiesel production using oil from vegetables and fats from animals
India	Non-identifiable target. Promotion in the use of jatropha biodiesel	Ethanol: 5% blending in gasoline required in designated states in 2008 and to increase to 20% by 2027	Ethanol: Excise duty concession promoted. Ethanol and biodiesel: Minimum encouraging charges for buying of ethanol by advertising corporations provided
Indonesia	Application of biofuels domestically: Projections of 2% of energy mix by 2020, 3% by 2022, and 5% by 2025 Considerations on the use of jatropha and cassava products	Biodiesel: Optional biodiesel blending but with plans to increase biodiesel blend to 10% by 2030	Biodiesel: Subsidies (at the same level as fossil fuels)
Japan	Plans to replace 500 ML/year of transportation petrol with biofuels. Promotion of transportation fuels obtained from biomass	No blending mandates. The upper limits for blending are 3% for ethanol and 5% for biodiesel	Ethanol: Subsidies for production and tax exemptions
Malaysia	No targets identified Promotion of jatropha, Nipa, etc.	Diesel: Blending of 5% palm oil in diesel	Diesel: Plans to subsidize prices for blended diesel
Philippines	Non identifiable targets. Studies and jatropha pilot projects	Ethanol: 5% by 2008; 10% by 2010 Diesel: 1% coconut blend; 2% by 2009	Ethanol and diesel: Tax exemptions and priority in financing.
Thailand	Replacement plans for 20% of vehicle fuel consumption with biofuels and natural gas by 2012	Ethanol: 10–20% by 2008 (gasohol 95) Diesel: 5% (B5) mix in 2007 and 10% (B10) by 2011	Ethanol: Pricing incentives through tax exemptions

Source: Adopted from Agoramoorthy et al. 2009; Li and Chan-Halbrendt 2009; Matsumoto et al. 2009; Silalertruksa et al. 2009; Yan and Lin 2009; Zhou and Thomson 2009; Silitonga et al. 2011; Mofijur et al. 2012a, b; Hassan and Kalam 2013

frameworks, financial and fiscal incentives, enabling mechanisms, and institutional interventions (Saladini et al. 2016; Hajilary et al. 2019; Jeswani et al. 2020). These policies have since been upgraded across most Asian countries as part of their biofuels and bioenergy development initiatives.

The biofuel policies across Asian countries differ from one country to another depending on the purpose for which the policy was meant (Oh et al. 2010; Shafie et al. 2012; Bakar et al. 2014; ERIA 2014). Although the different policies are meant to develop the biofuel sector, there are several differences based on individual countries' priorities. Table 12.6 shows the different policy drivers as they relate to current biofuel-related policies across Asian countries. In the case of the Philippines and Indonesia, the policies are mandatory and geared towards the introduction of bioethanol and biodiesel. In South Korea, the introduction of biodiesel as B2 is mandatory, while in other Asian countries, the biofuel promotion policies are based on the development of biofuel program rather than mandatory schemes. These features of biofuel policies across Asian countries include rural development, energy security, industry strategy, and environmental policy initiatives (Singh 2013; Yosiyana 2015; ADB 2015; IRENA and ACE 2016).

Table 12.7 presents the policy and targets for biofuel development across different Asian countries. Analysis of the table revealed that apart from Brunei Darussalam and Cambodia, all other Asian countries presented development programs and/or targets. The application of biodiesel and bioethanol is mandatory at domestic levels in countries such as the Philippines and Indonesia. The application of B2 in South

Table 12.6 Policy drivers for biofuel promotion across Asian countries

Country	Main policy drivers
China	Reduction in oil consumption; adjustment in food stock
India	Reduction in oil imports; exploration of clean fuel
Indonesia	Reduction in oil imports; improvement in energy security; development of rural communities
Cambodia	Reduction in oil imports; exploration of clean fuels
Brunei Darussalam	None
Japan	Reduction in CO_2 emission; supply of clean energy
Lao PDR	Improvement in energy security
Malaysia	Development of local industries
Myanmar	Improvement in energy security; policy development on agriculture
Philippines	Reduction in oil importation; improvement in energy security; development of rural communities
South Korea	Diversification of energy sources; reduction in oil importation; supply of clean energy
Singapore	Development of local industries
Thailand	Reduction in oil imports; improvement in energy security; development of rural communities
Vietnam	Reduction in imports; improvement in energy security

Sources: Oh et al. (2010); Shafie et al. (2012); Bakar et al. (2014); ERIA (2014)

Table 12.7 Policy and target for biofuel development across Asian countries. Sources: ERIA (2014); Tiziana et al. (2018); Tun et al. (2019)

Country	Status	Main policy, development program, and planning	Target
China	Not mandatory	1. The act for ethanol testing and expansion and blending of ethanol with gasoline for vehicles in 6 provinces and 27 cities (E10) 2. 500 yuan; which is equivalent to 6 cents/litre for the production of bioethanol 3. Renewable ethanol (RE) development in the 12 5-year plan	Bioethanol: Four million tons Biodiesel: One million tons
India	Not mandatory	1. National policy on biofuels in about 9 states and 4 union territories as well as E5 blending mandates across the nation	Biofuels: 20–40% and beyond 2027
Indonesia	B10 policies beyond 2013; and development of E3 options	1. Development of National Energy Policy beyond 2006 2. Presidential instruction no. 1/2006 3. Mandatory regulation DEMR no. 32/2008	5% national energy mix by 2025
Japan	Not mandatory	1. Voluntary target of 500,000 kiloliters (kL) of bioethanol by E3 (ETBE)	Total 500,000 kL bioethanol >2017
Lao PDR	Not mandatory	1. Strategies in renewable energy development strategy beyond 2010	10% of total transport fuels in 2020; bioethanol: 106 million liters; biodiesel: 205 million liters
Malaysia	Not mandatory	1. Development of national biofuel policy beyond 2006. 2. Promotion of B5 biofuel programs with effect from June 2011 and to be completed July 2014	B7 after July 2014
Myanmar	Not mandatory	1. Development of jatropha tree-planting project within the period 2006–2008	Approximately four million hectares (ha) by 2015 and beyond.
Philippines	E10, B2	1. Biofuel Act 2006 (RA 9367) 2. National Biofuel Program	2030: E20/E80; B20
South Korea	B2	1. Development of a second biodiesel long-term supply plan by 2010 to complement the first plan which had commenced in 2006	B5:Vehicle B20: Bus and trucks

(continued)

Table 12.7 (continued)

Country	Status	Main policy, development program, and planning	Target
Singapore	Not mandatory	1. Industry strategy on worldwide biofuel terminal and as a hub for biofuel trade	No target
Brunei Darussalam	None	None	None
Cambodia	None	None	None
Thailand	Not mandatory	1. Alternative energy development plan (AEDP:2013–2025) 2. Incentives to growing palm oil in large hectares of about 2.5 million rais 3. Price subsidy to biofuels and feedstock 4. Expansion of feedstock development program on land use from 2010–2025	2025: Bioethanol: Nine million liter/year 2025: Biodiesel: 7.2 million liter/year
Vietnam	Not mandatory	1. Prime Minister's decision no. 177 2. Mandatory from 2014: E5 (seven provinces and cities) 3. Expanding financial support for biofuel program to the tune of VND 259.2 billion within the time frame of 2007–2015	2025: E5 and B5 (1.8 million tons total of biofuels

Korea was made compulsory while the government of Thailand and Malaysia introduced various promotion programs at national scale levels. Vietnam mandated the use of E5 in 2014 (ERIA 2014; Tiziana et al. 2018; Tun et al. 2019).

As shown in Table 12.7, it is evident that the biofuel's target for individual countries across Asia is quite different. Careful observation of the table shows that Japan, China, and Thailand display targets that are quantitative in nature. However, South Korea, Vietnam, the Philippines, and Malaysia use blending share as a means of estimating their targets. India, Indonesia, and Lao People's Republic (Lao PDR) use biofuel's share in total fuel consumption of their nations as a means for individual target estimation.

Across Asian countries, the biofuels trade is not that active. Bioethanol trade volume is about 565,000 kL/year according to data (ERIA 2014; ADB 2015; Chang et al. 2016). Thailand is the major exporter of biofuels in Asia while bioethanol is mainly imported by the Philippines according to the report (ERIA 2014; IRENA and ACE 2016; Tun 2019). In 2013 alone, biodiesel trading volume across Asian countries was estimated to be 1.6 million kiloliters with Indonesia accounting for 1.55 million kiloliters as the leading exporter of biodiesel in 2013. Malaysia was next as the leading exporter of biodiesel in 2013 with 34,000 kiloliters according to the report (ERIA 2014; Yosiyana 2015; Tiziana et al. 2018). Unfortunately, production of biodiesel in both Indonesia and Malaysia is majorly marketed to Europe with very minute quantity sold to/exported to other Asian countries (ERIA 2014; ASEAN-MARD 2019).

Based on statistics and available data/information, it is apparent that at the developmental stages of biofuel utilization, each country across Asia introduces its policy on biofuel production and utilization. In China and India, the local governments are principally responsible for the introduction of biofuel policies, while in Vietnam and Myanmar, biofuel policies are still at the planning level in the developmental stage of the program. Other countries across Asia are at different levels or stages in their biofuel policy or program on the introduction of biofuels into their energy mix (ACE 2015; ASEAN-MARD 2019).

12.6 Present Biofuels and Bioenergy Production Activities across Asia

Presently, the leading producers of biofuels in Asia are Thailand, Indonesia, Malaysia, People's Republic of China (PRC), India, and the Philippines (Zhou and Thomson 2009; Hassan and Kalam 2013). However, India and China with South East Asian countries are the only countries actively participating in the biofuels and bioenergy industry at the moment. While Southeast Asian countries have concentrated their energies on biofuel exports, India and China are actively engaged in biofuel programs for the rapid development of their economies and a reduction in import dependency for petroleum-based fuels (Zhou and Thomson 2009; Mofijur et al. 2012a, b; Karwat et al. 2014).

According to Mofijur et al. (2012a, b), the first and second largest oil palm producers in the world are Malaysia and Indonesia. Both countries share and provide 85% of the world's oil palm requirement. The high production rate is mainly due to high potential and extremely favorable production factors such as climatic conditions and relatively cheap labor (Latner et al. 2006; Tan et al. 2009). Indonesia is also involved in the massive biodiesel production (Zhou and Thomson 2009) with about eight plants producing biodiesel as of 2011 and still counting. Thailand has replaced 10% of diesel with biodiesel since year 2012 with an appreciable number of gas stations marketing B5 mixtures since 2007 according to a study by Yan and Lin (2009). Accessibility of several forms of biofuel feedstocks in Thailand and Indonesia has positively influenced the biofuel activities in these two countries. The major biodiesel raw material of Thailand is palm oil, although coconut oil, jatropha oil, and waste cooking oils are also used to produce the biofuels (biodiesel). Other biofuel activities include bioethanol production, using molasses, cassava, and sugarcane. However, Indonesia uses only cassava and sugarcane for its bioethanol production. On the other hand, Malaysia produces biofuels from mainly palm oil making the country susceptible to market force (price) fluctuations and variations from petroleum and palm oil prices occasionally. The versatile nature of palm oil which finds applications in food and production of biodiesel has further exaggerated the price fluctuation and volatile nature of using palm oil in production activities of biofuel across Asian countries.

India and China both account for nearly 45% of world's energy demand for biofuels because of their populace and size (Hassan and Kalam 2013). The major raw material source for energy needs of India still remains to be coal with the importation of crude petroleum to supplement and cope with the increasing demand for transportation fuel. The bioethanol market has grown in limps and bounds over time much more than the biodiesel market. India's biodiesel feedstocks are mainly from Karanja oil, jatropha oil, mahua oil, and neem oil because there are no surplus edible oils to be utilized in producing biodiesel. India's policy direction has in the last decades encouraged 5% ethanol blending in gasoline across the different counties and regions of the country using the EBP (ethanol blended petrol) initiative (Arshad et al. 2018; Anto et al. 2020). In the People's Republic of China (PRC), feedstocks in biofuel production are mainly from corn and wheat. Fuel-grade ethanol in China is presently obtained from corn and accounts for about 80% of its biofuel (Ou et al. 2009; Hassan and Kalam 2013). Cassava and sorghum are mainly used for laboratory-scale experimental activities and not for large-scale commercial production activities. The biofuel products in China include corn-derived ethanol, cassava-derived ethanol, sweet sorghum-derived ethanol, soybean-derived biodiesel, jatropha oil-derived biodiesel, and used cooking oil-derived biodiesel (Ou et al. 2009; Hassan and Kalam 2013). The world's largest importers of vegetable oils is China and majority of the factories producing biodiesels function on a small-scale inefficient operation due to the nonavailability of feedstocks which is largely due to non-provision of incentives by the government (Latner et al. 2006; Hassan and Kalam 2013). In China, the ethanol production from available feedstocks is comparatively higher than biodiesel production from available feedstocks. Biodiesel production is majorly from fats from animals and wasted oils from plant sources in China, and this amounts to about 300,000 metrics according to Hassan and Kalam (2013).

The promotion and production of biofuel activities across Asian countries has grown tremendously in the last few years (IREP 2016). Thailand is utilizing the potentials inherent in palm oil, tapioca, and cassava as feedstocks for biofuel production, while in Indonesia, oil from jatropha and palm oil are utilized for different biofuel production. The extensive utilization of *Jatropha curcas* oil in India for biofuel production has been reported (NIE 2014). The volatility in the oil prices and supply inconsistency in the recent past has necessitated the need to create alternate means of biofuel production from sustainable sources, and this has prompted most Asian countries to develop the bioenergy sector to complement the fossil fuel-derived energy sources (Tun 2019).

12.7 Regimes and State Policies on Biofuel Production and Related Bioenergy Activities

Various researchers (Demirbas 2009a, b; Araújo et al. 2017; Khan et al. 2021; Saini et al. 2021) reported that State policies on biofuel and bioenergy across Asian countries were aimed at gradual but sustained application of biofuels in the transportation and power sectors in the coming years to supplement existing energy mix. The studies further stated that the policy was also aimed at the utilization, development, and promotion of local and indigenous feedstocks for biofuel production thereby improving upon domestic energy security, mitigation of environment modification, generating innovative occupation prospects, and providing means to compliment nonrenewable fuels with biofuels in the energy sector. The policy also aims to encourage the utilization of advanced production technologies for biofuel generation across Asia such that the market availability would be increased to meet up with the blending percentage.

India's countrywide biofuel policy was issued in 2018, but in June 2022, this policy was further amended to accommodate newer activities and broaden the scope of application (Canabarro and Silva-Ortiz 2023; IEA 2023). These amendments included an upward review of the blending target percentage and year from 20% bioethanol in petrol diesel from 2025/26 to 2030 and an additional increase in feedstock types and variety for biofuel production. The amended national policy on biofuels provided the requisite roadmap for ethanol blending in petrol diesel in India within the period 2020–2025 and supersedes the 2009 National Biofuel Policy. India also possesses a National Bioenergy Program which was in place since 2022 and effectively coordinates all energy-related activities within the country. There was also a scheme to enhance ethanol distribution in India since 2020, while ethanol demonstration schemes existed in Japan since 2007 but have been effectively terminated due to feedstock challenges.

Presently, Kern et al. (2017) and Khan et al. (2021) noted that only 2% ethanol blending percentage is used in petrol while less than 0.1% biodiesel blending percentage is applicable in diesel on an average across Asian countries. Based on State policy, it is envisaged that there should be 20% blending of ethanol in petrol and 5% blending of biodiesel in petrol diesel by 2030 (Kern et al. 2017; Khan et al. 2021). This policy is planned to be achieved using different regimes which include the deployment of new technologies for biomass conversion to biofuel, new feedstock development for biofuels, aggressive cultivation of second-generation (2G) feedstocks, and biorefineries establishment for 2G feedstocks, integration of main different fuels with biofuels while also creating favorable environment for biofuels, and increasing domestic production and supply of biofuels especially ethanol and biodiesel through reinforcement approach on existing regimes (Carriquiry et al. 2011; Van Uytvanck et al. 2014; Castañeda-Ayarza and Cortez 2016; Foster et al. 2017). The long-term contributions of biofuels to the development of the economy, stability of the environment, and assurance of energy safety and availability have made governments of several Asian countries enact policies that will enhance biofuel production and make them cost-effective while taking cognizance of the possible issues of balance between food crops and biofuels. A handful of Asian nations including China, Indonesia, Malaysia, and India are also encouraging the production of biofuels domestically by employing trade regulations, enhancing the supply and demand of biofuels from several feedstocks like sorghum, cassava, *Jatropha* sp. (Physic nut), sugar cane, palm oil, coconut oil, and tapioca and providing support for emerging industries (Zhou and Thomson 2009). These activities reflect diverse approaches to bioenergy across Asia and are influenced by resource availability, economic factors, policy frameworks, and environmental considerations. Some fundamental policies geared towards reducing reliance on imported oil and tackling environmental concerns in Asia include blending mandates, incentives, tariffs, research, and development (Arima and Schor 2017).

12.7.1 Blending Mandates

Blending mandates call for the blending of a specific proportion of biofuels with regular fossil fuels like fossil diesel and PMS (premium motor spirit) or gasoline. China is a substantial player in bioenergy production. They also possess large-scale biomass power plants employing agricultural residues and forestry waste. The cultivation of energy crops and the environmentally friendly and sustainable production of biological fuel are both supported by the Chinese government. There are several regional and national mandates for biofuel blending in transport fuels given by the government in China. An example is the February 2024 mandate (E10) in which five provinces were directed to incorporate 10% bioethanol into their petrol supply. In China, 16% of all passenger cars are in these areas. This mandate was also extended to 27 cities in four other provinces (Li et al. 2023). Additionally, on first January 2006, the law establishing renewable energy component of the energy policy came into effect. This law was meant to enhance the biofuel effort and generally boost renewable energy production. In particular, the law promotes the production and application of biofuels where gas-selling businesses must incorporate biofuel that complies with the national standard into their fuel delivery system (Qin et al. 2018).

India has been gradually increasing its focus on biofuels to enhance energy security and reduce emissions. The blending mandate became operational in 2006 and was effective in 20 states. South Korea and Japan are investing substantially in advanced biofuels and bioenergy technologies through research and development activities and focusing on sustainability and reducing greenhouse gas emissions. Producing ethanol as well as employing it in automobiles is an initiative that the Japanese government is embracing. Thus, as much as 3% of petrol can lawfully be blended with ethanol. The Philippines has policies aimed at promoting biofuels as an integral renewable energy strategy. A blending mandate for incorporating varied percentages of coconut oil in diesel was achieved in 2007 and 2009 and ethanol in other gasoline products in 2009 and 2011 in the Philippines. However, implementing blending mandates in Thailand and Indonesia was met with some obstacles and hence achieved partial success (Akashi et al. 2019).

12.7.2 Incentives and Tariffs

In China, biofuel production facilities were supported by the government through the provision of subsidies and grants. Accordingly, bioethanol companies got subsidies to offset economic losses from the high cost of feedstock. In 2008, the government boosted the subsidy for bioethanol production since feedstock costs were high due to the spike in global grain prices during that period (Lang et al. 2009). Additionally, Wang (2011) stated that these factories no longer had to pay value-added tax of 17%, and consumption tax of 5% as bioethanol was exempted with each stock receiving a preferential guarantee of roughly 100 Yuan as profit. Indonesia is a major producer of palm oil used for production biodiesel. The country also explores biomass power from agricultural residues. The government included a rapid depreciation and payment of amortization as an incentive against operational shortfalls for a period of 5–10 years and an exclusion from value-added tax for 15 commercial sectors including bioenergy (Slette and Wiyono 2011). Export taxes are also being considered in Indonesia, to promote the domestic use of biofuels rather than their exportation (OECD/FAO 2023). Malaysia, like Indonesia, utilizes palm oil for biodiesel production. The country also explores biomass power generation from palm oil residues and other agricultural waste. The high amount of subsidy made the production of biodiesel uncompetitive (Mustapha 2008).

The generation of ethanol and biodiesel from sugarcane, coconut, and oil palm is the main emphasis of the Philippines' expanding bioenergy sector. Rice husks, coconut shells, and other agricultural wastes are used in biomass power plants. The Philippines' Biofuels Act of 2006 provided several benefits. These include zero taxation on domestic and imported biofuel components, no wastewater costs on any water effluent, exemptions from value-added tax on items used in the manufacture of biofuels, and priority eligibility for financial help (Philippine Department of Energy 2020). The 2018 National Biofuel Policy promotes the production of ethanol in India by offering a range of incentives from tax breaks to subsidies to secured purchase agreements with oil marketing companies (OMCs). To shield home producers against lower-priced imports, biofuel import levies were also implemented. The tariffs are contingent on the type of biofuel and are periodically modified in response to changes in the market and governmental goals (Arora 2018).

The Thai government offered rebates on taxes to promote investment in the manufacturing of biofuels. This covers import charges on machinery and equipment used in the manufacture of biofuels, discounts or exemptions from corporate income tax, and other relevant tax benefits. Additionally, it levied import taxes on biofuels, especially biodiesel, to shield domestic producers from cheaper import competition. Furthermore, the goal of social and environmental incentives was to encourage biofuel usage. Among them are cutting back on greenhouse gas emissions, improving energy security, and promoting rural development by raising agricultural output and diversifying sources of income. (Asian Development Bank 2019). To promote the production and use of renewable energy, particularly biofuels, the Japanese government implemented the Feed-in-Tariff (FIT) program, which fixed purchase prices and provided subsidies (Koizumi 2011).

12.7.3 Research and Development

China has been making significant investments in biofuel research and development (R&D) as part of a larger plan to improve energy security, decrease dependency on imported fossil fuels, and minimize environmental effects. In this regard, China is providing leadership and direction on a number of far-reaching biofuel developmental initiatives that use diversity of raw materials including non-food crops, agrarian leftovers, and technical advancements. A few particular instances of biofuel-related R&D projects and initiatives in China include the development of technology to make ethanol from biomass that is rich in cellulose, algae strains, leftover cooking oil, animal fats, and sweet sorghum. The fulfillment of international aviation fuel criteria is another focus of Chinese research (e.g., ASTM D7566) and scaling up production to supply airports and airlines with SAF blends (Li et al. 2020; Ma et al. 2020). Research and development programs in India include the creation of enzymatic processes and technologies for converting farming residues like rice straw and sugarcane bagasse into ethanol, investigations into how to cultivate algae more effectively and efficiently to produce biofuel, and breeding programs for yield enhancement and biodiesel quality improvement from *jatropha curcas* seeds (Banerjee and Sharma 2020). Indonesia's biodiesel from palm oil and bioethanol from cassava are research and development programs of the government.

The Malaysian's government R and D programs focused on converting waste biomass from oil palm such as empty fruit bunches and palm kernel shells into biofuels by employing pretreatment and conversion technologies, while the Japanese government encourages the development of sustainable aviation biofuels (SAFs) from various feedstocks, including algae and waste oils, to meet aviation industry standards and reduce carbon emissions (Yokochi and Honda 2019). Additionally, the 2006 revision of the Japanese biomass strategy places more focus on the value of cellulosic biofuels, gives unutilized resources such as thinning timber priority, and encourages the development of "biomass towns," or settlements where biomass is used entirely.

12.8 Effects of Demand and Supply on Biofuel and Bioenergy Activities

Across most Asian countries, various governments are adopting several multi-pronged approaches to the promotion, encouragement, and deployment of renewable fuels (Silitonga et al. 2011; Anto et al. 2020). Within the next few years, there are projections that the biofuel markets across Asian countries are likely to see massive expansion based on several pivotal factors that are driving the market. The demand for biofuels would surge due to deliberate government initiatives targeted at greenhouse gas (GHG) emissions reduction in many Asian nations. In most of these Asian countries, policies have been implemented that incentivize biofuel use in transportation, heating, and industrial activities thereby creating the right environment for the growth and development of the biofuel market. Additionally, biofuel production technologies have seen rapid advancement which has engineered greater efficiency and cost reduction thereby making biofuels more competitive as against fossil-based fuels. Of special interest is the use of second- and third-generation biofuel feedstocks for production purposes since the process entailed use of non-food biomass and algae as feedstocks, and these are rapidly gaining traction in the market. Other factors include the growing investment in biofuel infrastructure and production facilities as well as enlarged investment in the biofuel sector from local and international/foreign players thereby creating a potentially beneficial and profitable venture. Similarly, there has been improved awareness in environmental conservation and sustainability, and this has prompted a steady shift to biomass sources such as biofuels. Thus, the development is projected to linger going forward into the future which would further drive the biofuel market.

Several studies (Uckun et al. 2016; Sajid et al. 2018; Dahman et al. 2019; Anto et al. 2020) reported that different programs have been adopted across Asian nations to encourage the demand and supply of biofuels. They include ethanol blending in petrol through ethanol blended petrol (EBP) programs where the ethanol is derived from multiple feedstocks, ethanol technologies, and commercialization via the development of second-generation (2G) biorefineries, biodiesel blending in petrol diesel through biodiesel blending programs (Scarlat et al. 2015). These were achieved through the exploration of multiple feedstocks; the development of drop-in fuels obtained from municipal solid wastes, industrial wastes, and biomass; and the development of advanced biofuels such as bio-compressed natural gas (Bio-CNG), bio-methanol, bio-hydrogen, and bio-jet fuels (Lamers et al. 2012; Junginger et al. 2014; Abdullah et al. 2019). The effect of demand and supply is to ensure the availability of biofuels from the local and indigenous feedstocks through the appraisal of the requisite biomass across the different countries. In ensuring a balance between the demand and supply of biofuel in-country and for exports, Asian countries are undertaking critical interventions as and when required to enhance domestic production of biofuels, provide adequate and extensive storage and distribution facilities for biofuels, and adopt a consultative approach with various stakeholders in biofuel value chain utilization (Gonzalez-Salazar et al. 2016; Dafnomilis et al. 2017; Ruan et al. 2019). Other reported interventions include adopting periodic appropriate financial and fiscal measures to support the growth and development of the biofuel sector; involvement and funding of research, development, and demonstration activities in support of feedstock cultivation and biofuel processing for a variety of applications; and development of advanced biofuels and newer feedstock variants with high yielding potentials for production of biofuel.

Maeda et al. (2018) and Sharma et al. (2019) reported that price for biofuel at present is determined by the various governments based on defined criteria, and these have effects on the demand and supply of the different biofuels. It was opined that the different governments need to continue to incentivize biofuels by administering prices or market-determined prices which are dependent on several factors such as market conditions, availability of biofuels in the domestic or local market, and import substitution requirement (Oumer et al. 2018). Advanced biofuel development has to be given a differential pricing mechanism to further incentivize them to encourage greater participation by investors (Jeswani et al. 2020). Using appropriate incentives, local production of biofuels in Asian countries could be encouraged through an emphasis on development of domestic biofuel industries and feedstock mechanisms. Arshad et al. (2018) and Jeswani et al. (2020) thus opined that allowing free and unrestricted importation of biofuels from outside Asia would adversely affect the domestic biofuel market. However, Mizik and Gyarmati (2021) stated that depending on availability of domestic raw materials and blending requirements, importation of raw materials for biofuel production could be permitted to the extent of necessity. Thus, with the knowledge that the domestic biofuel market has lower volume and quantity/quality across most of the Asian countries, export of biofuels from Asia to other parts of the world would be restricted so as to meet up with local demand and supply mandate (Hassan and Kalam 2013; Saladini et al. 2016; Datta et al. 2019; Hafeez et al. 2020; Mahapatra et al. 2021; Adama and Anani 2023). Biofuels and bioenergy play significant roles in Asia's energy landscape, driven by diverse factors including concerns for energy security, ecological sustainability goals, and monetary development strategies such as greater employment and income-generating opportunities for farmers.

12.8.1 Drivers of Demand and Supply for Biofuels in Selected Asian Countries

12.8.1.1 Energy Security

China, India, Japan, and South Korea are among the majority of Asian nations where the quest for energy are predominantly fulfilled through imports of fossil fuels. Notwithstanding, energy security can be improved by biofuels since they provide a domestically generated, renewable alternative (Yan and Lin 2009). In this respect, Indonesia achieved the status of been major exporter of crude oil as against importer in the mid-2000s. As a consequence, other Asian nations are beginning to seek how to augment, swap out, and diversify their traditional energy sources with renewable alternatives. Biofuels are viable since they blend well with fuels from nonrenewable sources (International Renewable Energy Agency 2020). Diversification of sources of energy away from nonrenewable fuels and escalating crude prices to minimize reliance on oil imports are among the variables that drive a country toward energy security.

12.8.1.2 Ever-Increasing Oil Prices

Crude oil prices have risen globally over time. Additionally, the changing oil prices grow in tandem with subsidies on fossil fuels thereby increasing a government's yearly expenditure on fossil fuels. A typical example is the investment of higher funds in subsidies by the Indonesian government than in social security, health care, education, defense, and education combined in 2012. This outcome exerted pressure on the government's finances and resources (International Institute for Sustainable Development 2012). Hence, developing renewable energy sources like biofuels can assist Asian nations in becoming less reliant on imports and less vulnerable to rising oil costs.

12.8.1.3 Decreasing Reliance on Imported Oil and Diversification

Reducing reliance on imports is one way to boost domestic petroleum and natural gas production amid the current trend of rising global oil costs and disruptions in the fuel supply. One crucial criterion for assessing the likelihood of energy security is the percentage of imported oil consumed in the nations or the degree of imported oil reliance. Kumar et al. (2013) reviewed the policy drivers of demand in Indonesia, Malaysia, the Philippines, and Thailand. According to their report, before 2004, about 40% of Thailand's main energy source came from imports of petroleum products and crude oil. There has been a decrease in reliance, which they suggested, may be related to strict bio-based energy strategies like the unconventional energy initiative could encourage a larger percentage of different bio-based energy avenues that would include biofuels. Additionally, Kumar et al. (2013) further reported that in the Philippines, a similar trend was noted as the country's reliance on the importation of foreign oil declined over the past 10 years due to the replacement of renewable fuels with conventional fuels. Furthermore, their investigation revealed that Indonesia and Malaysia's exports and dependence on oil decreased over the past 10 years indicating that their fossil oil reserves are dwindling. This further makes it imperative that they employ alternative sources of oil such as biofuel to ensure their energy security. Another report from the Asia-Pacific Economic Cooperation (2013) on the proportion of major energy source mix of the abovementioned Asian nations' (Indonesia, Malaysia, the Philippines, and Thailand) at intervals of 3 years showed that petroleum and crude oil were the main components of the energy mix. Renewable sources of energy including solar, wind, and hydropower constitute a minute proportion of cumulative energy mix. The countries' huge reliance on fuels from nonrenewable sources catalyzes diversifying their energy sources. Thus, diversifying the sources of their energy by moving beyond fossil fuels and towards more alternative and sustainable options like bioenergy is one approach for such nations to adapt to declining energy security induced by fossil fuels.

12.8.2 Economic and Societal Impacts

The economic and social implications of biofuels on one nation might differ greatly from another depending on the market dynamics, region, climate, and legislative environment of that nation. The following socioeconomic variables are determinants of the development of biofuels in any country: feedstock availability, greater revenue generation opportunities, entrance to innovative markets, and the need for correction of imperfections in foreign exchange, global balance of trade, and resource prospective.

12.8.3 Feedstock Availability

The type and availability of biomass feedstocks impact the supply chain and production costs of biofuels. In Southeast Asia, countries like Indonesia and Malaysia are engaged in large-scale production of biodiesel from oil palm and are driven by their extensive palm oil plantations. China and India have significant agricultural sectors and are exploring biofuels to lessen their reliance oil importation while also addressing issues from environmental concerns. Japan and South Korea have advanced technology sectors and are investing in research on 2G (second-generation) biofuels from non-edible sources (Garg and Kaur 2020). Therefore, it is feasible to promote biofuels in these areas due to the large volume reserve potential, including access to vast cultivable hectares of land and inexpensive cost of feedstock.

12.8.4 Revenue Creation and Job Opportunities for Farmers

The necessity to promote agricultural economies is a major policy incentive for government's encouragement of biofuels given that it has the propensity to reduce poverty and create jobs in villages while reducing rural-urban migration. This is as a result of the labor-intensive propensity of feedstocks' cultivation and harvesting processes. Furthermore, it encourages smallholder involvement in developing biofuel crops and provides income-generating avenues to farmers. For instance, the palm oil industry employed 1.4 million people in allied industries in 2007 and contributed approximately 5–6% of Malaysia's GDP (Malaysia Palm Oil Board and

American Palm Oil Council 2010). In the same vein, in a study conducted in Thailand, every milliliter of cassava-based ethanol produced adds almost 500,000 US dollars to the country's GDP. Comparably, 1 ML of ethanol derived from molasses, sugar cane, and biodiesel derived from palm oil each contribute approximately US$ 411,000, US$ 604,000, and US$ 632,000, respectively, to the GDP. This originates as a direct consequence of effects of the agronomy subsector, due to unforeseen effects of utilizing chemicals and energy (Silalertruksa et al. 2012).

According to a research report investigating biofuel's effects in Greater Mekong Subregion of Southeast Asia, advancement of biofuels globally will modify trade in agricultural products in the region and around the world, as well as dramatically raise prices for agricultural produce. Additionally, farmers' earnings will rise as a consequence of escalating agrarian costs of production and the concomitant upsurge in land rates and labor costs of farming activities. The development of biofuels thus has a great deal of potential to sustain a rural, agrarian economy (Yang et al. 2009).

12.8.4.1 Introduction of New Markets to Farmers

Rural farmers and other agricultural workers have the opportunity to enter new markets, thanks to the development of biofuels. By expanding their crop variety to include more biofuel-producing plants, farmers can reduce the danger of declining demand for crops used as food and potentially obtain better pricing for their produce in the new market. Growth in the supply and exploration of biofuels could be encouraged by the potential for more revenue and free market activities that come with it. Ewing and Msangi (2009) believe that commercial activities in biofuel and biomass are expected to rise from developing to advanced nations, as the Southeast Asian nations produce more biomass than industrialized nations due to their favorable environment as well as low labor costs. Consequently, developing nations may be inspired to produce biofuel if it has the potential to reach both domestic and foreign markets.

12.8.4.2 Maintenance of Foreign Exchange and Trade Balance

Governments create rules for biofuels for diverse reasons, including the need to preserve the equilibrium between trade and foreign exchange. Many Asian nations experienced unfavorable balance of trade, situations, and foreign exchange outflows due to their reliance on imports of petroleum, and the savings from these outflows are a major factor in the development of local biofuel during periods of high oil prices (World Bank Group 2018). The overall import value for Malaysia, Indonesia, Thailand, and the Philippines shows rising import values between 1990–2012. This showed that imports' high value has favorable effects on the development of biofuels (Chai et al. 2019).

12.9 Concerns Regarding the Environment

Increasing awareness of environmental change and air pollution drives motivation for cleaner source of energy like biofuels that releases minimal greenhouse gases and toxins than fossil fuels. Biofuels are often regarded as carbon-free since the carbon they emit during combustion is eliminated from the atmosphere, which helps the plant thrive. Thus, reduction of greenhouse gas emissions during biofuel production forms the major crux of the environmental goal (Tan and Daim 2017). In cases where ethanol was used instead of regular petrol, the greenhouse gas decline was approximately 13%, up to 90% when sugarcane was used for the production of the ethanol, and 40–60% when oil seed was employed to produce biodiesel (Quadrelli and Peterson 2007). Similarly, in Thailand, using petrol blended with ethanol reduced net greenhouse gas emissions by 2.0 million tons of CO_2 equivalent every year, whereas using diesel substituted with biodiesel reduced CO_2 emissions by 0.8 million tons of CO_2 equivalent annually (Bell et al. 2011). According to a 1990–2005 report on prospects of bioenergy and biomass in Southeast Asia, utilizing woody materials for energy instead of fossil fuels reduces gaseous emissions by 202–336.7 Tg/year (Sasaki et al. 2009).

12.9.1 Impact of Market Forces on Demand and Supply of Biofuel

A wide range of market forces including policy, economics, technology, sustainability, and global trade dynamics shape the demand and supply of biofuels in Asia. These forces interact in a dynamic way to impact investment decisions, market growth trajectories, and the sustainability of biofuel adoption in the region as a whole. Some of these forces are elucidated below.

12.9.1.1 Policy Frameworks and Regulatory Environment

Policy frameworks are sacrosanct in shaping biofuel markets across Asia. Governments enact policies to moderate reliance on fossil fuel, encourage sources of energy that are renewable, and ease environmental influences. Vital aspects include subsidies, incentives, mandates, targets, and regulatory standards. Biofuel blending mandates and renewable energy targets create a steady demand for biofuels in transport and other sectors. Furthermore, regulations on sustainability criteria, such as land use and GHG (greenhouse gas emissions), influence market access and trade agreements (Gulati and Satija 2018).

12.9.1.2 Economic Factors and Market Dynamics

Economic conditions and market dynamics shape the competitiveness and profitability of biofuels. Examples include oil prices, feedstock costs, market integration, and investment climate. Prices of feedstocks (e.g., palm oil, sugarcane, corn) affect biofuel production costs and profitability. Additionally, biofuel as an alternative source of energy is impacted by fluctuations in prices of crude petroleum, thus affecting the pricing and competitiveness. Furthermore, integration into global markets and trade policies influence biofuel prices, export opportunities, and market growth. Economic stability, availability of financing, and investor confidence also impact biofuel project development and expansion (Meng and Senauer 2009).

12.9.1.3 Technological Advancements and Innovation

Technological advancements drive efficiency improvements and cost reductions in biofuel production processes including conversion technologies, feedstock diversification, and scaling up. Advances in biochemical and thermochemical processes enhance biofuel yield and reduce production costs. In addition, research into new feedstocks and agricultural practices expands biofuel feedstock options and improves sustainability. Likewise, commercial-scale deployment of advanced biofuel technologies reduces pilot plant risks and enhances market competitiveness (Nizami et al. 2017).

12.9.1.4 Sustainability and Environmental Considerations

Environmental sustainability is increasingly important in biofuel markets. For example, concerns over deforestation and habitat loss due to biofuel feedstock cultivation influence market acceptance and regulatory scrutiny. Additionally, policies and certifications require biofuels to demonstrate lower lifecycle GHG (greenhouse gas emissions) in comparison to nonrenewable liquid and gaseous fuels. Certification schemes like compliance with sustainability certification standards (e.g., Roundtable on Sustainable Biomaterials, ISCC) enhance market access and brand reputation (Dufey 2006).

12.9.1.5 Global Market Dynamics and Trade Relationships

Global trade dynamics and geopolitical factors such as trade policies, international collaborations, and competitive landscape influence Asia's biofuel market. Tariffs, quotas, and trade agreements affect biofuel imports and exports, influencing regional supply and demand. International collaborations such as bilateral and multilateral agreements promote technology transfer, investment, and market expansion as competition from biofuel producers in other regions impacts Asia's market share and pricing strategies (Sorda et al. 2010).

12.10 Challenges and Considerations

In Asia, there are several obstacles and restrictions to biofuel development despite continuous initiatives to support renewable energy sources. The following are some major obstacles:

12.10.1 Feedstock Availability and Competition

Several Asian nations primarily rely on particular feedstocks (such as sugarcane in India and oil palm in Indonesia and Malaysia) for manufacture of biofuels, which might put food production and environmental problems (such as deforestation) in competition. In addition, seasonal availability and yield variations in feedstocks derived from agriculture might impact the consistency of biofuel production. Furthermore, biofuels may face competition from low-cost fossil fuels, particularly during periods of low oil prices, affecting market competitiveness (Bleischwitz et al. 2022).

12.10.2 Technological Limitations

Poor conversion efficiency and high production cost are two issues that some biofuel production technologies like cellulosic algae and ethanol biofuels are presently working to overcome. There are still many technological and financial obstacles that must be overcome before biofuel production systems can be expanded from laboratory or pilot stages to commercial levels (Zhang et al. 2010).

12.10.3 Policy and Regulatory Frameworks

Investors and industry stakeholders may experience uncertainty due to inconsistent or shifting government policies and laws regarding biofuels. Additionally, a strong reliance on incentives and subsidies to boost the production of biofuels may not be sustainable over time, necessitating the creation of balanced policy frameworks (Smith and Brown 2022).

12.10.4 Public Awareness and Acceptance

Demand in the market and governmental decisions can be influenced by public opinion on biofuels, which includes worries about food security, environmental effects, and technical dependability (Lee and Kang 2019).

12.11 Conclusions and Future Perspectives

The need for concerted and targeted development of biofuels and bioenergy across different Asian countries cannot be overemphasized. Energy challenges across Asian countries encourage the use of an amalgamation of diverse options geared towards decarbonizing the transportation sector including the use of bioenergy. Bioenergy employs solar energy via biological photosynthesis for capturing CO_2 from the atmosphere and converting the stored energy into renewable energy. Several forms of bioenergy include bioethanol, biodiesel, biogas, green diesel, biomethane, bio-hydrogen, biomass heat, and bioelectricity. Across Asian countries, biofuel blending mandates have been introduced to meet with expanding population's need for energy and transportation fuel. Thus, the biofuel market across Asia is rapidly opening up at rates higher than those observed for the conventional fossil fuel market. The energy security concerns and continuous energy demand are the reasons why local economies across Asia are constantly seeking local solutions to mitigate the pecuniary and shared responsibilities of citizens while ensuring that the environment is used sustainably. Thus, the development of bioenergy has ensured sustainable development strategies for energy security and environmental protection. Biofuel production and bioenergy activities have seen rapid growth and development within the last few decades in Asia occasioned by the deliberate increase in feedstock production and utilization across the value chain. In this regard, to mitigate a free-for-all situation without appropriate regulation, policies to guide the production and application of renewable energy activities and protocols against negative influences of market forces have been enunciated across different Asian countries. This chapter has effectively assessed the Asian market and policies regarding biofuels and bioenergy activities. An overview of the market and policy issues in biofuels and bioenergy was presented and discussed.

The benefits and market potentials of the policies for biofuels and bioenergy across Asian countries are numerous. The policy envisages biofuels of different categorizations to include 1G (first-generation) and 2G (second-generation) biofuels which would enable appropriate financial and social incentives to be extended in the development of the sector. The policy further provides and expands the scope of the different feedstocks for biofuel production which are readily available nonedible and waste biomass. The policy also encourages the development of demand and supply chain mechanisms for biofuel production from biomass and waste solid materials. It inspires a synergistic approach by all stakeholders to the development of different biofuels across various Asian countries. Across Asian countries, different policy mechanisms were highlighted as promoting biofuels and bioenergy. A variety of domestic agriculture, energy, fuel, and trade policies have been instituted to influence biofuel sector development. These guidelines were employed in forms of biofuel blends, research, development, and demonstration and financial incentives like production and blending tax credits on biofuel, subsidies on investment like reduced import tariffs, subsidies on export, direct subsidies, and loans with low interest. The biofuel policies across Asian countries are targeted at advancing social objectives ranging from transportation fuel price reduction, energy security improvement, and revitalization of rural communities, air pollution reduction, and greenhouse gas emission reduction.

An economic analysis of biofuel production taking the market price as a sole indicator cannot give an informed and objective assessment of the desirability of biofuels to the entire Asian society. To present an informed public policy about the market, it is necessary to present a private market perspective complemented with or by a social welfare perspective that should be able to evaluate biofuel influences across Asia. From the societal viewpoint, biofuel's fiscal successes can be taken to be a function of inclusive impact on the Asian country's welfare. Essentially, where there is food prices increase due to biofuel production policies, the country's social well-being tends to be lowered because of high production vagaries linked to the production activities. The shared aids of the manufacture of biofuels are difficult to evaluate and measure when compared to the commercial profitability of the production. In determining biofuels' impact on the social well-being of Asian citizens, there would be a need to evaluate their impact on the market inputs such as food yields, acreage, labor, and chemicals; the ecosystem, energy, and food security; budgets of government and trade balances; and sharing of revenue. The investigation within this context is intricate because of nonavailability of defined standards and values for comparison for certain shared resources like conservational facilities and energy safety nets.

The cost of biofuel production is somewhat expensive when compared to that of fossil fuels. As a result of this cost variation, agronomic and investment guidelines influencing demand, supply, and agricultural costing are major factors in the expansion and sustainability of biofuel activities across Asian countries. Even with the policy interventions of the various governments, such as agricultural price subsidies, import and export regulations still influence trade opportunities thereby affecting the market. Though the demand for biofuels, technology advancements, and policy support in the production of biofuels is building up across Asian countries, the supply of the much-needed feedstocks remains a practical challenge for sustainable growth of the biofuel and bioenergy segment across Asian countries. The absence or dearth of validated and good quality data on agricultural and agronomic practices, potential yields of the different feedstocks, pest and disease control measures, and water requirements are serious concerns in the development of the sector. Most developmental programs across Asian countries have not achieved the desired targets because of the choice of inappropriate species, climatic conditions, and feedstock varieties. Thus, choosing appropriate species is important to the success of the programs and also helps in the understanding of the climatic and edaphic factors. There is also the need to develop extensive research programs that would serve as a means of providing alternative and cheap feedstocks for biodiesel production. Research needs to be undertaken on different aspects of biofuel feedstock development including the development of an appropriate legislative framework. Additionally, it is pertinent that individual countries across Asia be deeply involved in undertaking concise identification of existing renewable resources and country-specific feedstocks that would provide additional resources for biofuel production. There should be a deliberate policy to localize imported technologies from outside Asia such that the citizens can fully be in charge of the production process with very minimal foreign interventions. Biofuel feedstock plantations could be utilized to improve the citizens' livelihood on a sustainable basis as they are potential sources for employment generation from plantation activities, harvesting, and processing activities and as a means of reducing migration. Essentially, intensive research and organized planning are key components for any sustainable biofuel program, and this must be incorporated in biofuels and bioenergy policies across Asian countries.

In summary, this chapter emphasizes the benefits of assessing the market and policies for biofuels/bioenergy across Asian countries to include reduction in input reliance, improved and better environment, well-being and wellness of their citizens, waste administration, investment in rural communities through provision of infrastructures, generation of job opportunities, and extra earnings to agriculturalists engaged in feedstock production activities for biofuel development. The way forward is to ensure a reduction in the crude oil and petroleum product import bill into Asian countries through advancement of continuous and sustained biofuel utilization in transportation, production of new and improved crops for biofuel development for rural and agricultural development, using wasteland and municipal wastes to produce sustainable biofuels, designing and implementing biofuel solution to provide both food and energy, and ensuring a community-based biofuel distribution program that benefits local economies including the farmers, businesses, and distribution outlets.

Acknowledgments The authors gratefully acknowledges their various institutions for providing the enabling environment to carry out research and for academic interactions among them.

References

Abdullah B, Muhammad SFA, Shokravi Z, Ismail S, Kassim AK, Mahmood NA, Maniruzzaman M, Aziz A (2019) Fourth-generation biofuel: a review on risks and mitigation strategies. Renew Sust Energ Rev 107:37–50

ACE (2015) ASEAN plan of action for energy cooperation (APAEC) 2010-2025. Phase I:2016–2020

Achten W, Muys B, Mathij E, Singh VP, Verchot L (2007) Life-cycle assessment of biodiesel from jatropha curcas L: energy balance, impact on global warming, land use impact. In: 5th international conference on LCA in foods, 25–26 April, Goteborg, Sweden

Adama KK (2021) Physicochemical composition and functional properties of jatropha seed oil and jatropha biodiesel: an agro-renewable product. J Sci Technol Res 3(1):153–161

Adama KK, Anani OA (2023) Experimental and theoretical assessment of phenomena linked with separation and purification of biodiesel from Ricinus communis seed oil. Heliyon 9:e16536. https://doi.org/10.1016/j.heliyon.2023.e16536

Adama KK, Aluyor EO, Audu TOK (2021) Component distribution associated with phase separation and purification of tropical almond biodiesel at different temperatures. Renew Energy 165:67–75

Adama KK, Ukhurebor UK, Pal K, Hossain I (2024) Effect of neem oil biodiesel on the surface and structural integrity of carbon steel alloy: chromatographic, spectroscopic and morphological investigations. Int J Biol Macromol 269:132199. https://doi.org/10.1016/j.ijbiomac.2024.132199

Agoramoorthy G, Hsu MJ, Chandhary S, Shieh PC (2009) Can biofuel crop alleviate tribal poverty in India's drylands? Appl Energy 86(Suppl 1):118–124

Akashi M, Rude J, Wada K (eds) (2019) Biofuels in Asia: sustainable technologies and policy. Elsevier

Alam M, Hasan M (2017) Feasibility study of biogas energy in Bangladesh. Doctoral dissertation, Daffodil International University

Anto S, Mukherjee SS, Muthappa R, Mathimani T, Deviram G, Kumar SS et al (2020) Algae as green energy reserve: technological outlook on biofuel production. Chemosphere 242:125079

Araújo K, Mahajan D, Kerr R, Silva M (2017) Global biofuels at the crossroads: an overview of technical, policy, and investment complexities in the sustainability of biofuel development. Agriculture 7(4):32. https://doi.org/10.3390/agriculture7040032

Arima TH, Schor T (2017) The effectiveness of biofuel policies in Asia: a comparative analysis. Renew Sust Energ Rev 70:826–834. https://doi.org/10.1016/j.rser.2016.11.231

Arora S (2018) Incentives for biofuel production and use in India. In: Ravindranath NH, Hall JR (eds) Bioenergy for sustainable development and international competitiveness: the role of sugar cane in Africa. Elsevier, Amsterdam, pp 273–284

Arshad M, Zia AM, Shah AF, Ahmad M (2018) An overview of biofuel. In: Arshad M (ed) Perspective on water usage for biofuel production. Springer, Cham, pp 1–38. https://doi.org/10.1007/978-3-319-66408-8_1

ASEAN-MARD (2019) Workshop report on biomass for growth: potential and challenges of bioenergy in agricultural communities and rural development in ASEAN region. Bangkok, Thailand

Asia Pacific Energy Research Center (APERC) (2006) APEC energy demand and supply outlook 2006. Asia Pacific Energy Research Center

Asian Development Bank (2006) Energy efficiency and climate change considerations for on-road transport in Asia. ADB/Clean air initiatives for Asian cities

Asian Development Bank (2015) Renewable energy developments and potentials in the greater Mekong sub-region. Asian Development Bank, Mandaluyong City

Asian Development Bank (2019) Biofuel developments and impacts in Thailand. https://www.adb.org/publications/biofuels-developments-impacts-thailand

Asian Development Bank (ADB) (2009) Rural renewable energy initiative in the Greater Mekong sub-region

Asia-Pacific Economic Cooperation (2013) Energy database. http://www.ieej.or.jp/egeda/database/

Axelsson L, Franzen M (2010) Performance of jatropha biodiesel production and its environmental and socio-economic impacts: a case study of southern India

Bakar MH, Ahmad N, Islam MN, Ani FN (2014) Characterization of biomass solid wastes of Brunei Darussalam for their pyrolytic conversion into liquid oil. In: Proceedings of the 5th Brunei international conference of engineering and technology (BICET 2014), Bandar Seri Begawan, Brunei, 1–3 November

Banerjee A, Sharma MP (2020) Jatropha curcas: a sustainable source for biofuel production. Renew Sust Energ Rev 121:109680. https://doi.org/10.1016/j.rser.2020.109680

Bell DR, Silalertruksa T, Gheewala SH, Kamens R (2011) The net cost of biofuels in Thailand: an economic analysis. Energy Policy 39(2):834–843. https://doi.org/10.1016/j.enpol.2010.11.002

Benites-Lazaro LL, Giatti L, Sousa Junior WC, Giarolla A (2020) Land-water-food nexus of biofuels: discourse and policy debates in Brazil. Environ Develop 100491:100491

Bennett A, Anex RP (2009) Production, transportation, and milling costs of sweet sorghum as a feedstock for centralized bioethanol production in the upper Midwest. Bioresour Technol 100:1595–1607

Blanco-Canqui H (2016) Growing dedicated energy crops on marginal lands and ecosystem services. Soil Sci Am J 80:845–858. https://dl.sciencesocieties.org/publications/sssaj/abstracts/80/4/845

Bleischwitz R, Yang M, Huang B, Xu X, Zhou J, McDowall W, Andrews-Speed P, Liu Z, Yong G (2022) The circular economy in China: achievements, challenges, and potential implications for decarbonisation. Res Conserv Recycl 183:106350

Buddenhagen E, Chimera C, Clifford P (2009) Assessing biofuel crop invasiveness: a case study. PLoS One 4(4):e5261

Canabarro NI, Silva-Ortiz P (2023) Sustainability assessment of ethanol and biodiesel production in Argentina, Brazil, Colombia, and Guatemala. Renew Sust Energ Rev 171:113019

Carriquiry MA, Du X, Timilsina GR (2011) Second generation biofuels: economics and policies. Energ Policy 39:4222–4234

Castañeda-Ayarza JA, Cortez LAB (2016) Final and B molasses for fuel ethanol production and some market implications. Renew Sust Energ Rev 95:107–129

Chai SK, Tan KH, Ramasamy K (2019) Drivers and barriers to biofuel development in ASEAN countries. Renew Sust Energ Rev 104:343–351. https://doi.org/10.1016/j.rser.2018.12.029

Chang Y, Fang Z, Li Y (2016) Renewable energy policies in promoting financing and investment among the East Asia summit countries: quantitative assessment and policy implications. Energy Policy 95:427–436

Collins K (2008) The role of biofuels and other factors in increasing farm and food prices: a review of recent developments with a focus on feed grain markets and market prospects. Kraft Foods Global, Inc

Creutzig F, Ravindranath NH, Berndes G, Bolwig S, Bright R, Cherubini F, Chum H, Corbera E, Delucchi M, Faaji A, Fargione J (2015) Bioenergy and climate change mitigation. GCB Bioenergy 7:916–944

Dafnomilis I, Hoefnagels R, Pratama WY, Schott LD, Lodewijks G, Junginger M (2017) Review of solid and liquid biofuel demand and supply in Northwest Europe towards 2030: a comparison of national and regional projections. Renew Sust Energ Rev 78:31–45

Dahman Y, Syed K, Begum S, Roy P, Mohtasebi B (2019) Biomass, biopolymer-based materials, and bioenergy. In: Biomass, biopolymer-based materials, and bioenergy. https://doi.org/10.1016/B978-0-08-102426-3.00014-X

Dale BE, Anderson JE, Brown RC, Csonka S, Dale VH, Herwick G (2014) Take a closer look: biofuels can support environmental, economic, and social goals. Environ Sci Technol 48:7200–7203

Datta A, Hossain A, Roy S (2019) An overview on biofuels and their advantages and disadvantages. Asian J Chem 31(8):1851–1858

Demirbas MF (2007) Current technologies for biomass conversion into chemicals and fuels. Energy Source A: Recovery, Util Environ Eff 28(13):1181–1188

Demirbas A (2009a) Biofuels securing the planet's future energy needs. Energy Convers Manag 50:2239–2249

Demirbas A (2009b) Political, economic and environmental impacts of biofuels: a review. Appl Energy 86:S108–S117. https://doi.org/10.1016/j.apenergy.2009.04.036

Demirbas A, Bafail A, Ahmad W, Sheikh M (2016) Biodiesel production from non-edible plant oils. Energy Explor Exploit 34(2):290–318

Department of Alternative Energy Development and Efficiency, Ministry of Energy (DAEDEME) (2012) Executive summary report; Developing biomass database potential in Thailand, pp 1–45

Divakara BN, Upadhyaya HD, Wani SP, Gowda CL (2010) Biology and genetic improvement of Jatropha curcas L: a review. Appl Energy 87:732–742
Dufey A (2006) Biofuel production, trade, and sustainable development: emerging issues. International Institute for Environment and Development, London
Elbehri A, Segerstedt A, Liu P (2013) Biofuels and the sustainability challenge: a global assessment of sustainability issues, trends, and policies for biofuels and related feedstock. FAO, Rome
ERIA (2014) Biofuel promotion policies and development status in east Asian countries. In: Yamaguchi K (ed) Study on Asian potential of biofuel market. ERIA research project report 2013–2020, pp 7–21
Ewing M, Msangi S (2009) Biofuels production in developing countries: assessing tradeoffs in welfare and food security. Environ Sci Pol 12:520–528
Fabian K (2008) Projections of air pollutants (PM) and greenhouse gas emissions (CO_2) from transport in Asia (2005–2035). In: Presentation at better air quality for Asian cities conference, Bangkok
Foster E, Contestabile M, Blazquez J, Manzano B, Workman M, Shah N (2017) The unstudied barriers to widespread renewable energy deployment: fossil fuel price responses. Energy Policy 103:258–264
Garg A, Kaur A (eds) (2020) Biofuels in Asia: sustainable and climate-friendly alternatives. CRC Press. https://doi.org/10.1201/9780429279824
German Technical Cooperation (2010) Jatropha reality check: a field assessment of the agronomic and economic viability of jatropha and other oilseed crops in Kenya
Gonzalez-Salazar MA, Venturini M, Poganietz WR, Finkenrath M, Kirsten T, Acevedo H (2016) Development of a technology roadmap for bioenergy exploitation including biofuels, waste-to-energy, and power generation and CHP. Appl Energy 180:338–352
Gulati M, Satija V (2018) Biofuel policies and food grain prices: a comparison of the expected impacts in India, China, and the United States. Renew Energy 116(Part A):34–51. https://doi.org/10.1016/j.renene.2017.09.067
Hafeez S, Al-Salem SM, Manos G, Constantinou A (2020) Fuel production using membrane reactors: a review. Environ Chem Lett 18:1477–1490
Hajilary N, Shrazian S, Rezakazemi M (2019) Biofuel types and membrane separation. Environ Chem Lett 17:1–18
Hassan MJ, Kalam MA (2013) An overview of biofuel as a renewable energy source: development and challenges. 5th BSME international conference on thermal engineering. Proc Eng 56:39–53
Ho DP, Ngo HH, Guo W (2014) A mini review on renewable sources for biofuel. Bioresour Technol 169:742–749
Hoff H (2011) Understanding the nexus. In: Background paper for the Bonn 2011 Nexus conference. Stockholm Environment Institute, pp 1–52. https://www.sei.org/publications/understanding-the-nexus
Institute of Renewable Energy Promotion (IREP) (2016) Ministry of energy and mines renewable energy data in Lao PDR. In: Proceedings of the East and Southeast Asia Renewable Energy Statistic Training Workshop, Bangkok, Thailand, 12–14 December
Intelligent Energy Systems (IES) (2019) Towards 100% renewable electricity by 2050. In: Cambodia report; power sector vision; Greater Mekong region, pp 1–138
International Energy Agency (IEA) Bioenergy (2023) Biofuels in emerging markets: potentials for sustainable production and consumption. IEA Bioenergy: 39, Technology Collaboration Program
International Institute for Sustainable Development (2012) A citizen's guide to subsidies in Indonesia. Global Subsidies Initiative and the Institute for Essential Services Reform, Manitoba
International Panel on climate Change (IPCC) (2014) Climate change 2014-impacts, adaptation, and vulnerability: regional aspects. Cambridge University Press
International Renewable Energy Agency (IRENA) (2020) Renewable Energy Statistics 2020. https://www.irena.org/publications/2020/Sep/Renewable-Energy-Statistics-2020

IRENA and ACE (2016) Renewable energy outlook for ASEAN: a REmap analysis. International renewable energy agency (IRENA), Abu Dhabi and ASEAN Center for Energy (ACE), Jakarta
Jeswani HK, Chilvers A, Azapagic A (2020) Environmental sustainability of biofuels: a review. Philos Trans R Soc Lond Ser B Biol Sci 476:2020035
Jonker JGG, Van Der Hilst F, Junginger HM, Cavalett O, Chagas MF, Faaij APC (2015) Outlook for ethanol production costs in Brazil up to 2030, for different biomass crops and industrial technologies. Appl Energy 147:593–610
Junginger M, Faaji A, Goh CS (2014) International bioenergy trade-history, status, and outlook on securing sustainable bioenergy supply, demand, and markets. Springer, Cham
Karwat DM, Eagle WE, Wooldridge MS (2014) Are there technological problems that technology cannot solve? Water scarcity, dams, climate change, and biofuels. Int J Environ Sci Dev 3:7–25
Kern JD, Hise AM, Characklis GW, Gerlach R, Viamajala S, Gardner RD (2017) Using life cycle assessment and techno-economic analysis in a real options framework to inform the design of algal biofuel production facilities. Bioresour Technol 225:418–428
Khan N, Sudhakar K, Mamat R (2021) Role of biofuels in energy transition, green economy, and carbon neutrality. Sustain For 13(22):12374. https://doi.org/10.3390/su132212374
Kibler KM, Reinhart D, Hawkins C, Motlagh AM, Wright J (2018) Food waste and the food-energy-water nexus: a review of food waste management alternatives. Waste Manag 74:52–62
Koizumi T (2011) The Japanese biofuel program-developments and perspectives. J Clean Prod 2497:57
Koizumi T, Ohga K (2007) Biofuels policies in Asian countries: impact of the expanded biofuels program on world agricultural markets. J Agric Food Ind Organ 5(2):8
Kumar A, Sharma S (2011) Potential non-edible oil resources as biodiesel feedstock: an Indian perspective. Renew Sust Energ Rev 15:1791–1800
Kumar S, Shrestha P, Salam PA (2013) A review of biofuel policies in the major biofuel-producing countries of ASEAN: production, targets, policy drivers, and impacts. Renew Sust Energ Rev 26:822–836
Lamers P, Junginger M, Hamelinck C, Faaji A (2012) Developments in international solid biofuel trade—an analysis of volumes, policies, and market factors. Renew Sust Energ Rev 16:3176–3199
Lang X, Zheng F, Cui H (2009) Evolution of fuel ethanol policy in China. Forestry Econ 29:29–33
Latner KC, O'Kray C, Jiang J (2006) The PRC biofuels: an alternative future for agriculture USDA Gain report CH6049
Lee SY, Kang HY (2019) Factors influencing public acceptance of biofuels in South Korea: implications for policy development. Energy Policy 126:196–204. https://doi.org/10.1016/j.enpol.2018.11.016
Li SZ, Chan-Halbrendt C (2009) Ethanol production in the PRC: potential and technologies. Appl Energy 86(Suppl 1):162–169
Li X, Xie Y, Chai M, Yang H, Qiu X (2020) Progress in the development of biodiesel in China. Renew Sust Energ Rev 133:110225. https://doi.org/10.1016/j.rser.2020.110225
Li Y, Wen Y, Chen B, Fu X, Wu Y (2023) The dilemma and potential development of biodiesel in China—given production capacity and policy. Energy Sustain Dev 75:60–71
Ma Y, Wang X, Wu Y (2020) Recent advances in biodiesel production from algae: a review. J Clean Prod 261:121106. https://doi.org/10.1016/j.jclepro.2020.121106
Maeda Y, Yoshino T, Matsunaga T, Matsumoto M, Tanaka T (2018) Marine microalgae for production of biofuels and chemicals. Curr Opin Biotechnol 50:111–120
Mahapatra S, Kumar D, Singh B, Sachan KP (2021) Biofuels and their sources of production: a review on cleaner sustainable alternative against conventional fuel, in the framework of the food and energy nexus. Energy Nexus 4:100036
Malaysia Palm Oil Board, American Palm Oil Council (2010) Palm oil development and performance in Malaysia: presentation to United States international trade commission, Washington, DC. http://www.americanpalmoil.com/pdf/USITCpre-PublicHearing-V2.pdf

Matsumoto N, Sano D, Elder M (2009) Biofuel initiatives in Japan: strategies, policies, and future potential. Appl Energy 86(Suppl 1):69–76

Meng X, Senauer B (2009) Measuring consumer preferences for ethanol blends using a conjoint experiment. Energy Econ 31(3):343–349. https://doi.org/10.1016/j.eneco.2008.12.004

Mizik T, Gyarmati G (2021) Economic and sustainability of biodiesel production-a systematic literature review. Clean Technol 3:19–36

Mofijur M, Masjuki HH, Kalam AM, Hazrat AM, Liaquat AM, Shahabuddin M, Varman M (2012a) Prospect of biodiesel from jatropha in Malaysia. Renew Sust Energ Rev 16:5007–5020

Mofijur M, Masjuki HH, Kalam AM, Hazrat AM, Liaquat AM, Shahabuddin M, Varman M (2012b) Palm oil methyl ester and its emulsions effect on lubricant performance and engine component wear. Energy Procedia 14:1748–1753

Moioli E, Salvati F, Chiesa M, Siecha RT, Manenti F, Laio F, Rulli MC (2018) Analysis of the current world biofuel production under a water-food-energy nexus perspective. Adv Water Resour 121:22–31

Mustapha N (2008) Top news in Malaysia. In: The Institute for Global Environmental Strategies (IGES) (ed) 2007 top news on the environment in Asia. IGES, Hayama

Netherland Enterprise Agency (NIE) (2014) Biomass market opportunities-Thailand. Utrecht, Netherland Enterprise Agency, pp 1–12

Nizami AS, Rehan M, Waqas M, Naqvi M, Ouda OK, Shahzad K, Pant D (2017) Waste biorefineries: enabling circular economies in Asia. Bioresour Technol 241:1101–1117. https://doi.org/10.1016/j.biortech.2017.05.193

OECD (2008a) Biofuels support policies. Organization for Economic Cooperation and Development, Paris

OECD (2008b) Economic assessment of biofuels support policies. Organization for Economic Cooperation and Development, Paris

OECD/FAO (2019) OECD-FAO Agricultural Outlook, Agriculture statistics (database), OECD https://doi.org/10.1787/agr-outl-data-en

OECD/Food and Agriculture Organization of the United Nations (2023) Biofuels. In: OECD-FAO agricultural outlook. OECD Publishing, Paris, pp 2023–2032

OECD-FAO (2008) OECD-FAO Agricultural Outlook 2008–2017

Oh TH, Pang SY, Chua SC (2010) Energy policy and alternative energy in Malaysia: issues and challenges for sustainable growth. Renew Sust Energ Rev 14:1241–1252

Önal H, Núñez HM (2014) An economic analysis of transportation fuel policies in Brazil

Orsi F, Muratori M, Rocco M, Colombo E, Rizzoni G (2016) A multi-dimensional well-to-wheels analysis of passenger vehicles in different regions: primary energy consumption, CO_2 emissions, and economic cost. Appl Energy 169:197–209

Ou X, Zhang X, Chang S, Guo Q (2009) Energy consumption and GHG emissions of six biofuel pathways by LCA in the People's Republic of China. Appl Energy 86(Suppl 1):S197–S208

Oumer AN, Hasan MM, Baheta AT, Mamat R, Abdullah AA (2018) Bio-based liquid fuels as a source of renewable energy: a review. Renew Sust Energ Rev 88:82–98

Philippine Department of Energy (2020) Biofuels Act of 2006: Implementing rules and regulations. https://www.doe.gov.ph/national-policies/biofuels-act-2006-implementing-rules-and-regulations

Prueksakorn K, Gheewala S (2006) Energy and greenhouse gas implications of biodiesel production from jatropha curcas L. In: Proceedings of the second joint international conference on "sustainable energy and environments (SEE2006)", Bangkok, Thailand, November 21–23

PTI (2016) India Will Soon be Zero Petroleum Import Country. India Times. 6 September 2016. http://economictimes.indiatimes.com/industry/energy/oil-gas/india-will-soon-be-zero-petroleum-import-country-nitin-gadkari/articleshow/54031397.cms. Accessed 6 Sept 2016

Qin Z, Zhuang Q, Cai X, He Y, Huang Y, Jiang D, Lin E, Liu Y, Tang Y, Wang MQ (2018) Biomass and biofuels in China: toward bioenergy resource potentials and their impacts on the environment. Renew Sust Energ Rev 82(3):2387–2400

Quadrelli R, Peterson S (2007) The energy–climate challenge: recent trends in CO_2 emissions from fuel combustion. Energy Policy 35:5938–5952

Raju VK (2006) Biofuels in South Asia: an overview. Asian Biotechnol Dev Rev 8(2):1–9

Ruan R, Zhang Y, Chen P, Liu S, Fan L, Zhou N, Ding K, Peng P, Addy M, Cheng Y, Anderson E, Wang Y, Liu Y, Lei H, Li B (2019) Biofuels: alternative feedstock and conversion processes for the production of liquid and gaseous biofuels. In: Biomass, biofuels, and biochemical. Elsevier, pp 3–43

Saini R, Osorio-Gonzalez CS, Brar SK, Kwong R (2021) A critical insight into the development, regulation, and future prospects of biofuels in Canada. Bioengineered 12(2):9847–9859. https://doi.org/10.1080/21655979.2021.1996017

Sajid Z, Khan F, Zhang Y (2017) Integration of interpretive structural modelling with Bayesian network for biodiesel performance analysis. Renew Energy 107:194

Sajid Z, Khan F, Zhang Y (2018) A novel process economics risk model applied to biodiesel production system. Renew Energy 118:615–626

Saladini F, Patrizi N, Pulselli FM, Marchettini N, Bastianoni S, Fyoor G (2016) Energy evaluation of first, second, and third generation biofuels. Renew Sust Energ Rev 66:221–227

Sarin R, Sharma M, Sinharay S, Malhotra RK (2006) Jatropha-palm biodiesel blends: an optimum mix for Asia. Fuel 86:1365–1371

Sasaki N, Knorr W, Foster DR, Etoh H, Ninomiya H, Chay S et al (2009) Woody biomass and bioenergy potentials in Southeast Asia between 1990 and 2020. Appl Energy 86:140–150

Scarlat N, Dallemand JF, Monforti-Ferrario F, Nita V (2015) The role of biomass and bioenergy in a future bioeconomy: policies and facts. Environ Dev 15:3–34

Shafie SM, Mahlia TM, Masjuki HH, Ahmad-Yazid A (2012) A review on electricity generation based on biomass residue in Malaysia. Renew Sust Energ Rev 16:5879–5889

Sharma D, Singh D, Soni SI, Sharma S, Sharma KP, Jhalani A (2019) A review on feedstocks, production processes, and yield for different generations of biodiesel. Fuel 10:874–889

Silalertruksa T, Gheewala S, Sagisaka M (2009) Impacts of Thai bioethanol policy on land use and greenhouse gas emissions. Appl Energy 86(Suppl 1):170–177

Silalertruksa T, Gheewala SH, Hünecke K, Fritsche UR (2012) Biofuels and employment effects: implications for socio-economic development in Thailand. Biomass Bioenergy 46:409–418. https://doi.org/10.1016/j.biombioe.2012.07.019

Silitonga AS, Atabani AE, Mahlia TMI, Masjuki HH, Badruddin AI, Mekhilet S (2011) A review on the prospect of jatropha curcas for biodiesel in Indonesia. Renew Sust Energ Rev 15(8):3733–3756

Singh BK (2013) South Asia energy security: challenges and opportunities. Energy Policy 63:458–468

Slette J, Wiyono IE (2011) Indonesia biofuels annual: 2011. Global agricultural information network (GAIN), US Department of Agriculture. http://gain.fas.usda.gov/Recentpercent20GAIN per cent20Publications/biofuelspercent20Annual_Jakarta_Indonesia_8-19-2011.pdf. Accessed 23 Sept 2011

Smith K (2013) Biofuels, air pollution and health: a global review. Springer, Cham

Smith J, Brown A (2022) Policy implications for biofuels in Asia. Energy Policy 45(2):123–145. https://doi.org/10.1016/j.enpol.2022.01.002

Sorda G, Banse M, Kemfert C, Löschel A (2010) An overview of biofuel policies across the world. Energy Policy 38(11):6977–6988. https://doi.org/10.1016/j.enpol.2010.07.043

Tan Y, Daim TU (2017) Drivers and challenges for the development of biofuels in ASEAN countries. Energy Procedia 143:534–539. https://doi.org/10.1016/j.egypro.2017.12.664

Tan KT, Lee KT, Mohammed AR, Bhata S (2009) Addressing issues and towards sustainable development. Renew Sust Energ Rev 13:420–427

Tiziana P, Andrea R, Constance M (2018) Sustainability of biogas and cassava-based ethanol value-chain in Vietnam. FAO WP, Roma

Tun MM (2019) An overview of renewable energy sources and their energy potential for sustainable development in Myanmar. Eur J Sustain Dev Res 1:em0071

Tun WW, Juchelkova D, Win MM, Thu MA, Puchor T (2019) Review biomass energy: an overview of biomass sources, energy potentials, and management in Southeast Asia countries, vol 8. Resource MDPI, Basel
Uckun E, Stamatelatan K, Antonopoulou G, Lyberatos G (2016) Production of biogas via anaerobic digestion. In: Handbook of biofuels production, vol 2. Elsevier, Amsterdam, pp 259–301
USAID/ASIA (2009) Biofuels in Asia: An analysis of sustainability options. Report of International Resources Group (IRG)-ECO-Asia Clean Development and Climate Program, EPP-1-100-03-00013-00, 9
Van Eijck J, Batidzirai B, Faaji A (2014) Current and future economic performance of first-and second-generation biofuels in developing countries. Appl Energy 135:115–141
Van Uytvanck PP, Hallmark B, Haire G, Marshall PJ (2014) Impact of biomass on industry: using ethylene derived from bioethanol within the polyester value chain. ACS Sust Chem Eng 2:1098–1105
Walker G, Simcock N, Day R (2016) Necessary energy uses and a minimum standard of living in the United Kingdom: energy justice or escalating expectations? Energy Res Soc Sci 18:129–138
Wang QA (2011) Time for commercializing non-food biofuels in China. Renew Sustain Energy Rev 15(1):621–629
World Bank Group (2018) Sustainable biofuels development in ASEAN: Potential, challenges, and policies. https://openknowledge.worldbank.org/handle/10986/30600
Yan J, Lin T (2009) Biofuels in Asia. Appl Energy 86:S1–S10
Yang J, Huang J, Qiu H, Rozelle S, Sombilla MA (2009) Biofuels and the greater Mekong subregion: assessing the impact on prices, production, and trade. Appl Energy 86:37–46
Yokochi R, Honda K (2019) Development of sustainable aviation biofuels: past, present, and future perspectives. Renew Sust Energ Rev 115:109364. https://doi.org/10.1016/j.rser.2019.109364
Yosiyana B (2015) Renewable energy development in ASEAN. ASEAN Center for Energy. In: proceedings of the 2nd Asia renewable energy workshop, Jakarta, Indonesia, 2–4 December
Zhang J, Smith KR, Ma Y (2010) Hurdles to promoting biofuels: pricing and technological risks in China. Renew Energy 35(1):29–35. https://doi.org/10.1016/j.renene.2009.05.001
Zhou A, Thomson E (2009) The development of biofuels in Asia. Appl Energy 86(1):S11–S20

Chapter 13
Role of Machine Learning and Artificial Intelligence in Biofuel/Bioenergy Productions

Saira Mansab, Saima Nasreen, and Kousar Parveen

Abstract Bioenergy is utilized as a sustainable substitute of the fossil energies. It is a fact that applications of bioenergy products on a large scale are limited because of various challenges of the complexity and nonlinearity of the biofuel production process, feedstock variability, handling, optimization of conversion economics, and supply chain reliability. These challenges can be overcome using rapid and accurate data-based technologies of artificial intelligence (AI) that aid in design optimization, monitoring, and future forecasting. Artificial intelligence systems comprise the areas of symbolic AI (fuzzy logic, case-based reasoning), machine learning approaches (interconnected nodes utilization of artificial neural network (ANN) function, support vector machine (SVM), Bayesian network, random forest), heuristics (optimization through algorithm of particle swarm, genetic algorithm), and hybrid systems (adaptive neuro-fuzzy interference system, agent-based modeling). This chapter provides an understanding of how artificial intelligence systems operate and how they are set up to be used in the production life cycle of biofuel. It also sheds light on the previous research based on machine learning techniques in predicting biomass properties, biomass conversion performance, bioenergy consumption systems, supply chain optimization, and modeling beginning with the collection of feedstock to end use, engine performance, and emission phase. Thus, AI and machine learning has a function in overcoming the challenges of bioenergy production and improving yield and quality.

Keywords Machine learning · Artificial intelligence algorithms · Support vector machine · Yield optimization · Neural networking

S. Mansab (✉) · S. Nasreen · K. Parveen
Department of Environmental Sciences, The Women University Multan, Multan, Pakistan
e-mail: saira.mansab@wum.edu.pk

M. A. Dar et al. (eds.), *Recent Trends in Lignocellulosic Biofuels and Bioenergy*, Clean Energy Production Technologies,
https://doi.org/10.1007/978-981-96-4636-4_13

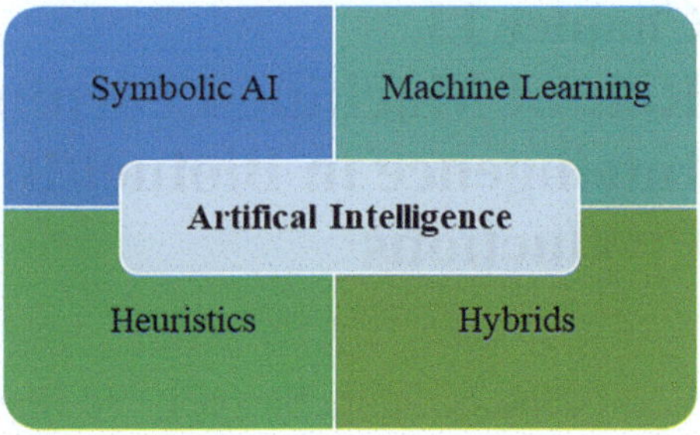

Fig. 13.1 Major branches of the artificial intelligence

13.1 Introduction

It was not more than a century ago when British computer scientist and mathematician Alan Turing predicted that computers would soon duplicate human intelligence. He performed a practical test of computer intelligence known as the Turing test; since then, various scientists have been intrigued by developing machines with the ability to think. In 1951, Christopher Strachey, a computer programmer, created the first AI program with the ability to play checker games. Later, various aspects of AI technology, including evolutionary computing and logical programming, were introduced, transforming different aspects of the fourth Industrial Revolution, including cognitive computing, artificial intelligence, etc. (Thakur et al. 2023). Artificial intelligence (AI) is a developing area that involves machines being able to carry out activities that mimic human cognitive function. It focuses on creating intelligent systems that demonstrate human-like characteristics in behavior, language, learning, reasoning, and problem-solving, although still lacking true cognitive abilities. It consists of four main branches (Fig. 13.1).

13.1.1 Symbolic Artificial Intelligence

It is also known as classical AI that was used until the late 1980s. Symbols are used to depict cognition, and logical deduction is employed to mimic human cognitive functions. In this branch, a system operates following a sequence of logic-based steps on language-type representations that are usually propositional. The system asserts relations between objects and computes additional relations based on established ones, following a defined set of inference rules (Garnelo and Shanahan 2019). Fuzzy logic (FL), case-based reasoning, and expert systems (ES) have been developed with a symbolism-based top-down strategy (Venkatasubramanian 2019). A challenge of symbolic AI is the symbol grounding problem, which questions whether its representations are created manually or learned from data, unlike machine learning which excels at identifying features in complex data independently. Process system engineering frequently utilizes symbolic AI (Uraikul et al.

2007), with application in bioenergy systems like fuzzy logic control for power plants using biomass (Jurado and Saenz 2002) and boilers (Romeo and Gareta 2009).

13.1.2 An Overview of Machine Learning

The main focus of machine learning (ML) is on examining a significant volume of data in order to obtain valuable insights. It includes various techniques (artificial neural networks) and statistical methods (support vector machine and Bayesian network) (Dounis 2010; Russell 2010). It offers datasets and data-centric approaches. ML is beneficial as it can analyze large datasets and understand their structure in a way beyond human capabilities. ML methodology (Fig. 13.2) involves generating datasets by conducting experiments, using simulations, or extracting information where variables such as biofuel composition, viscosities, and densities data are used as input separately. ML constructs prediction models based on dataset correlations, using supervised and unsupervised algorithms. Clustering models are used in unsupervised learning, while regression and classification models are used in supervised learning. Following best practices is essential for reliable outcomes in AI and ML methods, involving issue formulation, data collection, tuning, partitioning, and managing under–/over-fitting.

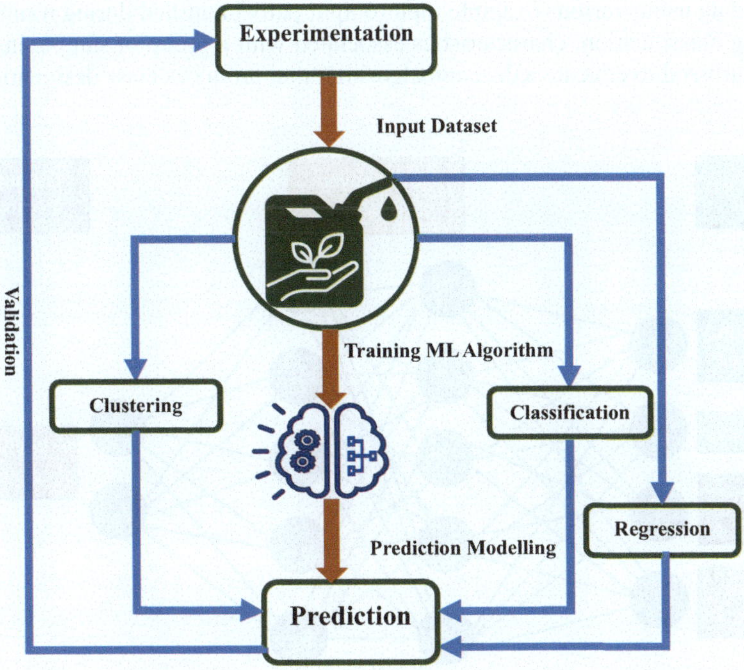

Fig. 13.2 Methodology involved in the machine learning processes

Artificial neural network (ANN) functions are groups of processing units that are connected to each other individually to pass information (Fig. 13.3). A new connection contains two different values (input layer and hidden layer) that generate a third value (output layer) as a sum of the input layer multiplied by the hidden layers. ANN is similar to the human brain in two ways: firstly, it gains knowledge through learning and stores it through robust interneuron connections. It deals with intricate systems containing numerous interconnected parameters, placing emphasis solely on input parameters. Neural networks can be constructed without any preexisting knowledge of the data distribution due to their nonparametric characteristic. It can also uncover nonlinear connections between variables and enhance model accuracy in a way that is nearly unattainable using conventional approaches (Mansab and Rafique 2024). The end result of ANN structure can be represented as

$$y_i = f\left(\sum_{i=1}^{I} X_i W_{ij} + b_j\right) \tag{13.1}$$

where i and j are representing neurons, W_{ij} represents the weight, X_i is the input data vector, b_j is the bias value of the output jth neuron, and y_i is the output of the ith neuron (Mazloom et al. 2020).

Support vector machine (SVM) is a group of effective supervised learning techniques primarily used for classification, but can also be utilized for regression purposes. Classification involves supervised learning, where input data is paired with output data using various example input-output pairs identified during training. By utilizing classification, characteristics associated with a group of sample instances can be utilized to educate a decision algorithm that produces class designations (or

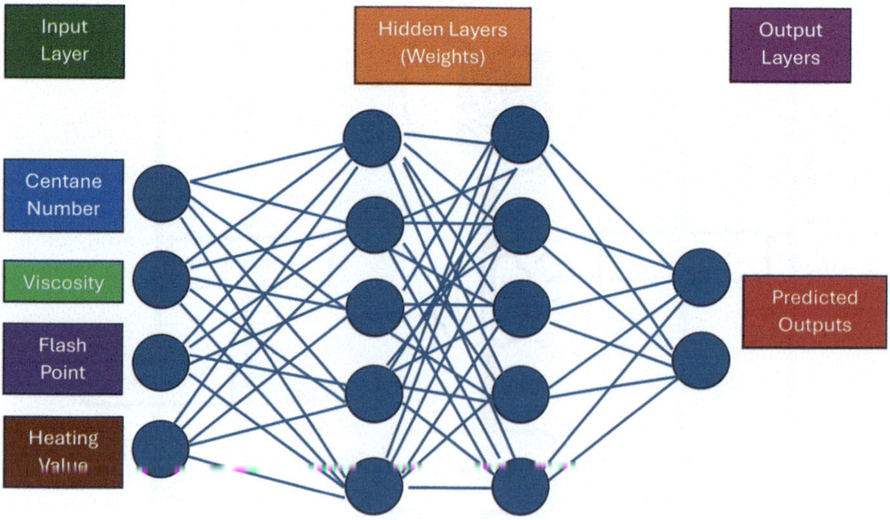

Fig. 13.3 A schematic diagram representing the artificial neural network models

"labels") with a specified level of precision (Pisner and Schnyer 2020). The SVM algorithm identifies the best hyperplane to separate data points in a multidimensional space into distinct classes. These data points, known as the support vectors, are located closest to the hyperplane and have a significant impact on its positioning and orientation. Removing points positioned further away from the hyperplane will not affect the model since they do not affect the loss function used for fitting the model (Bhavsar and Panchal 2012). This is the reason why it is named a support vector machine since it depends on these support vectors. It deals with linear and nonlinear classification. Kernel functions convert the data into a higher-dimensional space when it is not linearly separable. The utilization of different kernel functions depends on both data traits and the intended application (Awad et al. 2015). The parameters of the kernel function can be utilized to evaluate the performance of the SVR model. The anticipated outcome of the SVR model can be expressed in the following manner (Mansab and Rafique 2024).

$$y(x)_{pre} = \sum_{i=1}^{n} \alpha_i K(x_i, x_j) + b \tag{13.2}$$

K(xi, xj) may take the form of radial, gaussian, linear, or polynomial basis function kernel. *αi* and *b* represent the Lagrange multiplier and threshold parameter, respectively. The most commonly used function for representation is the radial basis function (RBF) kernel.

$$K(x_i, x_j) = \exp\left(\frac{-x_j - x_i^2}{\sigma^2}\right) \tag{13.3}$$

where σ is considered a variable for decision-making and is optimized using an external optimization method throughout the calculations (Bouchelkia et al. 2023).

13.1.3 Heuristic Method of AI

Heuristics is a stochastic search method involving learning-based techniques and experience (Zhang et al. 2022), with evolutionary algorithms like genetic algorithms and swarm intelligence algorithms (Chakraborty and Kar 2017; Vikhar and Karde 2017). It utilizes readily accessible, loosely applicable information to control problems. It is not certain that a particular solution or heuristic will be optimal, but heuristics are employed to expedite the search for satisfactory solutions when optimal ones are not feasible. The heuristic genetic algorithms is a method for efficiently arranging the sequence of data downloads. Its structure is a mix of heuristic rules and genetic algorithm, designed for a particular problem context to improve scheduling and planning outcomes. In genetic algorithm, selection, crossover, and mutation operations are applied to produce offspring (Song et al. 2019). Heuristic algorithms have been utilized in the production of biodiesel to determine the optimal reaction parameters. Additionally, a multiobjective genetic algorithm is used to ensure a strong optimization (Goharimanesh et al. 2016).

13.1.4 Hybrid Method of AI

Hybrid AI is an approach that combines rule-based systems (symbolic AI) and statistical learning methods (nonsymbolic AI) to organize, analyze, and present data or knowledge in a machine-readable format. An instance of a hybrid AI system could employ machine learning to examine extensive datasets and recognize patterns, followed by utilizing a rule-based expert system to make decisions. It is a quickly expanding domain in the artificial intelligence field, with possible applications in various industrial sectors (Azizi and Azizi 2019). When the rules governing a system are unclear or ambiguous, predicting the parameter values that influence the system's behavior can be challenging. In this scenario, it has been discovered that combining connectionist and symbolic techniques in a hybrid artificial intelligence model is more effective for making predictions than using them individually (Fig. 13.4). The hybrid prediction system created combines a case-based reasoning system having RBF, ANN, and different optimization algorithms (Corchado and Aiken 2002).

13.2 Role of ML and AI in the Process of Biofueled/ Bioenergy

Production processes of bioenergy have aroused a lot of interest with the ability of machine learning to estimate crop yields as well as improve technology. The use of these modern computational techniques makes it possible to study a large amount of data and causal relationships, which are difficult to achieve by conventional methods (Saju et al. 2025). In turn, for microalgae-based biofuel production, it is suggested that ML algorithms could assist in optimizing parameters of microbial fermentation, enhancing the output while minimizing input costs. AI techniques are capable of simulating different situations in bioenergy systems, which in turn promotes effective resource utilization and energy management decision-making. Modeling and forecasting by AI techniques finds numerous applications, as for instance in supply chain management processes and forecasting. Moreover, AI can also reduce the crisis, preselect feedstock, and determine the conversion measures and track the production processes, achieving sustainability and reduction of the environmental impact (Oguntola et al. 2024). The development of biofuel and bioenergy industries is essential for the transition to green energy sources, and AI and ML enhance their productivity and cost effectiveness. In different research articles (Fangfang et al. 2021; Farobie and Matsumura 2015; Tan et al. 2019), the author

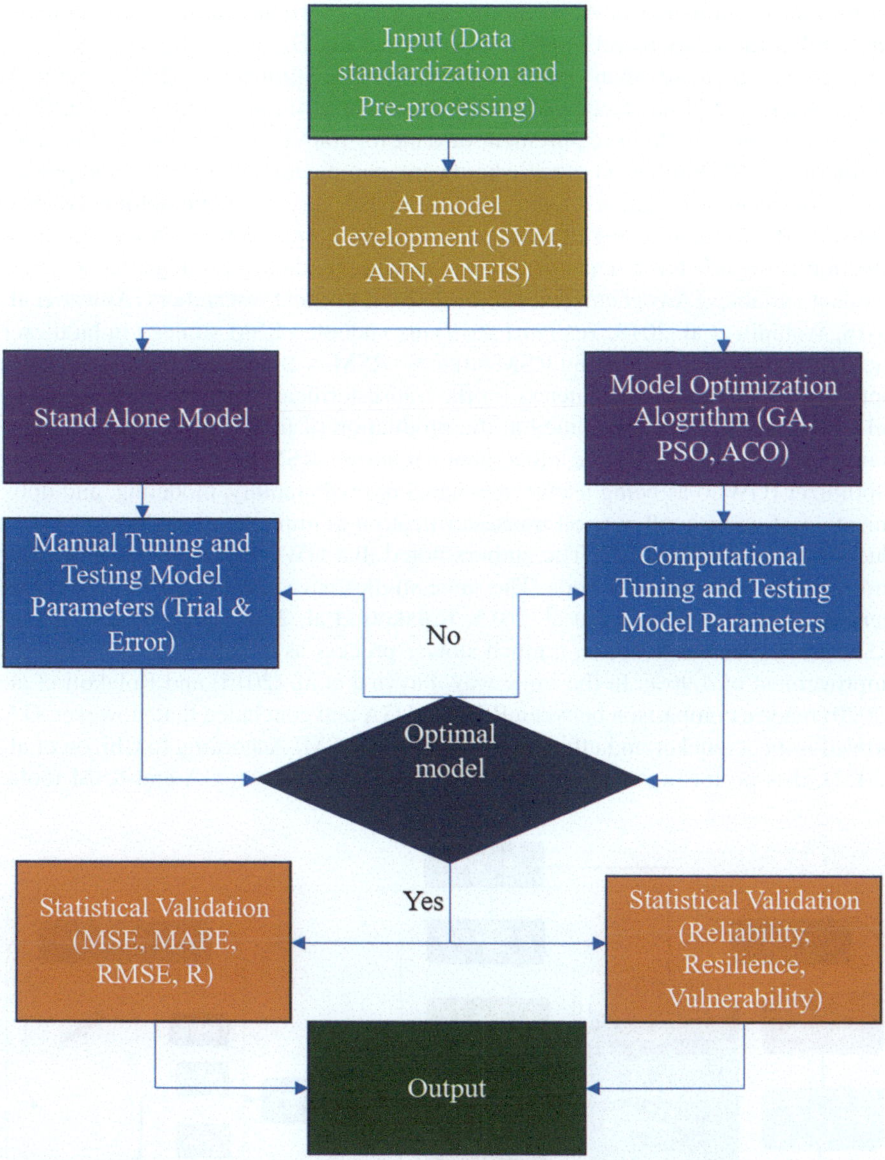

Fig. 13.4 Individual versus hybrid AI-optimized model. (Source: Ibrahim et al. (2022))

utilized ANN to forecast biodiesel output by utilizing brassica oil, *Jatropha* oil, and waste cooking oil as raw materials. The anticipated outcomes were found to be in line with the actual results, and implementing ANN model reduced material waste and time consumption. Other scholars have found that the application of ANN for modeling, forecasting, and optimizing variables that impact biodiesel production is reliable and able to store data throughout the network (Hoang et al. 2021). The performance evaluation of ANN model predicted by Hoang et al. (2021) demonstrated that the ANN-based model outperformed the GA model by 91% and was 76% more effective than the particle swarm optimization (PSO) method. Nevertheless, ANN has faced criticism for its lack of interpretability, difficulty in explanation, and the high requirement of data for training, validation, and testing. Furthermore, ANN relies on specific hardware, computational expertise, and power. These problems are tackled using the response surface methodology (RSM). Consider the frame working of RSM as an optimization technique. A measure of a function is regarded as a dependent variable while predictors are regarded as independent variables (Awogbemi and Kallon 2023). Different researchers (Anwar et al. 2018; Wahidin et al. 2018; Yesilyurt 2019) also adopted RSM strategy in biodiesel yield predictions and modeling RSM strategy. RSM is employed in the design and optimization of process parameters for the transesterification of Canadian rapeseed oil and stone fruit seed oil aimed at the production of high-quality biodiesel oils. Samuel et al. (2020), on the other hand, reported RSM along with Grey Wolf Optimizer (GWO) as being useful in enhancing predictability, modeling, and optimization of waste sunflower oil transesterification to improve on the biodiesel production process (Fig. 13.5). The authors noted that GWO was more accurate and more usable than other methods. The same studies have also compared RSM with genetic algorithm (Fayyazi et al. 2015; Kolakoti et al. 2020). They observed that RSM results were attained in a much slower process as compared to GA with an improvement of 4.96%. In the same way, Fayyazi et al. (2015) and Kolakoti et al. (2020) made a comparison between RSM and GA and concluded that, however, GA proved to be a quicker and effective method than RSM. According to Chizoo et al. (2022), it is promising to integrate ANN-based systems with GA and RSM tools.

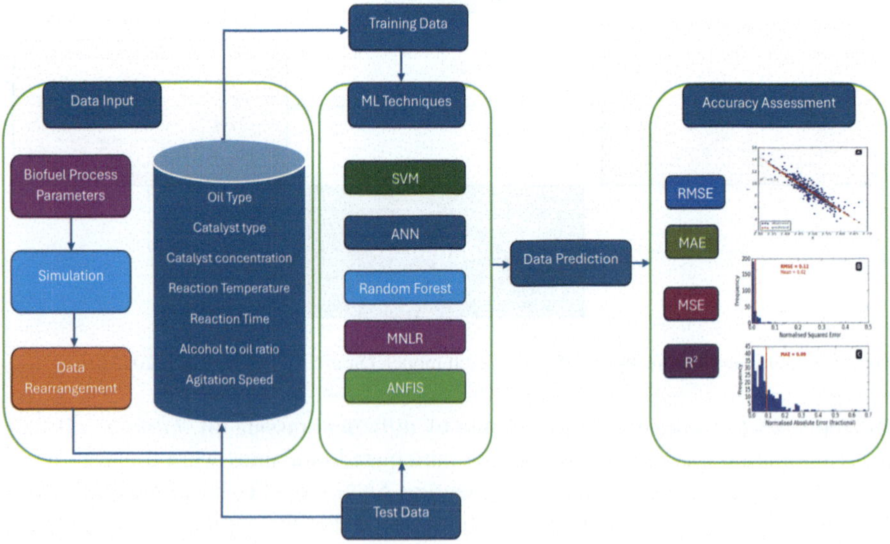

Fig. 13.5 Schematic sketch of ML approaches used in the biofuel process

The results showed that ANFIS-GA performed better in predicting results than RSM models. Similarly, findings of Jisieike et al. (2023) and Ishola et al. (2019) proposed that ANN and RSM can be utilized for forecasting of biodiesel yield. They observed that ANFIS is more dependable, precise, and efficient compared to ANN and RSM.

13.3 Life Cycle Analysis: Optimization of Parameters and Production Process

The use of AI and ML has been crucial in the development of biofuel technology, with various applications such as predicting product yield, optimizing process conditions, and assessing the impacts on the economy and the environment of biomass to biofuel technologies (Okolie 2024). The process of producing biofuel is intricate and consists of various stages like choosing feedstock, pretreating, fermenting, and purifying. AI-driven optimization can greatly improve outcomes at every stage.

13.3.1 AI Applications in Biofuel Crop Production Stage

Machine learning like any other domain in artificial intelligence (AI) assists in the growing of biofuel crops and has been applied in picking the best crop types, tracking the crops in the field, and forecasting output (Ahmad et al. 2021). These innovations are working towards improving efficiency to be more eco-friendly and financially viable. Machine learning techniques analyze large database sets for the determination of the crop varieties necessary for the biofuel. Additionally, the ML models can use genetic and other phenotypic and environmental characteristics to estimate the biomass or oil content of hybrids, or GM crops that are most suitable as biofuel (Godbole et al. 2021). Research conducted by Cacho et al. (2023) has utilized machine learning approaches to analyze cultivation practices coupled with climate data to identify high yielding switchgrass varieties. The predictive fidelity of the concerned models was gauged through themselves by the regression coefficient of determination (R2) and mean absolute error (MAE). The ML model after it processed the data, so to say, was able to discern important environmental parameters along with the cultivation conditions that improved biomass yield. Several studies have dealt with the machine learning of the plot and tree instances at the soil stage of biofuel life cycle. The most commonly used constitutive modules in soil studies include ML components of random forest (RF), SVM, and Gaussian process model (GPM). Sorghum has dual advantage as production of nutrient-rich food from seeds, animal feed, and production of renewable fuels. Huntington et al. (2020) applied the RF technique to evaluate the trends of *Sorghum bicolor* yield harbored in the prospects of four greenhouse emission scenarios and two watering system

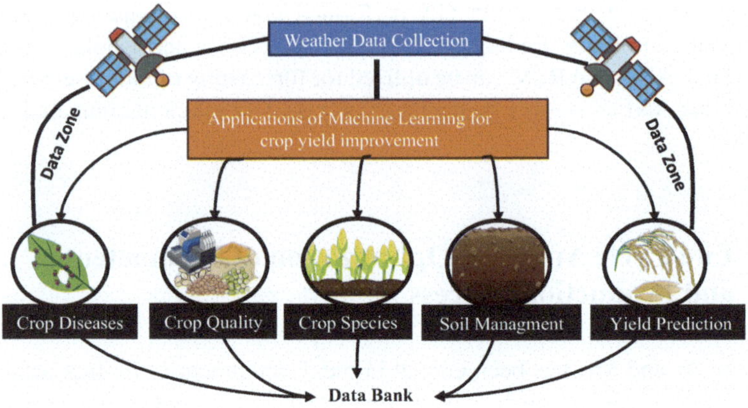

Fig. 13.6 Applications of ML in crop production stage

variations. The major factors affecting the productivity of *Sorghum* are concluded as vapor pressure deficit, irrigation modes, and timing. It was possible to predict *Sorghum* yield through RF model with reasonable accuracy by year and country due to the unique process of model training and sample classification. Habyarimana et al. (2019) utilized satellite images of broomcorn fields to forecast *Sorghum* biomass yield through various ML approaches such as SVM, SVM-R, GBL, GBD, and GBT, with GBT showing superior performance. Lee et al. (2020) managed to make use of a modeling algorithm of Boosted Regression Tree (BRT) to study four different scenarios and estimate the climatic effects for the production of corn from 2022 to 2100. The used model exhibited a correlation coefficient of 0.78 for global warming and 0.82 for eutrophication impact prediction (Fig. 13.6). In a different case, a Bayesian inference technique of GPM was adopted for a two-phase machine learning procedure which was actually a post-processing step that was used to refine the estimation. At first, a GPM downscaling of crop yields is done, and afterward a RF model is used to estimate the yield. Input parameters for future life cycle environmental assessments are generally soil attributes and climatic parameters (solar radiation, average precipitation data, wind speed, and temperature variability) (Yang et al. 2020).

13.3.2 *Applications of AI in Feedstock Selection*

The utilization of artificial intelligence in biofuel feedstock selection is changing the biofuel sector. In the website www.example.com, by machine learning algorithms, predictive modeling, and data analytics, AI is able to help in identifying the most suitable raw materials for producing biofuels by considering their possible productivity, cost-effectiveness, and eco-friendliness. This shades a more efficient and sustainable approach to biofuel production. AI's predictive modeling allows the simulation of different scenarios according to the different environment and taking into account the crop and farm practices. AIS can predict the performance of feedstocks in different environmental conditions. Apart from this, it is very much recommendable for the prediction of how these feedstocks may become more resilient and deploy high-yielding options which are what climate change entails. ANN and linear, statistical, and nonlinear regression models are additionally anticipated to be the most widely used methods in feedstocks phase studies which involve machine learning as a rule. The common input of these processes involves blending, temperature, and the time of processing and mixing duration while higher heating values, density, and viscousity, oxidation stability, heat of vaporization, methane and cetane number are used as common output parameters. In their research, the authors examined different biodiesels made from various sources including *Juglans* oil, helianthus oil, peanut oil, brassica oil, hydrogenated coconut oil, hydrogenated coconut palm oil, and beef tallow (Mairizal et al. 2020). They made use of multiple linear regressions for predicting the density, higher heating value, flashpoint, biodiesel's oxidative stability, and viscosity. The outcomes proposed that the accuracy of the forecast is higher when PU/MU (ratio of poly to mono unsaturated fatty acids) is used as a separate variable. Tchameni et al. (2019) employed various ANN and multiple nonlinear regression (MNLR) models to forecast the rheological properties of used cooking oil. Results showed that the ANN model outperformed the MNLR method. The correlation between biofuel yield in biomass structural components and chemical composition was found to be significant when using both single and multiple linear regressions for estimation. Machine learning models can also incorporate genetic data to enhance feedstock selection. By analyzing genetic traits, these models can predict which feedstocks are likely to exhibit desirable characteristics such as high biomass yield, drought resistance, or pest resistance. In a study by Squance (2014), machine learning algorithms were used to analyze genetic data from *Miscanthus* breeding programs. The researchers employed random forest models to predict the performance of different genetic variants under various growing conditions. The model identified specific genetic markers associated with high biomass yield and environmental resilience. This information was used to guide breeding programs, resulting in new *Miscanthus* varieties optimized for biofuel production.

13.3.3 Catalyst Choice for Biofuel Production

Machine learning is being more and more used to improve catalyst selection in biofuel production. Catalysts are essential in increasing reaction rates and enhancing yields in biofuel production methods like transesterification, hydrodeoxygenation, and pyrolysis. ML models can examine large amounts of data, such as catalyst properties, feedstock characteristics, and reaction conditions, to pinpoint the most successful catalysts. Transesterification is a key process in biodiesel production. Selecting the optimal catalyst is critical for efficiency and yield. Ali et al. (2023)

prepared biodiesel under a vacuum atmosphere utilizing waste cooking oil as a feedstock and a dolomite via Oscillatory Baffled Reactor (OBR). The assessment criteria included a methanol-to-oil ratio of 6:1; temperatures of 323, 333, and 343 K; residence times ranging from 5 to 40 min; various amplitudes of oscillation from 2 to 8 mm with interval of 2 mm; and frequencies of oscillation varying from 1 to 4.3 Hz. ANN model employed for forecasting the ongoing transesterification in the OBR. The structure consisted of 20 neurons in a single hidden layer. The simulated results closely matched the experimental results with determination regression coefficient (R^2) of 0.998 and a mean square error (MSE) of 0.0712 achieved. In another study, Soltani et al. (2019) employed ANN to simulate the impact of different reaction parameters like metal ratio, calcination temperature, reaction time, and temperature on the extraction of esters from palm fatty acid using a sulfonated mesoporous zinc oxide as a catalyst. The optimum conditions were estimated using 0.004 mole of zinc at 433 K and a calcining temperature of 973 K with a experimentation time of 18 min. The most effective parameter is zinc concentration, while the least effective parameter is reaction time. Hydrodeoxygenation (HDO) is employed for the generation of renewable diesel through the elimination of oxygen from molecules derived from biomass. The selection of a catalyst has a big influence on the effectiveness and specificity of this process. ML can help manage intricate data connections and enhance process variables in bio-oil catalytic HDO.

13.4 Estimation of Biofuel Properties, Yield, and Quality

When producing different types of biofuels like biodiesel or biohydrogen, the choice of machine learning technique in the manufacturing process is crucial. According to research, machine learning can be categorized into four main areas when applied to studying biodiesel: optimization of fuel quality and production yield, quality estimation, yield estimation, and process efficiency and condition optimization.

13.4.1 Quality Prediction

ANN created through a regression model utilizes factors such as experimentation temperature and time, flow rate pressure, and calcination temperature as inputs and viscosity, fatty acid methyl ester (FAME) content, chemical composition, product quantity, density, and cetane number as outputs. Sarve et al. (2015a, b) utilized ANN and RSM with a CCD to forecast FAME content in biodiesel made from *Sesamum indicum* oil, using barium hydroxide as the primary catalyst. A FAME content of 98.6% was received by using a methanol-to-oil molar ratio of 6.69:1 with a catalyst concentration of 1.79 wt.%, which is the best combination of optimal condition values. The research found that the amount of catalyst is the primary factor affecting the FAME levels in the end product. ANN is more effective in forecasting FAME content than RSM based on higher correlation coefficient, R2, SEP, and RPD values. Ahmad et al. (2018) employed a combination of ensemble learning techniques, specifically least squares boosting (LSB), with the polynomial chaos expansion (PCE) method to forecast characteristics of product quantity and quality, biofuel flow rate, FAME cetane number, and composition during the production of household cooking oil based biodiesel. The proposed model exposed excellent accuracy in the result predication and quantification of uncertainty in the process, and projected values suggest 1% uncertainty in all the parameters of the process through mean absolute deviation percent (MADP). To calculate the percentage conversion of vegetable oil toward methyl esters during the process of biodiesel production from vegetable oil process, PCA was applied. The method of utilizing PCA is considered an effective approach for not only differentiating but also classifying pure biodiesel from its blend.

13.4.2 Yield Estimation by AI

Bioethanol is the most common biofuel produced from biomass and other feedstocks which contain lignocellulosic matter. The bioethanol yield can be projected by ML models if given data on feedstock structure, pretreatment processes, and fermentation conditions are available for data analysis and prediction. Rivera et al. (2010) applied ML models to the optimization of the enzymatic hydrolysis of lignocellulosic biofuel. The scientists utilized a combination of ANN and GA to foresee the best circumstances for the enzyme to work at optimal levels. The program confirmed the significance of the enzyme concentrations, substrate kinds, as well as the reaction period to the improvement of the sugar yield and recommended the conditions that will generate a 25% increase in product yield which is fundamental for the acquiring of bioethanol. RSM and ANN were utilized to model the biodiesel synthesis process from waste goat tallow with high levels of FFAs and to find the best parameter values for maximum conversion of FAs. Both models exhibited the same reliability as reported by Chakraborty et al. (2014) when the experiment was done in ideal conditions. Kumar (2020) utilized the ANN model for forecasting biodiesel production during transesterification with *Jatropha curcas* and algae oil mixture as the input variables. It was validated by the high regression value of 0.9976 when compared to the experimental results. Kanat and Saral (2009) applied the ANN technique and a multilayer neural network structure to simulate and forecast the production rate of biodiesel and biogas prepared in a non-aerobic high temperature upflow sludge blanket digester. Assessed experimental data that was trained and tested in steady and abnormal conditions displayed a strong correlation coefficient indicating the effectiveness of using ANN for real-time monitoring of thermophilic reactors. Banerjee et al. (2017) utilized the ANN and CCD model to forecast the percentage of fatty acid methyl ester content in ricinus oil and biofuel transesterification with an acid catalyst. They developed a kinetic model based on both experimental and computed data. ANN and experimental results are further utilized for determination of rate constants for a kinetic model. The ANN model forecasted an 8% variance in the yield of biofuel. In a different research, a linear regression and ANN model utilizing optimization approach of Levenberg–Marquardt learning algorithm were created to forecast the transesterification yield of glycine oil-based biodiesel, with the ANN outperforming LR (Kumar 2020). Different factors affecting the transesterification process of glycine oil into biodiesel have been examined to forecast the yield of biodiesel (Moradi et al. 2013). This research utilizes kinetic models and ML approaches of ANN along with a multilayer feedforward neural network. The findings demonstrated that the ANN model outperformed the kinetic modeling method in terms of accuracy and clarity. Guo and Baghban (2017) employed an ANFIS method, to predict the biodiesel production yield in the SCM method without using a catalyst. The prediction was based on reaction parameters. The ANFIS model's significant impact on biodiesel yield forecasting is demonstrated by the high R-squared results. Maran and Priya 2015) conducted a comparison between ANN and RSM effectiveness in forecasting and modeling muskmelon oil-based biodiesel output. Central composite rotatable design, known as CCRD, was used to analyze the ANN model and compare it with the response surface methodology (RSM) model. The conversion of FAME by Multilayer Perceptron (MLP) neural network and RSM is influenced by the reaction parameters. The R^2 value was 0.869 for the RSM model and 0.991 for ANN models, demonstrating the superior performance of the ANN model.

13.4.3 Role of AI in Quality and Yield Optimization for Biofuels

Optimization of biofuel quality and its yield is dependent on the production procedure. Pyrolysis is an advanced method for producing biofuel from biomass through thermochemical conversion. ML approaches can be used to optimize the determining procedure for the best operational parameters, biomass characteristics, and types. Many research studies have concentrated on improving the quality and maximizing the yield of biodiesel. Rahimi et al. (2023) applied the RBF and GA based on RBF to improve the projecting performance of three products including bio-oil, bio-char, and syngas (Fig. 13.7). The optimal yields achieved 45, 36.04, and 54.16%, respectively, providing high feasibility of the RBF algorithm in prediction procedure. In another study, Corral Bobadilla et al. (2018) utilized SVM using multiple kernel functions and LR techniques to forecast and enhance biodiesel yield by focusing on specific characteristics such as turbidity and higher heating value. Utilizing GA in the regression models led to improved optimization scenarios for determining the optimal mix of variables. Ighose et al. (2017) concentrated using an RSM tool in conjunction with the ANFIS model to forecast and enhance the yield in the transesterification process of Thevetia peruviana seed oil. Using genetic

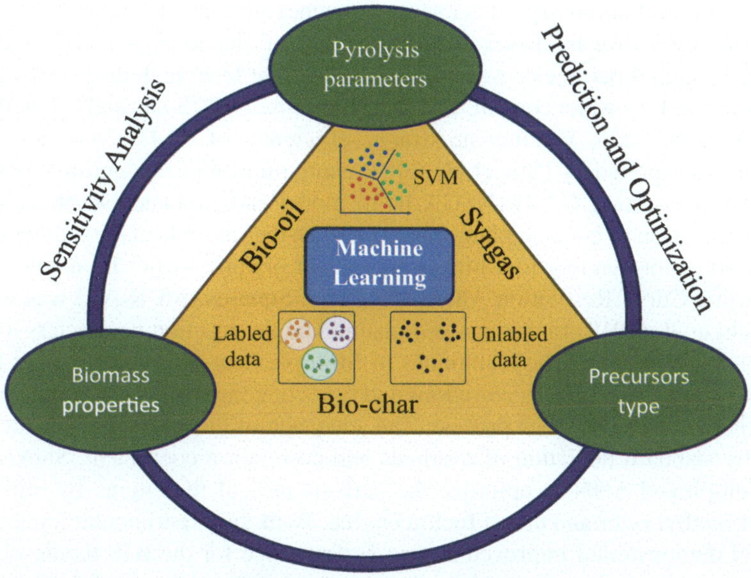

Fig. 13.7 ML applications for biofuel yield prediction and optimization

algorithms along with the ANFIS and RSM models led to an increased production of *Thevetia peruviana* methyl esters (TPME) in a shorter amount of time. The findings established ANFIS's superior prediction capability compared to the RSM model. Sivamani et al. (2019) employed GA-ANN and RSM models to forecast and enhance biodiesel production in *Simarouba glauca* through transesterification.

13.5 Optimization of Process Conditions and Efficiency by AI

Chain reactions such as transesterification, hydrodeoxygenation, and pyrolysis are involved in the production of biofuels. For improving process and product efficiency, ML has become a crucial tool to either identify the optimal process conditions or predict the outcome of a given process. Detailed data, which include multiple reaction parameters, the properties of feedstocks, and the composition of reaction products and intermediates, are typically used to train ML models. When provided with these process data, ML models can be used to identify optimal process parameters, to predict outcomes, and so on. Overall, optimizing a given process is a crucial and necessary step to maximize yield and/or efficiency. Many types of research have leveraged ML to optimize inputs or process conditions to improve efficiency. Karimi et al. (2016) used a hybridized method of RSM and ANN for a multi-criteria evaluation to predict energy efficiency in transesterification process of waste cooking oil for biodiesel manufacturing. In order to perform AI modeling, each experimental parameter was adjusted to optimal level including 35% catalyst concentration, 12% water content, with a reaction time of 20 hours and 6.7 methanol-to-WCO molar ratio, yielding an exergy efficiency of 80.1%. Another hybrid approach was applied by Patle et al. (2014) using multiobjective optimization with nondominated sorting GA-II (NSGA-II) to model and evaluate esterification and transesterification of coconut palm waste oil. With the rise in heat duty, there was an increase in profit, leading to a higher amount of organic waste. In another study, Multivariate Curve Resolution Alternative Least Squares (MCR-AL) was utilized by Rouchi et al. (2019) to manage and manipulate the reaction parameters towards the intended trajectory. The techniques of Multiple Scatter Correction and MCR-ALS are used to analyze concentrations and types of components to obtain biodiesel production from the soybean process. The adequacy of the MCR-ALS method was shown by standard deviation of residuals and correlation coefficient. Shukri et al. (2015) employed ANN to optimize the performance of the engine by utilizing a blend of methyl ester and diesel fuel in engine. Both the experimental findings and the ANN demonstrated improved engine performance for the B10 biodiesel blend and palm oil blend due to their higher heating value and cetane number. Sarve et al. (2015a, b) investigated the sensitivity, predictivity, and generalization capability of ANN and RSM in optimizing production, as well as the effects of parameters. The highest yield of 97.42% of FAEE was achieved by adjusting reaction parameters. The ANN model outperformed the RSM in predicting mahua oil FAEE content and fitting the data. Di Nicola et al. (2010) utilized a multiobjective GA optimization in their biodiesel production from vegetable oil to enhance the purification of key compounds and reduce energy consumption through parameter optimization in the process. The optimized configuration that was found meets the minimum specific energy consumption and meets the required standards for biodiesel quality. Others employed an Extended Kalman Filter (EKF) along with virtual sensors for measuring and predicting operational parameters, controlling performance, and overseeing the response (López-Zapata et al. 2017). The evaluation of *Jatropha* oil-based biodiesel involved analyzing alcohol, triglycerides (TG), methyl ester, diglycerides (DG), glycerol (GL), and monoglycerides (MG) concentrations, as only a few measurable variables such as pH and temperature were considered in the performance analysis.

13.6 Machine Learning for Characterization and Classification of Biodiesel

Characterization includes FAME composition, viscosity, density, and calorific value of biodiesel. Using spectral data can be a complex process and ML models easily help predict these properties accurately. Since then, most of the research focused on energy aspects and the classification or characterization properties have been documented through ML algorithms that are based on various machine learning features. The ML algorithms are developed to analyze the complex data, estimate fuel properties, and differentiate biodiesels according to raw materials or production manners with quality characteristics. A new near-infrared (NIR) method for the analysis of alcohols in diesel fuel was reported by Liu et al. (2022). By combining NIR spectroscopy with GAF image encoding and deep CNN, the proposed approach realized high accurate qualitative classification of different contents of diesel-alcohol mixtures where some similar collinearity occurs in the spectra. In order to check the effectiveness of new method, it has been evaluated against CNN, SVM, and ANN using the same data for a short time. We find that our proposed GAF-CNN method can discriminate well among different diesels. Results were shown to successfully predict the methanol and ethanol content dosage using a variety of quantified metrics, with low MSE/MAE, within high R2 on all cases. A study conducted by Saldana et al. (2012) utilized SVM to predict the viscosity and density of biodiesel at various temperatures. The SVM model accurately predicted engine performance and emissions by utilizing various learning algorithms and descriptors (molecular descriptors and functional group count descriptors). The models have demonstrated superior performance across a variety of chemical compounds and temperatures. Moreover, the models are used to make precise predictions of kinematic viscosities (ν) for molecules that resemble those in the training dataset.

13.7 Estimation of Engine Performance by AI in Biofuel Sector

ML is transforming how we estimate and optimize engine performance specifically in the case of biofuels. This enables ML models to predict both key performance metrics efficiency, power output, fuel consumption, and emissions by analyzing the complex datasets containing engine parameters like combustion properties, structure of fuels, as well as operational conditions. Rate of burn and volume will determine efficiency of the engine as power. These metrics can be predicted by ML models using input variables such as fuel type, injection timing, air-fuel ratio, and engine speed. The performance of an engine (or a powertrain) is measured in terms of its efficiency, often based on the total energy put into and used by it. For combustion engines, this can be related to its thermal or brake thermal efficiency. These models can review thousands of different elements seen with pushing for the right combination to predict engine efficiency. In a study (Panithasan and Venkadesan 2023), the test engine ran on lemon peel oil, varying water injection, rice husk nanoparticles, and proportions of Di tertiary butyl peroxide (DTBP). Six engine performance indicators including smoke, NO_x, BSFC, BTE, HC, and CO were evaluated for various combinations of DTBP ratios, engine load, RH nanoparticle levels, and water injection rates. All experimental values are obtained using a direct injection diesel engine with a single cylinder and constant speed. A total of 70 experimental values were utilized to both train and test the neural network created based on the results obtained. Each correlation between the predicted and experimental values was used to calculate the corresponding R^2 value. The obtained R^2 values range from 0.98209 to 0.99744, which is nearly identical to one, demonstrating the network's accurate prediction of engine performance and emissions. Power output, including brake power and indicated power, is another critical performance metric. ML models can predict power output based on input parameters such as fuel type, injection timing, and cylinder pressure. Research (Togun and Baysec 2010) used ANN to forecast brake-specific fuel consumption and torque of a gasoline engine. A specific ANN formulation was created to forecast the torque and brake-specific fuel consumption of a gasoline engine based on gas position, spark advance, and engine speed. The suggested ANN function is rooted in experimental findings. Research experiments were conducted to collect data for training and testing purposes. Out of a total of 81 datasets, there were 63 sets randomly chosen for training and 18 sets chosen for testing. A back-propagation learning algorithm was used to create an ANN model for the engine. The results presented that ANN is highly effective in predicting engine torque and brake-specific fuel consumption.

13.8 Biofuel Consumption-Emission Mechanism Estimate by AI

The consumption-emission mechanism deals with fuel consumption rates and emissions that result from combustion. The most important ones are NOx, CO, hydrocarbons, and particulate matter. These emissions are mainly determined by the fuel type, combustion completeness, engine load, and operational conditions. It is evident that ML is actively developing the approaches towards the understanding and optimization of the consumption-emission mechanism in biofuel engines. Fuel consumption and emission levels are important parameters that ML models can predict and optimize through the analysis of numerous datasets related to engine, fuel, and emissions, which in turn can make biofuel use more effective and ecofriendly. ML models use a variety of input parameters to estimate emissions and as a result utilize the knowledge to change engine settings to optimize harmful emissions while keeping performance targets. A study (Patnaik et al. 2024) employed ANN to predict emission characteristics of biofuel from waste cooking oil in compression ignition engines. The optimal ANN architectures for forecasting performance metrics like brake specific consumption, energy consumption, thermal efficiency, and exhaust gas temperature were found. The values for the MSE, RMSE, MAD, and MAP error in predicting brake thermal efficiency were 0.0397, 0.1993, 0.1234, and 0.5599, respectively. Sensitivity analysis indicated that ANN created to forecast fuel consumption is greatly influenced by torque. Likewise, the model for energy consumption is affected by brake power, and the models for efficiency of braking and temperature of exhaust gas are impacted by fuel consumption. The findings from this research support the increased use of biodiesel as a practical alternative fuel, reducing greenhouse gas emissions and encouraging ecofriendly energy behaviors for a more sustainable future. Gokalp et al. (2010) effectively used the MLPNN model to forecast the emissions from engines running on jet to biodiesel, marine to biodiesel, and diesel to biodiesel blends based on engine conditions and fuel properties. In their study, Tosun et al. (2016) evaluated the MLPNN model's performance in forecasting CO and NOx level of an engine using a blend of diesel-biodiesel-alcohol, compared to the LR model trained on engine speed performance and fuel physicochemical attributes. The neural model outperformed the LR method in accurately predicting the analyzed exhaust emissions. In other study, Menon and Krishnasamy (2018) effectively linked the smoke and NOx level of heavy and light duty engines to the FAME composition and load conditions through the MLPNN approach. The research effectively predicted the exhaust emissions from diesel/biodiesel blend-powered engines based on the physicochemical properties of fuel blends.

13.9 Conclusions

The need for cheap, readily available, and eco-friendly energy sources is expected to rise shortly. All stakeholders (such as fuel refiners, energy producers, researchers, etc.) of our society will face increased pressure to ramp up energy production to meet the growing global energy demand. Hence, studies focused on increasing the availability of renewable energy will remain a top priority. Biodiesel, a significant entity in renewable fuel, will maintain a key position in the renewable energy sector. Considering the ongoing use of cutting-edge technologies across various sectors, such as research and development, incorporating machine learning will streamline and develop technology-driven methods for advancing biodiesel research. These technologies can help make biodiesel production more accessible and lead to greater energy self-sufficiency. That is why the application of ML and other advanced technologies to model, predict, estimate, and optimize biodiesel production become significant.

13.10 Challenges and Future Prospects

Machine learning technologies will remain the top choice for predicting biodiesel yield over mathematical and numerical models. Continuous utilization of ML technologies in biodiesel research will result in the discovery and application of additional convenient and innovative technologies for further advancement in biodiesel research. Using tools like ANN, ML, mathematical, multiphysics, etc. for simulation, optimization, and modeling is not easy and demands great effort and focus to navigate the high levels of uncertainty and complexity. Choosing raw materials, production processes, and reaction conditions, optimizing input variables, improving procedures, and scaling up production and purification facilities are essential for advancing biodiesel production. A comprehensive examination and enhancement of various factors associated with the biodiesel value chain, such as economic, infrastructural, manpower, and environmental factors, as well as policy formulation and implementation, biodiesel mandates, and standards, is required. The needs for infrastructure for modeling, forecasting, and improving the biodiesel production process need to be addressed and enhanced to align with modern demands.

There are clear inherent weaknesses that need to be addressed directly. For example, difficulties may arise when interpreting results from ANN, RSM, and other statistical analyses. Some results are incomplete, cannot be independent, and are hard to understand. Some of these must be adjusted, examined, and adjusted based on the training data. ML technologies are created to oversee real-world biodiesel utilization processes. Additional research is needed to enhance ML technologies and similar tools for monitoring and controlling biodiesel quality to meet standards. Additional research is necessary to fully grasp the impact of biodiesel production and utilization on the environment, ecosystem, society, economy, and biodiversity.

References

Ahmad I, Ayub A, Ibrahim U, Khattak MK, Kano M (2018) Data-based sensing and stochastic analysis of biodiesel production process. Energies 12(1):63

Ahmad I, Sana A, Kano M, Cheema II, Menezes BC, Shahzad J, Ullah Z, Khan M, Habib A (2021) Machine learning applications in biofuels' life cycle: soil, feedstock, production, consumption, and emissions. Energies 14(16):5072

Ali MM, Gheni SA, Ahmed SM, Hmood HM, Hassan AA, Mohammed HR, Mohammed ST, Karakullukcu NT (2023) Catalytic production of biodiesel from waste cooking oil in a two-phase oscillatory baffled reactor: deactivation kinetics and ANN modeling study. Energy Conversion Manage X 19:100383

Anwar M, Rasul MG, Ashwath N, Rahman MM (2018) Optimisation of second-generation biodiesel production from Australian native stone fruit oil using response surface method. Energies 11(10):2566

Awad M, Khanna R, Awad M, Khanna R (2015) Support vector machines for classification. In: Efficient learning machines: theories, concepts, and applications for engineers and system designers, pp 39–66

Awogbemi O, Kallon DVV (2023) Application of machine learning technologies in biodiesel production process—a review. Front Energy Res 11:1122638

Azizi A, Azizi A (2019) Hybrid artificial intelligence optimization technique. Appl Artif Intell Techn Industry 4:27–47

Banerjee A, Varshney D, Kumar S, Chaudhary P, Gupta VK (2017) Biodiesel production from castor oil: ANN modeling and kinetic parameter estimation. Int J Ind Chem 8:253–262

Bhavsar H, Panchal MH (2012) A review on support vector machine for data classification. Int J Adv Res Comput Eng Technol 1(10):185–189

Bouchelkia N, Tahraoui H, Amrane A, Belkacemi H, Bollinger JC, Bouzaza A, Zoukel A, Zhang J, Mouni L (2023) Jujube stones based highly efficient activated carbon for methylene blue adsorption: kinetics and isotherms modeling, thermodynamics and mechanism study, optimization via response surface methodology and machine learning approaches. Process Safety Environ Protect 170:513–535

Cacho JF, Feinstein J, Zumpf CR, Hamada Y, Lee DJ, Namoi NL, Lee D, Boersma NN, Heaton EA, Quinn JJ (2023) Predicting biomass yields of advanced switchgrass cultivars for bioenergy and ecosystem services using machine learning. Energies 16(10):4168

Chakraborty A, Kar AK (2017) Swarm intelligence: a review of algorithms. In: Nature-inspired computing and optimization: theory and applications, pp 475–494

Chakraborty R, Das S, Pradhan P, Mukhopadhyay P (2014) Prediction of optimal conditions in the methanolysis of mustard oil for biodiesel production using cost-effective mg-solid catalysts. Ind Eng Chem Res 53(51):19681–19689

Chizoo E, Augustine SC, Chukwu EG, Gerald U (2022) Adaptive neuro-fuzzy inference system-genetic algorithm versus response surface methodology-desirability function algorithm modelling and optimization of biodiesel synthesis from waste chicken fat. J Taiwan Inst Chem Eng 136:104389

Corchado JM, Aiken J (2002) Hybrid artificial intelligence methods in oceanographic forecast models. IEEE Trans Syst Man Cybern C 32(4):307–313

Corral Bobadilla M, Fernández Martínez R, Lostado Lorza R, Somovilla Gómez F, Vergara González EP (2018) Optimizing biodiesel production from waste cooking oil using genetic algorithm-based support vector machines. Energies 11(11):2995

Di Nicola G, Moglie M, Pacetti M, Santori G (2010) Bioenergy II: modeling and multi-objective optimization of different biodiesel production processes. Int J Chem React Eng 8(1):1

Dounis AI (2010) Artificial intelligence for energy conservation in buildings. Adv Build Energy Res 4(1):267–299

Fangfang F, Alagumalai A, Mahian O (2021) Sustainable biodiesel production from waste cooking oil: ANN modeling and environmental factor assessment. Sustain Energy Technol Assess 46:101265

Farobie O, Matsumura Y (2015) A comparative study of biodiesel production using methanol, ethanol, and tert-butyl methyl ether (MTBE) under supercritical conditions. Bioresour Technol 191:306–311

Fayyazi E, Ghobadian B, Najafi G, Hosseinzadeh B, Mamat R, Hosseinzadeh J (2015) An ultrasound-assisted system for the optimization of biodiesel production from chicken fat oil using a genetic algorithm and response surface methodology. Ultrason Sonochem 26:312–320

Garnelo M, Shanahan M (2019) Reconciling deep learning with symbolic artificial intelligence: representing objects and relations. Curr Opin Behav Sci 29:17–23

Godbole V, Pal MK, Gautam P (2021) A critical perspective on the scope of interdisciplinary approaches used in fourth-generation biofuel production. Algal Res 58:102436

Goharimanesh M, Lashkaripour A, Akbari A (2016) Optimization of biodiesel production using multi-objective genetic algorithm. Tamkang J Sci Eng 19(2):117–124

Gokalp B, Metin Ertunc H, Hosoz M, Ibrahim Sarac H (2010) Performance prediction of a CI engine using artificial neural network for various SME and diesel fuel blends. Int J Vehicle Design 54(2):156–176

Guo J, Baghban A (2017) Application of ANFIS strategy for prediction of biodiesel production using supercritical methanol. Energy Sour Part A Recov Util Environ Effects 39(17):1862–1868

Habyarimana E, Piccard I, Catellani M, De Franceschi P, Dall' Agata M (2019) Towards predictive modeling of sorghum biomass yields using fraction of absorbed photosynthetically active radiation derived from sentinel-2 satellite imagery and supervised machine learning techniques. Agronomy 9(4):203

Hoang AT, Nižetić S, Ong HC, Tarelko W, Le TH, Chau MQ, Nguyen XP (2021) A review on application of artificial neural network (ANN) for performance and emission characteristics of diesel engine fueled with biodiesel-based fuels. Sustain Energy Technol Assess 47:101416

Huntington T, Cui X, Mishra U, Scown CD (2020) Machine learning to predict biomass sorghum yields under future climate scenarios. Biofuels Bioprod Biorefin 14(3):566–577

Ibrahim KSMH, Huang YF, Ahmed AN, Koo CH, El-Shafie A (2022) A review of the hybrid artificial intelligence and optimization modelling of hydrological streamflow forecasting. Alex Eng J 61(1):279–303

Ighose BO, Adeleke IA, Damos M, Junaid HA, Okpalaeke KE, Betiku E (2017) Optimization of biodiesel production from Thevetia peruviana seed oil by adaptive neuro-fuzzy inference system coupled with genetic algorithm and response surface methodology. Energy Conversion Manage 132:231–240

Ishola NB, Okeleye AA, Osunleke AS, Betiku E (2019) Process modeling and optimization of sorrel biodiesel synthesis using barium hydroxide as a base heterogeneous catalyst: appraisal of response surface. Neural Comput Applic 31:4929

Jisieike CF, Ishola NB, Latinwo LM, Betiku E (2023) Crude rubber seed oil esterification using a solid catalyst: optimization by hybrid adaptive neuro-fuzzy inference system and response surface methodology. Energy 263:125734

Jurado F, Saenz JR (2002) Possibilities for biomass-based power plant and wind system integration. Energy 27(10):955–966

Kanat G, Saral A (2009) Estimation of biogas production rate in a thermophilic UASB reactor using artificial neural networks. Environ Model Assess 14:607–614

Karimi M, Jenkins B, Stroeve P (2016) Multi-objective optimization of transesterification in biodiesel production catalyzed by immobilized lipase. Biofuels Bioprod Biorefin 10(6):804–818

Kolakoti A, Jha P, Mosa PR, Mahapatro M, Kotaru TG (2020) Optimization and modelling of mahua oil biodiesel using RSM and genetic algorithm techniques. Math Models Eng 6(2):134–146

Kumar S (2020) Comparison of linear regression and artificial neural network technique for prediction of a soybean biodiesel yield. Energy Sources Part A 42(12):1425–1435

Lee EK, Zhang WJ, Zhang X, Adler PR, Lin S, Feingold BJ, Khwaja HA, Romeiko XX (2020) Projecting life-cycle environmental impacts of corn production in the US Midwest under future climate scenarios using a machine learning approach. Sci Total Environ 714:136697

Liu S, Wang S, Hu C, Bi W (2022) Determination of alcohols-diesel oil by near infrared spectroscopy based on gramian angular field image coding and deep learning. Fuel 309:122121

López-Zapata BY, Adam-Medina M, Álvarez-Gutiérrez PE, Castillo-González JP, Hernández-de León HR, Vela-Valdés LG (2017) Virtual sensors for biodiesel production in a batch reactor. Sustain For 9(3):455

Mairizal AQ, Awad S, Priadi CR, Hartono DM, Moersidik SS, Tazerout M, Andres Y (2020) Experimental study on the effects of feedstock on the properties of biodiesel using multiple linear regressions. Renew Energy 145:375–381

Mansab S, Rafique U (2024) Adsorption simulation of 2,4-D pesticide on novel zinc–based 2-amino-4-(1H-1,2,4-triazole-4-yl)benzoic acid coordination complexes using machine learning approach. Environ Sci Pollut Res 31:36814. https://doi.org/10.1007/s11356-024-33668-1

Maran JP, Priya B (2015) Modeling of ultrasound assisted intensification of biodiesel production from neem (Azadirachta indica) oil using response surface methodology and artificial neural network. Fuel 143:262–267

Mazloom MS, Rezaei F, Hemmati-Sarapardeh A, Husein MM, Zendehboudi S, Bemani A (2020) Artificial intelligence based methods for asphaltenes adsorption by nanocomposites: application of group method of data handling, least squares support vector machine, and artificial neural networks. Nano 10(5):890

Menon PR, Krishnasamy A (2018) A composition-based model to predict and optimize biodiesel-fuelled engine characteristics using artificial neural networks and genetic algorithms. Energy Fuel 32(11):11607–11618

Moradi G, Dehghani S, Khosravian F, Arjmandzadeh A (2013) The optimized operational conditions for biodiesel production from soybean oil and application of artificial neural networks for estimation of the biodiesel yield. Renew Energy 50:915–920

Oguntola O, Boakye K, Simske S (2024) Towards leveraging artificial intelligence for sustainable cement manufacturing: a systematic review of AI applications in electrical energy consumption optimization. Sustain For 16(11):4798

Okolie JA (2024) Introduction of machine learning and artificial intelligence in biofuel technology. Curr Opin Green Sustain Chem 100928:100928

Panithasan MS, Venkadesan G (2023) Evaluating the outcomes of a single-cylinder CRDI engine operated by lemon peel oil under the influence of DTBP, rice husk nano additive and water injection. Int J Engine Res 24(2):308–323

Patle DS, Sharma S, Ahmad Z, Rangaiah G (2014) Multi-objective optimization of two alkali catalyzed processes for biodiesel from waste cooking oil. Energy Convers Manag 85:361–372

Patnaik S, Khatri N, Rene ER (2024) Artificial neural networks-based performance and emission characteristics prediction of compression ignition engines powered by blends of biodiesel derived from waste cooking oil. Fuel 370:131806

Pisner DA, Schnyer DM (2020) Support vector machine. In: Machine learning. Elsevier, Amsterdam, pp 101–121

Rahimi M, Mashhadimoslem H, Vo Thanh H, Ranjbar B, Safarzadeh Khosrowshahi M, Rohani A, Elkamel A (2023) Yield prediction and optimization of biomass-based products by multi-machine learning schemes: neural, regression and function-based techniques. Energy 283:128546. https://doi.org/10.1016/j.energy.2023.128546

Rivera EC, Rabelo SC, dos Reis GD, Filho RM, da Costa AC (2010) Enzymatic hydrolysis of sugarcane bagasse for bioethanol production: determining optimal enzyme loading using neural networks. J Chem Technol Biotechnol 85(7):983–992

Romeo LM, Gareta R (2009) Fouling control in biomass boilers. Biomass Bioenergy 33(5):854–861

Rouchi MB, Khorrami MK, Garmarudi AB, de la Guardia M (2019) Application of infrared spectroscopy as process analytics technology (PAT) approach in biodiesel production process utilizing multivariate curve resolution alternative Least Square (MCR-ALS). Spectrochim Acta Part A 213:347–353

Russell PN (2010) In: Davis E et al (eds) Artificial intelligence: a modern approach by Stuart. Russell and Peter Norvig contributing writers. EPDF

Saju L, Selvaraj D, Vairaperumal T (2025) Artificial intelligence and machine intelligence: modeling and optimization of bioenergy production. In: Computer vision and machine intelligence for renewable energy systems. Elsevier, Amsterdam, pp 163–176

Saldana DA, Starck L, Mougin P, Rousseau B, Ferrando N, Creton B (2012) Prediction of density and viscosity of biofuel compounds using machine learning methods. Energy Fuel 26(4):2416–2426

Samuel OD, Okwu MO, Oyejide OJ, Taghinezhad E, Afzal A, Kaveh M (2020) Optimizing biodiesel production from abundant waste oils through empirical method and grey wolf optimizer. Fuel 281:118701

Sarve A, Sonawane SS, Varma MN (2015a) Ultrasound assisted biodiesel production from sesame (Sesamum indicum L.) oil using barium hydroxide as a heterogeneous catalyst: comparative assessment of prediction abilities between response surface methodology (RSM) and artificial neural network (ANN). Ultrason Sonochem 26:218–228

Sarve AN, Varma MN, Sonawane SS (2015b) Response surface optimization and artificial neural network modeling of biodiesel production from crude mahua (Madhuca indica) oil under supercritical ethanol conditions using CO_2 as CO-solvent. RSC Adv 5(85):69702–69713

Shukri MR, Rahman M, Ramasamy D, Kadirgama K (2015) Artificial neural network optimization modeling on engine performance of diesel engine using biodiesel fuel. Int J Automot Mech Eng 11:2332

Sivamani S, Selvakumar S, Rajendran K, Muthusamy S (2019) Artificial neural network–genetic algorithm-based optimization of biodiesel production from Simarouba glauca. Biofuels 10(3):393–401

Soltani S, Rashid U, Roodbar Shojaei T, Nehdi IA, Ibrahim M (2019) Modeling of the nanocrystalline-sized mesoporous zinc oxide catalyst using an artificial neural network for efficient biodiesel production. Chem Eng Commun 206(1):33–47

Song YJ, Zhang ZS, Sun K, Yao F, Chen YW (2019) A heuristic genetic algorithm for regional targets' small satellite image downlink scheduling problem. Int J Aerospace Eng 1:1371852

Squance MG (2014) Computational modelling of the relationship between Miscanthus genotype, phenotype and environment. Aberystwyth University, Methodology, neural network and neuro-fuzzy system. Neural Comput & Applic 31:4929–4943

Tan YH, Abdullah MO, Kansedo J, Mubarak NM, San Chan Y, Nolasco-Hipolito C (2019) Biodiesel production from used cooking oil using green solid catalyst derived from calcined fusion waste chicken and fish bones. Renewable Energy 139:696–706

Tchameni AP, Zhao L, Ribeiro JX, Li T (2019) Predicting the rheological properties of waste vegetable oil biodiesel-modified water-based mud using artificial neural network. Geosystem Eng 22(2):101–111

Thakur L, Vasudev H, Singh J, Prashar G (2023) Artificial Intelligence and Machine Learning in the Thermal Spray Industry: Practices, Implementation, and Challenges: CRC Press.

Togun NK, Baysec S (2010) Prediction of torque and specific fuel consumption of a gasoline engine by using artificial neural networks. Appl Energy 87(1):349–355

Tosun E, Aydin K, Bilgili M (2016) Comparison of linear regression and artificial neural network model of a diesel engine fueled with biodiesel-alcohol mixtures. Alex Eng J 55(4):3081–3089

Uraikul V, Chan CW, Tontiwachwuthikul P (2007) Artificial intelligence for monitoring and supervisory control of process systems. Eng Appl Artif Intell, 20(2):115–131

Venkatasubramanian V (2019) The promise of artificial intelligence in chemical engineering: is it here, finally? AIChE J, 65(2): 466–478

Vikhar P, Karde P (2017) Exploring differential evolution algorithm for content based image retreival system. Paper presented at the 2017 2nd International Conference for Convergence in Technology (I2CT)

Wahidin S, Idris A, Yusof NM, Kamis NHH, Shaleh SRM (2018) Optimization of the ionic liquidmicrowave assisted one-step biodiesel production process from wet microalgal biomass. Energy Convers Manag 171:1397–1404

Yang P, Zhao Q, Cai X (2020) Machine learning based estimation of land productivity in the contiguous US using biophysical predictors. Environ Res Lett, 15(7):074013

Yesilyurt MK (2019) The effects of the fuel injection pressure on the performance and emission characteristics of a diesel engine fuelled with waste cooking oil biodiesel-diesel blends. Renewable Energy 132:649–666

Zhang Y, Bai R, Qu R, Tu C, Jin J (2022) A deep reinforcement learning based hyper-heuristic for combinatorial optimisation with uncertainties. Eur J Oper Res, 300(2):418–427

Chapter 14
Application of Computational and In Silico Strategies for Sustainable Future Biofuels and Bioenergy

Anjuman Ayub, Atif Khurshid Wani, Reena Singh, Anjuvan Singh, Juliana Heloisa Pinê Américo-Pinheiro, Showkeen Muzamil Bashir, Mudasir A. Dar, and Farida Rahayu

Abstract In modern society, the use of computational modeling has increased significantly in a variety of industries. This increase is especially noticeable because of growing concerns about the rapid depletion of fossil fuels and their harmful effects on the environment. As a result, there has been a lot of interest in researching alternative energy sources like biofuels. It becomes clear that computational modeling is a useful and affordable tool that can maximize output and financial sustainability in the processes involved in the production of biofuels. Notably, developments in artificial intelligence and related fields have made it possible to predict a number of parameters that are essential to the production of biofuels, such as substrate compositions, process yields, and ideal operational conditions for sugar fermentations. It can be used in a wide range of bioprocesses, including the pretreatment of lignocellulosic biomass and different phases of biofuel synthesis. The incorporation of

A. Ayub · A. K. Wani (✉) · R. Singh · A. Singh
School of Bioengineering and Biosciences, Lovely Professional University, Jalandhar, Punjab, India

J. H. P. Américo-Pinheiro
Department of Forest Science, Soils and Environment, School of Agricultural Sciences, São Paulo State University (UNESP), Botucatu, São Paulo, Brazil

Graduate Program in Environmental Sciences, Brazil University, São Paulo, São Paulo, Brazil

S. M. Bashir
Biochemistry and Molecular Biology Lab, Division of Veterinary Biochemistry, Faculty of Veterinary Sciences and Animal Husbandry, Sher-e-Kashmir University of Agricultural Sciences and Technology, Srinagar, Jammu and Kashmir, India

M. A. Dar
School of Environment and Safety Engineering, Biofuels Institute of Jiangsu University, Zhenjiang, Jiangsu, China

F. Rahayu
Research Center for Genetic Engineering, National Research and Innovation Agency, Bogor, Indonesia

M. A. Dar et al. (eds.), *Recent Trends in Lignocellulosic Biofuels and Bioenergy*, Clean Energy Production Technologies,
https://doi.org/10.1007/978-981-96-4636-4_14

computational modeling exhibits potential to mitigate the dependence on preliminary experiments and, in certain cases, eliminate the necessity for lab-scale trials, consequently optimizing the process of research and development. This chapter examines potential trends of biofuels and bioenergy and emphasizes the importance of computational modeling for maximizing biofuel yields while providing extensive explorations of computational and in silico strategies employed for biofuel productions, emphasizing their pivotal role in advancing sustainable bioenergy production.

Keywords Bioenergy · Bioinformatics · Stimulation · Enzyme engineering · Process optimization · System biology

14.1 Introduction

In modern times, human civilization encompasses several issues like global warming, overreliance on fossil fuels, and concerns regarding food and energy security. The rapid expansion of populations and industrial activities has surged the demand for energy, thereby exerting immense pressure on finite reserves of nonrenewable fossil fuels (Carbonell et al. 2020; Dar et al. 2024). The combustion of these fossil fuels not only releases greenhouse gases but also contributes to acid rain and climate change, posing a grave environmental threat (Heger et al. 2021). To address these demanding concerns and meet the rising energy demands sustainably, the adoption of renewable energy sources is imperative. Biofuels represent one such alternative energy source that holds promise in mitigating these challenges (Pahima et al. 2019). Derived from biological sources such as plants, microorganisms, animals, and waste, biofuels encompass a diverse array of options including bioethanol, biodiesel, biogas, and biohydrogen. Industrially produced biodiesel and bioethanol currently dominate the biofuel markets, accounting for over 90% of its total share (Godbole et al. 2021). These biofuels are obtained from both first-generation feedstocks like sugar beets and corn, as well as second-generation sources such as lignocellulosic biomass (Fig. 14.1) (Ranganathan et al. 2022). While first-generation biofuels offer higher production yields at lower costs, their reliance on arable land raises concerns about food security (Rulli et al. 2016). In contrast, second-generation biofuels, derived from nonfood sources, improve these food security concerns but still require cultivable land, water, and pesticides for production (Wani et al. 2024c). However, the utilization of lignocellulosic biomass in second-generation biofuels poses challenges due to the complex structure of lignin, requiring additional chemical processes and enzymes, thereby increasing production costs (Nanda et al. 2015; Renzaho et al. 2017).

In recent years, third-generation biofuels, particularly those derived from microalgae and cyanobacteria, have gathered significant attention. Microalgae, characterized by their rapid growth rate and minimal resource requirements, hold tremendous

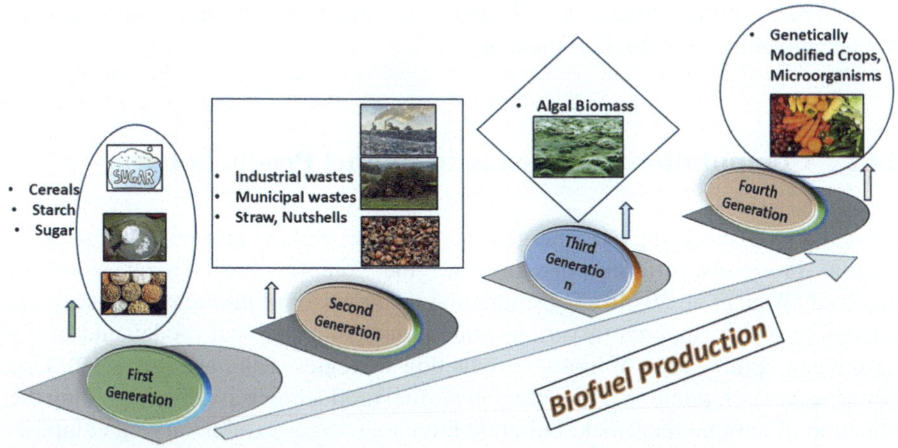

Fig. 14.1 The evolution of biofuel production: Harnessing crop-based biofuels in the first generation, advancing to waste biomass in the second, optimizing algae and microorganisms in the third, and exploring synthetic biology and genetic engineering for the fourth generation. A journey towards cleaner, more sustainable energy solutions for the future

potential as a sustainable biofuel source (Ananthi et al. 2021; Bajpai 2019; Nanda et al. 2015). Their cultivation presents advantages such as year-round growth and minimal impact on arable land and food supplies. Similarly, seaweed, or macroalgae, presents another viable option for bioenergy production, boasting a short cultivation time and high carbohydrate, lipid, and protein content. Despite the promise of microalgae-based biofuels, challenges such as cultivation contamination, harvesting difficulties, and complex downstream processing persist (Hebbale et al. 2017; Twigg et al. 2024). However, recent advancements in genetic engineering and molecular biology have revolutionized the field, enabling the development of novel strains optimized for biofuel production (Bharadwaj et al. 2020; Ng et al. 2017). Computational tools, including next-generation sequencing and genome-scale metabolic reconstruction, facilitate the identification of new targets and pathways for strain development, streamlining the experimental process and reducing time and capital expenditure (Hari et al. 2024; Rong 2023). The integration of genetic engineering, synthetic biology, and computational methodologies heralds the advent of fourth-generation biofuels, characterized by genetically modified organisms with enhanced biofuel yield. By leveraging these interdisciplinary approaches, researchers aim to design strains with heightened lipid productivity and improved cost-effectiveness, paving the way for a sustainable future powered by biofuels (Hari et al. 2024; Zengler et al. 2017). This book chapter aims to explore the potential of computational modeling in enhancing the production of biofuels. The objective is to assess how computational techniques, including artificial intelligence and in silico simulations, can optimize various bioprocesses involved in biofuel production. Specifically, it seeks to identify how modeling can predict key parameters such as substrate composition and process yields and optimal operational conditions, with the goal of improving efficiency, reducing experimental dependence, and advancing the sustainability of biofuel production.

14.2 Computational Modeling in Biofuel Production

Computational modeling has emerged as a powerful tool in optimizing biofuel production processes, offering insights into the complex biochemical pathways involved (Wani et al. 2024b). By simulating the behavior of biological systems and predicting their responses to various conditions, computational models aid in the design and optimization of biofuel production strategies (Pahima et al. 2019). One key aspect of computational modeling in biofuel production is its role in guiding the selection of optimal feedstock. Different biomass sources exhibit varying compositions and properties, influencing their suitability for biofuel productions (Sewsynker-Sukai et al. 2017). Computational models can analyze the biochemical composition of potential feedstocks and predict their performance in biofuel conversion processes. This enables researchers to identify feedstocks with the highest biofuel yield and minimal environmental impact, thus optimizing resource utilization (Banerjee et al. 2018). Furthermore, computational models play a crucial role in the design of enzymatic pathways for biofuel synthesis. Enzymes catalyze the conversion of biomass into biofuels through complex biochemical reactions (del Rio-Chanona et al. 2019). Computational modeling allows researchers to predict the behavior of enzymes under different conditions and optimize their activity for maximum efficiency (Garg et al. 2021). By simulating enzyme kinetics and substrate interactions, computational models help in the selection and engineering of enzymes with enhanced catalytic activity and stability, thereby improving overall biofuel production yields (Banerjee et al. 2018). Additionally, computational modeling facilitates the design and optimization of bioreactor systems for biofuel production. Bioreactors provide controlled environments for microbial fermentation or enzymatic reactions, allowing for efficient conversion of biomass into biofuels (Jones et al. 2024). Computational models can simulate fluid dynamics, mass transfer phenomena, and microbial growth kinetics within bioreactor systems, optimizing reactor design parameters such as reactor geometry, operating conditions, and feeding strategies. This enables researchers to maximize biofuel production rates and minimize production costs by improving process efficiency and scalability (Wani et al. 2023c). Overall, computational modeling serves as a valuable tool in biofuel production, enabling researchers to optimize feedstock selection, enzyme design, and bioreactor operation (Kavitha et al. 2024). By providing insights into the complex biochemical processes involved in biofuel production, computational models contribute to the development of sustainable and economically viable biofuel production technologies (Wani et al. 2019, 2023a, b).

14.3 In Silico Design of Biomass Feedstocks

The in silico design of biomass feedstocks represents a transformative approach within bioenergy research, aiming to revolutionize the selection and optimization of raw materials for sustainable fuel production (Huang et al. 2024). This innovative method employs computational tools and modeling techniques to predict and customize the properties of biomass feedstocks, offering significant advantages over traditional trial-and-error methods. By utilizing computational models, researchers can simulate the characteristics and behaviors of various biomass feedstock candidates, integrating data from genetic databases, biochemical pathways, and environmental factors to predict their performance under different conditions (Asolkar et al. 2024). Through these simulations, feedstocks with desirable traits such as high energy content, fast growth rates, and resistance to environmental stressors can be identified, expediting the discovery process and enabling the exploration of unconventional feedstock sources. Through these simulations, the process of identifying and optimizing feedstocks with desirable traits becomes significantly more efficient. By simulating various biological and environmental conditions, computational models allow for the rapid screening of feedstocks with traits such as high energy content, fast growth rates, and resistance to environmental stressors. For instance, computational models can analyze the molecular composition of feedstocks to identify those with high carbohydrate or lipid content, which can be converted into energy-dense biofuels. Algae like *Chlorella vulgaris* or yeasts like *Yarrowia lipolytica* may be prioritized for further study due to their lipid storage capabilities essential for biodiesel production (Sari and Utama 2024). Additionally, artificial intelligence and machine learning techniques can predict the growth rates of microorganisms like *Saccharomyces cerevisiae* and *Zymomonas mobilis* under varying environmental conditions, helping to select fast-growing strains for ethanol production, enhancing overall efficiency (Zhang et al. 2020).

Computational simulations also play a pivotal role in assessing the resilience of feedstocks to environmental stressors. By modeling how different strains respond to factors such as drought, salinity, or temperature fluctuations, researchers can identify promising candidates like *Rhodobacter sphaeroides* and *Pseudomonas putida* that can thrive in harsh conditions, making them ideal for biohydrogen production (Bharadwaj et al. 2020). The use of computational tools extends beyond conventional feedstocks, allowing the exploration of unconventional sources such as algae, fungi, or genetically engineered microorganisms like *Micractinium* sp. or *Sporomusa ovata*. These simulations reduce the need for extensive field testing, cutting down on time and costs associated with research and development. By simulating thousands of scenarios quickly, computational models expedite the discovery of optimal feedstocks like *Clostridium thermocellum* for cellulosic ethanol and *Rhodococcus opacus* for lipid-based biofuels, streamlining biofuel production and reducing the dependency on lengthy lab trials (Clomburg and Gonzalez 2010).

Moreover, in silico design facilitates the optimization of biomass feedstocks for specific applications and production processes (Fig. 14.2) (Bhushan et al. 2023).

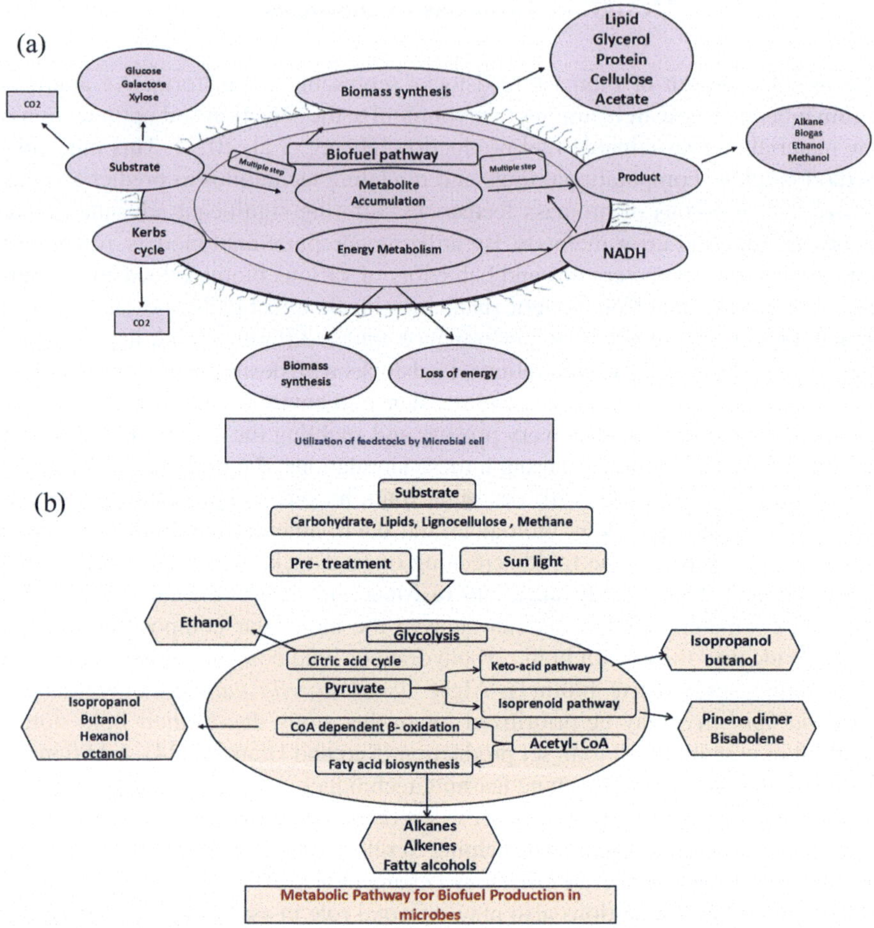

Fig. 14.2 (**a**) Utilization of feedstocks and (**b**) mechanism involved in production of biofuel by microbial cell

Through virtual modifications such as genetic engineering or chemical treatments, researchers can fine-tune feedstock properties to enhance their suitability for bioenergy production. This precision allows for the creation of tailored feedstock blends optimized for different conversion technologies, ultimately improving overall efficiency and sustainability. Additionally, the rapid screening capabilities of in silico design enable researchers to prioritize the most promising feedstock options for further experimentation and validation, saving both time and resources (Kaladhar et al. 2023). This streamlined approach not only accelerates the discovery and development of novel biomass feedstocks but also expands the pool of available options for bioenergy production, contributing to a more efficient and environmentally friendly bioenergy industry. The in silico design of biomass feedstocks holds immense potential for advancing bioenergy research (Gupta et al. 2023). By harnessing the power of computational modeling and predictive analytics, researchers can expedite the discovery, optimization, and utilization of biomass resources, paving the way for sustainable and environmentally friendly fuel production. This innovative approach offers a pathway towards more efficient and tailored bioenergy solutions, addressing the growing demand for renewable energy sources and contributing to global efforts towards a greener future (Samantaray et al. 2023).

14.4 Genome-Scale Metabolic Modeling

In recent years, the study of microbial ecosystems has undergone a significant transformation due to advances in genome sequencing technologies. Initially trusting on amplicon sequencing methods, the field has transitioned towards next-generation shotgun sequencing, resulting in the generation of extensive omics datasets (Banerjee et al. 2018). Understanding these datasets requires the application of systems biology tools to unravel the functional capacity of microbiomes and their interactions with the environment and potential hosts. Consequently, a diverse array of mathematical and computational approaches has emerged to elucidate the underlying mechanisms governing microbial interactions, notably including the rapid expansion of genome-scale modeling of metabolism (Banerjee et al. 2024). This modeling approach involves the large-scale mathematical formalization of knowledge consolidated from genomic, physiological, and biochemical evidence, enabling the simulation of microbial metabolism at a systems level (Vasileiadis et al. 2024). While early stages of genome-scale modeling depend heavily on manual knowledge accumulation, recent advancements have facilitated a shift towards automating these processes. Concurrently, emerging technologies for microbial isolation and the rapid generation of metagenomes housing uncharacterized microbial species present both opportunities and challenges in the reconstruction process (Wani et al. 2024a). Therefore, there is a critical need for rapid and automated reconstruction tools to support experts by automating tasks such as genome annotation, draft model generation, and gap filling. As genome-scale metabolic models continue to expand in scope, it becomes imperative to rigorously evaluate reconstruction tools to assist researchers in selecting suitable methods and software, especially when constructing multispecies models.

14.4.1 Enzyme Engineering and Metabolomics

Microbial biosynthesis presents a promising avenue for the sustainable production of various chemicals, including fuels, natural plant products, and plastics. Achieving cost-effective industrial-scale production necessitates the engineering of microbial strains and cultivation methods. Advances in genetic engineering and screening techniques have facilitated the construction of highly engineered strains, often guided by the Design–Build–Test–Learn (DBTL) cycle (Tanaka et al. 2024). This iterative workflow involves strain design, construction, evaluation, and learning from analysis data to inform subsequent design iterations. Metabolomics, a powerful tool within the DBTL cycle, enables the identification of bottleneck reactions in metabolic pathways and facilitates the enhancement of production yield and productivity through enzyme engineering (Holtz et al. 2024). Despite its utility, challenges persist, including the lack of metabolic information hindering productivity improvements and the need for multiple iterations of the DBTL cycle, which can be time-consuming (Gu et al. 2024). Metabolomics plays a crucial role not only in optimizing culture conditions and scale-up for industrial-scale production but also in enhancing mathematical modeling efforts. It enables the characterization of metabolic effects using various omics approaches and facilitates the development of mathematical models based on omics data to perform rational strain design (Kaur and Gupta 2024). Recent advancements in machine learning have further expanded the utility of metabolome data, allowing for accurate predictions of metabolome changes and maximizing the productivity of target compounds (Van Lent et al. 2023). However, to precisely control metabolic flux and increase productivity through metabolic engineering, a thorough understanding of metabolic mechanisms is essential. Integration of multi-omics data and advanced analytical techniques, such as supercritical fluid chromatography and imaging mass spectrometry, streamlines the DBTL workflow and accelerates systems biology model development, ultimately driving advancements in microbial chemical production (Table 14.1; Liu et al. 2024).

14.4.2 Molecular Stimulation and Enzyme Engineering

The quest to engineer microbial metabolism has improved in on fatty acyl compounds, showcasing their pivotal role. Scientists are continuously striving to bioengineer the fatty acid metabolism of microbes, redirecting it towards alkane/alkene biosynthesis through the incorporation of heterologous enzymatic modules (Liu et al. 2019). While pathways for alkane synthesis have been explored, achieving efficient production of structurally similar alkenes, particularly medium- and short-chain ones, remains challenging (Amer et al. 2020). Early attempts at biosynthesizing these alkenes have shown promise but suffer from low substrate conversion efficiencies (Bauer et al. 2020), necessitating further pathway optimization for potential commercialization. Improving enzymatic properties, such as stability and modified substrate specificity, is crucial but traditionally resource-intensive and time-consuming. Studies targeting specific residues in enzymes, like *Synechococcus elongatus* cyanobacterial aldehyde-deformylating oxygenase (cADO), have shown promising results (Bao et al. 2016). By engineering residues near the active sites of enzymes like *Chlorella variabilis* NC64A fatty acid photo decarboxylase (CvFAP) and *Jeotgalicoccus* species, researchers have enhanced substrate specificity for short-chain substrates, leading to increased propane and propene production titers. Despite successes, generating robust enzyme variants often requires multiple

Table 14.1 A list of the microorganisms that have been bioengineered through computational strategies for the production of biofuel from lignocellulose

Microorganisms	Biofuel	References
Saccharomyces cerevisiae	Ethanol, butanol, fatty acids, organic acids	Turner et al. (2018)
Zymomonas mobilis	Bioethanol	Zhang et al. (2019)
Pichia stipitis	Ethanol	Selim et al. (2018)
Lactobacillus plantarum	Biodiesel	Khan et al. (2020)
Bacillus licheniformis	Iso-butanol	Zhan et al. (2021)
Escherichia coli	Ethanol	Majidian et al. (2018)
Clostridium sp.	Syngas	Zhang et al. (2020)
Acinetobacter baylyi	Biodiesel	Luo (2022)
Pseudomonas putida	Butanol, ethanol, biodiesel	Sarwar et al. (2022)
Trichoderma reesei	Ethanol	Druzhinina and Kubicek (2017)
Synechococcus elongates	Isopropanol	Hirokawa et al. (2017)
Rhodococcus opacus	Lipid-based biofuels	Mondal et al. (2024)
Sporomusa ovata	Acetate for electrofuels	Li et al. (2024)
Clostridium thermocellum	Cellulosic ethanol	Pech-Canul et al. (2024)
Aspergillus niger	Bioethanol	Bendaoud et al. (2024)
Yarrowia lipolytica	Lipid-based biodiesel	Sari and Utama (2024)
Mucor circinelloides	Biodiesel	Zhang et al. (2024)
Micractinium sp.	Biodiesel	Khosravinia et al. (2023)

mutations, with destabilization effects sometimes counteracting improvements in substrate binding or catalysis.

To streamline enzyme engineering, computational tools have emerged, leveraging structural bioinformatics and modeling. These tools accelerate the identification of enzyme variants with enhanced properties, offering a pipeline for computational enzyme engineering and design. The pipeline involves structural analysis, enzyme-substrate complex modeling, identification of design positions, stability engineering, and activity/specify engineering, culminating in computational screening of designed variants (Arshad et al. 2023). Understanding enzyme structure-function relationships is fundamental. Crystal structures or predictive modeling provide insights into enzyme dynamics and substrate interactions. Molecular dynamics (MD) simulations can reveal hidden binding pockets and catalytic mechanisms. Subunit assembly and cofactor interactions are also critical considerations. Modeling enzyme-substrate complexes aids in activity and specificity engineering (Arshad et al. 2023). Molecular docking methods predict substrate-binding poses, evolving from rigid-body to flexible and MD-based approaches. Identification of design positions relies on understanding enzyme-substrate interactions and evolutionary conservation analysis (Barghout et al. 2023).

Stability engineering enhances enzyme robustness for industrial applications. Strategies include phylogenetic analysis and Rosetta-based screening for mutations.

The MD simulations and dynamic cross-correlation network analysis inform multipoint mutant design (Ferreira et al. 2022). Activity and specificity engineering focus on residues linked to catalytic centers or binding pockets. Iterative design algorithms like IPRO and FuncLib optimize enzyme-substrate interactions. Computational directed evolution tools, such as Computer-Aided Directed Evolution of Enzymes (CADEE), facilitate mutation screening based on enzymatic activation barriers (Damborsky and Brezovsky 2014). Computational screening accelerates variant identification, complementing experimental efforts. Predictive tools assess stability and binding affinity changes upon mutations. Empirical Valence Bond (EVB) approaches aid in transition state modeling for multipoint mutants (Sharma et al. 2021). Overall, computational enzyme engineering offers a systematic approach to accelerate biocatalyst development, crucial for industrial biofuel production and beyond.

14.4.3 System Biology and Pathway Engineering

Systems biology and pathway engineering represent powerful approaches in synthetic biology for understanding and manipulating biological systems to achieve desired outcomes. Systems biology aims to comprehend the complex interactions within biological systems by integrating computational modeling, data analysis, and experimental techniques. Pathway engineering, on the other hand, focuses on modifying metabolic pathways within organisms to produce valuable compounds, ranging from pharmaceuticals to biofuels (Clomburg and Gonzalez 2010). At the core of systems biology is the holistic view of biological systems, recognizing that individual components function within intricate networks and feedback loops. By leveraging high-throughput omics technologies such as genomics, transcriptomics, proteomics, and metabolomics, researchers can capture comprehensive snapshots of cellular processes (Sundarsingh et al. 2024). Computational tools then analyze these datasets to construct mathematical models that simulate the behavior of biological systems, helping to predict how they respond to various stimuli or genetic perturbations (Minnaar et al. 2024). Pathway engineering builds upon the insights gained from systems biology to reform metabolic pathways for specific purposes. This might involve introducing new enzymes, modifying existing ones, or altering regulatory mechanisms to optimize the production of desired compounds (Subramanian et al. 2024). Rational design approaches use computational modeling and protein engineering techniques to predict and implement changes in metabolic pathways, while directed evolution methods rely on iterative rounds of mutation and selection to evolve enzymes with improved properties.

One of the key applications of pathway engineering is the production of biofuels and biochemicals using microbial cell factories. By engineering microorganisms like bacteria, yeast, or algae, researchers can harness their metabolic capabilities to convert renewable resources such as sugars or fatty acids into fuels like ethanol, biodiesel, or hydrogen, for example, microbial bioengineering, driven by rational design and pathway engineering, has brought about remarkable advancements in biotechnology. Rational design focuses on making precise modifications to microbial genomes, leveraging known genetic and metabolic pathways to achieve desired outcomes. A notable example is the engineering of *Escherichia coli* to produce human insulin by incorporating the insulin gene into its genome, facilitating large-scale production of this vital hormone (Sarwar et al. 2022). Similarly, *Saccharomyces cerevisiae* (yeast) has been genetically modified to enhance ethanol production, a key biofuel, by improving its fermentative pathways. These targeted approaches have laid the foundation for more efficient and productive microbial systems. Pathway engineering, which involves modifying the entire metabolic routes to enhance the synthesis of specific compounds, has also seen considerable success (Selim et al. 2018). In *Corynebacterium glutamicum*, the shikimate pathway has been reengineered to boost the production of aromatic amino acids, while in *Pseudomonas putida,* carbon flux has been redirected to optimize bioplastic production, offering sustainable alternatives to traditional plastics. Additionally, genome-editing technologies such as CRISPR/Cas9 have been instrumental in advancing these efforts, allowing precise genetic modifications to improve antibiotic production in *Streptomyces* and enhance biofuel synthesis in *Clostridium* (Xie et al. 2024). Complementary techniques like metabolic flux analysis and synthetic biology circuits have further refined microbial engineering, optimizing production rates and controlling microbial behavior for more efficient industrial applications (Table 14.2). This process often involves optimizing metabolic pathways for higher yields, substrate utilization, and tolerance to environmental stresses (Erdrich et al. 2014). Another area of pathway engineering is the development of novel therapeutics through the biosynthesis of complex molecules like pharmaceuticals or natural products. By reconstructing or augmenting biosynthetic pathways within host organisms, researchers can produce valuable compounds that are difficult or costly to obtain through traditional chemical synthesis methods (Singh et al. 2024). This approach offers the potential for more sustainable and efficient production of drugs, as well as the discovery of new bioactive molecules through combinatorial biosynthesis strategies. Systems biology and pathway engineering represent complementary approaches in synthetic biology, enabling researchers to gain a deeper understanding of biological systems and manipulate them for various applications (Kang et al. 2024). By integrating computational modeling, experimental techniques, and engineering principles, these disciplines offer powerful tools for designing and optimizing biological systems for a wide range of purposes, from biotechnology and medicine to environmental remediation and beyond.

14.5 Challenges and Future Outlook

In the realm of biofuel production, computational modeling encounters several hurdles. Data integration proves arduous due to the diverse formats, quality, and accessibility of genomics, transcriptomics, proteomics, and metabolomics datasets

Table 14.2 "Comprehensive overview of microbial bioengineering: rational design and pathway engineering approaches"

Microorganisms	Renewable resource	Target biofuel	Metabolic pathway/ process	Engineering strategy	Example application	References
Bacteria (*E. coli*)	Sugars (e.g., glucose)	Ethanol	Glycolysis and fermentation	Overexpression of key enzymes in glycolytic pathways Deletion of competing pathways	Increased ethanol production from lignocellulosic biomass	Zhang et al. (2019)
Bacteria (*Pseudomonas putida*)	Aromatic compounds (e.g., lignin)	Bioplastics	Aromatic degradation pathways	Engineering pathways for aromatic compound utilization – Enhanced polymer production	Production of biodegradable plastics from lignin-based biomass	Minnaar et al. (2024)
Yeast (*Saccharomyces cerevisiae*)	Sugars (e.g., sucrose, glucose)	Bioethanol	Fermentation (ethanol production)	Introduction of alternative enzymes – Optimized fermentation conditions	Commercial bioethanol production for fuel blending	Xie et al. (2024)
Algae (*Chlorella* spp.)	Fatty acids	Biodiesel	Lipid accumulation pathways	Enhancement of lipid biosynthesis genes – Increase in CO2 fixation efficiency	Biodiesel production using microalgae grown in photobioreactors	Ananthi et al. (2021)
Cyanobacteria	CO2 (via photosynthesis)	Hydrogen	Photosynthetic hydrogen production pathways	Genetic modification to optimize hydrogenase activity – Improved light-harvesting complexes	Hydrogen production as a renewable energy source	Erdrich et al. (2014)
Bacteria (*Clostridium acetobutylicum*)	Cellulose, starch	Butanol	Acetone-butanol-ethanol (ABE) fermentation	Enhanced butanol tolerance through stress response genes – Increased conversion efficiency	Butanol as a gasoline alternative for internal combustion engines	Singh et al. (2024)
Yeast (*Yarrowia lipolytica*)	Waste cooking oil	Biodiesel	Lipid metabolism and β-oxidation pathways	Overexpression of lipid biosynthesis genes – Optimization of waste oil conversion processes	Conversion of industrial waste oils into biodiesel for sustainable fuel	Selim et al. (2018)

(Tanaka et al. 2024). While essential for model validation, the scarcity of high-quality experimental datasets restricts the process. Additionally, transitioning from lab-scale models to industrial-scale processes necessitates robust scaling algorithms and computational infrastructure. Accurately estimating model parameters, such as reaction kinetics and enzyme activities, poses a challenge due to limited experimental data availability. Incomplete or inaccurate genomic and biochemical data also hinders the model reconstruction and validation in genome-scale metabolic modeling (Bai et al. 2021). Managing the complexity of large-scale metabolic networks demands efficient algorithms and computational resources. Integrating multiple microbial species into metabolic models increases complexity and requires advanced computational techniques. Although automating the reconstruction process is vital for scalability, achieving it necessitates robust algorithms and tools. The complexity of metabolic networks necessitates advanced computational and experimental techniques for understanding and manipulation. Integrating multi-omics data for strain design and optimization demands sophisticated computational methods (Subramanian et al. 2024). Bridging the gap between computational predictions and experimental validation remains challenging. Ensuring the accuracy of computational predictions for enzyme properties and metabolic flux distributions is crucial for effective strain design.

Ensuring the accuracy of computational models for predicting biomass properties and performance under varying conditions is vital. Integrating diverse data sources, including genetic, biochemical, and environmental data, requires robust computational frameworks. Validating in silico predictions through experimental testing is necessary but challenging due to resource constraints (Samantaray et al. 2023). Developing efficient optimization algorithms for biomass feedstock design and modification is essential for scalability. Managing the computational complexity of molecular simulations and enzyme design algorithms necessitates advanced computational resources. Experimentally validating computationally designed enzyme variants remains challenging due to resource and time constraints. Improving the accuracy of computational predictions for enzyme properties and stability is crucial for successful enzyme engineering (Huang et al. 2024). Integrating structural and functional data for enzyme design and optimization requires sophisticated computational methods. Managing the complexity of large-scale biological networks demands efficient algorithms and computational resources (Sharma et al. 2021). Experimentally validating computational predictions for biological systems remains challenging due to the complexity of biological processes. Integrating diverse omics data for systems biology modeling requires robust computational frameworks. Scaling up computational models to larger biological systems and networks necessitates efficient algorithms and computational infrastructure.

Advancements in multi-omics technologies and computational methods will enable more comprehensive and accurate modeling of biological systems. Similarly, the integration of machine learning and artificial intelligence techniques will enhance the predictive accuracy and scalability of computational models (Damborsky and Brezovsky 2014). Automation of experimental workflows and high-throughput screening techniques will accelerate the validation of computational predictions and strain design. The development of collaborative platforms and open-access databases will facilitate data sharing and collaboration among researchers. Interdisciplinary collaborations between computational biologists, experimental biologists, and engineers will drive innovation and advancements in biofuel production and microbial biosynthesis (Kang et al. 2024). Increased adoption of computational modeling and systems biology approaches by the biotechnology industry will lead to the development of more efficient and sustainable bioenergy and biochemical production processes. The development of regulatory frameworks and standards for computational modeling and systems biology applications in biotechnology will ensure safety, reliability, and scalability of bio-based products and processes.

14.6 Conclusions

The integration of computational modeling, systems biology, and pathway engineering has revolutionized various aspects of bioenergy research and microbial biosynthesis. Through computational modeling, researchers can optimize biofuel production processes by simulating complex biochemical pathways, predicting the behavior of biological systems, and guiding the selection of optimal feedstocks (Bhushan et al. 2023). This approach enables the design and optimization of enzymatic pathways for biofuel synthesis and facilitates the development of efficient bioreactor systems for scalable production. Moreover, genome-scale metabolic modeling has transformed our understanding of microbial ecosystems by elucidating their functional capacity and interactions with the environment (Banerjee et al. 2024). By leveraging mathematical and computational approaches, researchers can simulate microbial metabolism at a systems level, paving the way for the design of novel metabolic pathways and the optimization of microbial strains for industrial applications. Enzyme engineering and metabolomics play crucial roles in enhancing microbial biosynthesis for the sustainable production of various chemicals, including fuels and pharmaceuticals (Sundarsingh et al. 2024). Metabolomics enables the identification of bottleneck reactions in metabolic pathways, guiding enzyme engineering efforts to improve production yield and productivity. The in silico design of biomass feedstocks offers a transformative approach to bioenergy research, allowing researchers to predict and customize the properties of biomass feedstocks for sustainable fuel production (Tanaka et al. 2024). By utilizing computational models, researchers can expedite the discovery and optimization of biomass resources, ultimately contributing to a more efficient and environmentally friendly bioenergy industry.

Furthermore, molecular simulation and enzyme engineering are pivotal in engineering microbial metabolism for the production of fatty acyl compounds, including alkanes and alkenes. Computational tools accelerate the identification of enzyme variants with enhanced properties, offering a systematic approach to accelerate biocatalyst development for industrial biofuel production (Damborsky and Brezovsky

2014). Overall, systems biology and pathway engineering represent powerful approaches in synthetic biology, enabling researchers to understand and manipulate biological systems for diverse applications. By integrating computational modeling, experimental techniques, and engineering principles, these disciplines offer valuable tools for designing and optimizing biological systems for a wide range of purposes, from bioenergy production to pharmaceutical development.

Conflict of Interest The authors declare that they have no known competing financial interests or personal relationships that could have appeared to influence the work reported in this chapter.

References

Amer M, Wojcik EZ, Sun C, Hoeven R, Hughes JM, Faulkner M, Yunus IS, Tait S, Johannissen LO, Hardman SJ (2020) Low carbon strategies for sustainable bio-alkane gas production and renewable energy. Energ Environ Sci 13:1818–1831

Ananthi V, Raja R, Carvalho IS, Brindhadevi K, Pugazhendhi A, Arun A (2021) A realistic scenario on microalgae based biodiesel production: third generation biofuel. Fuel 284:118965

Arshad I, Ahsan M, Zafar I, Sajid M, Sehgal SA, Yousaf W, Noor A, Rashid S, Garai S, Moovendhan M (2023) Bioinformatics approaches in upgrading microalgal oil for advanced biofuel production through hybrid ORF protein construction. In: Biomass conversion and biorefinery, pp 1–21

Asolkar SS, Anju M, Kumar R, Deshmukh A, Ghosalkar A, Kumbhar P (2024) In silico screening and validation of different dehydrogenases to produce 2, 3-butanediol in Bacillus subtilis. Biotechnol Bioprocess Eng 29:1–20

Bai D-P, Lin X-Y, Hu Y-Q, Chen Z-Z, Chen L, Huang Y-F, Huang X-H, Li J (2021) Metagenomics approach to identify lignocellulose-degrading enzymes in the gut microbiota of the Chinese bamboo rat cecum. Electron J Biotechnol 50:29–36

Bajpai P (2019) Third generation biofuels. Springer, Berlin

Banerjee A, Kumar N, Varjani SJ, Guria C, Bandopadhyay R, Shukla P, Banerjee C (2018) Computational modelling and prediction of microalgae growth focused towards improved lipid production. In: Biosynthetic technology and environmental challenges, pp 223–232

Banerjee D, Yunus IS, Wang X, Kim J, Srinivasan A, Menchavez R, Chen Y, Gin JW, Petzold CJ, Martin HG (2024) Genome-scale and pathway engineering for the sustainable aviation fuel precursor isoprenol production in Pseudomonas putida. Metab Eng 82:157

Bao L, Li J-J, Jia C, Li M, Lu X (2016) Structure-oriented substrate specificity engineering of aldehyde-deformylating oxygenase towards aldehydes carbon chain length. Biotechnol Biofuels 9:1–14

Barghout RA, Xu Z, Betala S, Mahadevan R (2023) Advances in generative modeling methods and datasets to design novel enzymes for renewable chemicals and fuels. Curr Opin Biotechnol 84:103007

Bauer D, Zachos I, Sieber V (2020) Production of propene from n-butanol: a three-step cascade utilizing the cytochrome P450 fatty acid decarboxylase OleTJE. ChemBioChem 21:3273–3281

Bendaoud A, Belkhiri A, Hmamou A, Tlemcani S, Eloutassi N, Lahkimi A (2024) Efficient bioethanol production from lignocellulosic biomass using diverse microbial strains. J Ecol Eng 25:353–362

Bharadwaj SV, Ram S, Pancha I, Mishra S (2020) Recent trends in strain improvement for production of biofuels from microalgae. In: Microalgae cultivation for biofuels production. Elsevier, Amsterdam, pp 211–225

Bhushan S, Jayakrishnan U, Shree B, Bhatt P, Eshkabilov S, Simsek H (2023) Biological pretreatment for algal biomass feedstock for biofuel production. J Environ Chem Eng 11:109870

Carbonell P, Le Feuvre R, Takano E, Scrutton NS (2020) In silico design and automated learning to boost next-generation smart biomanufacturing. Synth Biol 5:ysaa020

Clomburg JM, Gonzalez R (2010) Biofuel production in Escherichia coli: the role of metabolic engineering and synthetic biology. Appl Microbiol Biotechnol 86:419–434

Damborsky J, Brezovsky J (2014) Computational tools for designing and engineering enzymes. Curr Opin Chem Biol 19:8–16

Dar MA, Xie R, Zabed HM et al (2024) Termite microbial symbiosis as a model for innovative design of lignocellulosic future biorefinery: current paradigms and future perspectives. Biomass 4:180–201. https://doi.org/10.3390/biomass4010009

del Rio-Chanona EA, Wagner JL, Ali H, Fiorelli F, Zhang D, Hellgardt K (2019) Deep learning-based surrogate modeling and optimization for microalgal biofuel production and photobioreactor design. AICHE J 65:915–923

Druzhinina IS, Kubicek CP (2017) Genetic engineering of Trichoderma reesei cellulases and their production. J Microbial Biotechnol 10:1485–1499

Erdrich P, Knoop H, Steuer R, Klamt S (2014) Cyanobacterial biofuels: new insights and strain design strategies revealed by computational modeling. Microb Cell Fact 13:1–15

Ferreira P, Fernandes PA, Ramos MJ (2022) Modern computational methods for rational enzyme engineering. Chem Catal 2:2481–2498

Garg S, Devi S, Sharma A, Wani AK (2021) An insight into therapeutic profile of lectins in cancer and inflammation. Int J Res Edu Sci Methods 9:647–654

Godbole V, Pal MK, Gautam P (2021) A critical perspective on the scope of interdisciplinary approaches used in fourth-generation biofuel production. Algal Res 58:102436

Gu S, Zhu F, Zhang L, Wen J (2024) Mid–long chain dicarboxylic acid production via systems metabolic engineering: progress and prospects. J Agric Food Chem 72:5555–5573

Gupta KM, Aitipamula S, Chow PS (2023) In-silico design of covalent organic framework membranes for efficient water/ethanol separation. J Membr Sci 687:122083

Hari A, Doddapaneni TRKC, Kikas T (2024) Common operational issues and possible solutions for sustainable biosurfactant production from lignocellulosic feedstock. Environ Res 251:118665

Hebbale D, Chandran MS, Joshi N, Ramachandra T (2017) Energy and food security from macroalgae. J Biodivers 8:1–11

Heger S, Brendt J, Hollert H, Roß-Nickoll M, Du M (2021) Green toxicological investigation for biofuel candidates. Sci Total Environ 764:142902

Hirokawa Y, Dempo Y, Fukusaki E, Hanai T (2017) Metabolic engineering for isopropanol production by an engineered cyanobacterium, Synechococcus elongatus PCC 7942, under photosynthetic conditions. J Biosci Bioeng 123:39–45

Holtz M, Acevedo-Rocha CG, Jensen MK (2024) Combining enzyme and metabolic engineering for microbial supply of therapeutic phytochemicals. Curr Opin Biotechnol 87:103110

Huang J, Xie X, Zheng W, Xu L, Yan J, Wu Y, Yang M, Yan Y (2024) In silico design of multipoint mutants for enhanced performance of Thermomyces lanuginosus lipase for efficient biodiesel production. Biotechnol Biofuels Bioprod 17:33

Jones GB, Sims RC, Zhao J (2024) Computational modeling of microalgal biofilm growth in heterogeneous rotating algal biofilm reactors (RABRs) for wastewater treatment. App Math Model 131:487

Kaladhar DS, Kenar SK, Meesala S, Sahu S (2023) Biofuel production, phytochemical analysis and in silico analysis for active compound valorization from Jatropha Gossypifolia seeds. J Pharm Negat Results:1555–1565

Kang M-K, Yoon S-H, Kwon M, Kim S-W (2024) Microbial cell factories for bio-based isoprenoid production to replace fossil resources. Curr Opin Syst Biol 37:100502

Kaur D, Gupta S (2024) Biosynthetic pathways involved in the synthesis of next-generation biofuels in microbial cells. Process Biochem 143:248

Kavitha S, Ravi YK, Kumar G, Nandabalan YK (2024) Microalgal biorefineries: advancement in machine learning tools for sustainable biofuel production and value-added products recovery. J Environ Manage 353:120135

Khan I, Ganesan R, Dutta JR (2020) Probiotic lipase derived from lactobacillus plantarum and lactobacillus brevis for biodiesel production from waste cooking olive oil: an alternative feedstock. Int J Green Energy 17:62–70

Khosravinia S, Malekzadeh-Shafaroudi S, Bagheri A, Behdad A, Moshtaghi N (2023) Bioprospecting of ten microalgae species isolated from saline water lake for evaluation of the biodiesel production. Bioenergy Res 17:1090–1103. https://doi.org/10.1007/s12155-023-10707-2

Li Y, Cao M, Gupta VK, Wang Y (2024) Metabolic engineering strategies to enable microbial electrosynthesis utilization of CO_2: recent progress and challenges. Crit Rev Biotechnol 44:352–372. https://doi.org/10.1080/07388551.2023.2167065

Liu Q, Xun G, Feng Y (2019) The state-of-the-art strategies of protein engineering for enzyme stabilization. Biotechnol Adv 37:530–537

Liu Y, Zhang D, Yang T, Chen R, Luo X (2024) Pathway engineering of plant-derived bioactive compounds in microbes. In: Engineering biology for microbial biosynthesis of plant-derived bioactive compounds. Elsevier, Amsterdam, pp 73–87

Luo J (2022) Engineering Acinetobacter baylyi ADP1: new tools and strategies for enhanced production of chemicals using non-conventional substrates. Tampere University

Majidian P, Tabatabaei M, Zeinolabedini M, Naghshbandi MP, Chisti Y (2018) Metabolic engineering of microorganisms for biofuel production. Renew Sustain Energy Rev 82:3863–3885

Minnaar LS, Kruger F, Fortuin J, Hoffmeester LJ, den Haan R (2024) Engineering Saccharomyces cerevisiae for application in integrated bioprocessing biorefineries. Curr Opin Biotechnol 85:103030

Mondal S, Halder SK, Mondal KC (2024) State-of-art engineering approaches for ameliorated production of microbial lipid. Syst Microbiol Biomanuf 4:20–38. https://doi.org/10.1007/s43393-023-00195-y

Nanda S, Azargohar R, Dalai AK, Kozinski JA (2015) An assessment on the sustainability of lignocellulosic biomass for biorefining. Renew Sustain Energy Rev 50:925–941

Ng I, Tan S, Kao P, Chang Y, Chang J (2017) Recent developments on genetic engineering of microalgae for biofuels and bio-based chemicals. Biotechnol J 12:1600644

Pahima E, Hoz S, Ben-Tzion M, Major DT (2019) Computational design of biofuels from terpenes and terpenoids. Sustain Energy Fuels 3:457–466

Pech-Canul A, Hammer SK, Ziegler SJ, Richardson ID, Sharma BD, Maloney MI, Bomble YJ, Lynd LR, Olson DG (2024) The role of AdhE on ethanol tolerance and production in Clostridium thermocellum. J Biol Chem 300:107559

Ranganathan S, Mahesh S, Suresh S, Nagarajan A, Sen Z, Yennamalli RM (2022) Experimental and computational studies of cellulases as bioethanol enzymes. Bioengineered 13:14028–14046

Renzaho AM, Kamara JK, Toole M (2017) Biofuel production and its impact on food security in low and middle income countries: implications for the post-2015 sustainable development goals. Renew Sustain Energy Rev 78:503–516

Rong Y (2023) Development of synthetic biology tools for growth decoupled production and protein expression. Technical University of Denmark

Rulli MC, Bellomi D, Cazzoli A, De Carolis G, D'Odorico P (2016) The water-land-food nexus of first-generation biofuels. Sci Rep 6:22521

Samantaray B, Behera RR, Mishra RR, Thatoi H (2023) In silico studies of fungal xylanase enzymes: structural and functional insight towards efficient biodegradation of lignocellulosic biomass. Syst Microbiol Biomanuf 4:1–19

Sari MP, Utama GL (2024) The potential of Yarrowia lipolytica in converting bioenergy resources: a preliminary review. In: E3S web of conferences, vol 495. EDP Sciences, Les Ulis, p 01006

Sarwar A, Nguyen LT, Lee EY (2022) Bio-upgrading of ethanol to fatty acid ethyl esters by metabolic engineering of pseudomonas putida KT2440. Bioresour Technol 350:126899

Selim KA, El-Ghwas DE, Easa SM, Abdelwahab Hassan MI (2018) Bioethanol a microbial biofuel metabolite; new insights of yeasts metabolic engineering. Fermentation 4:16

Sewsynker-Sukai Y, Faloye F, Kana EBG (2017) Artificial neural networks: an efficient tool for modelling and optimization of biofuel production (a mini review). Biotechnol Biotechnol Equip 31:221–235

Sharma A, Gupta G, Ahmad T, Mansoor S, Kaur B (2021) Enzyme engineering: current trends and future perspectives. Food Rev Int 37:121–154

Singh VK, Jha S, Rana P, Soni R, Lalnunpuii R, Singh PK, Sinha RP, Singh G (2024) Cyanobacteria as a biocatalyst for sustainable production of biofuels and chemicals. Energies 17:408

Subramanian B, Basak S, Thirumurugan R, Saleena LM (2024) Metabolic flux analysis and metabolic engineering for polyhydroxybutyrate (PHB) production. Polym Bull 81:1–20

Sundarsingh TJA, Ameen F, Ranjitha J, Raghavan S, Shankar V (2024) Engineering microbes for sustainable biofuel production and extraction of lipids–current research and future perspectives. Fuel 355:129532

Tanaka K, Bamba T, Kondo A, Hasunuma T (2024) Metabolomics-based development of bioproduction processes toward industrial-scale production. Curr Opin Biotechnol 85:103057

Turner TL, Kim H, Kong II, Liu J-J, Zhang G-C, Jin Y-S (2018) Engineering and evolution of Saccharomyces cerevisiae to produce biofuels and chemicals. Synth Biol Eng 162:175–215

Twigg G, Fedenko J, Hurst G, Stanley MS, Hughes AD (2024) A review of the current potential of European brown seaweed for the production of biofuels. Energy Sustain Soc 14:21

van Lent P, Schmitz J, Abeel T (2023) Simulated design–build–test–learn cycles for consistent comparison of machine learning methods in metabolic engineering. ACS Synth Biol 12:2588–2599

Vasileiadis V, Kyriklidis C, Karayannis V, Tsanaktsidis C (2024) Application of evolutionary computation to the optimization of biodiesel mixtures using a nature-inspired adaptive genetic algorithm. Algorithms 17:181

Wani MH, Baba SH, Baht A (2019) Economic appraisal of water-ecosystem in Jammu and Kashmir: India. Int J Environ Clim Change 9:193–203

Wani AK, Akhtar N, Chopra C, Américo-Pinheiro JHP, Quadir M, Yadav KK, Kumar P, Indrayani I, Krismawati A, Herwati A (2023a) Exploring the world hot springs: a bibliometric analysis of global trends in metagenomics research. Curr Res Biotechnol 6:100161

Wani AK, Akhtar N, Prakash A, Garg SS, Singh R (2023b) Karrikins: molecular association with strigolactones: role in plant growth regulation and quorum sensing modulation. In: Strigolactones, alkamides and Karrikins in plants. CRC, Boca Raton, FL, pp 209–224

Wani AK, Chopra C, Singh R, Ahmad S, Américo-Pinheiro JHP (2023c) Mining microbial tapestry using high-throughput sequencing and in silico analysis of Trehalose synthase (TreS) derived from hot spring metagenome. Biocatal Agric Biotechnol 52:102829

Wani AK, Chopra C, Ansari MA, Dar MA, Américo-Pinheiro JHP, Singh R (2024a) Characterization of thermostable carboxypeptidase from high-altitude hot spring metagenome. Int J Biol Macromol 276:133974

Wani AK, Khan Z, Sena S, Akhtar N, Alreshdi MA, Yadav KK, Alkahtani AM, Rahayu F, Tafakresnanto C, Latifah E (2024b) Carbon nanotubes in plant dynamics: unravelling multifaceted roles and phytotoxic implications. Plant Physiol Biochem 210:108628

Wani AK, Rahayu F, Ben Amor I, Quadir M, Murianingrum M, Parnidi P, Ayub A, Supriyadi S, Sakiroh S, Saefudin S (2024c) Environmental resilience through artificial intelligence: innovations in monitoring and management. Environ Sci Pollut Res 31:18379–18395

Xie R, Danso B, Sun J et al (2024) Biorefinery and bioremediation strategies for efficient management of recalcitrant pollutants using termites as an obscure yet promising source of bacterial gut symbionts: a review. Insects 15:908

Zengler K, Palsson B, Lewis N (2017) Next-Gen3: sequencing, modeling, and advanced biofuels-final technical report. University of California, San Diego, CA

Zhan Y, Xu Y, Lu X, Zhou F, Zheng P, Wang D, Cai D, Yang S, Chen S (2021) Metabolic engineering of bacillus licheniformis for sustainable production of isobutanol. ACS Sustain Chem Eng 9:17254–17265

Zhang K, Lu X, Li Y, Jiang X, Liu L, Wang H (2019) New technologies provide more metabolic engineering strategies for bioethanol production in Zymomonas mobilis. Appl Microbiol Biotechnol 103:2087–2099

Zhang L, Zhao R, Jia D, Jiang W, Gu Y (2020) Engineering clostridium ljungdahlii as the gas-fermenting cell factory for the production of biofuels and biochemicals. Curr Opin Chem Biol 59:54–61

Zhang Y, Liu Z, Sun Y, Du Y, Zhao Z, Liu Q, Song Y (2024) Lipid production from corn straw by cellobiohydrolase and delta-6 desaturase engineered Mucor circinelloides strains under solid state fermentation. Sci Rep 14:18784

Chapter 15
Fourth-Generation Biofuels: Genetic Engineering for Sustainable Energy

Sunil P. Bhagat, Balasaheb D. Wagh, Savita M. Satpute, Mahesh B. Gunjal, Shreya K. Kasar, and Praveen B. Roylawar

Abstract Increasing costs of crude oil act as a driving force for developing novel methods to produce biofuels from feedstocks. The genetic engineering of the plants that provide stock for biofuels will also be required to make biofuels viable and cost-effective. The fourth-generation biofuels hold the key for renewable energy from feedstocks that do not compete with the food chain. Bioenergy production by cellulose hydrolysis and microbial synthesis calls for unique optimization of sugars and metabolic intermediates. Today, plant biologists and microbiologists are creating transgenic plants with changed, energy storing cell walls and changed reproductive organs. This type of functional reorganization for directing carbon flow through specific pathways or for directly constructing the final bioenergy compounds is the focus for this new generation of dedicated bioenergy crops. In the present chapter, an attempt was made to overview the fourth-generation biofuels under the following lines: (1) to enlist key characteristics of fourth-generation biofuels, (2) to discuss the genetic engineering techniques for biofuel production, (3) to highlight the biofuel production from genetically engineered microorganisms, (4) to focus on ligno-

Sunil P. Bhagat and Balasaheb D. Wagh have contributed equally and are both first authors.

S. P. Bhagat
Department of Botany, ADTs Sharadabai Pawar Mahila Arts, Commerce and Science College, Shardanagar Baramati, Maharashtra, India

B. D. Wagh · M. B. Gunjal
Department of Botany, Sahakar Maharshi Bhausaheb Santuji Thorat Arts, Science and Commerce College, Sangamner, Ahilyanagar, Maharashtra, India

S. M. Satpute
Department of Botany, Shankarrao Mohite Mahavidyalaya, Akluj, Solapur, Maharashtra, India

S. K. Kasar · P. B. Roylawar (✉)
Department of Botany, S.N. Arts, D.J.M. Commerce and B.N.S. Science College (Autonomous), Sangamner, Ahilyanagar, Maharashtra, India
e-mail: roylawar@sangamnercollege.edu.in

M. A. Dar et al. (eds.), *Recent Trends in Lignocellulosic Biofuels and Bioenergy*, Clean Energy Production Technologies,
https://doi.org/10.1007/978-981-96-4636-4_15

cellulosic biofuel enhancement through genetic engineering, (5) to overview the genetic engineering for lipid-based biofuels, (6) to discuss the potential role of synthetic biology and artificial pathways for fourth-generation biofuel production, and (7) to suggest future prospective of fourth-generation biofuels.

Keywords Genetic engineering · Fourth-generation biofuels · Synthetic biology · Microbial synthesis · Sustainable energy

15.1 Introduction

Biofuels have evolved significantly over the decades, progressing through multiple generations to address environmental, economic, and technological challenges (Liew et al. 2014). The journey started with the first-generation biofuels which are mainly produced from food crops such as corn, sugarcane, and soy beans (Alalwan et al. 2019). Bioethanol and biodiesel are some of the biofuels that were and are still seen as an effort to replace fossil fuels (Adewuyi 2020). However, with the competition with food production, use of land, and their general sustainability, their sustainability was an issue that kept the program from being sustainable (Renzaho et al. 2017). To address these challenges, second-generation biofuels developed as these biofuels rely on nonfood feedstocks such as agricultural and forestry residues and energy crops (Van-Eijck et al. 2014; Singh et al. 2020). Some second-generation biofuels, for instance, cellulosic ethanol, were able to avoid the food or fuel dilemma by utilizing waste materials and unused land (Guo and Lü 2021). All were considered offering a better approach towards energy production, but technological difficulties in biomass prospecting and processing, high production costs, and low conversion ratios hindered their use (Cheah et al. 2020; Asghar et al. 2022). The third generation of biofuels was a step ahead by concentrating on algae and other microorganisms for biofuels (Alaswad et al. 2015). Some species of algae yield high lipids, carbohydrates, and proteins suitable for biodiesel, bioethanol and biohydrocarbons. These algae can grow in places that are not suitable for agriculture and grow in, or use, wastewater, or CO2 emission is advantageous (Chowdhury and Loganathan 2019). As a result, third-generation biofuels have numerous advantages; however, there are some issues which hinder their commercialization, for instance, high cultivation costs, low efficiency in harvesting, plus the necessity of extensive infrastructure (Paravantis 2022; Maliha and Abu-Hijleh 2023). To tackle such hurdles, the solutions of fourth-generation biofuels have come into being using the know-how of genetic engineering and synthetic biology that help to redesign biofuel production (Mat Aron et al. 2020; Maliha and Abu-Hijleh 2023). In contrast to the prior generations of biofuels, the fourth generation of biofuels is to incorporate the employment of genetically modified organisms GMOs and state of the art biotechnological tools with the aim of eliciting enhanced biofuel yields rates, higher efficiency, and decreased impacts on the environment (Cavelius et al. 2023). These biofuels are intended to be carbon-neutral or carbon-negative through injecting carbon dioxide into biofuel production process (Yang et al. 2021; Moioli and Schildhauer 2022). The primary characteristic of fourth-generation biofuels is the concept of genetically engineered microorganisms, plants, and algae for improved biofuel yield (Shokravi et al. 2019). Changing metabolic pathways allows for producing more of the needed biofuels like the bioethanol, biobutanol, or biodiesel (Kim and Reed 2010; Immethun et al. 2016). For example, through genetic engineering, algae can contain improved levels of lipids (Radakovits et al. 2010), while yeasts strains have the ability to metabolize sugars to ethanol (Adebami et al. 2022). Such precision leads to increase in yields, less time taken during processing as compared to conventional methods and therefore cheaper. These capabilities are expanded upon by synthetic biology which creates brand new routes for biosynthesis or artificial organisms that could generate biofuels (Cavelius et al. 2023). Using synthetic biology, one can engineer microorganisms that would select CO2 from the atmosphere and converts it into hydrocarbons as a promising mitigation measure against greenhouse gases emissions (DeLisi et al. 2020; Shokravi et al. 2022; Onyeaka and Ekwebelem 2023). These leaps are making the biofuels to be more sustainable with the added advantage of being climate change fighters. The reason for fourth-generation biofuels is a necessity to solve more urgent, global problems (Shokravi et al. 2019). As the climate changes persistently, there is a burning question to shift from the use of fossil fuels to renewables (Arutyunov and Lisichkin 2017; Holechek et al. 2022). Fourth generation has provided the best answer for it provides renewable, scalable have striking characteristics of biofuels that make the fuel sources safe and sustainable (Thirunavukkarasu et al. 2023). It can decrease reliance on scarce and dwindling fossil resources, minimize greenhouse gas emissions, and advance energy security (Rial 2024). In addition, the usage of these biofuels builds a circular bioeconomy system, in which waste products and CO2 emissions are thus turned into useful fuels (Venkata Mohan et al. 2019; Narisetty et al. 2023). Fourth-generation biofuels utilize both genetic engineering and synthetic biology to offer a value added for efficient generation of energy while sticking to the concept of achieving net zero emissions in consonance with the international climate goals. Recently, a group of researchers (Malik et al. 2024) provided comparative account of the different generations of biofuels in Fig. 15.1 for better understanding.

In the present chapter, an attempt was made to overview the basic concept, the need of fourth-generation biofuels, and the role played by genetic engineering and synthetic biology to improve the microbes and plant for enhanced and sustainable production of biofuels.

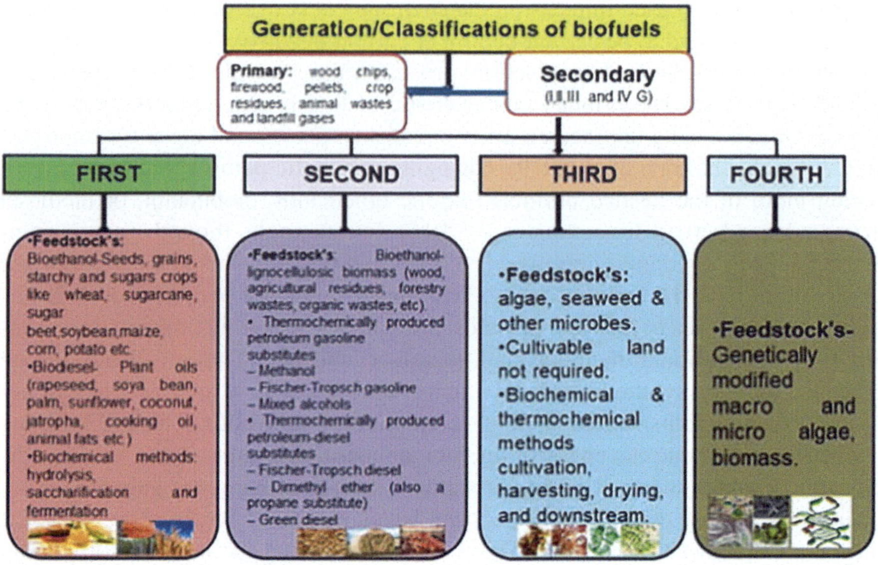

Fig. 15.1 A schematic representation of different generations of biofuels with feedstock profiles. (Source: Malik et al. (2024))

15.2 Fourth-Generation Biofuels

Carbon-neutral and carbon-negative potential: Fourth-generation biofuels are prospective and promising solutions on the path to creating sustainable energy sources to combat climate change (Kumar et al. 2021). Biofuels have the potential to produce no net CO_2 Energy Management System (EMS) or even negative CO_2 EMS to the environment (Abdullah et al. 2016). This means they can sequester more CO_2 from the atmosphere than can be emitted from its fabrication and utilization. It includes microorganisms that have been bioengineered; algae and bacteria are some examples of microorganisms that navigate through the system to capture the CO_2 in forms of carbon sinks through photosynthesis (Zahed et al. 2021). The ability to store carbon dioxide also makes fourth-generation biofuels an energy source that helps decrease climate change and emissions of carbon dioxide (Rahman et al. 2017). Currently, methods in genetic engineering and modification of the organism have developed complex ways of creating GMOs used for producing traits that are suitable for biofuels (Shokravi et al. 2022). Application of biotechnology, particularly the use of genetically modified organisms, is central to the manufacture of fourth-generation biofuels (Shokravi et al. 2019; Cavelius et al. 2023). Biotechnological techniques, such as CRISPR-Cas9, have enabled scientists to create GMOs with enhanced traits specifically tailored for biofuel production (Garg et al. 2023). For instance, the genetically modified algae and cyanobacteria have high lipid content which is required for biofuel production (Khoo et al. 2023). These modifications do not only enhance yield of biofuels but also enhance productivity of the production process. When the metabolic pathways of these organisms are fine-tuned, then desired higher growth rates, increased biomass production, and enhanced efficiency of feedstock conversion to usable energies can be obtained (Johnson and Wen 2009).

15.2.1 Higher Production Efficiency and Lower Impact on the Environment

The improvement of production efficiency is the primary benefit of the fourth-generation biofuels (Mat Aron et al. 2020). Genetic engineering has the potential of altering the metabolic pathways of organisms, in a bid to ensure optimum production of biofuels with minimal by-product (Adegboye et al. 2021). They also get to provide better energies for every unit feedstock enhancing the overall efficiency and likelihood of the production process (Sticklen 2008). In the same way, the use of GMOs enables the use of undeveloped or marginal land in the desert or saline soils for biofuel production (Adenle et al. 2013). This in turn minimizes competition with food crops and also ensures sane levels of genetic variance (Baumber 2016). Furthermore, the fourth-generation biofuels are relatively more friendly to the environment as compared to the fossil fuels (Mat Aron et al. 2020). These burn with less emissions of undesirable gases like sulfur as well as nitrogen oxides when compared to other fossil fuels hence improving the quality of biofuels and impose relatively small impact on the health of human beings and other living organisms (Neupane 2023).

15.2.2 Feedstocks Suitable for Production of Fourth-Generation Biofuels

Fourth-generation biofuels take feedstocks that are chosen based on sustainability and efficiency (Lin and Lu 2021). Some typical feedstocks are algae, cyanobacteria, and other photosynthetic microorganisms (Lalman et al. 2017). These organisms have efficient photosynthetic potential and are known to be raised in water, freshwater, seawater, and wastewater treatment systems (Abdullah et al. 2019). Moreover, plants genetically modified and agricultural residues are also likely to be utilized as feedstocks of fourth generations of biofuel (Shaah et al. 2021). For instance, the switchgrass and miscanthus that are common as modified energy crops can be grown on prodigal lands and generate biomass in huge qualities (Howe et al. 2023). The leftover matter from plants like corn and wheat known as corn stover and wheat straw, respectively, can be used to produce biofuels, generate value for waste, and create an extra stream of wealth for the farmers (Bakala et al. 2023).

15.3 Application of Genetic Engineering Tools for Biofuel Production

15.3.1 CRISPR-Cas9 Technology for Biofuel Production

Clustered regularly interspaced short palindromic repeats (CRISPR)-CRISPR-associated protein 9 (Cas9) is one of the most exciting tools in genetic engineering to edit genes in microbes and plants to enhance biofuel production and has many other uses (Lakhawat et al. 2022; Garg et al. 2023). With these techniques, it is possible to get hold of the modification protocol in an organism's DNA and eradicate any undesirable facet or improve on the favorable ones (Potts et al. 2023). For example, genes can be added to algae that increase its lipid content, improve the rate of fermentation a yeast, or make bacterium capable of producing biofuels from other substances (Radakovits et al. 2010). The feature of high precision and a significantly lower level of off-target effects of CRISPR-Cas9 allows for the use of the method in the development of new strains of microorganisms capable of producing advanced biofuels (Garg et al. 2023; Hemalatha et al. 2023).

15.3.2 Synthetic Biology Tools and Pathways for Biofuel Production

Synthetic biology is the process of building new biological parts, devices, and systems from the scratch with a view of achieving useful objectives, or modifying functioning natural biological systems to put them into other useful uses (Ausländer et al. 2017; Yue et al. 2023). In the context of biofuel synthesis, actual synthetic biology tools and pathways are used to modify microorganisms having designed metabolic pathways (Jagadevan et al. 2018). This can include creating artificial switch that controls generation of biofuels, designing new pathways that improve the rate at which biofuels are synthesized, or developing new gene modules from parts of different organism to produce new types of organisms with better performance (Nath 2024). Synthetic biology allows the researchers to design microorganisms that are faster, more adaptable, and stable in the synthesis of sustainable biofuels (Yang et al. 2022).

15.3.3 Biochemical Engineering for Biofuel Production

Metabolic engineering is defined as the enhancement of some cellular processes to enhance generation of a compound like biofuels (Majidian et al. 2018). This is achieved by changing the flux distribution of substrates within microorganisms to the desired biofuels (Banerjee et al. 2016). For instance, the chemical pathways by which algae's lipids are used in making biodiesel are known, and they can be synthesized humanely by using efficient and ethical means that ensure least or no harm to the environment or other systems (Dickinson et al. 2017). In bacteria, metabolic engineering can enhance the efficiency of conversion of sugar to bioethanol (Zhang et al. 2019), whereas in yeast it can further enhance the efficiency of fermentation (Adebami et al. 2022). Thus, metabolic engineering strategies include techniques like overexpression of certain genes, knockouts of certain genes, and the addition of new biosynthesis pathways, all of which are designed to drive the efficiency of biofuel production (Das et al. 2020).

15.3.4 Microbial Strain Improvement for Bacteria, Algae, and Yeasts for Biofuel Production

The microbial strain enhancement process aims to invest in and design microorganisms solely for biofuel applications (Adegboye et al. 2021). CRISPR-Cas9, synthetic biology, and metabolic engineering are integrated in this process to design biofuel production strains of high efficiency. The modification of bacterial strains is reported to convert a variety of substrates, including lignocellulosic biomass or industrial waste, into biofuels (Kumar et al. 2017; Ashokkumar et al. 2022). Organisms such as algae can undergo genetic manipulation to enable the organism to have a high ability to perform photosynthesis, contain high lipids, and be actively resistant to stress (Ranjbar and Malcata 2022; Sreenikethanam et al. 2022). Through recombinant DNA, yeast can be modified to metabolically target diverse sugar types and to resist toxicity of biofuels (Adebami et al. 2022). Through the microbial engineering, researchers aimed to produce a durable biofuel manufacturing process that is effective under different circumstances and feedstock.

15.4 Biofuel Generation Using Microorganisms

15.4.1 Modification in Bacteria for Biofuels Production

Bioethanol and biobutanol are prepared from the biomass by the usage of genetically engineered bacteria in processes referred to as fermentation (Verardi et al. 2019). Bioethanol is normally produced through the conversion of sugar by ferments like *Escherichia coli* (Wagner et al. 2022) and *Zymomonas mobilis* (Zhang et al. 2019). Lignocellulosic biomass conversion has been improved by genetically modifying these bacteria to efficiently metabolize the different feedstocks by employing the CRISPR-Cas9 technology for the enhance uptake of sugars and ethanol production (Joshi et al. 2022). Biobutanol production is also reported by *Clostridium acetobutylicum* (Ranjan et al. 2013) and *Clostridium beijerinckii* (Tekin et al. 2023). These bacteria engage in acetone-butanol-ethanol (ABE) fermentation process. Biological engineering concepts have enabled specification of these pathways in order to improve butanol production. There are new modifications that enhance the activities of butanol synthesizing enzymes and decrease the secretion of secondary metabolites, such as acetone (Veza et al. 2021). Furthermore, biobutanol produced with these bacteria has been produced in a more tolerated genetic modification in order to accommodate high concentrations of the chemical compound in anticipation of marketing the chemical to the market since compared to bioethanol; biobutanol is toxic to microbial cells (Mitra et al. 2024).

15.4.2 Production of Higher Yield Biofuel Through Genetic Modification of Yeast Strains

Saccharomyces cerevisiae (yeast) plays a big role in making biofuels because it ferments stuff and converts it into biofuel (Miah et al. 2022). *S. cerevisiae* was modified using genetic engineering with enhanced potential to breakdown sugars like xylose and arabinose (Sonderegger and Sauer 2003; Naghshbandi et al. 2019). These by-products are usually produced from plant materials that normal yeasts not able to produce (Fehér 2018). Researchers reported that the use certain enzymes such as xylose isomerase by yeast to gobble up these sugars and spit out ethanol (Seike et al. 2019). Yeast, especially *S. cerevisiae*, is significant in preparation of biofuels as regards the competency in fermentation. Several genetically modified yeasts clone have already been designed to enhance their ability to transform biomass to biofuel (Sedlak and Ho 2004; Tai and Stephanopoulos 2013). A major aspect of genetic engineering is to improve the yeast to ferment sugars obtained from the hemicellulose, such as xylose and arabinose, which are not metabolized by the *S. cerevisiae* in-frame. Therefore, the scientists have been able to modify and acquire high marginal yeast strains that can ferment the sugars such as xylose into ethanol by optimizing the xylose isomerase and other related enzymes (Kwak and Jin 2017). In addition, metabolic engineering has potential in enhancing the yeast cells resistance to ethanol since this metabolite when accumulated in high concentration toxic to cellular processes (Naghshbandi et al. 2019).

Several strategies for the improvement of the overall efficiency of the pathway, which converts biomass into biofuels, involve optimizing the pathway (Jones et al. 2015). The creation of synthetic biology is a systemic approach toward the optimization of the metabolic routes in microorganisms as fuel producers (Jagadevan et al. 2018). One idea is to use synthetic biology to build artificial routes to guide substrate flux around these limitations and into biofuel formation (Keasling et al. 2021). For example, scientists can insert synthetic routes in a microorganism that directs more carbon flow toward the production of ethanol or butanol and less flow toward other unneeded routes that will produce waste products. The other strategy is to increase the gene expression of enzymes that can enhance the synthesis of biofuel (Singh et al. 2022). This can be done using such methods as the use of different strength promoters; promoter engineering is the use of promoters of different strengths to control the strength of genes. Moreover, an application of dynamic regulatory mechanisms that control the synthesis of enzymes dependent on the fluctuations in metabolic status can also improve the pathway's performance (Liu et al. 2018). Furthermore, two or more genetically different microorganisms can be cultured together where the abilities of one microorganism will complement the other (Bertrand et al. 2014). For instance, a system of engineered bacteria and yeasts can be applied for the conversion of composite structural biomass of lignocellulose to simple sugars and its subsequent conversion to biofuels. This synergistic mechanism can enhance the total conversion efficiency and the yields of the biosources into biofuels.

15.5 Lignocellulosic Biofuel Enhancement Through Genetic Engineering

15.5.1 Bioconversion of Biomass and the Challenge of Biomass Recalcitrance Through Engineered Enzymes

Biomass, made of cellulose, hemicellulose, and lignin, is a promising but difficult feedstock for biofuel production (Williams et al. 2016; Lin 2022). One of the biggest challenges in using this biomass is the biomass recalcitrance, the plant cell walls being very resistant to degradation (Kalluri et al. 2014). This challenge has however been tackled in the genetic engineering field through the development of engineered enzymes which favorably degrades such structural complexities into fermentable sugars (Bhatia et al. 2024). Among these enzymes, the cellulases and hemicellulases are of significant importance. These enzymes are thereby activated and stabilized under industrial circumstances through genetic manipulations. For instance, enzymes can be modified to tolerate high temperatures and fluctuating pH which is a norm in bioreactor systems (Mitra and Murthy 2022). The development of key enzymes like the multi-modular enzymes which include both cellulases and hemicellulases has been seen to improve the rate at which plant biomass can be degraded (Fan et al. 2021). Furthermore, the protein engineering methodologies such as directed evolution and rational design are employed to develop enzymes with enhanced binding capability and rate of catalytic reaction (Xiong et al. 2021). When these enzymes are improved, the transformation of lignocellulosic biomass to sugars is greatly enhanced, and less funding is required to buy hazardous chemicals for pretreatments. It not only increases the total production of biofuels, but the efficiency of its production makes them easy on the pocket, ecological, and sustainable.

15.5.2 Genetically Modified Plants for Reduced Lignin Content

Lignin, a complex polymer in plant cell walls, provides structural support and resistance to microbial attacks (Zhu et al. 2023). However, its presence poses a significant barrier to the efficient conversion of lignocellulosic biomass into biofuels (Ashokkumar et al. 2022). Genetic engineering offers a solution by creating genetically modified plants with reduced lignin content or altered lignin composition, making the biomass easier to process (Liu and Eudes 2022). Techniques such as RNA interference (RNAi) and CRISPR-Cas9 have been employed to knock down or knockout genes responsible for lignin biosynthesis in plants. For instance, reducing the expression of key enzymes like cinnamate-4-hydroxylase (C4H) and 4-coumarate-CoA ligase (4CL) leads to a significant reduction in lignin content(Li et al. 2022). Furthermore, modifying the composition of lignin by altering the ratio of its monomeric units (such as syringyl to guaiacyl lignin) can make the lignin more amenable to chemical and enzymatic degradation (Schutyser et al. 2018). The production of these genetically modified plants not only facilitates the easier breakdown of biomass but also increases the overall yield of fermentable sugars. These modifications have been successfully implemented in some energy crops, viz., switchgrass (Shen et al. 2013), and miscanthus (Malinská et al. 2021), demonstrating enhanced biofuel production efficiencies. The reduction in lignin content also results in lower energy and chemical input requirements for biomass processing, further promoting the sustainability of biofuel production (Zhu and Zhuang 2012).

15.6 Genetic Engineering for Lipid-Based Biofuels

15.6.1 Enhancing Capability of Lipid Storage in Engineered Microalgae and Plants

The engineering of algae and plants to enhance lipid formation has been identified as one of the most viable strategies for enhancing production of the lipid-based biofuels (Wang et al. 2022a). Fats, particularly triacylglycerols (TAG), are the major feedstock for biodiesel production (Andeden et al. 2021). Having high photosynthetic efficiency algae is most suitable for lipid productivity (Benvenuti et al. 2015). Genetic improvements in algae are targeting fatty acid synthesis while avoiding pathways that potentially divert carbon/energy into parallel sink pathways. To increase lipid yields, it is common for key synthetic genes such as ACC, DGAT, and GPDH to be overexpressed (Xue et al. 2017; Ranjbar and Malcata 2022). They include gene silencing by RNA interference or RNA interference simply known as RNAi which is used to suppress genes that are responsible for synthesizing starch and carbohydrates to produce more carbon for lipids. For illustration, suppression of the activity of ADP glucose pyrophosphorylase (AGPase) contributes to enhanced lipid content (Xu et al. 2019). Another advancement in algal genetic engineering is the enhanced CRISPR-Cas9 technique which adopts more precise as well as efficient genetic changes in algal genomes (Patel et al. 2023). In plants, genetic engineering strategies plan towards cell engineering to enhance lipid content from vegetative segments. Well-known oilseed crops such as soybean and rapeseed are oilseed crops where lipids are accumulated in the seeds (Savadi et al. 2017). To achieve desired levels of lipid accumulation in tissues not characteristic of seeds, such as leaves and stems, manipulation of lipid biosynthesis pathways is being pursued. This includes increasing the concentration of enzymes at the later step of the fatty acid biosynthesis pathway and enhancing the capacity of plants to produce higher amounts of oils in low lipid regions (De Bhowmick et al. 2015).

15.6.2 Avenues By Which Lipids Can Be Converted to Biodiesel

Lipids to biodiesel conversion involves transesterification process in which triglycerides which are a main content of lipids react with alcohol usually methanol in the presence of a catalyst to produce fatty acid methyl esters (FAME), the chemical biodiesel and glycerol as a spare product (Vyas et al. 2010). Organism improvement can be complementing this process using genetic engineering to design new organisms capable of producing good quality lipids for biodiesel production and possibly engineering microbes that could catalyze the transesterification process (Ranjbar and Malcata 2022). The desired fatty acid profile can be created by modifying pathways of engineered algae and plants, useful for biodiesel production specifically because they contain high levels of unsaturated fatty acids which improve fuel quality (Maheshwari et al. 2022). To improve biodiesel quality, lipid composition is changed by modulating the expression of desaturase genes, which are responsible for converting saturated fatty acids into unsaturated fatty acids. In addition, new developments in synthetic biology have made it possible to engineer microbial organisms to produce biodiesel feedstocks in a process that involves reprogramming microorganisms to create new synthetic metabolic pathways that use simple feedstocks such as glucose or CO_2 (Cravens et al. 2019). Such engineered pathways can avoid the conventional lipid separation and recovery process, which may not only simplify the processes involved in biodiesel production but also reduce cost.

15.7 Synthetic Biology and Artificial Pathways for Biofuels

15.7.1 Synthetic Biology to Reconstruct New Pathways for Synthesis of Biofuels

Synthetic biology has opened up new avenues for designing artificial biosynthetic pathways for synthesis of new-generation biofuels (Fig. 15.2) (Jagadevan et al. 2018).

This includes de novo synthesis of novel biocatalytic networks through the recombination of genetic building blocks from multiple organisms, effectively engineering biosynthetic routes that are unavailable in nature (Yi et al. 2021). Such pathways might build fuels and useful chemicals directly from light compounds

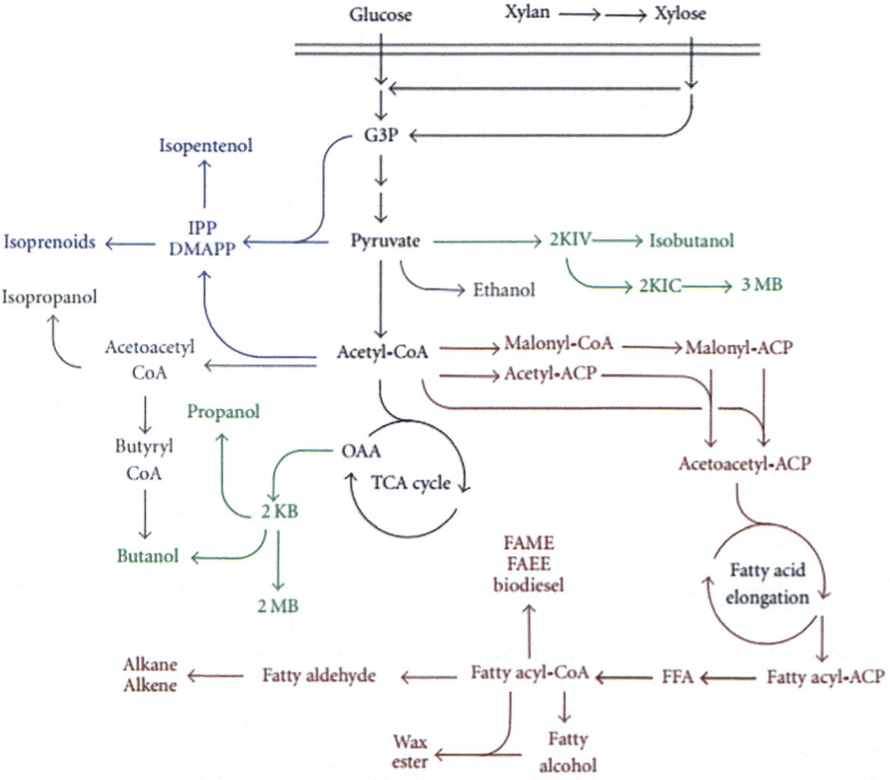

Fig. 15.2 Schematic representation of metabolic pathways for biofuel production. Central metabolic routes in the production of biofuels that can be made by conventional fermentative process (gray), higher chain alcohols (green), isoprenoid fuels (blue), and fatty acid fuels (red): *2 KB* 2-ketobutyrate, *2 KIV* 2-ketoisovalerate, *2KIC* 2-ketoisocaproate, *2 MB* 2-methyl-1-butanol, *3 MB* 3-methyl-1-butanol, *ACP* acyl carrier protein, *CoA* coenzyme A, *DMAPP* dimethylallyl diphosphate, *FAEE* fatty acid ethyl ester, *FAME* fatty acid methyl ester, *FFA* free fatty acid, *G3P* glyceraldehyde-3-phosphate, *IPP* isopentyl diphosphate, *OAA* oxaloacetate. (Source: Connor and Atsumi (2010))

known as substrates, which can be sugars, carbon dioxide, or other compounds, into higher energy density compounds like isoprenoids, fatty acid derivatives, and alcohols. A good example is genetically engineering of microorganisms like the production of advanced biofuel isobutanol with higher energy content than ethanol (Lakshmi et al. 2021). This is done by incorporating biosynthetic gene clusters (BGCs), i.e., genes that direct the synthesis of target product from various organisms into a single host such as *E. coli* (Li et al. 2023) or yeast (Xu et al. 2023).

15.7.2 Engineering Synthetic Microbes to Act as Hydrocarbons

The alkane and olefin hydrocarbons naphtha, distillate, and jet fuel are primary constituents of gasoline, diesel, and jet fuel (ASTR 1999). Microbial engineering in synthetic biology engineering also embraces the ability to synthesize these hydrocarbons using microbes that are produced directly from renewable resources (Khatib and Yassine 2019). However, cyanobacteria and algae are interesting candidates for such purpose because they are capable of photosynthesis and CO_2 fixation. Researchers have genetically modified cyanobacteria to synthesize alkanes using pathways borrowed from plants and other bacteria (Parveen and Yazdani 2022). These pathways transform fatty acids, which are in the lipid resources, into alkanes that can be collected and later channeled into biofuels. These cyanobacteria are engineered genetically, and the process they use involves light energy and CO_2 making the whole process carbon-friendly and renewable (Parveen and Yazdani 2022). Likewise, for hydrocarbon synthesis, yeast strains have also been redesigned for fatty acid synthesis and decarboxylation genes (Lu et al. 2022). These synthetic microbes can metabolize sugars from biomass into hydrocarbons; the method provides a flexible model of biofuel production (Rabinovitch-Deere et al. 2013).

15.8 Future Prospects of Fourth-Generation Biofuels

Fourth-generation biofuels are a paradigm shift in renewable energy by synthesizing the life science of synthetic biology and genetic engineering into sustainable solutions for global energy challenges (Shokravi et al. 2019). Unlike first- and second-generation biofuels which are based on food crops or nonfood biomass, fourth-generation biofuels directly convert renewable resources, such as CO2, agricultural residues, or sunlight using genetically engineered microorganisms such as *E. coli* or yeast, producing the biofuels (Padder et al. 2024). The microbes are "engineered" to maximize metabolic pathways, increase the efficiency of the enzymes, and increase the yield of high energy compounds such as biodiesel, bioethanol, or biobutanol. In practice, this technology hinges on a key synthetic biology capability of precisely editing and assembly of biosynthetic gene clusters (BGCs) for

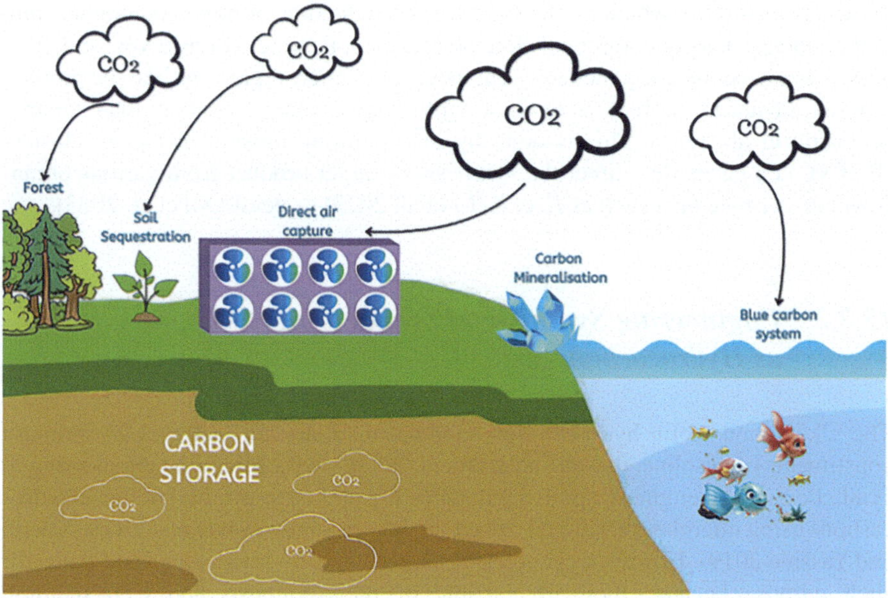

Fig. 15.3 Overview of the strategies used to capture CO_2 emissions from sources deemed to be significant greenhouse gas emitters. (Source: Bose et al. (2024))

hydrocarbon production (Wang et al. 2022b). The design of efficient strains capable of biofuel production under minimal resource inputs is hastened by advances in CRISPR-Cas9, metabolic modeling, and systems biology (Fabris et al. 2020). Secondly, carbon capture-based technologies could be coupled with these biofuel systems to cut greenhouse gas emissions significantly, providing a combined solution for climate change mitigation as well as energy sustainability (Fig. 15.3). Scaling up production systems, increasing the economic viability of engineered microbes, and making them more robust in industrial conditions are future prospects (Bose et al. 2024). Further developments in bioinformatics and artificial intelligence can further rein AIT strain designs to shorten development times and costs.

Transitioning these technologies to commercial scales also will rely upon making this work at the policymaking level and with industry collaboration (Wiesenthal et al. 2012). Genetic engineering and fourth-generation biofuels provide infinite potential to create a circular bioeconomy where waste is efficiently converted into energy that can lead to a cleaner, sustainable future (Garg et al. 2024).

15.9 Conclusion

The fourth-generation biofuels are being developed to achieve this while providing higher production efficiency and lower environmental impact. Optimal yields for these biofuels are achieved through diverse feedstocks including genetically engineered algae, bacteria, and plants. Advanced versions of genetic engineering tools like CRISPR-Cas9 and synthetic biology have also been used to boost the efficiency of microbial strains for producing biofuel. Enzymes engineered to improve saccharification processes and genetically modified plants that reduce biomass recalcitrance enhance lignocellulosic biofuel production. Lipid-based biofuels, the products of engineered organisms that store more lipids and biofuels that break to biodiesel, are conduits for conversion. As synthetic biology moves forward, and artificial pathways are developed to boost biofuel production, the prospects of fourth-generation biofuels are promising. Overall, these innovations continue the progress towards a sustainable and environmentally friendly energy future.

Acknowledgments PBR gratefully acknowledges the financial support provided by Shikshan Prasarak Sanstha, which has been instrumental in facilitating this work.

References

Abdullah M-A, Chew B-C, Hamid S-R (2016) Benchmarking key success factors for the future green airline industry. Procedia Soc Behav Sci 224:246. https://doi.org/10.1016/j.sbspro.2016.05.456

Abdullah B, Syed Muhammad SAF, Shokravi Z et al (2019) Fourth generation biofuel: a review on risks and mitigation strategies. Renew Sustain Energy Rev 107:37

Adebami GE, Kuila A, Ajunwa OM et al (2022) Genetics and metabolic engineering of yeast strains for efficient ethanol production. J Food Process Eng 45:e13798

Adegboye MF, Ojuederie OB, Talia PM, Babalola OO (2021) Bioprospecting of microbial strains for biofuel production: metabolic engineering, applications, and challenges. Biotechnol Biofuels 14:5

Adenle AA, Haslam GE, Lee L (2013) Global assessment of research and development for algae biofuel production and its potential role for sustainable development in developing countries. Energy Policy 61:182. https://doi.org/10.1016/j.enpol.2013.05.088

Adewuyi A (2020) Challenges and prospects of renewable energy in Nigeria: a case of bioethanol and biodiesel production. Energy Rep 6:77

Alalwan HA, Alminshid AH, Aljaafari HAS (2019) Promising evolution of biofuel generations. Subject review. Renew Energy Focus 28:127

Alaswad A, Dassisti M, Prescott T, Olabi AG (2015) Technologies and developments of third generation biofuel production. Renew Sustain Energy Rev 51:1446

Andeden EE, Ozturk S, Aslim B (2021) Evaluation of thirty microalgal isolates as biodiesel feedstocks based on lipid productivity and triacylglycerol (TAG) content. Curr Microbiol 78:775. https://doi.org/10.1007/s00284-020-02340-5

Arutyunov VS, Lisichkin GV (2017) Energy resources of the 21st century: problems and forecasts. Can renewable energy sources replace fossil fuels? Russ Chem Rev 86:777. https://doi.org/10.1070/rcr4723

Asghar A, Sairash S, Hussain N et al (2022) Current challenges of biomass refinery and prospects of emerging technologies for sustainable bioproducts and bioeconomy. Biofuels Bioprod Biorefin 16:1478. https://doi.org/10.1002/bbb.2403

Ashokkumar V, Venkatkarthick R, Jayashree S et al (2022) Recent advances in lignocellulosic biomass for biofuels and value-added bioproducts—a critical review. Bioresour Technol 344:126195

ASTR (1999) Overview of Total petroleum hydrocarbons (TPH). In: Toxicological profile for total petroleum hydrocarbons. Agency for Toxic Substances and Disease Registry (US), Atlanta, GA

Ausländer S, Ausländer D, Fussenegger M (2017) Synthetic biology—the synthesis of biology. Angew Chem Int Ed Engl 56:6396
Bakala HS, Devi J, Ankita et al (2023) Utilization of wheat and maize waste as biofuel source. In: Agroindustrial waste for green fuel application, pp 27–66
Banerjee C, Dubey KK, Shukla P (2016) Metabolic engineering of microalgal based biofuel production: prospects and challenges. Front Microbiol 7:432
Baumber A (2016) Bioenergy crops for ecosystem health and sustainability. Routledge, London
Benvenuti G, Bosma R, Cuaresma M et al (2015) Selecting microalgae with high lipid productivity and photosynthetic activity under nitrogen starvation. J Appl Phycol 27:1425. https://doi.org/10.1007/s10811-014-0470-8
Bertrand S, Bohni N, Schnee S et al (2014) Metabolite induction via microorganism co-culture: a potential way to enhance chemical diversity for drug discovery. Biotechnol Adv 32:1180
Bhatia T, Bose D, Sharma D, Patel D (2024) A review on cellulose degrading microbes and its applications. Ind Biotechnol 20:26
Bose D, Bhattacharya R, Kaur T et al (2024) Innovative approaches for carbon capture and storage as crucial measures for emission reduction within industrial sectors. Carbon Capture Sci Technol 16:100238
Cavelius P, Engelhart-Straub S, Mehlmer N et al (2023) The potential of biofuels from first to fourth generation. PLoS Biol 21:e3002063. https://doi.org/10.1371/JOURNAL.PBIO.3002063
Cheah WY, Sankaran R, Show PL et al (2020) Pretreatment methods for lignocellulosic biofuels production: current advances, challenges and future prospects. Biofuel Res J 7:1115
Chowdhury H, Loganathan B (2019) Third-generation biofuels from microalgae: a review. Curr Opin Green Sustain Chem 20:39
Connor MR, Atsumi S (2010) Synthetic biology guides biofuel production. J Biomed Biotechnol 2010:1–9
Cravens A, Payne J, Smolke CD (2019) Synthetic biology strategies for microbial biosynthesis of plant natural products. Nat Commun 10:2142
Das M, Patra P, Ghosh A (2020) Metabolic engineering for enhancing microbial biosynthesis of advanced biofuels. Renew Sustain Energy Rev 119:109562
De Bhowmick G, Koduru L, Sen R (2015) Metabolic pathway engineering towards enhancing microalgal lipid biosynthesis for biofuel application - a review. Renew Sustain Energy Rev 50:1239
DeLisi C, Patrinos A, MacCracken M et al (2020) The role of synthetic biology in atmospheric greenhouse gas reduction: prospects and challenges. BioDesign Res 2020:1016207
Dickinson S, Mientus M, Frey D et al (2017) A review of biodiesel production from microalgae. Clean Technol Environ Policy 19:637
Fabris M, Abbriano RM, Pernice M et al (2020) Emerging technologies in algal biotechnology: toward the establishment of a sustainable, algae-based bioeconomy. Front Plant Sci 11:279
Fan MZ, Wang W, Cheng L et al (2021) Metagenomic discovery and characterization of multifunctional and monomodular processive endoglucanases as biocatalysts. Appl Sci 11:5150
Fehér C (2018) Novel approaches for biotechnological production and application of L-arabinose. J Carbohydr Chem 37:251
Garg D, Samota MK, Kontis N et al (2023) Revolutionizing biofuel generation: unleashing the power of CRISPR-Cas mediated gene editing of extremophiles. Microbiol Res 274:127443
Garg A, Basu S, Shetti NP et al (2024) Biowaste to bioenergy nexus: fostering sustainability and circular economy. Environ Res 250:118503
Guo X, Lü X (2021) The need for biofuels in the context of energy consumption. In: Advances in 2nd generation of bioethanol production, pp 9–30
Hemalatha P, Abda EM, Shah S et al (2023) Multi-faceted CRISPR-Cas9 strategy to reduce plant based food loss and waste for sustainable bio-economy – a review. J Environ Manage 332:117382
Holechek JL, Geli HME, Sawalhah MN, Valdez R (2022) A global assessment: can renewable energy replace fossil fuels by 2050? Sustainability 14:4792. https://doi.org/10.3390/su14084792

Howe A, Stopnisek N, Dooley SK et al (2023) Seasonal activities of the phyllosphere microbiome of perennial crops. Nat Commun 14:1039. https://doi.org/10.1038/s41467-023-36515-y

Immethun CM, Henson WR, Wang X et al (2016) Engineering central metabolism for production of higher alcohol-based biofuels. In: Biotechnology for biofuel production and optimization. Elsevier, Amsterdam, pp 1–34

Jagadevan S, Banerjee A, Banerjee C et al (2018) Recent developments in synthetic biology and metabolic engineering in microalgae towards biofuel production. Biotechnol Biofuels 11:185

Johnson MB, Wen Z (2009) Microalgae as a feedstock for biofuel production. Virginia Cooperative Extension. p 11

Jones JA, Toparlak TD, Koffas MAG (2015) Metabolic pathway balancing and its role in the production of biofuels and chemicals. Curr Opin Biotechnol 33:52

Joshi A, Verma KK, Rajput D et al (2022) Recent advances in metabolic engineering of microorganisms for advancing lignocellulose-derived biofuels. Bioengineered 13:8135

Kalluri UC, Yin H, Yang X, Davison BH (2014) Systems and synthetic biology approaches to alter plant cell walls and reduce biomass recalcitrance. Plant Biotechnol J 12:1207

Keasling J, Garcia Martin H, Lee TS et al (2021) Microbial production of advanced biofuels. Nat Rev Microbiol 19:701

Khatib SE, Yassine NA (2019) Advances in synthetic biology and metabolic engineering in the production of biofuel. Int J Curr Microbiol App Sci 8:1762. https://doi.org/10.20546/ijcmas.2019.809.204

Khoo KS, Ahmad I, Chew KW et al (2023) Enhanced microalgal lipid production for biofuel using different strategies including genetic modification of microalgae: a review. Prog Energy Combust Sci 96:101071

Kim J, Reed JL (2010) OptORF: optimal metabolic and regulatory perturbations for metabolic engineering of microbial strains. BMC Syst Biol 4:53. https://doi.org/10.1186/1752-0509-4-53

Kumar D, Singh B, Korstad J (2017) Utilization of lignocellulosic biomass by oleaginous yeast and bacteria for production of biodiesel and renewable diesel. Renew Sustain Energy Rev 73:654

Kumar S, Singh NK, Kumar A, Shukla P (2021) Next generation biofuel production in the omics era: potential and prospects. In: Omics technologies for sustainable agriculture and global food security, vol 2, pp 293–311

Kwak S, Jin YS (2017) Production of fuels and chemicals from xylose by engineered Saccharomyces cerevisiae: a review and perspective. Microb Cell Fact 16:82

Lakhawat SS, Malik N, Kumar V et al (2022) Implications of CRISPR-Cas9 in developing next generation biofuel: a mini-review. Curr Protein Pept Sci 23:574. https://doi.org/10.2174/1389203723666220907110310

Lakshmi NM, Binod P, Sindhu R et al (2021) Microbial engineering for the production of isobutanol: current status and future directions. Bioengineered 12:12308

Lalman JA, Shewa WA, Gallagher J, Ravella S (2017) Biofuels production from renewable feedstocks. In: Lau PCK (ed) Quality living through Chemurgy and Green Chemistry. Springer, Berlin, pp 193–220

Li M, Guo L, Wang Y et al (2022) Molecular and biochemical characterization of two 4-coumarate: CoA ligase genes in tea plant (Camellia sinensis). Plant Mol Biol 109:579. https://doi.org/10.1007/s11103-022-01269-6

Li X, Gadar-Lopez AE, Chen L et al (2023) Mining natural products for advanced biofuels and sustainable bioproducts. Curr Opin Biotechnol 84:103003

Liew WH, Hassim MH, Ng DKS (2014) Review of evolution, technology and sustainability assessments of biofuel production. J Clean Prod 71:11. https://doi.org/10.1016/j.jclepro.2014.01.006

Lin CY (2022) The influences of promising feedstock variability on advanced biofuel production: a review. J Mar Sci Technol 29:714

Lin CY, Lu C (2021) Development perspectives of promising lignocellulose feedstocks for production of advanced generation biofuels: a review. Renew Sustain Energy Rev 136:110445

Liu CJ, Eudes A (2022) Lignin synthesis and bioengineering approaches toward lignin modification. In: Advances in Botanical Research. Elsevier, Amsterdam

Liu D, Mannan AA, Han Y et al (2018) Dynamic metabolic control: towards precision engineering of metabolism. J Ind Microbiol Biotechnol 45:535. https://doi.org/10.1007/s10295-018-2013-9
Lu R, Shi TQ, Lin L et al (2022) Advances in metabolic engineering of yeasts for the production of fatty acid-derived hydrocarbon fuels. Green Chem Eng 3:289
Maheshwari P, Haider MB, Yusuf M et al (2022) A review on latest trends in cleaner biodiesel production: role of feedstock, production methods, and catalysts. J Clean Prod 355:131588
Majidian P, Tabatabaei M, Zeinolabedini M et al (2018) Metabolic engineering of microorganisms for biofuel production. Renew Sustain Energy Rev 82:3863
Maliha A, Abu-Hijleh B (2023) A review on the current status and post-pandemic prospects of third-generation biofuels. Energy Syst 14:1185
Malik K, Capareda SC, Kamboj BR, Malik S, Singh K, Arya S, Bishnoi DK (2024) Biofuels production: a review on sustainable alternatives to traditional fuels and energy sources. Fuels 5(2):157–175. https://doi.org/10.3390/fuels5020010
Malinská HA, Vaněk M, Nebeská D et al (2021) Plant priming changes physiological properties and lignin content in Miscanthus x giganteus. Ind Crop Prod 174:114185. https://doi.org/10.1016/j.indcrop.2021.114185
Mat Aron NS, Khoo KS, Chew KW et al (2020) Sustainability of the four generations of biofuels—a review. Int J Energy Res 44:9266. https://doi.org/10.1002/er.5557
Miah R, Siddiqa A, Chakraborty U et al (2022) Development of high temperature simultaneous saccharification and fermentation by thermosensitive Saccharomyces cerevisiae and bacillus amyloliquefaciens. Sci Rep 12:3630. https://doi.org/10.1038/s41598-022-07589-3
Mitra S, Murthy GS (2022) Bioreactor control systems in the biopharmaceutical industry: a critical perspective. Syst Microbiol Biomanuf 2:91
Mitra S, Paliya S, Mandpe A (2024) Microbial engineering in biofuel production—a global outlook, advances, and roadmap. In: Shah M, Deka D (eds) Emerging sustainable technologies for biofuel production, 1st edn. Springer, Cham, pp 547–593
Moioli E, Schildhauer T (2022) Negative CO2 emissions from flexible biofuel synthesis: concepts, potentials, technologies. Renew Sustain Energy Rev 158:112120
Naghshbandi MP, Tabatabaei M, Aghbashlo M et al (2019) Progress toward improving ethanol production through decreased glycerol generation in Saccharomyces cerevisiae by metabolic and genetic engineering approaches. Renew Sustain Energy Rev 115:109353
Narisetty V, Reshmy R, Maitra S et al (2023) Waste-derived fuels and renewable chemicals for bioeconomy promotion: a sustainable approach. Bioenergy Res 16:16. https://doi.org/10.1007/s12155-022-10428-y
Nath S (2024) Biotechnology and biofuels: paving the way towards a sustainable and equitable energy for the future. Discov Energy 4:8
Neupane D (2023) Biofuels from renewable sources, a potential option for biodiesel production. Bioengineering 10:29
Onyeaka H, Ekwebelem OC (2023) A review of recent advances in engineering bacteria for enhanced CO2 capture and utilization. Int J Environ Sci Technol 20:4635
Padder SA, Khan R, Rather RA (2024) Biofuel generations: new insights into challenges and opportunities in their microbe-derived industrial production. Biomass Bioenergy 185:107220
Paravantis JA (2022) Socioeconomic aspects of third-generation biofuels. In: 3rd generation biofuels: disruptive technologies to enable commercial production, pp 869–917
Parveen H, Yazdani SS (2022) Insights into cyanobacterial alkane biosynthesis. J Ind Microbiol Biotechnol 49:kuab075
Patel VK, Das A, Kumari R, Kajla S (2023) Recent progress and challenges in CRISPR-Cas9 engineered algae and cyanobacteria. Algal Res 71:103068
Potts J, Jangra S, Michael VN, Wu X (2023) Speed breeding for crop improvement and food security. Crops 3:276. https://doi.org/10.3390/crops3040025
Rabinovitch-Deere CA, Oliver JWK, Rodriguez GM, Atsumi S (2013) Synthetic biology and metabolic engineering approaches to produce biofuels. Chem Rev 113:4611
Radakovits R, Jinkerson RE, Darzins A, Posewitz MC (2010) Genetic engineering of algae for enhanced biofuel production. Eukaryot Cell 9:486

Rahman FA, Aziz MMA, Saidur R et al (2017) Pollution to solution: capture and sequestration of carbon dioxide (CO2) and its utilization as a renewable energy source for a sustainable future. Renew Sustain Energy Rev 71:112

Ranjan A, Mayank R, Moholkar VS (2013) Process optimization for butanol production from developed rice straw hydrolysate using clostridium acetobutylicum MTCC 481 strain. Biomass Convers Biorefin 3:143. https://doi.org/10.1007/s13399-012-0062-2

Ranjbar S, Malcata FX (2022) Challenges and prospects for sustainable microalga-based oil: a comprehensive review, with a focus on metabolic and genetic engineering. Fuel 324:124567

Renzaho AMN, Kamara JK, Toole M (2017) Biofuel production and its impact on food security in low and middle income countries: implications for the post-2015 sustainable development goals. Renew Sustain Energy Rev 78:503

Rial RC (2024) Biofuels versus climate change: exploring potentials and challenges in the energy transition. Renew Sustain Energy Rev 196:114369

Savadi S, Lambani N, Kashyap PL, Bisht DS (2017) Genetic engineering approaches to enhance oil content in oilseed crops. Plant Growth Regul 83:207. https://doi.org/10.1007/s10725-016-0236-1

Schutyser W, Renders T, Van Den Bosch S et al (2018) Chemicals from lignin: an interplay of lignocellulose fractionation, depolymerisation, and upgrading. Chem Soc Rev 47:852

Sedlak M, Ho NWY (2004) Production of ethanol from cellulosic biomass hydrolysates using genetically engineered saccharomyces yeast capable of cofermenting glucose and xylose. Appl Biochem Biotechnol 113-116:403

Seike T, Kobayashi Y, Sahara T et al (2019) Molecular evolutionary engineering of xylose isomerase to improve its catalytic activity and performance of micro-aerobic glucose/xylose co-fermentation in Saccharomyces cerevisiae. Biotechnol Biofuels 12:139. https://doi.org/10.1186/s13068-019-1474-z

Shaah AHM, Hossain MS, Salem Allafi FA, Alsaedi A, Ismail N, Ab Kadir MO, Ahmad MI (2021) A review on non-edible oil as a potential feedstock for biodiesel: physicochemical properties and production technologies. RSC Adv 11:25018

Shen H, Poovaiah CR, Ziebell A et al (2013) Enhanced characteristics of genetically modified switchgrass (Panicum virgatum L.) for high biofuel production. Biotechnol Biofuels 6:71. https://doi.org/10.1186/1754-6834-6-71

Shokravi Z, Shokravi H, Aziz MA, Shokravi H (2019) The fourth-generation biofuel: a systematic review on nearly two decades of research from 2008 to 2019. In: Fossil free fuels, pp 213–251

Shokravi H, Heidarrezaei M, Shokravi Z et al (2022) Fourth generation biofuel from genetically modified algal biomass for bioeconomic development. J Biotechnol 360:23

Singh D, Sharma D, Soni SL et al (2020) A review on feedstocks, production processes, and yield for different generations of biodiesel. Fuel 262:116553

Singh R, Langyan S, Rohtagi B et al (2022) Production of biofuels options by contribution of effective and suitable enzymes: technological developments and challenges. Mater Sci Energy Technol 5:294

Sonderegger M, Sauer U (2003) Evolutionary engineering of Saccharomyces cerevisiae for anaerobic growth on xylose. Appl Environ Microbiol 69:1990. https://doi.org/10.1128/AEM.69.4.1990-1998.2003

Sreenikethanam A, Raj S, Rajesh Banu J et al (2022) Genetic engineering of microalgae for secondary metabolite production: recent developments, challenges, and future prospects. Front Bioeng Biotechnol 10:836056

Sticklen MB (2008) Plant genetic engineering for biofuel production: towards affordable cellulosic ethanol. Nat Rev Genet 9:433

Tai M, Stephanopoulos G (2013) Engineering the push and pull of lipid biosynthesis in oleaginous yeast Yarrowia lipolytica for biofuel production. Metab Eng 15:1. https://doi.org/10.1016/j.ymben.2012.08.007

Tekin N, Karatay SE, Dönmez G (2023) Optimization studies about efficient biobutanol production from industrial tea waste by Clostridium beijerinckii. Fuel 331:125763. https://doi.org/10.1016/j.fuel.2022.125763

Thirunavukkarasu M, Sawle Y, Lala H (2023) A comprehensive review on optimization of hybrid renewable energy systems using various optimization techniques. Renew Sustain Energy Rev 176:113192

Van-Eijck J, Batidzirai B, Faaij A (2014) Current and future economic performance of first and second generation biofuels in developing countries. Appl Energy 135:115. https://doi.org/10.1016/j.apenergy.2014.08.015

Venkata Mohan S, Dahiya S, Amulya K et al (2019) Can circular bioeconomy be fueled by waste biorefineries—a closer look. Bioresour Technol Rep 7:100277. https://doi.org/10.1016/j.biteb.2019.100277

Verardi A, Lopresto CG, Blasi A et al (2019) Bioconversion of lignocellulosic biomass to bioethanol and biobutanol. In: Lignocellulosic biomass to liquid biofuels, pp 67–125

Veza I, Said MFM, Latiff ZA (2021) Recent advances in butanol production by acetone-butanol-ethanol (ABE) fermentation. Biomass Bioenergy 144:105919

Vyas AP, Verma JL, Subrahmanyam N (2010) A review on FAME production processes. Fuel 89:1

Wagner E, Sierra-Ibarra E, Rojas NL, Martinez A (2022) One-pot bioethanol production from brewery spent grain using the ethanologenic Escherichia coli MS04. Renew Energy 189:717. https://doi.org/10.1016/j.renene.2022.03.014

Wang J, Singer SD, Souto BA et al (2022a) Current progress in lipid-based biofuels: feedstocks and production technologies. Bioresour Technol 351:127020

Wang L, Liu W, Liang J et al (2022b) Mining of novel secondary metabolite biosynthetic gene clusters from acid mine drainage. Sci Data 9:760. https://doi.org/10.1038/s41597-022-01866-6

Wiesenthal T, Leduc G, Haegeman K, Schwarz HG (2012) Bottom-up estimation of industrial and public R&D investment by technology in support of policy-making: the case of selected low-carbon energy technologies. Res Policy 41:116. https://doi.org/10.1016/j.respol.2011.08.007

Williams CL, Westover TL, Emerson RM et al (2016) Sources of biomass feedstock variability and the potential impact on biofuels production. Bioenergy Res 9:1

Xiong W, Liu B, Shen Y et al (2021) Protein engineering design from directed evolution to de novo synthesis. Biochem Eng J 174:108096

Xu X, Vanhercke T, Shrestha P et al (2019) Upregulated lipid biosynthesis at the expense of starch production in potato (Solanum tuberosum) vegetative tissues via simultaneous downregulation of ADP-glucose Pyrophosphorylase and sugar Dependent1 expressions. Front Plant Sci 10:1444. https://doi.org/10.3389/fpls.2019.01444

Xu Z, Park TJ, Cao H (2023) Advances in mining and expressing microbial biosynthetic gene clusters. Crit Rev Microbiol 49:18

Xue LL, Chen HH, Jiang JG (2017) Implications of glycerol metabolism for lipid production. Prog Lipid Res 68:12

Yang Y, Tian Z, Lan Y et al (2021) An overview of biofuel power generation on policies and finance environment, applied biofuels, device and performance. J Traffic Transport Eng 8:534

Yang Y, Chaffin TA, Ahkami AH et al (2022) Plant synthetic biology innovations for biofuels and bioproducts. Trends Biotechnol 40:1454

Yi D, Bayer T, Badenhorst CPS et al (2021) Recent trends in biocatalysis. Chem Soc Rev 50:8003

Yue K, Chen J, Li Y, Kai L (2023) Advancing synthetic biology through cell-free protein synthesis. Comput Struct. Biotechnol J 21:2899

Zahed MA, Movahed E, Khodayari A et al (2021) Biotechnology for carbon capture and fixation: critical review and future directions. J Environ Manage 293:112830

Zhang K, Lu X, Li Y et al (2019) New technologies provide more metabolic engineering strategies for bioethanol production in Zymomonas mobilis. Appl Microbiol Biotechnol 103:2087

Zhu JY, Zhuang XS (2012) Conceptual net energy output for biofuel production from lignocellulosic biomass through biorefining. Prog Energy Combust Sci 38:583. https://doi.org/10.1016/j.pecs.2012.03.007

Zhu D, Yang B, Wang H, Shahnawaz M (2023) Editorial: lignin valorization: recent trends and future perspective. Front Bioeng Biotechnol 11:1190128

The manufacturer's authorised representative in the EU is Springer Nature Customer Service Centre GmbH, Europaplatz 3, 69115 Heidelberg, Germany. If you have any concerns regarding our products, please contact ProductSafety@springernature.com

Printed and bound by CPI Group (UK) Ltd, Croydon, CR0 4YY
01/10/2026
02225929-0012